C0-AWP-532

# MICROWAVE CIRCUITS
# AND
# PASSIVE DEVICES

# MICROWAVE CIRCUITS AND PASSIVE DEVICES

M.L. SISODIA
*Department of Physics*
*Rajasthan University*
*JAIPUR*

G.S. RAGHUVANSHI
*Department of Physics*
*Engineering College, Kota*
*KOTA*

**JOHN WILEY & SONS**

NEW YORK CHICHESTER BRISBANE TORONTO SINGAPORE

First Published in 1987 by
WILEY EASTERN LIMITED
4835/24 Ansari Road, Daryaganj
New Delhi 110 002, India

**Distributors:**

*Australia and New Zealand*
JACARANDA WILEY LTD., JACARANDA PRESS
JOHN WILEY & SONS, INC.
GPO Box 859, Brisbane, Queensland 4001, Australia

*Canada:*
JOHN WILEY & SONS CANADA LIMITED
22 Worcester Road, Rexdale, Ontario, Canada

*Europe and Africa:*
JOHN WILEY & SONS LIMITED
Baffins Lane, Chichester, West Sussex, England

*South East Asia:*
JOHN WILEY & SONS, INC
05-05, Block B, Union Industrial Building
37 Jalan Pemimpin, Singapore 2057

*Africa and South Asia:*
WILEY EASTERN LIMITED
4835/24 Ansari Road, Daryaganj
New Delhi 110 002, India

*North and South America and rest of the world:*
JOHN WILEY & SONS, INC
605 Third Avenue, New York, NY 10158 USA



**Library of Congress of Cataloging-in-Publication Data**

Sisodia, M.L.
Microwave Circuits and Passive Devices

Includes bibliographies

1. Microwave Circuits. 2. Microwave devices.
I. Raghuvanshi, G.S. II. Title.
TK 7876. S 575 1985 621.381'3 85-26319

ISBN 0-470-20264-5 John Wiley & Sons, Inc.
ISBN 0-85226-856-4 Wiley Eastern Limited

Printed in India at Maharani Printers, Delhi, India.

# CONTENTS

## 1 INTRODUCTION



## 2 TRANSMISSION LINES



## 3 BASIC ELECTROMAGNETICS

## *SECTION A: MATRIX DESCRIPTION OF A MICROWAVE CIRCUITS*



## *SECTION B: WAVEGUIDE COMPONENTS AND TEE JUNCTIONS*

# 1

# INTRODUCTION

This chapter is intended to provide an introduction to the field of Microwave Electronics so that the reader can better appreciate the text that follows later in the book. It discusses the characteristic and applications as well as basic concepts of microwave electronics.

## 1.1 MICROWAVE BAND AND ITS APPLICATIONS

The term 'microwaves'* is used to designate electromagnetic oscillations with frequencies ranging from about 300 MHz to 300 GHz or with wavelengths in air ranging from 100 cm to 1 mm. Table 1.1 shows the whole electromagnetic spectrum divided into two regions, namely, the radio-spectrum extending from d.c. to 300 GHz and the optical-spectrum extending from 300 GHz to infinity.

**TABLE 1.1 The electromagnetic spectrum**

| Region | *Frequency limits* (GHz) | | *Wavelength limits* (cm) | | Remarks |
|---|---|---|---|---|---|
| | *Minimum* | *Maximum* | *Maximum* | *Minimum* | |
| (*a*) Radio-spectrum | d.c. | 300 | Infinity | 0.1 | VLF/LF/MF/HF/VHF/UHF SHF etc. |
| (*b*) Optical spectrum | | | | | |
| (*i*) Infra-red | 300 | $375\times10^3$ | 0.1 | $8\times10^{-5}$ | Heat and black light |
| (*ii*) Visible | $375\times10^3$ | $790\times10^3$ | $8\times10^{-5}$ | $38\times10^{-6}$ | Light red to violet |
| (*iii*) Ultra-violet | $790\times10^3$ | $225\times10^5$ | $38\times10^{-6}$ | $12\times10^{-7}$ | Chemically invisible |
| (*iv*) X-rays | $225\times10^5$ | $450\times10^8$ | $12\times10^{-7}$ | $6\times10^{-10}$ | |
| (*v*) $\gamma$-rays | $450\times10^8$ | $270\times10^9$ | $6\times10^{-10}$ | $1\times10^{-10}$ | Radioactive |
| (*vi*) Cosmic rays | $270\times10^9$ | Indefinite | $1\times10^{-10}$ | Indefinite | little known |

*Corrara in 1932 first used the term 'microwaves' to designate electromagnetic waves of 30 cm wavelength, which is also when the term appeared for the first time in the proceedings of the IRE.

The position of microwave bands in the entire radio-spectrum is shown in Table 1.2.

**TABLE 1.2 Position of microwave bands in the entire radio-spectrum**

| *Official F.C.C. band* | *Frequency band* | *Wavelength band* | *Remarks* |
|---|---|---|---|
| (*i*) Very long waves (VLF) | 0 to 30 kHz | $\infty$ to $10^4$ m | |
| (*ii*) Long waves (LF) | 30 to 300 kHz | $10^4$ to $10^3$ m | |
| (*iii*) Medium waves (MF) | 300 to 3000 kHz | $10^3$ to $10^2$ m | |
| (*iv*) Short waves (HF) | 3 to 30 MHz | 100 to 10 m | |
| (*v*) Very short waves (VHF) | 30 to 300 MHz | 10 to 1 m | |
| (*vi*) Ultra short waves (UHF) | 0.3 to 3 GHz | 1 to 0.1 m | Microwaves |
| (*vii*) Super short waves (SHF) | 3 to 30 GHz | 10 to 1 cm | —do— |
| (*viii*) Extreme short waves (EHF) | 30 to 300 GHz | 10 to 1 mm | —do— |

It can be seen that microwaves constitute only a small part of the whole radio-spectrum, but their uses have become increasingly important due to the rapid development of various branches of science and engineering such as radar, telecontrol and telemetry, telecommunications, television, industrial electronics, basic research and medicine, etc. Table 1.3 summarises the typical applications of microwaves.

**TABLE 1.3 Typical microwave applications**

| *Frequency band* | *Applications* |
|---|---|
| 0.3 to 3 GHz | TV, radar (troposcatter and meteorological), microwave point to point communications, telemetry, medicine, food industry (microwave ovens). |
| 3 to 30 GHz | Altimeter, air-and ship-borne radar, navigation, satellite communication. |
| 30 to 300 GHz | Radioastronomy, radiometeorology, space research, nuclear physics, nucleonics, radio spectroscopy. |

In fact, microwaves display certain specific physical properties and characteristics that make them distinct from the waves of adjacent bands and specifically suitable for certain applications.

In communications, microwaves occupy a relatively smaller percentage of the total band width of information channels where modulation rides upon the main carrier. This is because of the increase in band width. For instance, a frequency band $10^9$ to $10^{12}$ Hz (microwave region) contains thousand sections such as the entire frequency band from 0 to $10^9$ Hz. Thus, any one of these thousand sections may be used to transmit all the radio, television and other communication that is presently transmitted by the 0 to $10^9$ Hz band. This extremely high informational capacity of the microwave band facilitates multichannel telephone and television communication. The capability of transmitting a vast amount of imformation has been enhanced further with the development of quantum oscillators—masers and lasers. Therefore, there is currently a trend to use microwaves more and more in various communication applications such as telephone networks, television networks, space communication and telemetry and several other fields (such as defence and railways, etc.).

Because of the high frequency of the microwave band, the size of the antenna used for communication necessarily becomes larger than the operating wavelength. This allows the antenna to give an extremely directed beam, just the same way as an optical lens focuses light rays. Therefore, microwaves are said to possess *quasi-optical* properties. For instance, a 140 cm diameter parabolic reflector type antenna produces a beam of 1° beam width (angular beam width$=140°/(D/\lambda)$, where D is the diameter of the parabola and $\lambda$ is the wavelength at 30 GHz ($\lambda=1$ cm). At 300 MHz, i.e. at 100 cm wavelength the diameter of the parabolic antenna becomes 140 m to provide a beam width of 1°. This size is too large to be carried abroad in an aeroplane. Therefore, microwave is the only choice for directive signal transmission (navigation) and locating and ranging objects in space (radar). At one time microwave electronics was almost synonymous with radar electronics because of the great stimulus given to the development of microwave systems for use in high resolution radar capable of detecting and locating enemy ships and planes. Even today, radar in its many varied forms such as early warning radar, missile tracking and guidance radar, fire control radar, weather detecting radar, air traffic control radar, automobile speed controlling radar, etc. occupies more than 70 percent of the total microwave hardware.

Unlike long radio waves and optical infra-red waves, microwaves in the frequency band ranging from about 300 MHz to 10 GHz are capable of freely propagating through the ionized layers surrounding the earth as well as through the atmosphere. The presence of such a transparent "window" in a microwave band facilitates the study of microwave radiation from the sun and stars in radio astronomical research of space. Moreover, this transparency property also makes possible duplex communications and exchange of information between the ground stations and space vehicles.

During the last one and half decades, the industrial applications of microwaves have expanded greatly. Microwave heating has entered into commercial use in the form of microwave ovens, in which cooking is done quickly and uniformly by waves inside and outside simultaneously. Microwave drying machines are used in textile and paper industries. Industrial applications are particularly evident in the food processing industry, the rubber industry and foundries. Other industrial applications include the central and on-line measurement of expensive materials during processing procedures in environments hostile to opto-electronic techniques. Recently, microwaves have been used for non-destructive testing of metals.

Microwaves have provided a very powerful experimental probe for the study of basic properties of matter. Several molecular, atomic and nuclear systems have energy states such that the differences in their energy levels correspond to the quantum of energy at microwave frequencies. Consequently, microwaves (specially EHF) are capable of energetically interacting with matter. This feature is widely used in microwave and radio frequency spectroscopy for structural analysis. Microwave absorption spectra provide information about molecular structure and energy levels. Useful molecular resonances exist at microwave frequencies in the diodes of certain crystal materials. The resonant interaction of microwaves with crystals has been used for the generation of microwave power. Gunn, Read and Impact oscillators and amplifiers are some examples, the study of which form a separate branch of engineering, namely, quantum electronics. The non-reciprocal ferrite devices and masers are yet another example of microwave resonances in molecules.

The interaction of an electron beam with periodic slow-wave microwave structures has been used to design high power linear accelerators which are indispensable instruments in nuclear research.

The potential of microwaves for use in medicine is immense. The exact location of deep cancerous tissue, in particular, can be known by means of microwave radiometers. Microwave diathermy machines are used to remove rheumatic pains by producing heat inside the muscle without affecting the skin. Patients afflicted by uncontrollable pain or random muscle movements can be treated using microwave irradiation which creates thermal blocks in the nerve network.

Although the field of microwave engineering is already well developed, the scope of its applications in communication, industry and basic research is ever increasing. The extension of microwave techniques into the field of optics is one such example.

However, these many and varied uses of microwaves have increased microwave pollution and the risk to health and this has stimulated the study of the biological effects of microwaves in several large research projects.

## 1.2 BASIC MICROWAVE CONCEPTS

### (*i*) Microwave Transmission

The principles of microwave transmission cannot be derived by mere extensions of either low frequency radio or high frequency optical concepts although they are all based upon the same fundamental laws of electromagnetism. For instance, if microwave power is fed in a conventional two conductor line where the longitudinal and transverse dimensions of the line are comparable to the wavelength of the propagating signal, it leads to a series of interesting effects that fall outside the scope of problems examined by classical theory of long transmission lines. It turns out that such a line cannot be used for microwave transmission. One has, therefore, to use hollow metal tubes called waveguides. The energy propagation in these structures is basically a reflection phenomena. Conversely, a hollow-pipe waveguide cannot be treated by the rules of low-frequency, electricity, for by these rules opposite currents cannot flow in the same metallic conductor without coalescing into one net current, yet we see that opposite currents can flow in the same conductor in waveguides. Another class of waveguides of most recent origin is called surface waveguides which is absolutely uncommon to low frequency transmission.

(*ii*) **Microwave Circuit Elements**

It is a well-established fact that conventional circuit elements such as resistors, inductors and capacitors do not respond well at microwave frequencies. For instance, a coil of wire may be an excellent inductor at 1 MHz, but at 500 MHz it may be an equally good capacitor because of the predominating effect of inter-turn capacitance. However, this does not mean that energy dissipating (resistors) and storing (capacitors and inductors) elements cannot be constructed at microwave frequencies but their geometrical shape will be quite different. As will be seen a section of a microwave line (distributed parameters), offers reactances varying from $-\infty$ to $+\infty$ if its length is suitably chosen.

Similarly, conventional resonant and anti-resonant circuits are replaced by resonant microwave line sections known as resonant cavities. Often resonant cavities are used as circuit elements with varying properties.

When a number of such microwave circuit elements are connected together, we have a *microwave circuit*. The analysis of a microwave circuit can be carried out either in terms of equivalent transmission line voltage and current waves or in terms of amplitudes of the incident and reflected waves. The first approach is called the conventional equivalent *impedance description approach* while the second is known as the *scattering matrix approach*. The latter approach being closely related to the wave nature of fields, is discussed in detail in the book.

(*iii*) **Generation and Amplification of Microwaves**

The operation of conventional vacuum tubes and solid state devices is limited by transit time effects. However, the frequency range of operation of these devices can be extended to the lower edge of the microwave spectrum at the cost of power output and noise characteristics. Therefore, the development of new devices was essential to exploit this frequencies region. Fortunately, number of new principles of operation such as velocity modulation, interaction of space change waves with electromagnetic fields, avalanche breakdown, quantum mechanical, tunneling, transferred electron techniques etc. have enabled the generation microwaves.

(*iv*) **Power Coupling**

The waveguide transmission of microwaves is associated with a number of interesting problems such as coupling of power to another system say from generator to line, exciting of waves in a waveguide, etc. To overcome these, three basic coupling methods, viz. electrical coupling (probe), magnetic coupling (loop), and aperture coupling (waveguide to waveguide) have been evolved. The basic feature of these methods is that one can control the amount of coupling because these structures have small antennas that radiate into the waveguide to be coupled.

(*v*) **Microwave Component Hardware**

Microwave hardware may be divided into three basic catagories:

(1) *Active microwave components.* These include generation, amplification and detection of microwaves.

(2) *Passive microwave components.* This catagory includes microwave hardware used in attenuation, propagation and analysis of microwaves. The important components are couplers, tee junctions, isolators, circulators, resonant cavities, joints, switches or duplers, transformers and filters, etc.

(3) *Microwave measurement hardware.* This includes microwave measurement instrumentation.

Any book on the fundamentals of microwave electronics should therefore, discuss microwave propagation, circuit elements and their analysis, microwave measurements, and generation and amplification of microwaves. The present volume deals with the second category of microwave hardware. Volume II, *Basic Microwave Techniques and Laboratory Manual* deals with the third category and Volume III, *Generation and Amplification of Microwaves* deals with category one of the microwave hardware.

## 1.3 OUTLINE OF THE BOOK

Chapter 2 of this book deals with conventional transmission line theory using distributed parameters. The concepts of transmission line theory such as VSWR and reflection coefficient are discussed in great detail. A brief reference to non-uniform transmission line has also been made.

As mentioned earlier low frequency techniques are insufficient to explain microwave phenomena and one has to reformulate the structure of fundamental laws of electromagnetism. To fulfil this requirement, in Chapter 3 it has been included.

Chapters 4 and 5 deal with waveguide theory and microwave resonators respectively. Besides the general characteristics of microwave waveguides, the characteristics of practically used dominant mode are also discussed at length. A reference to the wall currents existing in waveguide walls has also been made so that one can appreciate the manner in which the location of the coupling hole and slot determines the characteristics of the new structure. To make the subject matter up-to-date optical waveguides and optical resonators have also been included at the end of the respective chapters.

Chapter 6 is divided into two sections. Section A deals with basic microwave circuit concepts while section B is devoted to various passive microwave components. The special feature that besides discussing wall currents and fields in various components, the derived scattering matrix has been used to study the characteristics of a particular network. Many applications of tee junctions and directional couplers have been included to emphasize their role in microwave engineering.

Chapter 7, besides giving a brief account of ferrite properties, discusses various microwave components such as circulator, isolator gyrator, phase shifter, etc.

The last chapter on microwave fitters and slow-wave structures has been included to serve two purposes. Firstly, filters are the basic elements of electronic regime. Secondly, slow-wave structures form the basis of the microwave generators commonly known as travelling wave tubes and backward wave oscillators.

# 2

# TRANSMISSION LINES

## 2.1 INTRODUCTION

In general transmission line is any structure used to guide the flow of energy from one point to another without permitting the radiation of energy. The two main features desired in transmission lines are: single mode of propagation over a wide band of frequencies, and small attenuation. A great variety of structures have been developed to satisfy the above requirements and can be grouped in one of the following three categories.

### (*i*) Transmission Lines

Examples are two-wire line, coaxial line, shielded two conductor line, and shielded strip line, on which the dominant mode of propagation is a Transverse Electromagnetic Wave (TEM), i.e. both electric and magnetic fields are transverse to the direction of propagation. These are in common use at audio, radio and lower microwave frequencies and are shown in Fig. 2.1(a).

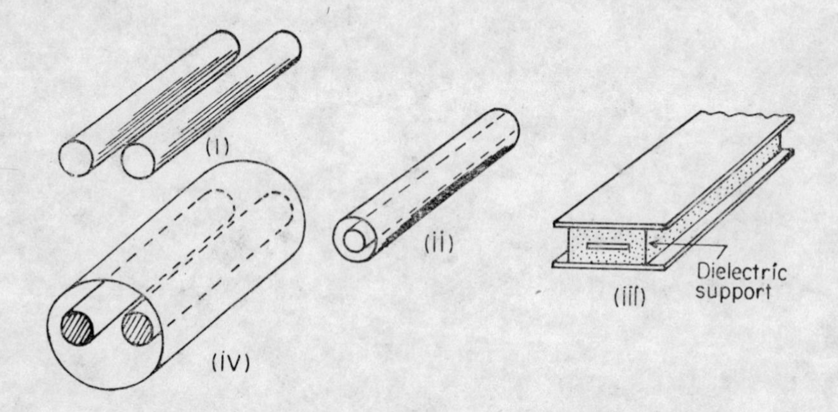


**Fig. 2.1** (*a*) (*i*) Two conductor line (*ii*) Coaxial line (*iii*) Shielded stripe line, (*iv*) Shielded two conductor lines

### (*ii*) Hollow Pipe Lines

These are generally called waveguides. They may be of any cross-section, e.g. rectangular, circular or elliptical. The dominant mode of propagation may be either Transverse Electric (TE) or Transverse Magnetic (TM) [Fig. 2.1(b)]. They are preferred to transmission lines at wavelengths below 10 cm, because of better electrical and mechanical properties. These lines are discussed in Chapter 4.

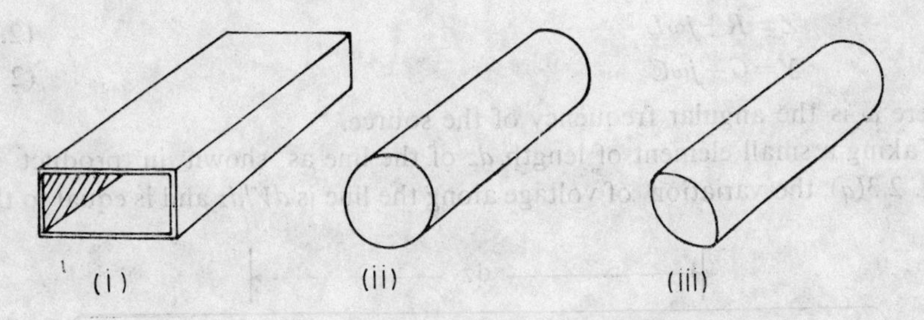


**Fig. 2.1 (*b*)** (*i*) Rectangular waveguide (*ii*) Circular cross-section waveguide (*iii*) Elliptical cross-section waveguide

(*iii*) **Surface Waveguides or Open Boundary Structures**

For example a dielectric rod [Fig 2.1(c)] which supports a surface wave of propagation and is preferred in the millimetre wavelength region.

In this chapter we shall present the basic transmission line theory by treating it as a network with distributed parameters and solving for the voltage and current waves that may propagate along the line.

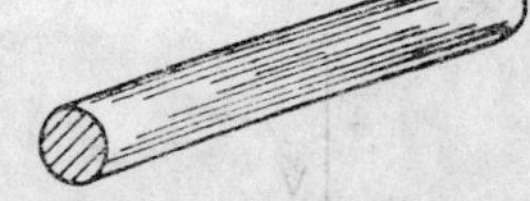

**Fig. 2.1 (*c*)** Dielectric rod

## 2.2 THE TRANSMISSION LINE EQUATIONS

Consider the familiar two-wire transmission line which consists of two parallel conductors of constant cross-sectional area and extending indefinitely in the $z$-direction. The spacing between the two parallel wires is uniform and small as compared to wavelength so that the effect is one wire of a change in current is instantaneous in the other wire [Fig. 2.2].

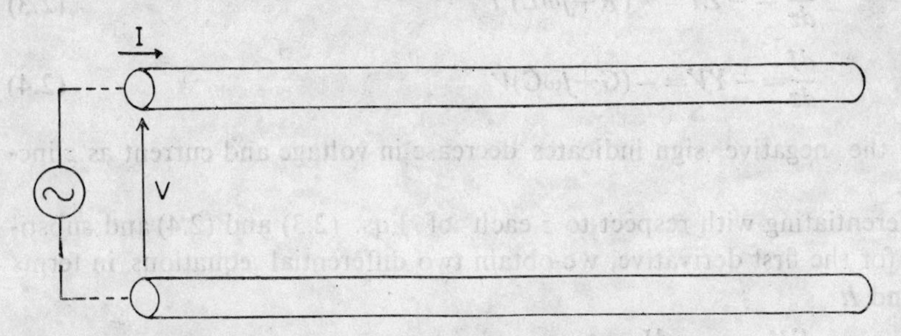


**Fig. 2.2** Current and voltage in two wire transmission line

The line may be assumed to have uniformly distributed parameters unlike the lumped constants in a network. Let the series resistance, series inductance, shunt capacitance and shunt conductance per unit length of the line be $R$, $L$, $C$ and $G$ respectively. Then the series impedance and shunt admittance of a unit length of the line are

$$Z=R+j\omega L \tag{2.1}$$

$$Y=G+j\omega C \tag{2.2}$$

where $\omega$ is the angular frequency of the source.

Taking a small element of length $dz$ of the line as shown in product of Fig. 2.3($a$), the variation of voltage along the line is $dV/dz$ and is equal to the

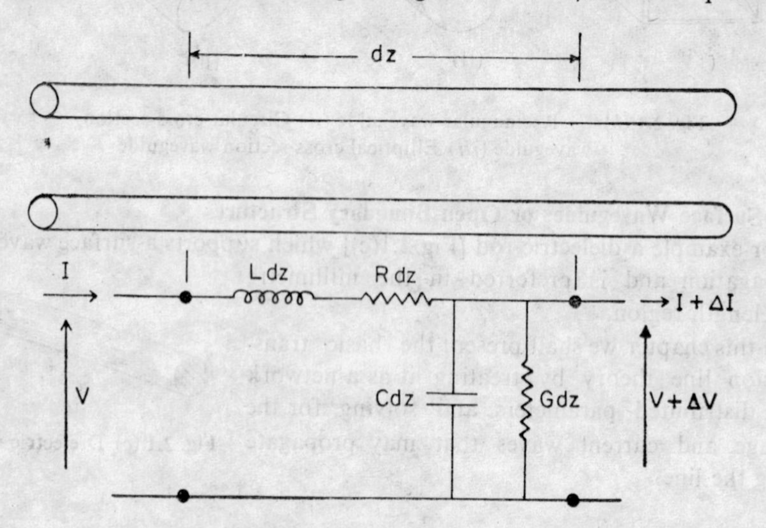


**Fig. 2.3** **Equivalenent circuit of a differential length of transmission line**

series impedance per unit length and current i.e. $Z$ times the current $I$. Similarly, the variation of current, along the line is $dI/dz$ and is equal to the product of the shunt admittance per unit length $Y$ and voltage $V$. Thus we have

$$\frac{dV}{dz}=-ZI=-(R+j\omega L)\,I \tag{2.3}$$

and

$$\frac{dI}{dz}=-YV=-(G+j\omega C)V \tag{2.4}$$

where the negative sign indicates decrease in voltage and current as $z$ increases.

Differentiating with respect to $z$ each of Eqs. (2.3) and (2.4) and substituting for the first derivative, we obtain two differential equations in terms of $V$ and $I$,

$$\frac{d^2V}{dz^2}=-Z\,\frac{dI}{dz}=+ZYV=\gamma^2V \tag{2.5}$$

and $$\frac{d^2I}{dz^2}=-Y\frac{dV}{dz}=+ZYI=\gamma^2 I \tag{2.6}$$

where $\gamma=\sqrt{YZ}=\sqrt{(R+j\omega L)(G+j\omega C)}$. This is known as the *propagation constant*. Equations (2.5) and (2.6) are the second order differential equations of voltage and current respectively. They may be recognized as differential equations characteristic of wave propagation, in this case waves of voltage and waves of current. The solutions of Eqs (2.5) and (2.6) which also satisfy Eqs (2.3) and (2.4) are:

$$V(z)=V_1\, e^{-\gamma z}+V_2\, e^{\gamma z} \tag{2.7}$$

$$I(z)=I_1\, e^{-\gamma z}+I_2\, e^{\gamma z} \tag{2.8}$$

where $V(z)$ and $I(z)$ refer respectively to the voltage and current existing at a distance $\epsilon$ from the source along the line. $V_1$, $I_1$ and $V_2$, $I_2$ are arbitrary constants but when evaluated from boundary conditions these respectively correspond to the amplitudes of voltage $V_1$ and current $I_1$ of the wave travelling in a positive $z$-direction i.e. the incident wave, and voltage $V_2$ and current $I_2$ of the wave travelling in a negative $z$-direction may be taken as the reflected wave. Differentiating Eq. (2.7) with respect to $z$ and combining it with Eq. (2.3), we have

$$\frac{dV}{dz}=-\gamma V_1 e^{-\gamma z}+\gamma V_2\, e^{\gamma z}=-\gamma\,(V_1\, e^{-\gamma z})-V_2 e^{\gamma z}=-ZI$$

or $$I=-\frac{1}{Z}\frac{dV}{dz}=\frac{\gamma}{Z}\,(V_1\, e^{-\gamma z}-V_2\, e^{\gamma z})$$

$$=\frac{1}{Z_0}\,(V_1\, e^{-\gamma z}-V_2\, e^{\gamma z}) \tag{2.9}$$

where we have introduced a new quantity

$$Z_0=\frac{Z}{\gamma}=\frac{Z}{\sqrt{ZY}}=\sqrt{\frac{Z}{Y}}=\sqrt{\frac{R+j\omega L}{G+j\omega C}}. \tag{2.10}$$

This quantity is known as *characteristic impedance* and its inverse as *characteristics admittance* $Y_0$,

$$Y_0=\sqrt{\frac{G+j\omega C}{R+j\omega L}}. \tag{2.11}$$

The physical significance of these new quantities, viz. $\gamma$, $Z_0$ and $Y_0$ shall be given in Sections 2.4 and 2.5.

In terms of the arbitrary amplitude constants $V_1$ and $V_2$ the solutions for the transmission line equations are

$$V(z)=V_1\, e^{-\gamma z}+V_2\, e^{\gamma z} \tag{2.12}$$

$$I(z)=\frac{1}{Z_0}\,[V_1\, e^{-\gamma z}-V_2 e^{\gamma z}]. \tag{2.13}$$

## 2.3 INTERPRETATION OF SOLUTIONS

The propagation constant

$$\gamma=\sqrt{ZY}=\sqrt{(R+j\omega L)(G+j\omega C)}$$

is a complex quantity and may be written as

$$\gamma=\alpha+j\beta=\sqrt{(R+j\omega L)(G+j\omega C)}. \tag{2.14}$$

Consequently, the complete solution of the wave Eq. (2.7) including the time factor, can be written as

$$V\, e^{j\omega t}=V_1 e^{-\alpha z}\, e^{j(\omega t-\beta z)}+V_2\, e^{\alpha z}\, e^{j(\omega t+\beta z)}. \tag{2.15}$$

Eq. (2.15) contains two terms, each of which has a phase factor of the type $e^{j(\omega t\pm\beta z)}$. This indicates that both the terms vary sinusoidally with time and space. Therefore, these two terms may be recognized as sinusoidal voltage waves travelling in positive and negative $z$-directions respectively. Further, the amplitude of each wave has a factor $e^{\pm\alpha z}$ indicating thereby that the amplitude of the wave is attenuated at the rate of $\alpha$, nepers* per metre. Similar interpretations follow for the terms in Eq. (2.8). $\alpha$, which is responsible for the attenuation, i.e. loss of power in a transmission line from various sources, is called the *attenuation constant* of the line.

To understand the significance of $\omega$, we may recall that the phase angle of the wave $(\omega t)$ increases by $2\pi$ radians when time $t$ increases by the periodic time $T=1/\nu$, $\nu$ being the frequency of the waves. That is

$$\omega\left(t+\frac{1}{\nu}\right)=\omega t+2\pi.$$

or

$$\frac{\omega}{\nu}=2\pi. \tag{2.16}$$

The factor $e^{j(\omega t-\beta z)}$ is a phase function in the first term of Eq. (2.15) and remains unaltered when seen by an observer moving with the speed of the wave. The constancy of this phase function implies that

$$\frac{d}{dt}\, j(\omega t-\beta z)=0.$$

Therefore, the time rate of change of $z$ is

$$\frac{dz}{dt}=\frac{\omega}{\beta}=v_p. \tag{2.17}$$

$v_p$ is the velocity of the equiphase plane of the wave and is called the *phase velocity* of the travelling wave. Similarly, in the second term of Eq. (2.15), the factor $e^{j(\omega t+\beta z)}$ represents a travelling wave with negative phase $v_p=-\omega/\beta$.

Further, to evaluate $\beta$ we may recall that the phase angle must increase by $2\pi$ radian between the successive equiphase points $z$ and $z+\lambda$, that is,

$$\beta(z+\lambda)=\beta z+2\pi$$

*Neper is the unit of attenuation (see Section 2.5).

or $\quad \beta = 2\pi/\lambda$ (2.18)

$\beta$, related to wavelength $\lambda$ as above, is the phase shift in radians per unit length of wave-travel and is generally called the *phase constant*. It should, however, be borne in mind that the wavelength of the waves used here is the wavelength (distance between two successive equiphase points) of the waves existing on the line and not that associated with free space.

We may note that both $\omega$ and $\beta$ are positive for ordinary transmission lines, hence, $v_p$ is positive. This confirms the statement that the two terms in Eq. (2.15), represent voltage waves travelling in positive and negative $z$-directions respectively.

The substitution of Eqs. (2.16) and (2.18) in Eq. (2.17) yields the fundamental equation of wave motion, i.e.

$$v_p = \frac{\omega}{\beta} = \frac{2\pi\nu}{2\pi/\lambda} = \lambda \tag{2.19}$$

which may be solved for $\lambda$

$$\lambda = \frac{v_p}{\nu}. \tag{2.20}$$

## 2.4 CHARACTERISTIC IMPEDANCE

Consider an infinite line extending in the negative $z$-direction, i.e. the receiving end is at $z=0$ and the generators is connected at $-\infty$. In such a case it is clear that the wave (signal) cannot reach the far end in finite time. Therefore, we have a wave travelling in the positive $z$-direction only and no wave in the negative $z$-direction. Hence $V_2 = I_2 = 0$, or from Eqs. (2.12) and (2.13) we have,

$$\frac{V_1}{I_1} = Z_0 = \sqrt{\frac{R + j\omega L}{G + j\omega C}} \tag{2.21a}$$

that is voltage $V_1$ applied across the conductors of an infinite line causes a current, $I_1$, to flow. By this observation, the line looks like an impedance which is designated $Z_0$.

Further, if the generator and receiver are interchanged we have only a wave travelling in the negative $z$-direction. For such a wave

$$\frac{V_2}{I_2} = Z_0 = -\sqrt{\frac{R + j\omega L}{G + j\omega L}}. \tag{2.21b}$$

We see that the amplitude of each voltage wave is related to its corresponding current wave amplitude by a factor $Z_0$, which is the characteristic constant of the line under discussion. Form Eqs. (2.21a and 2.21b) it is seen that the characteristic impedance of an infinite line is the impedance seen by the generator at any point on the line. Since the infinite line does not have any reflected wave, any line terminated in a similar infinite line or in its characteristic impedance will behave as an infinite line having input impedance $Z_0$, irrespective of the fact that the infinite line alone has an impedance $Z_0$. It may, therefore, be concluded that **a finite line of characteristic impedance $Z_0$ has an input impedance $Z_0$ when terminated in $Z_0$.** Furthermore, a line terminated in its characteristic impedance does not reflect any electrical power travelling towards the load ($Z_0$ in this case). Such lines are known as matched lines or **properly terminated lines. Thus the characteristic impedance $Z_0$ of a finite line is equal to the input impedance of that line when it is terminated by its characteristic impedance $Z_0$. It is that value of the impedance which when terminated to a finite length of line stops reflection from the distant end.**

Since the characteristic impedance expresses the ratio of voltage to current and in general the voltage and current have different phases, $Z_0$ is in general complex. This may also be verified from Eqs. (2.21a) and (2.21b). However, in a particular case when voltage and current are in phase $Z_0$ becomes a pure quantity, i.e. it is a pure resistance.

For a given line of known parameters, $Z_0$ can be calculated by substituting the values of the constants in Eq. (2.10) and simplifying them.

$$Z_0=\sqrt{\frac{R+j\omega L}{G+j\omega C}}=\left\{\frac{(R^2+\omega^2L^2)^{1/2}\,\Big/\tan^{-1}\left(\frac{\omega L}{R}\right)}{(G^2+\omega^2C^2)^{1/2}\,\Big/\tan^{-1}\left(\frac{\omega C}{R}\right)}\right\}^{1/2}$$

$$=\left\{\frac{R^2+\omega^2L^2}{G^2+\omega^2C^2}\right\}^{1/4}\Big/\frac{1}{2}\left(\tan^{-1}\frac{\omega L}{R}-\tan^{-1}\frac{\omega C}{R}\right). \tag{2.22}$$

We also see that $Z_0$ is frequency dependent.

## 2.5 PROPAGATION CONSTANT

The propagation constant is defined as

$$\gamma=\sqrt{ZY}=\sqrt{(R+j\omega L)(G+j\omega C)}.$$

This quantity $\gamma$ defines the propagation characteristics of a wave on the line. It is, in general, complex and its real part $\alpha$, specifies the rate at which the amplitude of the wave (current or voltage) is attenuated while the imaginary part $\beta$ specifies the rate of phase shift as the disturbance (wave) is propagatng along the line (see Eq. 2.15).

To express $\alpha$ and $\beta$ in terms of line parameters $\gamma$ in Eq. (2.12) may be eparated into real and imaginary parts.

$$\alpha+j\beta=\sqrt{(R+j\omega L)(G+j\omega C)}$$
$$=\pm\{(R+j\omega L)(G+j\omega C)\}^{1/2}$$

$$=\pm[(RG-LC\omega^2)+j\omega(LG+RC)]^{1/2} \tag{2.23}$$

or $$(\alpha+j\beta)^2 = (R+j\omega L)(G+j\omega C).$$

Equating the real and imaginary parts

$$\alpha^2-\beta^2=RG-\omega^2LC \tag{2.24}$$

$$2\alpha\beta=\omega(LG+RC). \tag{2.25}$$

Substituting the value of $\beta$ from Eq. (2.25) in Eq. (2.24) yields a quadratic equation in $\alpha^2$ which may be solved for $\alpha$ to give

$$\alpha=\sqrt{\tfrac{1}{2}[(RG-\omega^2LC)+\sqrt{(RG-\omega^2LC)^2+\omega^2(LG+RC)^2}]}. \tag{2.26}$$

Similarly,

$$\beta=\sqrt{\tfrac{1}{2}[(\omega^2LC-RG)+\sqrt{(RG-\omega^2LC)^2+\omega^2(LG+RC)^2}]}. \tag{2.27}$$

**Example 2.1.** For a coaxial loss-less line with outer and inner diameters $b$ and $a$ respectively, show that

$$Z_0=\frac{138.2}{\sqrt{\epsilon}}\log_{10}\frac{b}{a}, \qquad \lambda=\frac{\lambda_0}{\sqrt{\epsilon}} \text{ and power}$$

$$P=\frac{E_a^2a^2\sqrt{\epsilon}}{120}\ln\left(\frac{b}{a}\right)$$

where $\lambda_0$ is the vacuum wavelength and $\epsilon$ is the dielectric constant of the medium; $E_a$ is the radial electric field intensity at the centre.

*Solution*: The capacitance and inductance per unit length of a coaxial line are respectively given by

$$C=\frac{2\pi\epsilon_1}{\ln(b/a)}; \qquad L=\frac{\mu_1}{2\pi}\ln\left(\frac{b}{a}\right)$$

where $\epsilon_1$ and $\mu_1$ are the permittivity and permeability respectively of the medium between two conductors.

Therefore,

$$\beta=\sqrt{CL}=\sqrt{\frac{2\pi\epsilon}{\ln(b/a)}\times\frac{\mu_1}{2\pi}\ln\left(\frac{b}{a}\right)}$$

$$=\omega\sqrt{\mu_1\epsilon_1}=\frac{\omega}{C}\sqrt{\mu\epsilon}$$

where $\mu$ and $\epsilon$ are the relative permeability and relative permittivity (dielectric constant) of the medium

$$v_p=\frac{\omega}{\beta}=\frac{\omega C}{\omega\sqrt{\mu\epsilon}}=\frac{C}{\sqrt{\mu\epsilon}}.$$

Wavelength $\lambda=\dfrac{C}{\sqrt{\mu\epsilon}}=\dfrac{\lambda_0}{\sqrt{\epsilon}}.$

$$Z_0=\sqrt{\frac{L}{C}}=\sqrt{\frac{\mu_1}{2\pi}\ln\left(\frac{b}{a}\right)\Big/\frac{2\pi\epsilon_1}{\ln(b/a)}}$$

$$=\frac{1}{2\pi}\sqrt{\frac{\mu_1}{\epsilon_1}}\ln\left(\frac{b}{a}\right)=\frac{1}{2\pi}\sqrt{\frac{\mu\mu_0}{\epsilon\epsilon_0}}\ln\frac{b}{a}$$

$$=\frac{138.2}{\sqrt{\epsilon}}\ln g_{10}\left(\frac{b}{a}\right)\Omega$$

where $\sqrt{\frac{\mu_0}{\epsilon_0}}=120\,\pi=376.7\,\Omega$ and $\mu=1.$

Power, $P=\frac{1}{2}\frac{V^2}{Z_0}=\frac{\sqrt{\epsilon}\,V^2}{120\ln(b/a)}.$

Now, $V=\int_a^b\frac{E_a a}{r}\,dr=E_a\,a\ln\left(\frac{b}{a}\right)$

$$P=\frac{1}{2Z_0}E_a^2a^2\left(\ln\frac{b}{a}\right)^2$$

$$=\frac{E_a^2a^2\sqrt{\epsilon}}{120}\ln\left(\frac{b}{a}\right)W.$$

## 2.6 CONDITIONS FOR DISTORTIONLESS TRANSMISSION WITH MINIMUM ATTENUATION

If a signal received in a transmission line is an exact replica of the transmitted signal then such a line is known as a *distortionless line*. In general, there may be three sources of distortion in a line: (*i*) The characteristic impedance of the line may vary with frequency and if the line is terminated in an impedance then it may not vary in an identical manner. (*ii*) The attenuation constant ($\alpha$) may vary with frequency so that different frequency components of the signal suffer different attenuation, This type of distortion is referred to as *attenuation distortion*. (*iii*) The velocity of wave propagation may vary with frequency as a result of which different frequency components of the signal arrive at different times, in the load. This type of distortion is called *phase or velocity distortion*.

For a line to be distortionless each of $Z_0$, $\alpha$ and $\beta$ must be independent of frequency. The conditions for distortionless transmission are given below:

**(*i*) Distortion Due to $Z_0$ Being Frequency Dependent**

$$Z_0=\sqrt{\frac{R+j\omega L}{G+j\omega C}}=\sqrt{\frac{R\left(1+\frac{j\omega L}{R}\right)}{G\left(1+\frac{j\omega C}{G}\right)}}$$

$Z_0$ will be independent of frequency when

$$\frac{j\omega L}{R}=\frac{j\omega C}{G} \quad \text{or} \quad \frac{L}{R}=\frac{C}{G} \tag{2.28}$$

and the value of $Z_0$ is

$$Z_0=\sqrt{R/G}=\sqrt{L/C} \tag{2.29}$$

*(ii)* **Attenuation Distortion**

$$\gamma=\sqrt{(R+j\omega L)(G+j\omega C)}$$

or
$$\gamma^2=(R+j\omega L)(G+j\omega C)=RG+j\omega(CR+LG)-\omega^2 LC. \tag{2.30}$$

Taking $\quad CR=LG$

so that $\quad CR=LG=\sqrt{CRLG}$

or $\quad CR+LG=2\sqrt{CRLG}$

$$\gamma^2=RG+2j\omega\sqrt{RGCL}-\omega^2 LC$$
$$=(\sqrt{RG}+j\omega\sqrt{LC})^2$$
$$\alpha=\sqrt{RG} \quad \text{and} \quad \beta=\omega\sqrt{LC}. \tag{2.31}$$

Attenuation distortion is removed when $CR=LG$ or $L/R=C/G$. This condition is the same as in Eq. (2.28).

*(iii)* **Phase Distortion**

Velocity of propagation is

$$v=\omega/\beta.$$

From Eq. (2.31)

$$\beta=\omega\sqrt{LC}$$
$$v=\frac{1}{\sqrt{LC}}. \tag{2.32}$$

The velocity is independent of frequency and so there will be no phase distortion.

Besides for a line to be distortionless, it is necessary for the line to have minimum attenuation. The value of $L$ for minimum attenuation can be obtained by differentiating Eq. (2.26) with respect to $L$ and making it equal to zero, that is,

$$\frac{d\alpha}{dL}=0.$$

This leads to the condition

$$L=CR/G \quad \text{or} \quad L/C=R/G \tag{2.33}$$

which is the same as obtained in Eq. (2.28).

Similarly, the value of $C$ for minimum attenuation can be obtained by putting $d\alpha/dC=0$. This leads to the result of Eq. (2.33). However, by the differentiation of $\alpha$ with respect to $R$ and $G$ no minima can be found. But, by letting both $R$ and $G=0$, the attenuation $\alpha$ becomes zero; hence $R$ and $G$ should be kept as low as possible.

It is seen that the condition of distortionless transmission $L/C=R/G$ is also the condition for minimum attenuation. Hence the performance of a line may be improved by either increasing $L$ or decreasing $C$ so as to make the ratio $L/C=R/G$.

## 2.7 LOSSFREE TRANSMISSION LINE

For a line without losses, i.e. for which $R=G=0$, the propagation constant is $\gamma=j\beta=j\omega\sqrt{LC}$ or $\beta=\omega\sqrt{LC}$. The characteristic impedance is then a pure real quantity given by

$$Z_0=\sqrt{L/C}.$$

The velocity of propagation is

$$v_p=\omega/\beta=\frac{1}{\sqrt{LC}}. \tag{2.34}$$

Hence the wavelength of propagating waves is

$$\lambda=\frac{v_p}{\nu}=\frac{1}{\nu\sqrt{LC}}. \tag{2.35}$$

## 2.8 LOW-LOSS TRANSMISSION LINES

Most practical transmission lines are constructed so as to minimize the series resistance and shunt leakage conductance in order to reduce power losses in the line. Hence for low-loss lines at microwave frequencies, $R\ll\omega L$ and $G\ll\omega C$. When this is the case, the term $RG$ in the expression for $\gamma$ may be neglected

$$\begin{aligned}\gamma&=\sqrt{RG+j\omega(CR+LG)-\omega^2LC}\\&=\sqrt{j\omega(CR+LG)-\omega^2LC}\\&=[-LC\omega^2]^{1/2}\left[1-j\left(\frac{G}{\omega C}+\frac{R}{\omega L}\right)\right]^{1/2}.\end{aligned}$$

As $\quad \frac{G}{C}\ll1 \quad$ and $\quad \frac{R}{L}\ll1,$

a binomial expansion of the above equation gives

$$\gamma\simeq j\omega\sqrt{LC}+\frac{1}{2}\sqrt{LC}\left(\frac{R}{L}+\frac{G}{C}\right)=\alpha+j\beta. \tag{2.36}$$

Thus the phase constant of such a low-loss line is still $\beta=\omega\sqrt{LC}$ but the attenuation constant is

$$\alpha=\tfrac{1}{2}\sqrt{LC}\left(\frac{R}{L}+\frac{G}{C}\right)=\frac{1}{2}\left(R\sqrt{\frac{C}{L}}+G\sqrt{\frac{L}{C}}\right).$$

Since the first order characteristic impedance is still given by $Z_0=\sqrt{L/C}$, the expression for $\alpha$ reduces to

$$\alpha=\left(\frac{R}{2Z_0}+\frac{GZ_0}{2}\right)=\tfrac{1}{2}(RY_0+GZ_0). \tag{2.37}$$

$Y_0$ is the characteristic admittance. As $G$ is in general very small in the case of practical lines,

$$\alpha=\frac{R}{2Z_0}. \tag{2.38}$$

Equation (2.37) suggests that $\alpha$ for a low-loss line is made up of two parts: one depends on $R$ and hence represents conductor ohmic losses; while the other depends on $G$ and hence represents dielectric losses. So, in general, we can write $\alpha=\alpha_c+\alpha_d$

where $\alpha_c=\dfrac{R}{2Z_0}=\dfrac{1}{2}RY_0$ and $\alpha_d=\dfrac{GZ_0}{2}$.

The phase constant $\beta$ and corresponding wavelength may be expressed in terms of $\alpha_c$ and $\alpha_d$. Let $\beta_0$ and $\lambda_0$ represent parameters of a loss-less line, i.e.

$$\beta_0=\omega\sqrt{LC},\ \lambda_0=\frac{1}{\nu\sqrt{LC}};\ \text{then}$$

$$\beta=\beta_0\left[1+\frac{1}{2}\left(\frac{\alpha_c}{\beta_0}-\frac{\alpha_d}{\beta_0}\right)\right] \tag{2.39}$$

and

$$\lambda=\lambda_0\left[1-\frac{1}{2}\left(\frac{\alpha_c}{\beta_0}-\frac{\alpha_d}{\beta_0}\right)\right]. \tag{2.40}$$

These equations suggest that $\alpha$ introduces phase distortion. However, if the conductor loss is equal to the dielectric loss, this type of distortion may be removed. This condition also yields the same relation as expressed in Eq. (2.28).

The characteristic impedance is given by

$$Z_0=\sqrt{L/C}\left[1+\frac{1}{2}\left(\frac{\alpha_c}{\beta_0}+3\frac{\alpha_d}{\beta_0}\right)\left(\frac{\alpha_c}{\beta_0}-\frac{\alpha_d}{\beta_0}\right)-j\left(\frac{\alpha_c}{\beta_0}-\frac{\alpha_d}{\beta_0}\right)\right]. \tag{2.41}$$

## 2.9 TERMINATED TRANSMISSION LINE

The solutions for the transmission line equations are

$$V(z)=V_1e^{-\gamma z}+V_2e^{\gamma z} \tag{2.42}$$

$$I(z)=\frac{1}{Z_0}(V_1\, e^{-\gamma z}-V_2\, e^{\gamma z}) \tag{2.43}$$

where $Z_0$ is the characteristic impedance of the line. To have complete information about the solutions, the amplitude constants $V_1$ and $V_2$ have to be evaluated in terms of a pair of experimental values such as; the voltage and current $V_s$ and $I_s$ at the transmitting end of the line ($z=-z$) or the voltage and current $V_R$ and $I_R$ at the receiving end ($z=0$). We shall evaluate these in terms of the experimental values at the receiving end.

Consider a transmission line as shown in Fig. 2.4. The line is terminated at the receiving end at $z=0$ in an arbitrary load Impedance $Z_R$ and the line extends in the negative $z$-direction. Then at $z=0$

$$V=V_R,\quad I=I_R \quad \text{and} \quad \frac{V_R}{I_R}=Z_R.$$

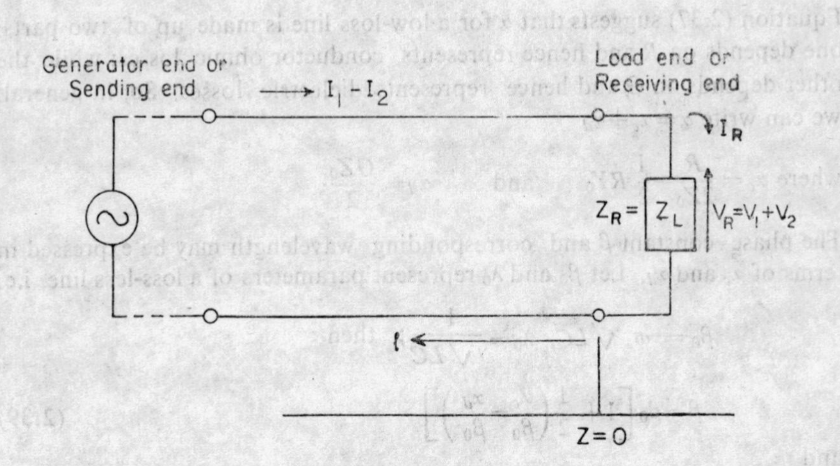


**Fig. 2.4** Terminated transmission line.

Equations (2.34) and (2.35) then become

$$V_R=V_1+V_2$$

$$I_R=\frac{1}{Z_0}(V_1-V_2)$$

from which we obtain

$$V_1=\frac{V_R}{2}\left(1+\frac{Z_0}{Z_R}\right),\quad V_2=\frac{V_R}{2}\left(1-\frac{Z_0}{Z_R}\right). \tag{2.44}$$

Substitution of these values of $V_1$ and $V_2$, and $Z_R=V_R/I_R$ in Eqs. (2.42) and (2.43) yields,

$$V(z)=\frac{V_R}{2}\left(1+\frac{Z_0}{Z_R}\right)e^{-\gamma z}+\frac{V_R}{2}\left(1-\frac{Z_0}{Z_R}\right)e^{\gamma z} \tag{2.45}$$

$$I(z)=\frac{I_R}{2}\left(1+\frac{Z_R}{Z_0}\right)e^{-\gamma z}+\frac{I_R}{2}\left(1-\frac{Z_R}{Z_0}\right)e^{\gamma z}. \tag{2.46}$$

In hyperbolic functions Eqs. (2.45) and (2.46) become

$$V(z)=V_R \cosh \gamma z-Z_0 I_R \sinh \gamma z \tag{2.47}$$

$$I(z)=I_R \cosh \gamma z-\frac{V_R}{Z_0} \sinh \gamma z. \tag{2.48}$$

The ratio of the voltage to current at any point on the transmission line is the impedance seen by the generator at that point due to the terminating impedance $Z_R$ at the receiving end.

$$Z(z)=\frac{V(z)}{I(z)}=\frac{V_R \cosh \gamma z-Z_0 I_R \sinh \gamma z}{I_R \cosh \gamma z-\frac{V_R}{Z_0}\sinh \gamma z}.$$

Putting $V_R=I_R Z_R$

$$Z(z)=Z_0\frac{Z_R \cosh \gamma z-Z_0 \sinh \gamma z}{Z_0 \cosh \gamma z-Z_R \sinh \gamma z}$$

or

$$Z(z)=Z_0\frac{Z_R-Z_0 \tanh \gamma z}{Z_0-Z_R \tanh \gamma z}. \tag{2.49}$$

## 2.10 INPUT IMPEDANCE

If the generator is connected at a distance $l$, left of the terminating end (or receiving end) at $z=0$, then $z=-l$. Equation (2.49) is then reduced to

$$Z_{in}=Z(-l)=Z_0\frac{Z_R+Z_0 \tanh \gamma l}{Z_0+Z_R \tanh \gamma l}. \tag{2.50}$$

$Z(-l)$ is the impedance seen by the generator as offered by the terminating line at its input and is called **input impedance**. For loss-less lines $\gamma=j\beta$ and Eq. (2.50) is simplified to

$$Z(-l)=Z_{in}=Z_0\frac{Z_R+jZ_0 \tan \beta l}{Z_0+jZ_R \tan \beta l}. \tag{2.51}$$

A considerable amount of information about the nature of the transmission line ($\gamma=\alpha+j\beta$) and a substantial part of microwave technique centres around Eq. (2.50), which in fact, is a relation between $Z_{in}$, $Z_0$ and the terminating or receiving end impedance $Z_R$, of a line of length $l$ and propagation constant $\gamma$. Some of its numerous implications* follow immediately from a simple substitution for certain cases of interest such as short-and open-circuit terminations.

When $Z_R=0$, i.e. the line is short-circuited at the receiving end, its input impedance is

$$Z_{sc}=Z_0 \tanh \gamma l. \tag{2.52}$$

*The most important implication is the Smith Chart discussed in Section 2.17.

When $Z_R=\infty$, i.e. the line is left open-circuited at the receiving end, then the input impedance is

$$Z_{oc}=\frac{Z_0}{\tanh \gamma l}=Z_0 \coth \gamma l. \tag{2.53}$$

Multiplying Eq. (2.52) with Eq. (2.53) yields

$$\sqrt{Z_{sc}\times Z_{oc}}=Z_0 \tag{2.54}$$

or in phasor form

$$Z_0=\sqrt{|Z_{sc}|e^{j\phi}|Z_{oc}|\epsilon^{j\psi}}=\sqrt{|Z_{sc}|\ |Z_{oc}|}\ \Big/\frac{\phi+\psi}{2} \tag{2.55}$$

where $\phi$ and $\psi$ are the phases of $Z_{sc}$ and $Z_{oc}$ respectively. Equation (2.55) provides a means of measuring the characteristic impedance of a given transmission line. To measure $Z_0$, measure the input impedances $Z_{sc}$ and $Z_{oc}$ (by measuring the ratio of voltage to current at the input, i.e. generator end) when the line is short-circuited and open-circuited and then find out the geometric mean. For experimental details refer to the literature.*

The short circuit, open circuit and characteristic terminations are of specific interest and will be examined in some detail.

**Short-Circuited and Open-Circuited Lines**

For a loss-less line (or a low-loss line at microwave frequencies), $\gamma=j\beta$ and so the input impedances $Z_{sc}$ and $Z_{oc}$ as given by Eqs. (2.52). and (2.53) may be written as

$$Z_{sc}=jZ_0 \tan \beta l \tag{2.56}$$

$$Z_{oc}=-jZ_0 \cot \beta l. \tag{2.57}$$

Equations (2.56) and (2.57) both indicate that the input impedance is pure reactance. Figure 2.5 which is the plot of Eqs. (2.56) and (2.57) shows that the reactance of each of the lines varies periodically between $-\infty$ and $+\infty$ as the length of the line is increased. It is noted that when $l<\lambda/4$ or $\beta l<\pi/2$ the reactance of the short-circuited line is positive and increases as the length of the line is increased while the reactance of the open-circuited line is negative but decreases in magnitude as the length of the line is increased.

When the length of the line happens to be

$$l=(2n-1)\lambda/4 \qquad \text{or} \qquad \beta l=(2n-1)\pi/2$$

where $n$ is an integer, then

$$Z_{sc}=\infty \qquad \text{and} \qquad Z_{oc}=0.$$

That is, the shorted quarter wave line behaves like an antiresonant line while the open-circuited quarter wave line behaves as a series resonant circuit.

*M.L. Sisodia and G.S. Raghuvanshi, *Basic Microwave Technique and Laboratory Manual*, Wiley, Eastern Limited, New Delhi.

When the length of the line is between $\lambda/4$ and $\lambda/2$, that is $\lambda/2>1>\lambda/4$ or $\pi>\beta l>\pi/2$, the shorted line has a negative decreasing reactance while the open-circuited line has a positive increasing reactance.

When the length of the line is $n\lambda/2$ or $\beta l/n\pi$,

$$Z_{sc}=0 \quad \text{and} \quad Z_{oc}=\infty.$$

That is, the shorted half wave line behaves as a resonant circuit while the half wave open-circuited line behaves as an antiresonant circuit. The similarity of these impedances (resonant and antiresonant lines) to those of resonant circuits may be seen by comparing Fig. 2.5 with Fig. 2.6 for resonant circuits. However, it is to be noted that lumped circuits are resonant at single frequencies, while the shorted and open lines are resonant at many frequencies. Moreover, if we consider $l$ to be constant and $\beta(=\omega/v;\ v=\text{constant})$ to be variable the results inferred above will be the same. It can, therefore, be concluded that the behaviour of the quarter and half wave lines with respect to change in the frequency of the source will be the same with respect to the variation in length.

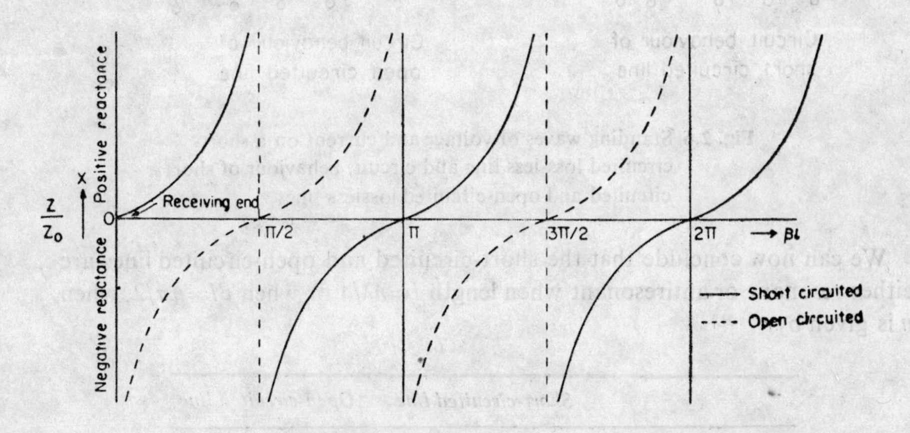


**Fig. 2.5** Impedance of short-circuited and open circuited loss-less line as a function of $\beta z$.

The current and voltage existing on such lines can be obtained from Eqs. (2.47) and (2.48). For a shorted line

$$V(-l)=jZ_oI_R \sin \beta l \tag{2.58}$$

$$I(-l)=I_R \cos \beta l. \tag{2.59}$$

Figure 2.6 depicts these variations as standing waves of voltage and current. It is noted that voltage and current are in space quadrature as evidenced by Fig. 2.6 and also in time quadrature and indicated by team $j$ in Eq. (2.58). In case of an open-circuited line the voltage and current standing waves are similar to those shown in Fig. 2.6, but with voltage and current interchanged. It follows from these plots that the reactance of the quarter and half wave line will be ∞ and 0 respectively in case of a shorted line and 0 and ∞ respectively in case of an open-circuited line.

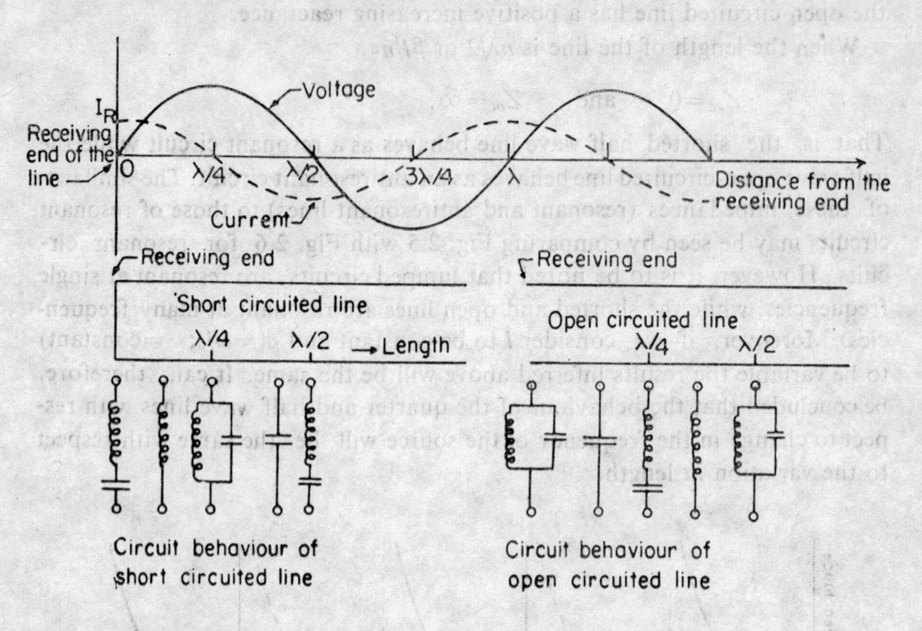


**Fig. 2.6** Standing waves of voltage and current on a short-circuited loss-less line and circuit, behaviour of short circuited and open-circuited lossless lines.

We can now conclude that the short-circuited and open-circuited lines are either resonant or antiresonant when length $l=n\lambda/4$ or when $\beta l=n\pi/2$, when $n$ is given by

| | *Short-circuited line* | *Open-circuited line* |
|---|---|---|
| Resonance | $n$=even integer | $n$=odd integer |
| Antiresonance | $n$=odd integer | $n$=even integer |

For a lossy line the impedance of a shorted line is given by Eq. (2.52); the corresponding admittance will be

$$Y_{sc}=\frac{1}{Z_{sc}}=\frac{\text{Coth}\,(\alpha+j\beta)l}{Z_0}.$$

Taking $Z_0$ as real*, the magnitude of this admittance is given by

$$Y_{sc}=\frac{1}{Z_0}\sqrt{\frac{\sin^2\alpha l\cosh^2\alpha l+\sin^2\beta l\cos^2\beta l}{(\sinh^2\alpha l\cos^2\beta l+\cosh^2\alpha l\sin^2\beta l)^2}}. \tag{2.60}$$

*The assumption holds good for most of the lines at very high frequencies.

In general, for a low-loss line at microwave frequencies $\alpha l \ll \beta l$. The plot of Eq. (2.60) under low-loss conditions is shown in Fig. 2.7. It is seen that the impedance periodically oscillates with a continuously decreasing peak value. The line behaves as a lossy antiresonant circuit when the line length is an odd multiple of $\lambda/4$ while it behaves as a lossy resonant circuit when its length is an even multiple of $\lambda/4$. Similar results may be drawn for a lossy open-circuited line.

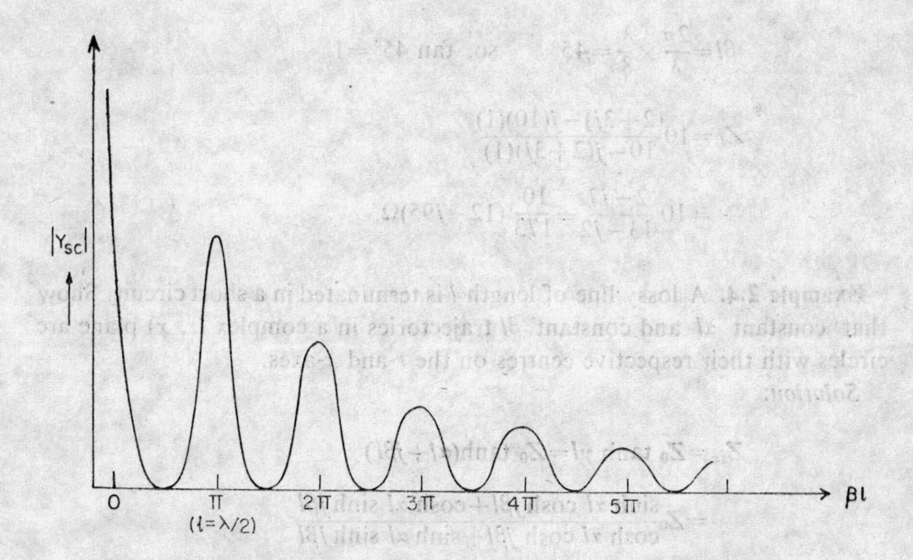


**Fig. 2.7** Magnitude if short-circuited lowless line admittance as a function of $\beta l$.

**Example 2.2.** A $\lambda/8$ section of a loss-less transmission line is terminated in a load $25+j50\ \Omega$. If the characteristic impedance of the line is $100\ \Omega$, determine the input impedance of the line.

*Solution*:

$$Z_{in}=Z_0\,\frac{Z_L+jZ_0\tan\beta l}{Z_0+jZ_L\tan\beta l}$$

$$\beta l=\frac{2\pi}{\lambda}\times\frac{\lambda}{8}=45^\circ \quad \text{so,} \qquad \tan\beta l=1$$

$$Z_{in}=100\,\frac{25+j50+j(100)(1)}{100+j(25+j50)(1)}$$

$$=100\,\frac{25+j150}{50+j25}$$

$$=100\,\frac{1+j6}{2+j}=(266.7+j166.7)\ \Omega.$$

**Example 2.3.** A $\lambda/8$ section of a loss-less transmission line is terminated in a load such that its input impedance is $2+3j\ \Omega$. If the characteristic impedance of the line be $10\ \Omega$ determine the load impedance terminating the line.

*Solution*:

$$Z_L=Z_0\frac{Z_{in}-jZ_0 \tan \beta l}{Z_0-jZ_{in} \tan \beta l}$$

$$\beta l=\frac{2\pi}{\lambda}\times\frac{\lambda}{8}=45^\circ \qquad \text{so, } \tan 45^\circ=1$$

$$Z_L=10\frac{(2+3j)-j(10)(1)}{10-j(2+3j)(1)}$$

$$=10\,\frac{2-j7}{13-j2}=\frac{10}{173}\,(12-j95)\Omega.$$

**Example 2.4.** A lossy line of length $l$ is terminated in a short circuit. Show that constant $\alpha l$ and constant $\beta l$ trajectories in a complex $(r, x)$ plane are circles with their respective centres on the $r$-and $x$-axes.

*Solution*:

$$Z_{sc}=Z_0 \tanh \gamma l=Z_0 \tanh(\alpha l+j\beta l)$$

$$=Z_0\frac{\sinh \alpha l \cosh j\beta l+\cosh \alpha l \sinh j\beta l}{\cosh \alpha l \cosh j\beta l+\sinh \alpha l \sinh j\beta l}$$

$$=R+jX$$

which gives

$$r=\frac{R}{Z_0}=\frac{\sinh \alpha l \cosh \alpha l}{\cosh^2 \alpha l \cos^2 \beta l+\sinh^2 \alpha l \sin^2 \beta l}$$

$$x=\frac{X}{Z_0}=\frac{\sin \beta l \cos \beta l}{\cosh^2 \alpha l \cos^2 \beta l+\sinh^2 \alpha l \sin^2 \beta l}.$$

Plots of the above equations are circles with equations

$$(r-r_1)^2+x^2=\rho_\alpha^2$$

$$r^2+(x-x_1)^2=\rho_\beta^2$$

where $\rho_\alpha$ and $\rho_\beta$ are radii of circles and

$$r_1=\tfrac{1}{2}(\tan \alpha l+\coth \alpha l)=\coth 2\alpha l$$

$$x_1=\tfrac{1}{2}(\coth \alpha l-\tanh \alpha l)=\operatorname{cosch} 2\alpha l$$

(Equations for the circles may be verified by substitution of values $r_1$, $\rho_\alpha$, $x_1$ and $\rho_\beta$.)

**Line Terminated in $Z_0$**

When a line is terminated at $Z_0$, i.e. $Z_R = Z_0$, it follows from Eq. (2.50) that the input impedance of the line $Z_{in}$ equals $Z_0$, i.e. $Z_{in} = Z_0$. It must, therefore, be concluded that a finite line of characteristic impedance $Z_0$ has an input impedance $Z_0$ when it is terminated in $Z_0$ irrespective of its length and the fact that the line alone has the characteristic impedance $Z_0$.

Substituting the values of $Z_R = Z_0$ in Eqs. (2.45) and (2.46), gives

$$V(z) = V_R\, e^{-\gamma z}$$

$$I(z) = I_R\, e^{-\gamma z}.$$

The voltage and current at the receiving end can be written as

$$V_R = V_s e^{-\gamma l} \tag{2.61}$$

$$I_R = I_s e^{-\gamma l}. \tag{2.62}$$

Where $V_s$ and $I_s$ are sending and voltage and current respectively.

Equations (2.61) and (2.62) suggest that the amplitudes of voltage and current waves are exponentially attenuated as the wave travels away from the generator (transmitting end). The variation of voltage and current wave amplitudes with the distance from the generator for such a properly terminated line, is shown in Fig. 2.8.

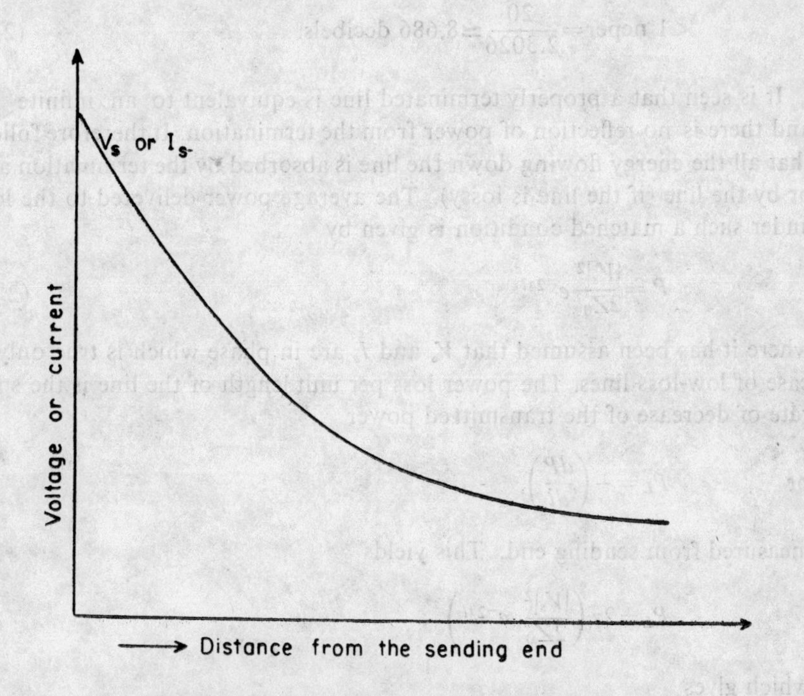


**Fig. 2.8** Magnitude of voltage and current as a function of distance along a line terminated in its characteristic impedance ($Z_0$).

Equations (2.61) and (2.62) may be used to evaluate a mathematical expression for $\alpha$ in terms of $V_A$ and $V_B$. Considering two points $A$ and $B$ ($A$ is near the generator) on such a properly terminated line, and following Eq. (2.61) we can write,

$$|V_B|=|V_A|\, e^{-\alpha l}$$

where $|V_B|$ and $|V_A|$ are the voltage wave amplitudes of the travelling wave at points $B$ and $A$ respectively at $l$ distance apart. Then

$$\alpha l=\log_e \left|\frac{V_A}{V_B}\right|$$

$$=2.3026 \ \log_{10}\left|\frac{V_A}{V_B}\right| \tag{2.63}$$

and $\alpha$ will be expressed in neper/metre. $\alpha$ can also be expressed in decibel/ metre, thus

$$\alpha l=10 \log \frac{P_A}{P_B}=20 \log\left|\frac{V_A}{V_B}\right| \ \ db \text{ (decibels)} \tag{2.64}$$

where $P_A$ and $P_B$ are the powers at points $A$ and $B$ respectively.

From Eqs. (2.63) and (2.64) it follows that

$$1 \text{ neper}=\frac{20}{2.3026}\simeq 8.686 \text{ decibels.} \tag{2.65}$$

It is seen that a properly terminated line is equivalent to an infinite line and there is no reflection of power from the termination. It therefore follows that all the energy flowing down the line is absorbed by the termination and/ or by the line (if the line is lossy). The average power delivered to the load under such a matched condition is given by

$$P=\frac{|V_s|^2}{2Z_0}\, e^{-2\alpha l} \tag{2.66}$$

where it has been assumed that $V_s$ and $I_s$ are in phase which is true only in case of low-loss lines. The power loss per unit length of the line is the space rate of decrease of the transmitted power

or $$P_L=-\left(\frac{dP}{dl}\right),$$

measured from sending end. This yields

$$P_L=2\alpha\left(\frac{|V_s|^2}{2Z_0}\, e^{-2\alpha l}\right)$$

which gives

$$\alpha=\frac{P_L}{2P}. \tag{2.67}$$

Thus, the attenuation constant is the ratio of the power loss per unit length of the line to twice the transmitted power. This is a very important relation and may be used to evaluate $\alpha$ for any uniform guiding structure.

**Example 2.5.** Using the general definition $\alpha = P_L/2P$, show that the total attenuation per unit length in an infinite loss-less line is $\alpha = \frac{1}{2}[GZ_0 + RY_0]$ neper/metre where symbols have their usual meanings.

*Solution*: Power loss per unit length

$$P_L = \frac{I_1^2 R}{2} + \frac{V_1^2 G}{2} = \frac{V_1^2}{2}\left[G + \frac{R}{Z_0^2}\right].$$

Power at the input

$$P = \tfrac{1}{2} V_1 I_1 = \frac{1}{2}\frac{V_1^2}{Z_0}$$

$$\alpha = \frac{P_L}{2P} = \frac{1}{2}\left[\frac{V_2}{2}\left(G + \frac{R}{Z_0^2}\right) \div \frac{1}{2}\frac{V_1^2}{Z_0}\right]$$

$$= \frac{GZ_0}{2} + \frac{RY_0}{2} \text{ neper/metre.}$$

## 2.11 REFLECTION COEFFICIENT

The ratio of the complex reflected voltage to the complex incident voltage at any point on the transmission line is called the voltage *reflection Coefficient* of that point and is denoted by $\Gamma$. From Eq. (2.42) the voltage reflection coefficient at a distance $l(Z = -l)$ from the termination is given by

$$\Gamma(-l) = \frac{V_2}{V_1} e^{-2\gamma l}. \tag{2.68}$$

It is seen that the voltage reflection coefficient at the termination ($l = z = 0$) is

$$\Gamma_0 = \frac{V_2}{V_1}$$

which on using Eq. (2.44) reduces to

$$\Gamma_0 = \frac{1 - Z_0/Z_R}{1 + Z_0/Z_R} = \frac{Z_R - Z_0}{Z_R + Z_0}. \tag{2.69}$$

It is seen from Eq. (2.69) that

(*i*) for $Z_R = Z_0$, i.e. matched termination, $\Gamma_0 = 0$;

(*ii*) for $Z_R = 0$, i.e. short-circuit termination, $\Gamma_0 = -1$;

(*iii*) for $Z_R = \infty$, i.e. open-circuit termination $\Gamma_0 = 1$.

Furthermore, since each of $Z_R$ and $Z_0$ are complex in general, $\Gamma_0$ is complex and can be written as

$$\Gamma_0 = |\Gamma_0|\, e^{j\phi} \tag{2.70}$$

$\phi$ is the phase of the voltage reflection coefficient at the termination, i.e. it is the ratio of the complex reflected voltage wave phase to the complex incident voltage wave phase and, therefore, is a measure of the reactance of the termination. It can be verified from Eqs. (2.69) and (2.70) that $\phi$ has a value $\pi$ for short-circuit termination and a value 0 for open-circuit and pure resistance terminations (for a loss-less line which has $Z_0$ as a real quantity).

Combining Eqs. (2.68) and (2.70), we obtain

$$\Gamma(-l)=\Gamma_0 \, e^{j\phi} \, e^{-2\alpha l} \, e^{-2j\beta l}. \tag{2.71}$$

Equation (2.71) indicates that the complex reflection coefficient at any distance $l$ from the load behaves as a vector whose value varies from $\Gamma_0$ at the load to zero as $l$ approaches infinity. The amplitude decreases exponentially with the distance while phase varies cyclically ($e^{-2j\beta l}$). The variation of $\Gamma(-1)$ with distance is shown in Fig. 2.9.

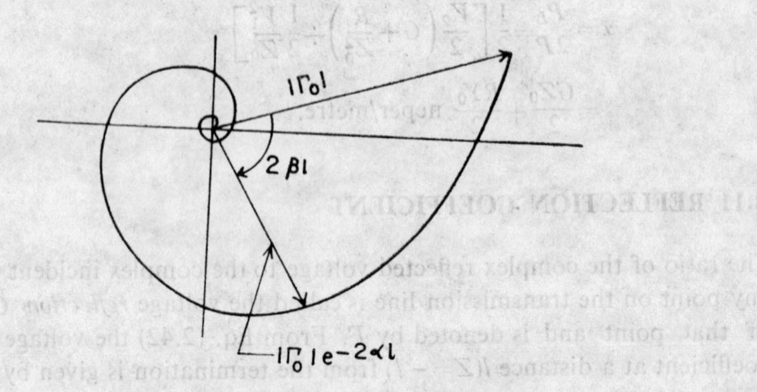


Fig. 2.9 Transformation of a voltage reflection coefficient along the line.

The current reflection coefficient may also be defined in a similar way. It is however less frequently used than the voltage reflection coefficient.

## 2.12 VOLTAGE AND CURRENT DISTRIBUTIONS ALONG A TERMINATED LINE

The voltage at a distance $l$ from the receiving end, i.e. load is obtained by substituting $z=-1$ in Eq. (2.42)

$$V(-l)=V_1 \, e^{\gamma l}+V_2 \, e^{-\gamma l}$$

which on using Eq. (2.71) reduces to

$$V(-l)=V_1 \, e^{\gamma l}[1+|\Gamma_0| \, e^{j\phi} \, e^{-2\alpha l} \, e^{-2j\beta l}] \tag{2.72}$$

where $\phi$ is the phase of the reflection coefficient at the termination, due to impedances $Z_0 < Z_R$.

**Lossless Line**

Assuming the line to be loss-less, i.e. $\gamma = j\beta$ and $Z_0$, the real Eq. (2.72) reduces to

$$V(-l) = V_1 e^{j\beta l}[1 + |\Gamma_0| e^{-j(2\beta l - \phi)}].$$

The magnitude of this voltage is given by

$$\begin{aligned} V(-l) &= V_1[\{1 + |\Gamma_0|\}^2 \cos^2(\beta l - \phi/2) \\ &\quad + \{1 - |\Gamma_0|^2\} \sin^2(\beta l - \phi/2)]^{1/2} \\ &= V_1[\{1 - |\Gamma_0|\}^2 - 4|\Gamma_0| \sin^2(\beta l - \phi/2)]^{1/2}. \end{aligned} \tag{2.73}$$

Similarly, the magnitude of the current at a distance $(-l)$ measured from the termination is given by

$$\begin{aligned} I(-l) &= \frac{V_1}{Z_0}[\{1 - |\Gamma_0|\}^2 \cos^2(\beta l - \phi/2) \\ &\quad + \{1 + |\Gamma_0|\}^2 \sin^2(\beta l - \phi/2]^{1/2} \\ &= \frac{V_1}{Z_0}[\{1 - |\Gamma_0|\}^2 + 4|\Gamma_0| \sin^2(\beta l - \phi/2)]^{1/2} \end{aligned} \tag{2.74}$$

Equations (2.73) and (2.74) indicate that $V(-l)$ and $I(-l)$ oscillate back and fourth between maximum values $V_1(1+|\Gamma_0|)$ and $V_1/Z_0(1+|\Gamma_0|)$, and minimum values $V_1(1-|\Gamma_0|)$ and $V_1/Z_0(1-|\Gamma_0|)$ respectively. Also the ratio $V_{max}/I_{max}$ and $V_{min}/I_{min}$ each equals $Z_0$, the characteristic impedance of the line regardless of the magnitude of the reflection coefficient, $|\Gamma_0|$. The voltage is maximum when $(\beta l - \phi/2) = n\pi$ and minimum when $(\beta l - \phi/2) = n\pi + \pi/2$, where $n$ is an integer. Furthermore, it can be seen by considering the time factor $e^{j\omega t}$ in Eq. (2.42) (from where we started), that the resultant amplitude of the voltage and current waves does not depend on time but varies periodically with distance. Such waves are commonly known as *standing waves* of voltage and current respectively. Equations (2.73) and (2.74) are, therefore, the equations of voltage and current standing waves. The positions of voltage maximum are called *antinodal points* or "*maxs*" while the positions of minimum voltage are called *nodal points* or "*mins*". The pattern as a whole is called the *standing wave pattern*. The ratio of maximum voltage to minimum voltage in the standing wave pattern is known as the *voltage standing wave ratio* (*VSWR*) and is denoted by $\rho$. It follows from Eq. (2.73) that

$$\text{VSWR} = \rho = \frac{V_{max}}{V_{min}} = \frac{1+|\Gamma_0|}{1-|\Gamma_0|} = \frac{|Z_R|}{|Z_0|}. \tag{2.75}$$

The last step follows from Eq. (2.69) which gives

$$|\Gamma_0| = \frac{\rho - 1}{\rho + 1}. \tag{2.76}$$

Since $|\Gamma_0|$ carries the information about the magnitude of mismatch at the termination, $\rho$ will also carry this information. The advantage of using $\rho$ over $|\Gamma_0|$ is that it is a well defined and easily measurable parameter.

Further the distance between successive maxima and minima is $d=\pi/\beta=\lambda/2$. Thus, measuring the distance between two successive minima or maxima which is $\lambda/4$, the wavelength* of the waves travelling on a transmission line (guide wavelength) may be obtained.

The distribution of voltage and current along an idealized transmission line when terminated in pure reactances of various magnitudes is shown in Fig. 2.10(a). These are obtained by applying the necessary boundary conditions on Eqs. (2.73) and (2.74), as follows:

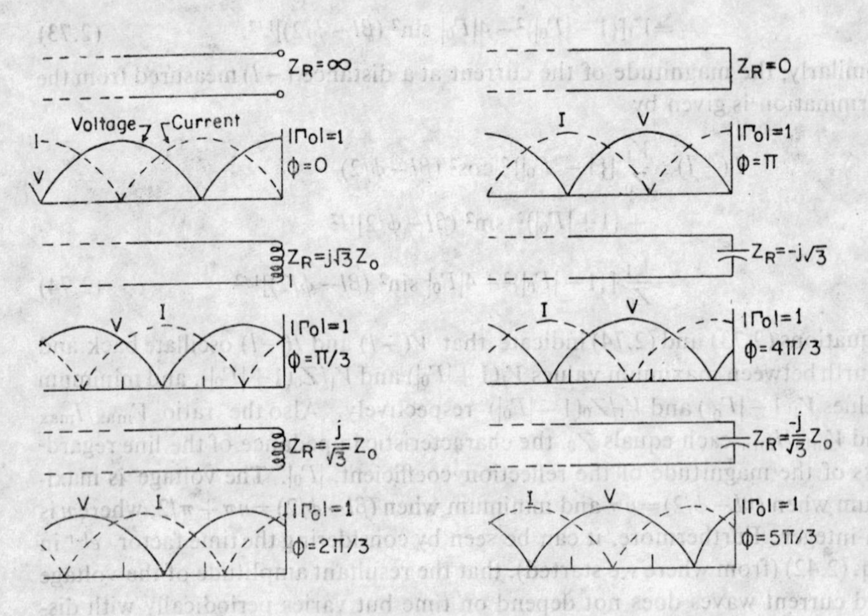


**Fig. 2.10(a)** Voltage and current standing waves corresponding to various reactive terminal.

1. When $Z_R=\infty$, i.e. the line is open-circuited, then we have $|\Gamma_0|=1$ and $\phi=0$. Hence

$$|V(-l)|=|V_m \cos \beta l|$$

$$|I(-l)|=|I_m \sin \beta l|$$

where $|V_m|$ and $|I_m|$ are the maximum values of voltage and current respectively. It is noted that $|V_m|=2V_1$ and $|I_m|=2V_1/Z_0$.

2. When $Z_R=i\sqrt{3}\, Z_0$, i.e. the load is purely inductive but less than $Z_0$, then, $|\Gamma_0|=1$, $\phi=\pi/3$. Hence

$$|V(-l)|=|V_m \cos(\beta l-\pi/6)|$$

$$|I(-l)|=|I_m \sin(\beta l-\pi/6)|$$

*This is the guide wavelengh and may be different from the free space wavelength (see Chapter 4).

3. When $Z_R = \frac{j}{\sqrt{3}} Z_0$, i.e. the load is purely inductive but less than $Z_0$, then

$$|\Gamma_0| = 1,\ \phi = 2\pi/3.$$

Hence
$$|V(-l)| = |V_m \cos(\beta l - \pi/3)|$$
$$|I(-l)| = |I_m \sin(\beta l - \pi/3)|.$$

4. When $Z_R = 0$, i.e. the line is short-circuited, then $|\Gamma_0| = 1$, and $\phi = \pi$. Hence

$$|V(-l)| = |V_m \sin \beta l|$$
$$|I(-l)| = |I_m \cos \beta l|.$$

5. When $Z_R = \frac{-jZ_0}{\sqrt{3}}$, i.e. the load is purely capacitive and greater than $Z_0$, then

$$|\Gamma_0| = 1 \text{ and } \phi = 4\pi/3$$
$$V(-l) = V_m \sin(\beta l - \pi/6)$$
$$I(-l) = I_m \cos(\beta l - \pi/6).$$

6. When $Z_R = \frac{-jZ_0}{\sqrt{3}}$, i.e. the load is purely capacitive and less than the characteristic impedance. Then

$$|\Gamma_0| = 1 \text{ and } \phi = (5\pi/3).$$

Hence
$$|V(-l) = |V_m \sin(\beta l - \pi/3)|$$
$$|I(-l)| = |I_m \cos(\beta l - \pi/3)|.$$

If the load is a pure resistance then $\phi$ will be zero and the voltage and current distribution given by Eqs. (2.73) and (2.74) may be reduced to

$$|V(-l)| = V_R \sqrt{\cos^2 \beta l + \left(\frac{Z_0}{Z_R}\right)^2 \sin^2 \beta l} \tag{2.77}$$

$$|I(-l)| = I_R \sqrt{\cos^2 \beta l + \left(\frac{Z_R}{Z_0}\right)^2 \sin^2 \beta l}. \tag{2.78}$$

Figure 2.10(b) shows the distribution of current and voltage along a transmission line terminated in pure resistances of different values.

It is noted that voltage maximum for the case $Z_R < Z_0$ occurs when $\sin \beta l = 1$ and has a value

$$V_{max} = V_R \frac{Z_0}{Z} = V_R \frac{R_0}{R}$$

$R_0 = Z_0$ is the characteristic impedance which is pure resistance as assumed for lowloss lines. Also the voltage minimum occurs when $\sin \beta l = 0$ and has a value $V_{min} = V_R$. The VSWR is given by

$$\rho = \frac{V_{max}}{V_{min}} = \frac{R_0}{R}. \tag{2.79}$$

For the termination $Z > Z_0$ (i.e., $R > R_0$) these relations are, however, reversed and so we have

$$\rho = R/R_0. \tag{2.80}$$

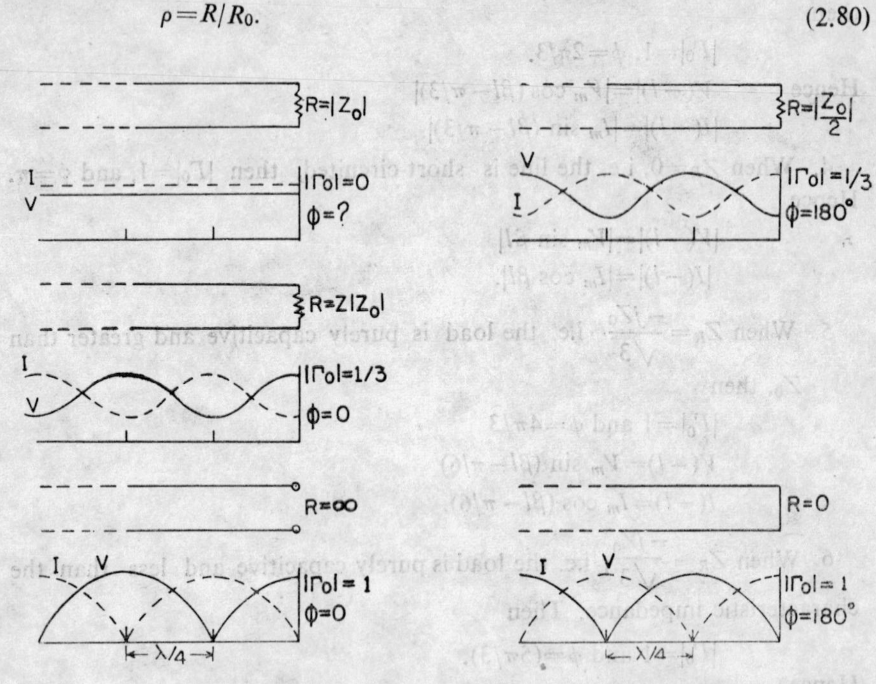


**Fig. 2.10(b)** Voltage and current standing waves corresponding to various resistive terminative.

**Lossy Line**

In case of lossy-lines $\gamma$ cannot be replaced by $j\beta$ and this makes mathematical computation somewhat tedious. It is therefore advisable to consider the hyperbolic functions. The voltage and current distribution along a lossy-line are then obtained from Eqs. (2.47) and (2.48) respectively by substituting $(z=-l)$. Thus

$$V(-l) = V_R \cosh \gamma l + Z_0 I_R \sinh \gamma l$$

$$I(-l) = I_R \cosh \gamma l + V_R/Z_0 \sinh \gamma l.$$

Only two cases are of interest, viz. the short-circuited line and the open-circuited line at the receiving end. In the first case $V_R=0$ while in the second $I_R=0$. Hence for a short-circuited lossy line the above equations can be written as

$$V(-l) = I_R Z_0 (\sinh \alpha l \cos \beta l + j \cosh \alpha l \sin \beta l)$$

$$I(-l) = I_R (\cosh \alpha l \cos \beta l + j \sinh \alpha l \sin \beta l)$$

where we have used $\sinh j\beta l = j \sin \beta l$, $\cosh j\beta l = \cos \beta l$

or
$$\left|\frac{V(-l)}{I_R Z_0}\right| = \sqrt{\sinh^2 \alpha l \cos^2 \beta l + \cosh^2 \alpha l \sin^2 \beta l} \tag{2.81}$$

$$\left|\frac{I(-l)}{I_R}\right|=\sqrt{\cosh^2 \alpha l \cos^2 \beta l+\sinh^2 \alpha l \sin^2 \beta l}. \tag{2.82}$$

Since $I_R$ and $Z_0$ are constants on the line, the parameters $|V(-l)/I_R Z_0|$ and $|I(-l)/I_R|$ will correspond to relative voltage and relative current amplitudes respectively existing at any point $(-l)$ on the line. As such Eqs. (2.81) and (2.82) represent the standing wave pattern of voltage and current respectively that exists on such a lossy line. The variations of the parameters $|V(-l)/I_R Z_0|$ and $|I(-l)/I_R|$ with $l$ are shown in Fig. 2.11(a). We see that voltage and current periodically oscillate but their amplitudes are exponentially

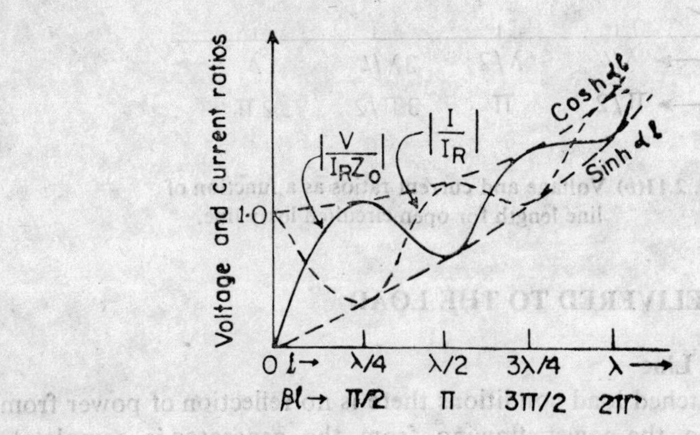


**Fig. 2.11(*a*)** Voltage ratio $|V/I_R Z_0|$ and current ratio $|I/I_R|$ as a function of line length for short-circuited lossy line.

attenuated due to the losses existing on the line. This can mathematically be understood from Eq. (2.81). We find that

$$\text{for } l=(2n-1)\frac{\lambda}{4}, \quad \cos \beta l=\cos \frac{2\pi}{\lambda}(2n-1)\frac{\lambda}{4}=0; \text{ and}$$

$$\text{for } l=2n\frac{\lambda}{4}, \ \sin \beta l=\cos\frac{2\pi}{\lambda}\cdot\frac{2n\lambda}{4}=0.$$

Therefore, $|V(-l)/I_R Z_0|$ has a value $\cosh \alpha l$ or $\sinh \alpha l$ depending upon whether the line length is an odd or even multiple of $\lambda/4$. Similarly, $|I(-l)/I_R|$ has values $\cosh \alpha l$ and $\sinh \alpha l$ depending upon even or odd integral values of $n$. The curves for $\cosh \alpha l$ and $\sinh \alpha l$, therefore, represent the decay in the amplitude of the standing wave pattern existing on a lossy line.

Similarly, the distribution of voltage and current along a lossy line terminated in an open-circuit can be studied. The results for such a line are shown in Fig. 2.11(*b*).

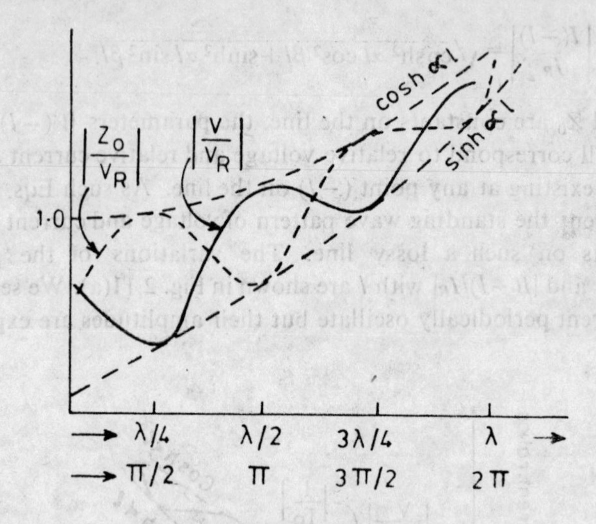


Fig. 2.11(b) Voltage and current ratios as a function of line length for open circuited lossy line.

## 2.13 POWER DELIVERED TO THE LOAD

**CASE I: Matched Line**

Since under matched load conditions there is no reflection of power from the termination, so the power flowing from the generator is completely absorbed in the termination (and the line if it is not lossless). The power delivered to the load is given by

$$P_o = \tfrac{1}{2}\, R_e\ (V_1 I_1^*) = \tfrac{1}{2}\ \frac{V_1^2}{Z_0} \tag{2.83}$$

where $V_1$ and $I_1$ are the complex voltage and current at the load ($z=0$).

**CASE II: Mismatched but Loss-less Line**

In case of a loss-less line terminated in a load $Z_L = Z_R$, the power delivered to the load is given by

$$P_L = \tfrac{1}{2}\, R_e(V_L\, I_L^*)$$

where $V_L$ and $I_L$ are the complex voltage and current at the load.

or $\quad P_L = \tfrac{1}{2}\, R_e\, [(V_1 + V_2)(I_1 + I_2)^*]$

$$= \frac{1}{2}\, R_e \left[ V_1(1+\Gamma_L)\ \frac{V_1}{Z_0}(1-\Gamma_L)^* \right]$$

$$= \frac{1}{2}\,\frac{V_1^2}{Z_0}\ [1 - |\Gamma_L|^2]$$

$\Gamma_L$ is the reflection coefficient at the load which on using Eq. (2.83), can be written as

$$P_L = P_0 [1 - |\Gamma_L|^2]. \tag{2.84}$$

Equation (2.84) represents the obvious result that the power delivered to the load is the difference between the incident and reflected powers.

**CASE III: Mismatched and Lossy-line**

The power flowing towards the load at any point $z = -l$ on the lossy line is given by

$$\begin{aligned} P(-l) &= \tfrac{1}{2} R_e [V(-l)\, I(-l)^*] \\ &= \tfrac{1}{2} R_e [V(-l) \{1+\Gamma\} \frac{V^*(-l)}{Z_0} \{1-\Gamma\}^*] \\ &= \frac{1}{2} \frac{|V(-l)|^2}{Z_0} [1 - |\Gamma_{(-l)}|^2] \\ &= \frac{1}{2} \frac{|V_1|^2}{Z_0} e^{2\alpha l} [1 - |\Gamma_{(-l)}|^2] \end{aligned} \tag{2.85}$$

where $|\Gamma_{(-l)}|$ is the magnitude of the reflection coefficient at a distance $l$ left of the load which is at $z = 0$. If the reflection coefficient at the load is $\Gamma_L$, then, from Eq. (2.71) it follows that

$$|\Gamma_L| = |\Gamma_{(l)}|^2 e^{2\alpha l}. \tag{2.86}$$

Using Eq. (2.86), Eq. (2.85) can be reduced to

$$P(l) = \frac{|V_1|^2}{2Z_0} (e^{2\alpha l} - |\Gamma_L|^2)$$

which on using Eq. (2.83) gives

$$P(l) = P_0 (e^{2\alpha l} - |\Gamma_L|^2). \tag{2.87}$$

Substracting Eq. (2.85) from Eq. (2.87) yields

$$P(l) - P_L = \frac{V_1^2}{2Z_0} (e^{2\alpha l} - 1) = P_0 (e^{2\alpha l} - 1).$$

Since $P(l) - P_L$ is nothing but the power dissipated in the line, the power dissipated in a lossy-line is given by

$$P_{\text{diss.}} = P_0 (e^{2\alpha l} - 1). \tag{2.88}$$

**Effect of Mismatch on Power Delivered to the Load**

To study the effect of mismatch on power delivered to the loan in case of a loss-less line, we divide power delivered to the load under mismatched conditions by the power delivered under matched conditions to yield

$$\frac{P_L}{P_0} = \frac{V_L^2}{2Z_L} \div \frac{V_L^2}{2Z_0} = \frac{Z_0}{Z_L}$$

which on using Eq. (2.75) yields

$$\frac{P_L}{P_0}=\frac{1}{\rho}. \tag{2.89}$$

This equation reveals that the greater the mismatch ($\rho$) the smaller is the relative power delivered to the load.

The mismatch also increases the attenuation constant $\alpha$.

Under mismatch conditions, Eq. (2.67) gives

$$\alpha=\frac{P_L}{2P}=\frac{|V_1|^2}{2Z_0}(1+|\Gamma|^2)\div\frac{2|V_1|^2}{2R}(1-|\Gamma|^2)$$

or
$$\alpha=\frac{R}{2Z_0}\left(\frac{1+|\Gamma|^2}{1-|\Gamma|^2}\right). \tag{2.90}$$

where $R$ is the resistance per unit length of the line

Under matched conditions $\alpha_m=R/2Z_0$.

Hence, the relative attenuation $\alpha_r$ is given by

$$\alpha_r=\alpha/\alpha_m=\frac{1+|\Gamma|^2}{1-|\Gamma|^2}=\frac{\rho^2+1}{2\rho}. \tag{2.90a}$$

It is seen from Eq. (2.90a) that as mismatch increases ($|\Gamma|$ increases) the attenuation constant of the line also increases. Therefore, in higher mismatch conditions more power is used to overcome the losses in the line. A detailed variation in $\alpha$ with mismatch is summarized in Fig. 2.12 which

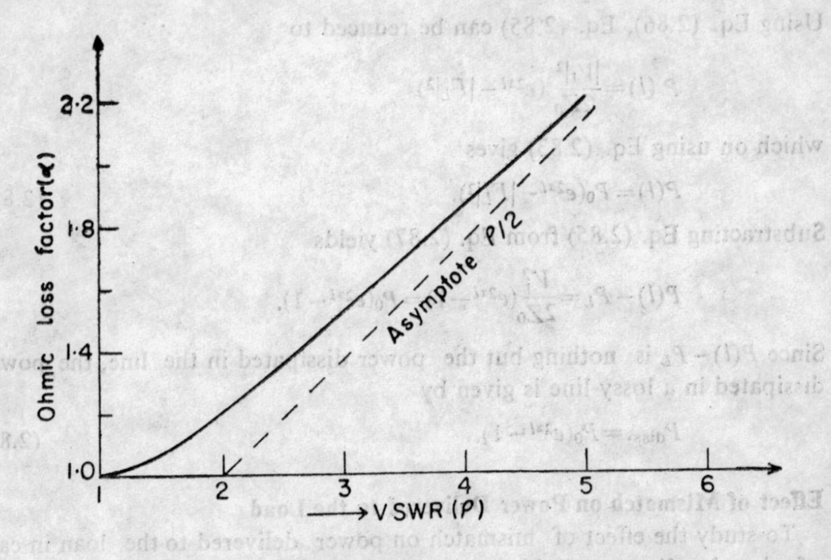


Fig. 2.12 Variation of ohmic loss factor with mismatch.

shows the variation of $\alpha$ with VSWR. It should also be noted that not only ohmic losses increase with mismatch but dielectric losses also increase owing to standing waves which result in higher values of mean square voltage across the line.

**Example 2.6.** Find the reflection coefficient, standing wave ratio and fraction of incident power delivered to the load when a transmission line of characteristic impedance 50 Ω is terminated in a load $(25+j25)$ Ω.

*Solution*: Normalized load is

$$z_L=\frac{25+j25}{50}=0.5+j\,0.5$$

$$\Gamma=\frac{z_L-1}{z_L+1}$$

$$=\frac{0.5+0.5j-1}{0.5+0.5j+1}=-\frac{0.5+j0.5}{1.5+j0.5}$$

$$=-0.2+0.4\,j$$

$$|\Gamma|=\sqrt{(0.2)^2+(0.4)^2}=0.447.$$

The angle of the reflection coefficient is given by

$$\tan\phi=\frac{0.4}{-0.2}=-2$$

$$\phi=\tan^{-1}(-2)$$

$$\text{VSWR}\ \rho=\frac{1+|\Gamma|}{1-|\Gamma|}=\frac{1+0.447}{1-0.447}\simeq 2.62.$$

Fraction of power delivered to the load is

$$\frac{P}{P_0}=1-|\Gamma|^2=1-0.20=0.80$$

i.e. 80 percent.

**Example 2.7.** Show that the time average power flowing down a loss-less transmission line is $|V_{max}|^2/2(\text{VSWR})\,Z_0$.

*Solution*:

$$P_{av}=\tfrac{1}{2}R_e(VI^*)$$

$$=\tfrac{1}{2}R_e\left[V_1\,e^{j\beta l}\{1+\Gamma_{(l)}\}\cdot\frac{V_1^*}{Z_0}\ e^{-j\beta l}\{1-\Gamma_{(l)}^*\}\right]$$

$$=\tfrac{1}{2}R_e\left[\frac{|V_1|^2}{Z_0}\{1-|\Gamma_{(l)}|^2+\Gamma_{(l)}-\Gamma_{(l)}^*\}\right]$$

$$=\frac{V_1^2}{2Z_0}[1-|\Gamma_{(l)}|^2]$$

$$=\frac{V_1^2}{2Z_0}[1-|\Gamma_0|^2]$$

$$=\frac{|V_1|^2[1+|\Gamma_0|][1-|\Gamma_0|]}{2Z_0}$$

$$=\frac{|V_{max}|\,|V_{min}|}{2Z_0}.$$

Since $\frac{|V_{max}|}{|V_{min}|}=\text{VSWR}=\rho$

$$P_{av}=\frac{|V_{max}|^2}{2\rho Z_0}$$

$$=\frac{|V_{max}|^2}{2(\text{VSWR})\times Z_0}.$$

## 2.14 RESONANT AND ANTIRESONANT LINES

While analyzing the input impedance characteristics of a lossless line in Section 2.9 we observed that short-circuited lines behaved as conventional resonant and antiresonant circuits depending upon whether their length was an even or odd multiple of quarter wavelength. The open-circuited lines are resonant and antiresonant when their lengths are odd or even multiples of $\lambda/4$. The mechanism of resonance is easy to visualize. Let us assume that a small voltage signal is induced into a shorted quarter wave ($l=\lambda/4$) line near its shorted end. There will be a voltage wave sent down the line and reflected without change of phase at the open end. This reflected wave travels back and is reflected again at the shorted end with reversal of phase. Because it is required that one half cycle travel up and down the line, this twice-reflected wave will now be in phase with the original induced voltage and so adds directly to it. These additions will continue to increase the voltage (and current) in the line until $I^2R$ loss is equal to the power being put into the line. Such a line will behave as an antiresonant line. A voltage step up of several hundred times is possible depending upon the quality of the line.

Since the voltage and current step up depends upon the quality of the line, it is therefore of interest to consider the order of magnitude of $Q$, the quality factor and impedance that can be obtained.

### Series Resonance Short-Circuited Line

Consider a short-circuited air-filled line* of length $l$ having parameters $R$, $L$, and $C$ per unit length. Let the length of the line be $l=\lambda/2$ at $f=f_0$, i.e. at $\omega=\omega_0$. The Line is resonant at frequency $f_0$. At frequency near $f_0$ say $f=f_0+\Delta f$ we can write

$$\beta l=\frac{2\pi fl}{c}=\frac{\pi\omega}{\omega_0}=\pi+\pi\frac{\nabla\omega}{\omega_0}. \tag{2.91}$$

*Air-filled line has been taken so that the losses in dielectrics ($G$) may be neglected. This is also usually the case.

The input impedance of a short-circuited line is given by Eq. (2.52) and can be written as

$$Z_{sc}=Z_0 \tanh(\alpha l+j\beta l)=Z_0\frac{\tanh \alpha l+j\tan \beta l}{1+j\tan \beta l \tanh \alpha l}. \tag{2.92}$$

For low-loss lines $\tanh \alpha l \simeq \alpha l \ll 1$ and

$$\tan \beta l=\tan\left(\pi+\pi\frac{\Delta\omega}{\omega_0}\right)=\tan \pi\frac{\Delta\omega}{\omega_0}.$$

Since $\Delta\omega/\omega_0$ is small and $\tan \pi\Delta\omega/\omega_0 \simeq \pi\Delta\omega/\omega_0$, Eq. (2.92) reduces to

$$Z_{sc}=Z_0\frac{\alpha l+j\pi\frac{\Delta\omega}{\omega_0}}{1+j\alpha l\cdot\pi\frac{\Delta\omega}{\omega_0}}.$$

Since the second term in the denominator of the above equation is the product of two small terms and may be neglected as compared to other terms,

$$Z_{sc}\simeq Z_0\left(\alpha l+j\pi\frac{\Delta\omega}{\omega_0}\right). \tag{2.93}$$

Using parametric expression for $\alpha$, $\omega$ and $Z_0$, i.e.

$$Z_0=\sqrt{L/C},\ \alpha=(R/2)\sqrt{C/L},\ \text{and}\ \beta l=\omega_0\sqrt{LC}l,=\pi \quad \text{Eq. (2.93)}$$

yields

$$Z_{sc}=\sqrt{L/C}\,(\tfrac{1}{2}R\sqrt{C/L}\,l+j\Delta\omega l\sqrt{LC})=\tfrac{1}{2}Rl+jlL\Delta\omega. \tag{2.94}$$

It is of interest to compare Eq. (2.94) with a series resonant $L_0C_0R_0$ circuit shown in Fig. 2.13.

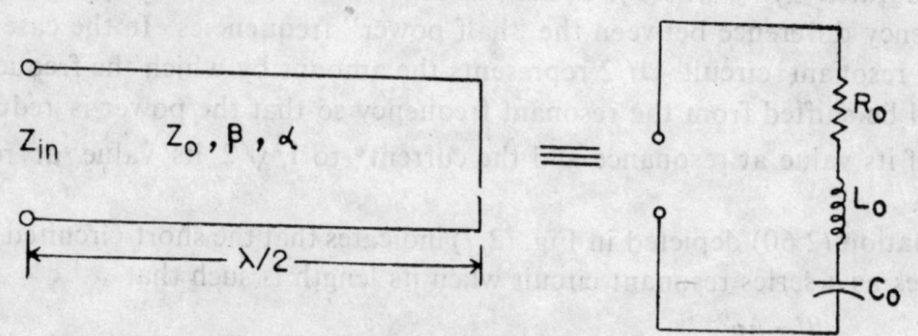


**Fig. 2.13** Comparison of a half wave shorted line with series resonant circuit.

The input impedance of such a circuit is given by

$$Z_{in}=R_0+j\omega L_0\left(1-\frac{1}{\omega^2L_0C_0}\right).$$

Taking the resonant frequency of such a circuit to be $f_0$

$$2\pi f_0=\omega_0=\frac{1}{\sqrt{L_0C_0}}$$

or $$Z_{in}=R_0+j\omega L_0\frac{\omega^2-\omega_0^2}{\omega^2}.$$

$$=R_0+j\omega L_0\frac{2\omega\Delta\omega}{\omega^2}.$$

Since $(\omega+\omega_0\simeq 2\omega$, and $\omega-\omega_0=\Delta\omega$.

$$Z_{in}=R_0+j\omega L_0\frac{2\omega\Delta\omega}{\omega^2}. \tag{2.95}$$

$$R_0+j2L_0\nabla\omega.$$

The comparison of Eq. (2.94) with Eq. (2.95) suggests that in the vicinity of frequency $f_0$ when $l=\lambda_0/2$, a short-circuited line behaves as a series resonant circuit with equivalent resistance $R_0=Rl/2$ and equivalent inductance $L_0=Ll/2$. The reducing factor $\frac{1}{2}$, appears because half the sinusoidal waves in voltage and current have been considered.

We know that for a series circuit of Fig. 2.13, the $Q$ value is defined as the ratio of inductive (or capacitive) reactance at resonance to the total series resistance. Similarly the merit or quality factor in the line is given by

$$Q=\frac{\omega_0 L_0}{R_0}=\frac{\omega_0 L}{R}=\frac{\beta}{2\alpha}. \tag{2.96}$$

Equation (2.96) is important because $Q$ measures the losses and selective properties of the line. This equation can alternatively be derived from the basic definition of the quality factor. The quality factor is the inverse of *selectivity* or the ability of a resonant circuit to pass some frequencies freely, but to discriminate against others. This is conveniently stated in terms of the ratio $\Delta f/f_0$, where $f_0$ is the resonant frequency and $\Delta f=f_2-f_1$ is the frequency difference between the "half power" frequencies. In the case of a series resonant circuit $\Delta f/2$ represents the amount by which the frequency should be shifted from the resonant frequency so that the power is reduced to half its value at resonance and the current* to $1/\sqrt{2}$ its value at resonance.

Equation (2.60) depicted in Fig. (2.7) indicates that the short-circuited line behaves as a series resonant circuit when its length is such that

$$\beta l=n\pi.$$

Let us assume the line to be a low-loss line so that $\alpha l\ll\beta l$ and $\sinh^2\alpha l\simeq(\alpha l)^2$ and $\cosh^2\alpha l\simeq 1$. Furthermore, moving off the resonance, we can write

$$\beta l=n\pi+\delta \tag{2.97}$$

where $\delta$ is a small phase shift.

Under these conditions, Eq. (2.60) reduces to

$$|Y_{sc}|=\frac{1}{Z_0}\frac{1}{\sqrt{(\alpha l)^2+\delta^2}}. \tag{2.98}$$

*A constant current source has to be used.

Since $\alpha l$ varies very slowly as compared to $\delta$ therefore from Eq. (2.97) $l \simeq n\pi/\beta$. Hence

$$\alpha l \simeq \alpha \cdot \frac{n\pi}{\beta}.$$

Substitution of the above result in Eq. (2.98) and squaring gives

$$|Y_{sc}|^2 = \frac{1}{Z_0^2}\left[\frac{1}{\delta^2 + \left(\frac{\alpha n\pi}{\beta}\right)^2}\right]. \tag{2.99}$$

We see from Eq. (2.97) that admittance is maximum and the line is resonant when $\delta = 0$. Therefore, the power will fall to half of its value when the admittance reduces to half of its value at resonance. This will be when

$$\delta = \pm \frac{\alpha}{\beta} n\pi. \tag{2.100}$$

Therefore, the condition of Eq. (2.100) shall correspond to half power frequency points. Since $\beta l$ is proportional to frequency, the ratio of the difference of two values of $\delta$ and $\beta l$ will correspond to the half power frequency band width. Thus we have

$$\frac{\Delta f}{f} = \frac{2\alpha}{\beta}$$

or

$$Q = \frac{f}{\Delta f} = \beta/2\alpha. \tag{2.101}$$

Using parameteric expressions for $\beta$ and $\alpha$ it can be varified that $Q = \omega_0 L/R$.

**Example 2.8.** Find the minimum length of the short-circuited line where the generator is to be connected in order that it behaves as a resonant line vhen operated at 9 kHz, 9 MHz and 9 GHz.

*Solution*: $Z_{sc} = Z_0 \tanh \beta l = 0$ (for resonance)

$\tan \beta l = \tan n\pi = \tan \pi (n = 1 \text{ for } l_{\min})$.

So $$l_{\min} = \frac{\pi}{\beta} = \frac{\lambda}{2} = \frac{C}{2f}$$

(*i*) At $f = 9$ kHz, $l_{\min} = \frac{C}{2f} = \frac{3 \times 10^8}{2 \times 9 \times 10^3} = 1.667 \times 10^4$ m

(*ii*) At $f = 9$ MHz, $l_{\min} = \frac{3 \times 10^8}{2 \times 9 \times 10^6} = 16.67$ m

(*iii*) At $f = 9$ GHz, $l_{\min} = \frac{3 \times 10^8}{2 \times 9 \times 10^9} = 1.667$ cm

**Example 2.9.** Calculate the respective lengths for an open-circuited line in Ex. 2.8.

$$Z_{oc}=Z_o \coth \beta l=0$$

$$\beta l=\pi/2(2n-1);\quad n=1, 2, 3$$

$$l_{\min}=\frac{\pi}{2\beta}=\frac{\lambda}{4}=\frac{C}{4f}$$

(i) $l_{\min}=\frac{3\times 10^8}{4\times 9\times 10^3}=8.33\times 10^3$ m

(ii) $l_{\min}=\frac{3\times 10^8}{4\times 9\times 10^6}=8.33$ m

(iii) $l_{\min}=\frac{3\times 10^8}{4\times 9\times 10^9}=0.833$ cm.

It may be noted that resonant line lengths are comparable to circuit dimensions at microwave frequencies (GHz) so these lines can suitably be used as resonant and antiresonant circuits.

The value of $Q$ for a resonant line can also be obtained from the general definition of $Q$, i.e.

$$Q=\omega_0\,\frac{\text{Time average energy stored in the resonant system}}{\text{Energy loss per second in the system}}\cdot \tag{2.102}$$

This is the general definition and is applicable to all resonant systems.

In case of a quarterwave shorted line the current on the line is a pure standing wave given by

$$I=I_m \cos \beta l\, e^{j\omega t}.$$

We know that in a low-less line, the total time average energy stored in the magnetic field is equal to the time average electric energy stored and hence total time average energy stored in the resonant circuit is

$$P_{\text{av}}=2\int_0^{\lambda_0/2}\frac{1}{4}LII^*\,dl=2\times\int_0^{\lambda_0/2}\frac{1}{4}\cdot LI_m^2\cos^2\beta l\,dl$$

$$=\frac{\lambda_0}{8}\,I_m^2L. \tag{2.103}$$

Assuming that losses do not modify the current distribution, the power loss per cycle is given by

$$P_{\text{loss}}=\frac{1}{2}\int_0^{\lambda_0/2}RII^*\,dl=\frac{1}{2}\int_0^{\lambda_0/2}I_m^2\cos^2\beta l\,dl$$

$$=\frac{\lambda_0}{8}RI_m^2. \tag{2.104}$$

Substitution of Eqs. (2.103) and (2.104) in Eq. (2.102) gives

$$Q=\frac{\omega_0P_{\text{av}}}{P_{\text{loss}}}=\frac{\omega_0L}{R}=\frac{\beta}{2\alpha}. \tag{2.105}$$

This follows from the fact that

$$\alpha=\frac{1}{2}\frac{R}{Z_0}=\tfrac{1}{2}R\sqrt{C/L}$$

$$\beta=\omega_0\sqrt{LC}.$$

**Open-Circuited Resonant Line**

We know that an open-circuited line behaves as a resonant circuit at a frequency when its length is an odd multiple of quarter wavelength. Let $l=\lambda_0/4$ at $f=f_0$, that is at $\omega=\omega_0$, and for $f$ near $f_0$ say $f=f_0+\Delta f$, $\beta l=\frac{2\pi fl}{c}=\frac{\pi}{2}\,\frac{\omega}{\omega_0}=\frac{\pi}{2}+\frac{\Delta\omega}{\omega_0}\,\frac{\pi}{2}$ since at $\omega_0$, $\beta l=\pi/2$. The input impedance of a line is given by Eq. (2.53).

$$Z_{oc}=Z_0 \coth \gamma l=Z_0 \coth(\alpha l+j\beta l)$$

since $$\cot \beta l=\cot\left(\frac{\pi}{2}+\frac{\Delta\omega}{2\omega_0}\,\pi\right)=\cot\frac{\Delta\omega}{2\omega_0}\,\pi.$$

Hence under low-loss conditions $\tanh \alpha l \simeq \alpha l \ll 1$ and $Z_{oc}$ will have a value

$$Z_{oc}=Z_0\frac{\alpha l+j\frac{\pi}{2}\,\Delta\omega/\omega_0}{1+j\alpha l\frac{\pi}{2}\,\Delta\omega/\omega_0}.$$

Since the second term in the denominator is small neglecting the same we have

$$Z_{oc}\simeq Z_0\left(\alpha l+j\frac{\pi}{2}\,\frac{\Delta\omega}{\omega_0}\right).$$

Since in an open-circuited case $\beta l=\omega_0\sqrt{LC}\,l=\pi/2$

so $$\frac{\pi}{2\omega_o}=l\sqrt{LC},\quad \alpha=\frac{1}{2}\,\frac{R}{Z_0}=\frac{R}{2}\sqrt{\frac{C}{L}}.$$

Thus $$Z_{oc}=\sqrt{L/C}(\tfrac{1}{2}R\sqrt{C/L}+j\,\Delta\omega l\sqrt{LC})=\tfrac{1}{2}Rl+jlL\Delta\omega. \tag{2.106}$$

Equation (2.106) is identical with Eq. (2.94). It therefore follows that an open-circuited quarter wave line behaves as a series resonant circuit similar to a short-circuited half wave line.

**Short-Circuited Antiresonant Line**

A short-circuited line behaves as an antiresonant circuit in the frequency range where it has a length which is close to an odd multiple of quarter wavelength.

Taking $l=\lambda_0/4$ at $\omega=\omega_0$ then at the frequency off the resonance we have

$$\beta l=\omega\sqrt{LC}\,l=\omega_0\sqrt{LC}\,l+\Delta\omega\sqrt{LC}\,l$$

and at $\omega$ the input admittance is given by

$$Y_{sc}=Y_0 \coth(\alpha l+j\beta l)$$

$$=Y_0\frac{1+j\tan\beta l\tanh\alpha l}{\tanh\alpha l+j\tan\beta l}.$$

Since $\tan(\omega_0\sqrt{LC}\,l+\Delta\omega\sqrt{LC}\,l)=\tan(\pi/2+\Delta\omega\sqrt{LC}\,l)$

$$=-\cot(\Delta\omega\sqrt{LC}\,l)$$

$$\simeq-(\Delta\omega\sqrt{LC}\,l)^{-1}.$$

For low-loss lines $\tanh\alpha l\simeq\alpha l$. Therefore,

$$Y_{sc}=Y_0\frac{j\Delta\omega\sqrt{LC}\,l+\alpha l}{1+j\Delta\omega\alpha l^2\sqrt{LC}}.$$

The second term in the denominator is the product of two small (i.e. $\ll 1$) terms. Neglecting the same, we have

$$Y_{sc}\simeq Y_0(j\Delta\omega\sqrt{LC}\,l+\alpha l)$$

$$Y_{sc}\simeq\left(\frac{RC}{2L}\,l+j\,\Delta\omega Cl\right) \tag{2.107}$$

where we have used $Y_0=\sqrt{C/L}$ and $\alpha=Y_0R/2$ comparing Eq. (2.107) with the input admittance of the antiresonant circuit shown in Fig. 2.14 wit' parameters $L_0C_0R_0$.

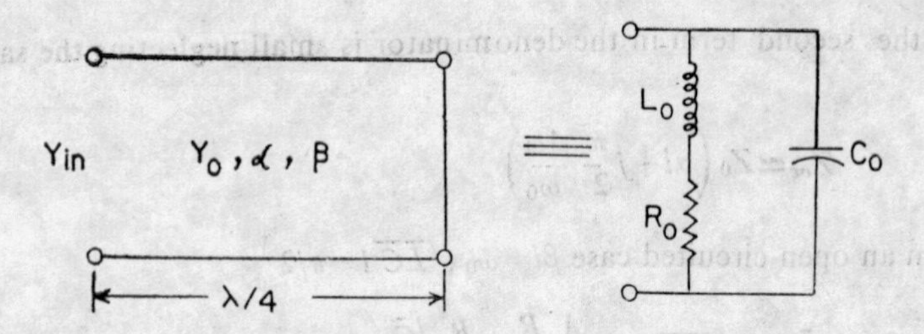


**Fig. 2.14** Comparison of a quarter short-circuited line with antiresonant circuit.

$$Y_{sc}=\frac{1}{R_0+j\omega L_0}+j\omega C_0$$

$$=\frac{j\omega C_0(R_0+j\omega L_0)+1}{R_0+j\omega L_0}.$$

Assuming low-loss conditions $R_0\ll j\omega L_0$ and defining $\omega_0^2L_0C_0=1$, we have

$$Y_{sc}=R_0\frac{C_0}{L_0}-jL_0C_0\frac{(\omega_0-\omega)(\omega_0+\omega)}{\omega L_0}$$

$$\simeq R_0\frac{C_0}{L_0}+jC_0\,2\Delta\omega. \tag{2.108}$$

Comparison of Eqs. (2.108) and (2.107) leads to the conclusion that a short-circuited line in the vicinity of a quarter wavelength is equivalent to an antiresonant circuit with

$$\frac{C_0R_0}{L_0}=\frac{RCl}{2L} \quad\text{and}\quad Cl=C_0,\ \therefore\ \frac{R_0}{L_0}=R/L.$$

The $Q$ value of the circuit is given by

$$Q=\frac{\omega_0 L_0}{R_0}=\frac{\omega L}{R}=\beta/2\alpha. \tag{2.109}$$

Similarly, it may be shown that even a multiple quarter wave open-circuited line behaves as an antiresonant circuit. $Q$ value may alternatively be derived for this case also.

Although, sections of transmission lines behave as resonant and antiresonant circuits, they are in reality a much more complicated network because every line has an infinite number of resonant and antiresonant frequencies and thus an exact equivalent circuit would consist of an infinite number of resonant circuits coupled together. $Q$ values also vary with frequency because equivalent parameters $R$, $L$ and $C$ vary with frequency unlike lumped constants.

## 2.15 TRANSMISSION LINE SECTIONS AS IMPEDANCE AND VOLTAGE TRANSFORMER

### Impedance Transformer

The input impedance of a loss-less line terminated at a distance $l$ from the generator is given by

$$Z_{in}=Z(-l)=Z_0\frac{Z_R\cos\beta l+jZ_0\sin\beta l}{Z_0\cos\beta l+jZ_R\sin\beta l}.$$

If the length of the line be such that at a given frequency, $\beta l=n\pi$, i.e. line length is an integral multiple of half wavelength, then $\cos\beta l=\cos n\pi=\pm1$, $\sin\beta l=\sin n\pi=0$. Then

$$Z_{in}=Z_0\,Z_R/Z_0=Z_R. \tag{2.110}$$

Thus an integral multiple half wavelength section of a transmission line has the effect of transforming the load normally at the remote end, i.e. source. The multisection half wave section behaves as a normal or 1 : 1 impedance transformer. The basic advantage of this transformer is that, using the section as a four terminal network, any desired load may be connected to a source and, the generator will match the load.*

If instead, we consider a transmission line section of length equal to an odd multiple of $\lambda/4$, i.e. $\beta l=(2n-1)\,\pi/2$, $n$ is an integer. Then $\cos\beta l=0$ and $\sin\beta l=\pm1$. The input impedance of such a section is given by

$$Z_{in}=Z_0\frac{jZ_0}{jZ_R}=\frac{Z_0^2}{Z_R}$$

or

$$Z_0=\sqrt{Z_{in}\times Z_R}. \tag{2.111}$$

It is seen from Eq. (2.111) that impedance $Z_R$ at the termination is transformed to an impedance $Z_0^2/Z_R$ at the generator. If $Z_R$ is very small, it will

*This is desired for maximum power transfer (see Chapter 6A).

appear as a large impedance and vice-versa at the generator end. If a quarter wave line is short-circuited at the receiving end ($Z_R=0$) it will appear at the generator as very high ($Z_{in}=\infty$) impedance and the line acts as a *quarter wave choke*. Moreover, whatever the terminating impedance may be, the impedance ($Z_0^2/Z_R$) shall appear at the input and in this way the quarterwave section is an impedance transformer with a turn ratio $\sqrt{Z_R/Z_{in}}$ or $Z_R/Z_0$ more correctly an *impedance inverter*. If the terminating impedance consisted of a resistance $R_R$ in *series* with an inductance $X_{LR}$, the input impedance would be given by a resistance $R_{in}$ *parallel* with a reactance $X_{C_{in}}$ where

$$R_{in}=\frac{R_0^2\,R_R}{Z_R^2} \quad \text{and} \quad X_{C_{in}}=\frac{R_0^2\,X_L}{Z_R^2}. \tag{2.112}$$

A pure resistance $R_R$ at the termination is transformed into a pure resistance $R_0^2/R_R$ at the input.

Relation $Z_0=\sqrt{Z_{in}\times Z_R}$ is of enormous practical importance. It provides a way to join together, without impedance mismatch, lines having different characteristic impedances say $Z_1$ and $Z_2$; it is only necessary to make the characteristic impedance of the quarterwave matching section equal to the geometric mean of the two $Z_{02}$ to be matched, i.e.

$$Z_0=\sqrt{Z_1Z_2}.$$

So far we have discussed half and quarter wave transformers. However, one may also be interested in the intermediate lengths of the connecting line which also behave as an impedance transformer. The transformation equations for the resistive and reactive components of the terminating load may be obtained by separating Eq. (2.51) into real and imaginary parts. This is done where exact results are required otherwise graphical methods serve the purpose for moderately accurate results. One such graphical method, the Smith Chart, will be discussed in section (2.17).

**Voltage Transformer**

The voltage transforming action of the quarter wave section is quite evident from its impedance transforming action the line is considered loss-less, the ratio of input to output voltages which will be the same as the square root of the ratio of input (sending end) and output (receiving end) impedances, i.e.

$$\frac{V_s}{V_R}=j\frac{I_R Z_0}{V_R}=j\frac{Z_0}{Z_R}=j\sqrt{Z_s/Z_R}$$

or

$$\frac{V_s}{V_R}=\sqrt{\frac{Z_s}{Z_R}}.$$

The turn ratio is $\sqrt{Z_s/Z_R}$.

The short circuit at the termination will result in an infinite voltage step up at the generator. This result is inconsistent with physical observations, the reason being that we have taken the line to be lossless. However, for real or low-loss lines, the voltage at the sending end is obtained from Eq. (2.47) by putting $z=-l$. For a quarter wavelength $\cos \beta l=0$ $\sin \beta l=1$, so

$$V_s = V_R \sinh \alpha l + jI_R Z_0 \cosh \alpha l.$$

Taking an open circuit at the termination, $I_R=0$

$$\frac{V_R}{V_s} = \frac{1}{\sinh \alpha l} \frac{1}{\alpha l} = \frac{2Z_0}{Rl}.$$

For a quarter wave section $(l=\lambda/4)$

$$\left|\frac{V_R}{V_s}\right| = \frac{2Z_0}{R\,\lambda/4} = \frac{8Z_0}{R\lambda} = \frac{8Z_0 f}{Rv}. \tag{2.113}$$

For a three-quarter wave section

$$\frac{V_R}{V_s} = \frac{8Z_0 f}{3Rv}.$$

It is noted from Eq. (2.113) that the voltage step up depends upon the frequency of operation and the characteristic impedance of the line.

## 2.16 IMPEDANCE MATCHING WITH REACTIVE ELEMENTS

In microwave power transmission circuits, if the load or terminating impedance happens to be different from $Z_0$ (characteristic impedance of the line) then reflections set in due to mismatch and result in undesirable standing waves. These standing waves impair maximum transfer of power from the generator to the load. To avoid this power loss and the possibility of insulation breakdown (wave discharge due to high voltage), transmission line (lossless) sections of length $l$, such that* $(2n+1)\,\lambda/4 > l > 2n\,\lambda/4$ where $n$ is an integer, are used. These sections produce the desired reactance required for the match. These lines are commonly known as *tuning stubs* or *stub transformers* or simply *stub*. These may be sections of either open-circuited or short-circuited lines but since an open circuit is difficult to obtain precisely so short-circuited line quarter wave stubs are commonly used. Moreover, these have better mechanical rigidity, less power radiation and easy adjustment.

The stubs may be placed in series or parallel to the load to be matched. We shall confine our discussion to the commonly used shunt (parallel) stubs; however, for series stubs refer to the literature.**

### Single Stub Tuner

Consider a line terminated in a pure resistive load of normalized*** admittance† $y_r$ (or $y_L$) as shown in Fig. 2.15. Let a shorted line of length $l_0$ be

*True for shorted stub.

**R.E., Collin, *Foundations of Microwave Engineering*, New York McGraw-Hill Book Co. (1966).

***Normalized load admittance is the ratio of normalized load admittance with the characteristic admittance $y_R = Y_R/Y_0$.

†While discussing shunt systems the analysis in terms of admittance greatly simplifies the mathematical procedure.

placed at a distance $d$ from the termination. The normalized input admittance at point $AA$, i.e. at the position of the stub is given by

$$\frac{Y_{\text{in}}}{Y_0}=y_{\text{in}}=1+jb \tag{2.114}$$

where $b$ is normalized input susceptance.

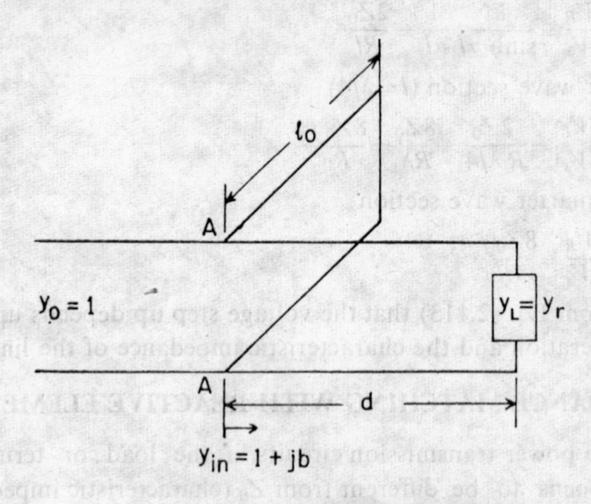


**Fig. 2.15** Single stub matching network.

For the line to be matched the stub connected at $AA$ must offer an input admittance such that $y_{\text{in}}=1$, i.e. its input normalized input susceptance must be $-jb$. Moreover, the stub should be connected nearest to the load to have minimum frequency sensitivity but offer an input susceptance $-jb$. To find the position of the stub $d$ one has to solve the following equation:

$$y_{\text{in}}=1+jb=\frac{y_L+j\tan\beta d}{1+jy_L\tan\beta d}. \tag{2.115}$$

As we have assumed pure resistive termination so $y_L=g$ (conductance). Thus

$$(1+jb)(1+jg\tan\beta d)=g+j\tan\beta d.$$

Equating real and imaginary parts we obtain

$$\left.\begin{aligned}1+bg\tan\beta d&=g\\ j(b+g\tan\beta d)&=j\tan\beta d\end{aligned}\right\}. \tag{2.116}$$

Equation (2.116) on solving give

$$\tan^2\beta d=\frac{1}{g}$$

or $$\frac{1-\cos^2\beta d}{\cos^2\beta d}=\frac{1}{g}$$

or $$d=\frac{\lambda}{2\pi}\cos^{-1}\sqrt{\frac{g}{1+g}} \tag{2.117}$$

where $\beta=2\pi/\lambda$.

It is seen from Eq. (2.117) that two principal values of $d$ are possible depending on which sign is chosen for the square root. An alternate relation is obtained if we replace $2\cos^2\beta d$ by $1+\cos 2\beta d$. Thus

$$1+\cos 2\beta d=\frac{2g}{1+g}$$

$$\cos 2\beta d=\frac{g-1}{g+1}$$

$$d=\frac{\lambda}{4\pi}\cos^{-1}\frac{g-1}{g+1}. \tag{2.118}$$

If $d_1$ is a solution of Eq. (2.118), then $\lambda/2-d_1$ is another principal solution since $\pm d_1 \pm n\lambda/2$ are all solutions of Eq. (2.118).

Now having located the position $d$ of the stub, the stub length $l_0$ has to be determined. The first of Eq. (2.116) gives

$$b=(1-g)\tan\beta d=\frac{(1-g)}{\sqrt{g}}. \tag{2.119}$$

Since the shorted stub of length $l_0$ must offer a normalized input susceptance $-jb$, we have

$$-jb=-j\cot\beta l_0$$

which on using Eq. (2.119) gives

$$\cot\beta l_0=\frac{1-g}{\sqrt{g}}$$

or $$l_0=\frac{\lambda}{2\pi}\tan^{-1}\left(\frac{\sqrt{g}}{1-g}\right). \tag{2.120}$$

However, the sign $\sqrt{g}$ must be chosen in such a way that it gives the correct sign to $b$ in Eq. (2.119). For $0<d<\lambda/4$, the positive square root should be used whereas in other solutions $\lambda/4<d<\lambda/2$, the negative square root must be used.

A similar analysis may be carried out when the load is complex but the analysis also becomes complicated. The following procedure is usually followed instead. First we locate the position of minimum voltage from the termination. Since at this position the reflection coefficient is a negative real quantity, the input admittance is *pure real* and is given by

$$Y_{\text{in}}=\frac{1-\Gamma}{1+\Gamma}=\frac{1+|\Gamma|}{1-|\Gamma|}=\rho\ (\text{VSWR}). \tag{2.121}$$

Referring to Fig. 2.16, let $d_0$ be the distance of the point from the minimum voltage position where normalized input admittance has a value $1+jb$. Then

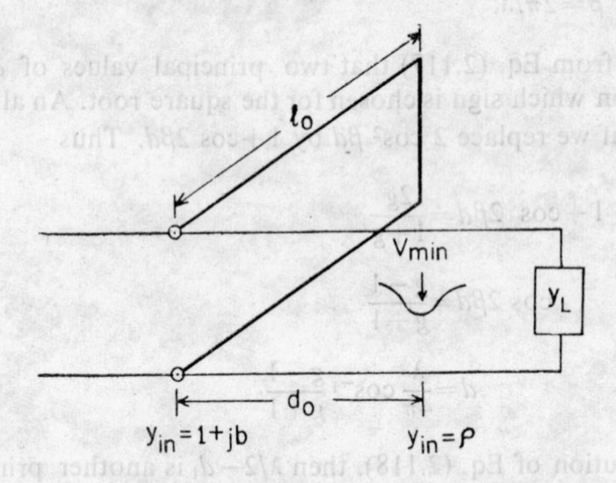


**Fig. 2.16** Location of stub relative to a voltage minimum.

the values of $d_0$ and $l_0$ will be the same as above but with $g$ replaced by $\rho$, i.e.

$$d_0=\frac{\lambda}{4\pi}\cos^{-1}\left(\frac{\rho-1}{\rho+1}\right)$$

$$l_0=\frac{\lambda}{2\pi}\tan^{-1}\left(\frac{\sqrt{\rho}}{\rho-1}\right). \tag{2.122}$$

The disadvantage of a single stub matching system is that every load requires a new stub position ($d$). This disadvantage is overcome in a system of two tuning stubs spaced by a fixed distance and located at a fixed distance from the load. Such a system is called a *double stub tuner*.

**Example 2.10.** A 50 Ω lossless transmission line is connected to a load of $30-j40$ Ω. A second identical but short-circuited line is connected at a point in parallel to the said line. Find out the location (distance from the load) and the length of the shorted line (stub) so that there is maximum transfer of power from the generator to the load.

*Solution*: Maximum transfer of power occurs when the load is well matched. As shown in Fig. Ex. 2.10, at plane *aa* the input admittance must be $Y_0$. Thus the location of the stub should be such that the real part of input admittance at *aa* be $Y_0$ and the stub length be such that it offers a susceptance equal and opposite to the input susceptance at *aa* on the line. To evaluate the desired parameters we proceed as follows:

(*i*) $\Gamma_0=\dfrac{Z_L-Z_0}{Z_L+Z_0}=\dfrac{30-j40-50}{30-j40+50}=0.5\, e^{-j\pi/2}$

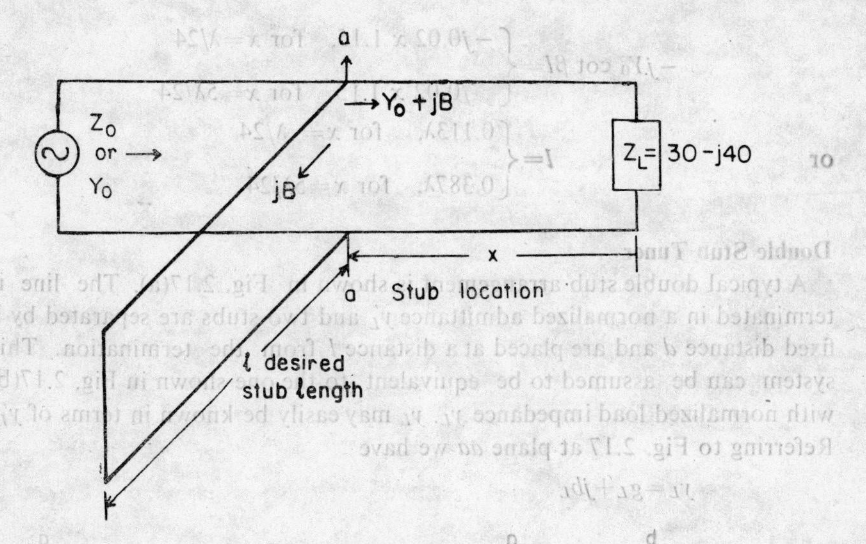


**Fig. Ex. 2.10**

(*ii*) $\Gamma(-x)=\Gamma_0\, e^{-j2\beta x}=0.5\, e^{-j(2\beta x+\pi/2)}$

(*iii*) Input admittance

$$Y_{in}=Y(-x)=\frac{1}{Z_0}\,\frac{1-\Gamma(-x)}{1+\Gamma(-x)}$$

$$=\frac{1}{50}\,\frac{1-0.5\,e^{j(2\beta x+\pi/2)}}{1+0.5\,e^{-j(2\beta x+\pi/2)}}$$

$$=0.02\,\frac{0.75+j\cos\beta x}{1.25-\sin 2\beta x}.$$

Since the real part of $Y_{in}=1/Z_0=0.02$, then

$$0.02\,\frac{0.75}{1.25-\sin 2\beta x}=0.02$$

$$\sin\beta x=0.5 \quad\text{or}\quad x=\frac{\lambda}{24} \quad\text{or}\quad \frac{5\lambda}{24}.$$

(There are two stub locations.)

The values of the susceptances at these stub locations are

$$B=\begin{cases} 0.02\ x\ 1.15, & \text{for } x=\dfrac{\lambda}{24} \\ 0.02\ x\ (-1.15), & \text{for } x=\dfrac{5\lambda}{24}. \end{cases}$$

If the corresponding length of the stub be $l$ then for the shorted stub,

$$Y_{in}=\frac{1}{jZ_0\tan\beta l}=-jY_0\cot\beta l$$

$$-jY_0 \cot \beta l = \begin{cases} -j0.02\ x\ 1.15, & \text{for } x=\lambda/24 \\ \ \ j0.02\ x\ 1.15, & \text{for } x=5\lambda/24 \end{cases}$$

or

$$l = \begin{cases} 0.113\lambda, & \text{for } x=\lambda/24 \\ 0.387\lambda, & \text{for } x=5\lambda/24 \end{cases}$$

**Double Stub Tuner**

A typical double stub arrangement is shown in Fig. 2.17(a). The line is terminated in a normalized admittance $y'_L$ and two stubs are separated by a fixed distance $d$ and are placed at a distance $l$ from the termination. This system can be assumed to be equivalent to the one shown in Fig. 2.17(b) with normalized load impedance $y_L$. $y_L$ may easily be known in terms of $y'_L$. Referring to Fig. 2.17 at plane $aa$ we have

$$y_L = g_L + jb_L$$

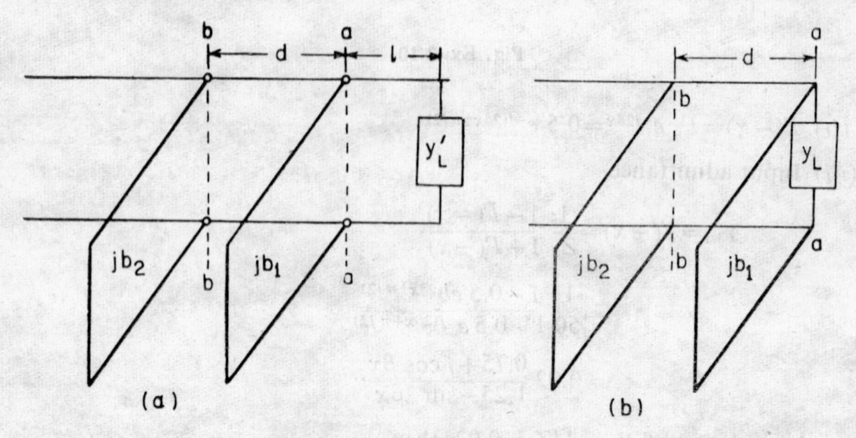


Fig. 2.17 The double stub tuner.

and just to the left of the first stub we have

$$y_a = g_L + jb_L + jb_1$$

where $jb_1$ is the admittance offered by the first stub at its input, i.e. at $aa$. The normalized admittance just to the right of the second stub is then given by

$$y_b = \frac{(g_L + jb_L - jb_1) + j \tan \beta d}{1 + j \tan \beta d (g_L + jb_L + jB_1)} \tag{2.123}$$

since for matching

$$y_b = 1 + jb. \tag{2.124}$$

Hence, equating Eqs. (2.123) and (2.124), and equating the real part we have

$$g_L = \frac{1+\tan^2 \beta d}{2 \tan^2 \beta d}\left[1 \pm \sqrt{1 - \frac{4 \tan^2 \beta d (1 - b_L \tan \beta d - b_1 \tan \beta d)^2}{(1+\tan^2 \beta d)^2}}\right] \tag{2.125}$$

Since $g_L$ is real, therefore the term under the radical sign must be positive, or zero. Hence, the value of the square root term lies between zero and one. The corresponding limits of $g_L$ are

$$0 < g_L < \frac{1+\tan^2 \beta d}{\tan^2 \beta d} = \frac{1}{\sin^2 \beta d}. \tag{2.126}$$

Equation (2.126) imposes a limit on the load admittances that can be matched. Of course, theoretically, $d$ may be chosen as small as possible, say, near zero. Then the range of non-matchable load admittances is negligibly small but in practice $d$ has a finite value.

The solution of Eq. (2.125) yields the normalized susceptance of the first stub, i.e.

$$b_1 = -b_L + \frac{1 \pm \sqrt{(1+\tan^2 \beta d)\, g_L - g_L^2 \tan^2 \beta d}}{\tan \beta d}. \tag{2.127}$$

where $b_L$, $g_L$ and $\tan \beta d$ are all known.

Equating the imaginary parts of Eq. (2.123), we obtain

$$b = \frac{(1 - b_L \tan \beta d - b_1 \tan \beta d)(b_L + b_1 + \tan \beta d) - g_L^2 \tan \beta d}{(g_L^2 \tan^2 \beta d) + (1 - b_L \tan \beta d - b_1 \tan \beta d)^2}.$$

Substituting the value of $b_1$ from Eq. (2.127), the above equation reduces to

$$b = \frac{\mp \sqrt{g_L(1+\tan^2 \beta d) - g_L^2 \tan^2 \beta d} - g_L}{g_L \tan \beta d}. \tag{2.128}$$

For a match the susceptance of the second stub has to be chosen as $-jb$. Also the lower and upper signs in Eqs. (2.127) and (2.128) go together.

The disadvantage of the double stub tuner in that it cannot match all impedances, is overcome by a *triple stub tuner* which is nothing but two double stub tuners placed in series. The analytical expression for the normalized stub admittances may be obtained in a similar manner.

The stub matching problems are greatly simplified with the aid of graphical methods such as the Smith impedance chart discussed in the next section.

## 2.17 SMITH CHART

Most of the microwave techniques are centred around Eqs. (2.50), (2.69) and (2.75). Two of these three equations, viz. Eqs. (2.50) and (2.75) may be combined into a single equation

$$Z(-l) = Z_{in} = Z_0 \frac{1 + j\, Z_0/Z_R \tan \beta l}{Z_0/Z_R + j \tan \beta l} = Z_0 \frac{1 + j\rho \tan \beta d_{min}}{\rho + j \tan \beta d_{min}}.$$

The last step follows form the fact that at the position of voltage minimum

$l=d_{min}$, $\rho=Z_0/Z_R$. The normalized* input impedance can be written as

$$z_{in}=\frac{Z_{in}}{Z_0}=\frac{1+j\rho\tan\beta d_{min}}{\rho+j\tan\beta d_{min}}. \tag{2.129}$$

Since $z_{in}$ is closely related to the reflection coefficient $[Z_R=Z_{in}\,(l)]$

$$\Gamma(l)=\frac{Z_{in}(l)-Z_0}{Z_{in}(l)+Z_0}=\frac{z_{in}-1}{z_{in}+1} \tag{2.130}$$

and both $\Gamma(l)$ and $\rho$ are well defined experimentally measurable quantities, these are independent of any guiding structure, so the necessity for making the calculations involved in Eqs. (2.50), (2.69) and (2.75) can by obviated by the use of a graphical device, namely, the *Smith chart transmission line diagram.*** This chart is nothing but a polar plot of Eq. (2.130). However, there are a large number of such polar plots for which reference can be made to the literature.***

To derive the Smith chart we substitute the value of $\Gamma(l)$ from Eq. (2.71) in Eq. (2.130). That is

$$|\Gamma_0|\,e^{j\phi}\,e^{-2\alpha l}\,e^{-2j\beta l}=\frac{z_{in}-1}{z_{in}+1}.$$

For a lossless line (extension of the chart for a lossy line be considered later)

$$\alpha=0 \quad \text{or} \quad e^{-2\alpha l}=1.$$

Hence, $$|\Gamma_0|\,e^{j(\phi-2\beta l)}=\frac{z_{in}-1}{z_{in}+1}$$

Or, in polar form

$$|\Gamma_0|\,\underline{/\phi-2\beta l}=\frac{z_{in}-1}{z_{in}-1}. \tag{2.131}$$

The LHS of Eq. (2.131) represents a family of circles of radius $|\Gamma_0|$ centred at the origin. The next step is to find the form of impedance coordinates which can be accomplished by the method of geometric inversion†. For this the RHS of Eq. (2.131) should be arranged in such a way that the coefficient of $1/z_{in}+1$ is a constant. This may be achieved by adding and subtracting 1 to the RHS of Eq. (2.131). That is

$$|\Gamma_0|\,\underline{/\phi-2\beta l}=\frac{z_{in}-1}{z_{in}+1}-1+1$$

$$=\frac{-2}{z_{in}+1}+1 \tag{2.132}$$

*When we normalize the impedance, quantities involving impedances become independent of the guiding structure; however, the impedances have to be properly interpreted.

**P.H. Smith "Transmission line calculator., *Electronics*" Vol. 12, Jan. (1939); also by the same author, "An improved transmission line calculator", *Electronics*, Vol. 17 Jan (1944.)

***W. Jackson and L.G.H. Huxley, "The solution of transmission line problems by use of circle diagram of impedance," Jour. *I.E.E.* (London), Vol. 91, Part III (1944).

†For details readers are referred to R.V., Churchill, *Introduction to Complex Variable and Applications*, New York: McGraw-Hill Book Co. (1948).

Now to map the RHS on a complex plane we require the following steps.

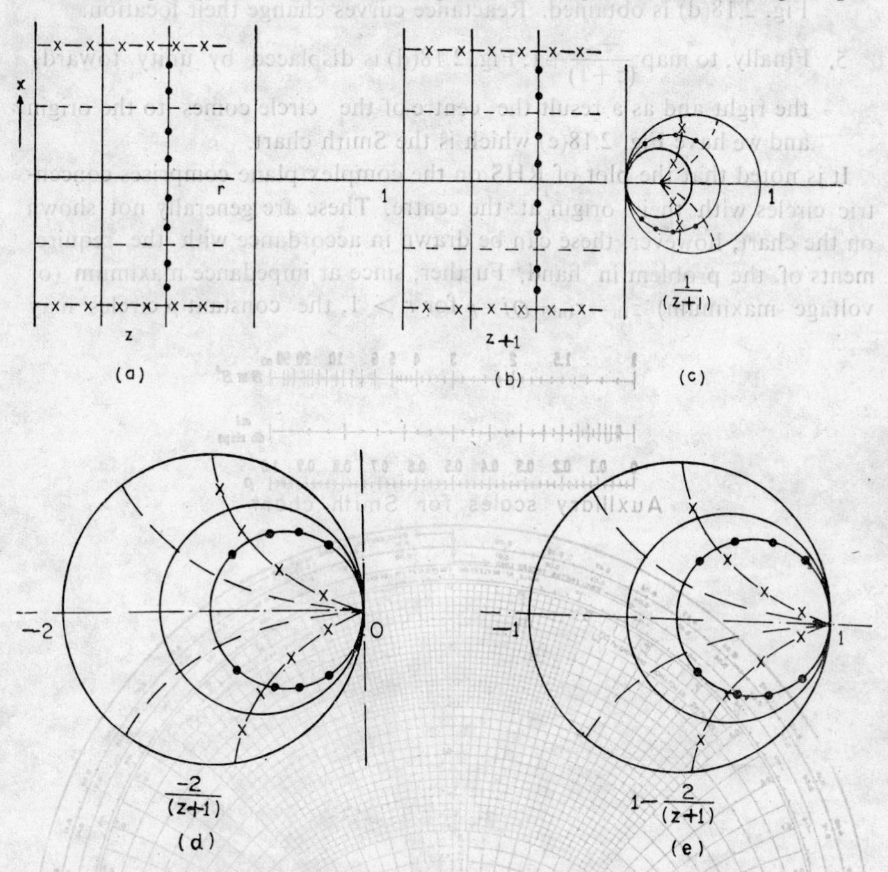


**Fig. 2.18** Development of Smith chart by maping Eq. (2.132) on complex plane.

1. $z_{in}$ may be complex in general, so we take a half $z$ plane as shown in Fig 2.18(a). The vertical lines correspond to the constant values of $r$ (resistance), while the horizontal lines to constant $x$ (reactance) and $r$ and $x$ are normalized values. Any point on the $z$ plane (any value of $z_{in}$, say, $z$) can be mapped on this plane. Consider some typical points on constant $r$ and constant $x$ lines.
2. To map $z+1$ points, the whole diagram is shifted to the right by one unit as depicted in Fig. 2.18(b).
3. Figure 2.18(c) shows the map of $1/z+1$. This is obtained by geometric inversion where lines are mapped into the circles. Two notable consequences of this inversion are (*i*) curves of negative reactance (positive susceptance) appear in the upper half circle and curves of positive reactance in the lower half circle, (*ii*) The entire infinite range of $z$ for positive values of $r$ has been mapped to the region inside the *unit circle*.

4. To map $-2/z+1$, Fig. 2.18(c) is multiplied by $-2$ and consequently Fig. 2.18(d) is obtained. Reactance curves change their location.
5. Finally, to map $\dfrac{-2}{(z+1)}+1$, Fig. 2.18(d) is displaced by unity towards the right and as a result the centre of the circle comes to the origin and we have Fig. 2.18(e) which is the Smith chart.

It is noted that the plot of RHS on the complex plane comprises concentric circles with their origin at the centre. These are generally not shown on the chart; however, these can be drawn in accordance with the requirements of the problem in hand. Further, since at impedance maximum (or voltage maximum) $z_{in}=r_{in}+Oj=\rho$ for $r>1$, the constant $\rho$ circles may

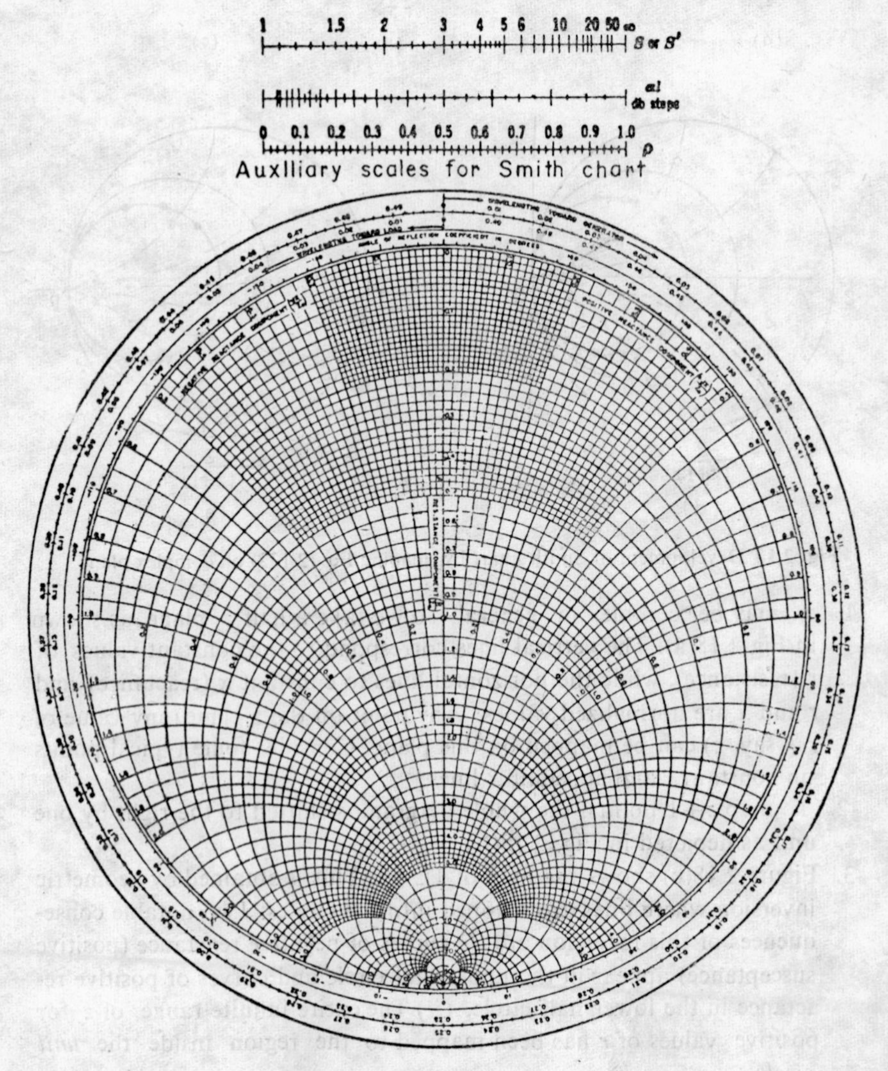


Fig. 2.19 Smith chart

conveniently be located by the intersections of the right hand portion of the real axis and the constant $r$ circles. The value of $|\Gamma_0|$ for any given value of VSWR $\rho$ can be determined from Eq. (2.76). The final plot is shown in Fig. 2.19. It consists of two families of circles, one corresponding to each reactive and resistive value of $z$ (or $z_{in}$). The circles of constant $r$ have their centres on the real axis while the circles of constant $x$ have their centres on an imaginary axis. The chart is calibrated around its periphery in terms of wavelengths rather than in degrees of $2\beta l$. Also the chart furnishes a direct calibration of the reflection coefficient angle $\phi$ in degrees. For a given value of the reflection coefficient, the corresponding input impedance ($z_{in}$) can be read directly from the chart and a movement of a distance $l$ along the line of corresponds to a change in the reflection coefficient by a factor $e^{-2j\beta d}$ only. This is represented by a simple rotation through an angle $2\beta l$, so the corresponding impedance point moves on a constant radius circle through this angle to its new value. The chart thus enables the transformation of impedance along a transmission line to be evaluated graphically in an efficient and direct manner.

If we wish to find explicit expressions for the contours of constant $r$ and constant $x$, we may let

$$z = r + jx$$

$$\Gamma = u + jv$$

and substitute these values in Eq. (2.130). Then on separating the equation so obtained into real and imaginary parts we obtain

$$r = \frac{1 - u^2 - v^2}{(1-u)^2 + v^2}, \quad x = \frac{2v}{(1-u)^2 + v^2}$$

which may be written as

$$\left(u - \frac{r}{r+1}\right)^2 + v^2 = \left(\frac{1}{r+1}\right)^2 \tag{2.133}$$

and

$$(u-1)^2 + \left(v - \frac{1}{x}\right)^2 = \frac{1}{x^2}. \tag{2.134}$$

Equations (2.133) and (2.134) provide explicit expressions for the circles of constant $r$ and constant $x$ on a complex plane.

To extend the chart for lossy lines, we may recall that for lossi lines the reflection coefficient has an additional multiplicative term $e^{-2\alpha l}$. As a result the value of $|\Gamma_0|$ does not remain constant. Hence the locus of $|\Gamma_0|$ with respect to line length is a spiral collapsing at the origin rather than a circle, as shown in Fig. 2.20. In practice we move on a constant $|\Gamma_0|$ circle first through an angle $2\beta l$ and then radially until we are at a distance $|\Gamma_0|\ e^{-2\alpha l}$ from the centre of the chart. Many practical charts have calibrated scales, scaled outside the chart to facilitate this inward displacement.

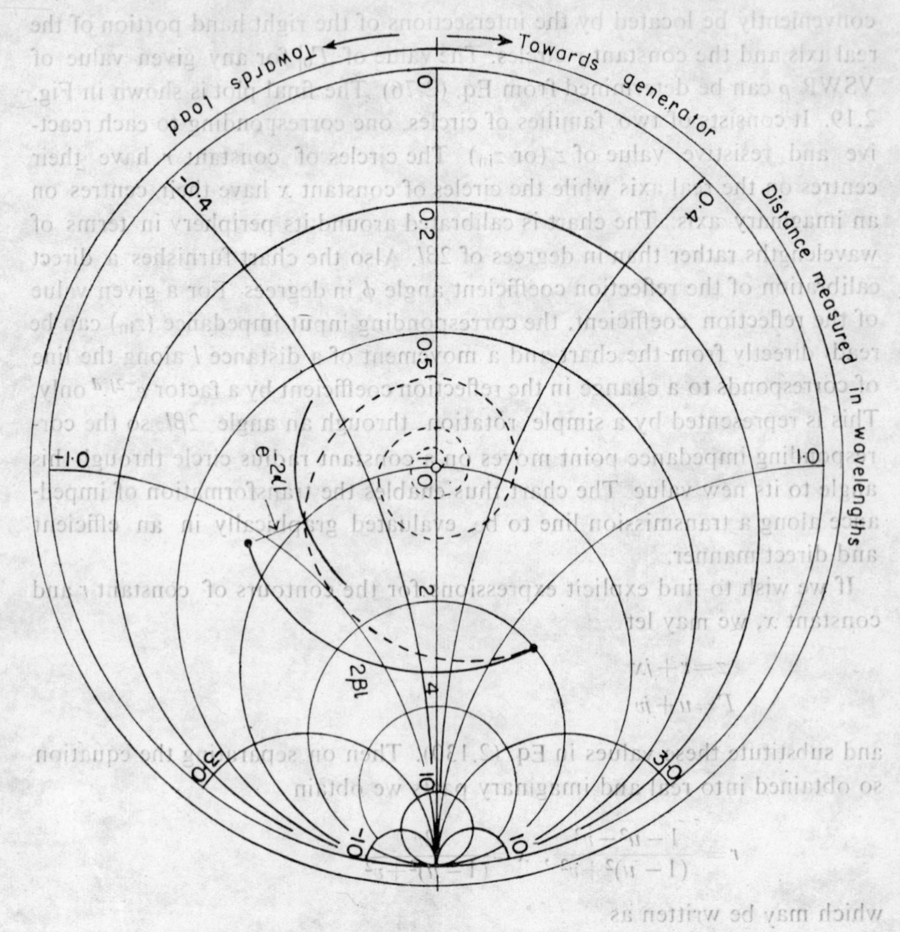


**Fig. 2.20** Smith chart showing effect of attention on input admittance of transmission line.

The applications of the Smith chart can best be understood with the following numerical examples.

It may be added that the same chart may be used as an admittance chart for admittance calculations. Any point $z$, if rotated through 180°, or moved to a circum distance $\lambda/2$, will correspond to $y$.

## 2.18 APPLICATIONS OF THE SMITH CHART

### 1. Study of Input Impedance

Let a transmission line be terminated in a normalized load impedance $r+jx$. This impedance is located on the Smith chart where two circles, one each corresponding to $r$ and $x$, intersect. The input impedance at a distance $l$ from the termination is obtained by rotating the point on constant $|\Gamma|$ circle or constant VSWR circle by a distance $l/\lambda$ in a clockwise (towards generator) direction.

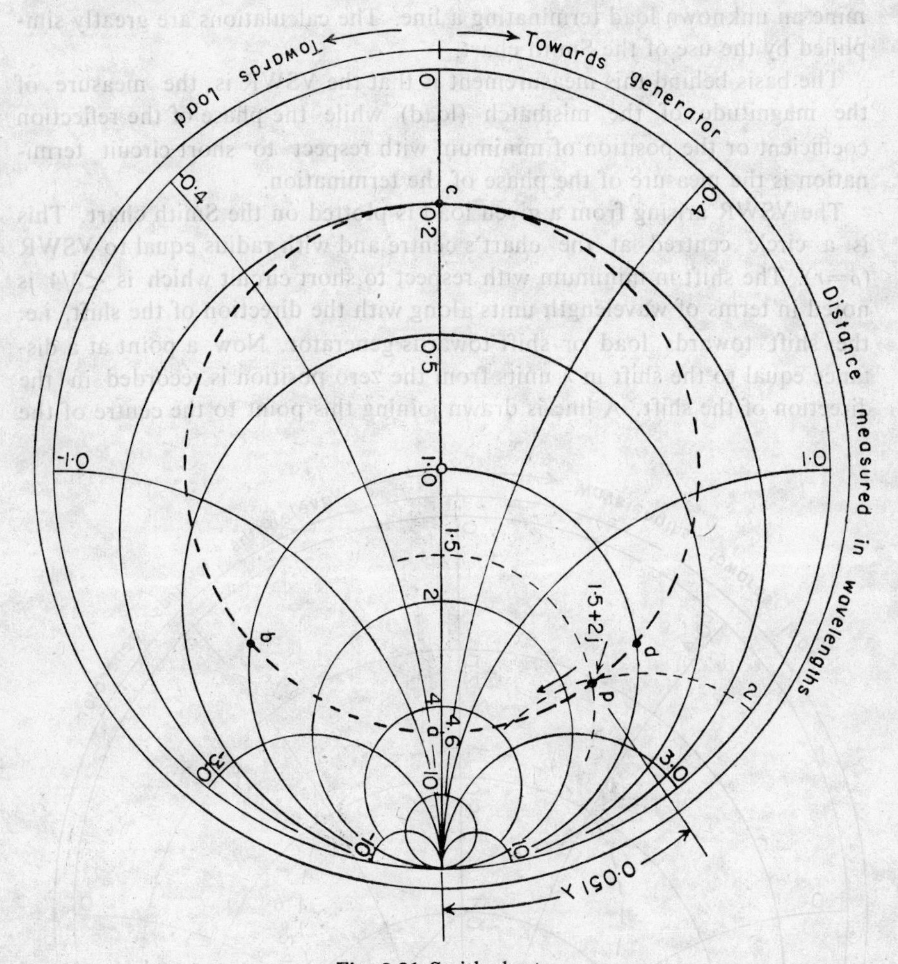


**Fig. 2.21** Smith chart

As a numerical example, consider a line terminated in a normalized load $z_R$ (or $z_L$)$=1.5+2j$. This impedance corresponds to point $P_\alpha$ on the Smith chart of Fig. 2.21. The VSWR is 5 so the VSWR circle is *bcd*. It is seen that if we move a peripheral distance *pa* which is approximately $0.051\lambda$, we reach point a which is on the resistance (real) axis. Thus the normalized impedance at a distance $0.051\lambda$ from the normalized load $1.5+2j$ is pure resistance having a value 4.6. If we move a distance *pb*, the normalized input impedance is no more purely resistive but has a negative reactive (capacitive) part. But the movement of $\lambda/4$ from *a*, i.e. to point *C*, leads to pure resistive input impedance of value 0.22 consistent with the transformation properties of a quarter wave line. The intermediate input impedances may similarly be obtained.

**2. Determination of Unknown Load**

The transformation properties of a transmission line may be used to determine an unknown load terminating a line. The calculations are greatly simplified by the use of the Smith chart.

The basis behind this measurement is that the VSWR is the measure of the magnitude of the mismatch (load) while the phase of the reflection coefficient or the position of minimum with respect to short circuit termination is the measure of the phase of the termination.

The VSWR arising from a given load is plotted on the Smith chart. This is a circle centred at the chart's centre and with radius equal to VSWR ($\rho=r$). The shift in minimum with respect to short circuit which is $<\lambda/4$ is noted in terms of wavelength units along with the direction of the shift, i.e. the shift towards load or shift towards generator. Now, a point at a distance equal to the shift in $\lambda$ units from the zero position is recorded in the direction of the shift. A line is drawn joining this point to the centre of the

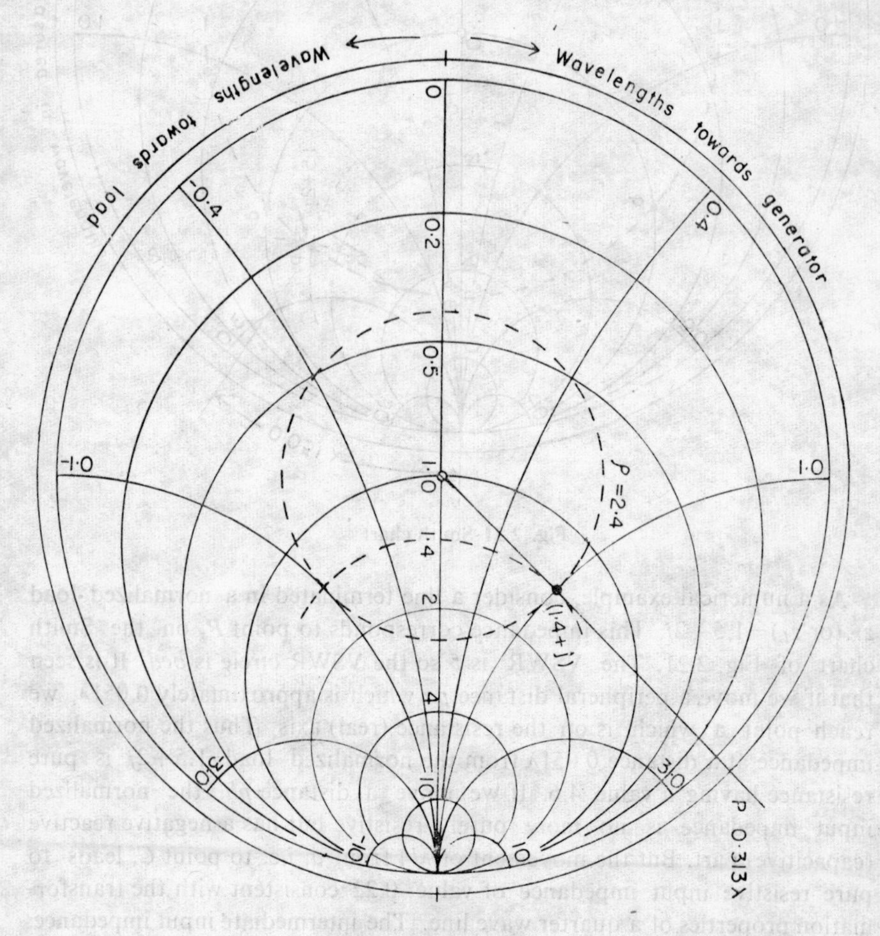


Fig. 2.22 Smith chart.

chart. The point of intersection of the VSWR circle and this line gives the impedance at the termination.

As a numerical example, consider a load terminating a given line of characteristic impedance 50 Ω. The VSWR measurements show its value to be 2.4 and the first minimum is located at a distance 0.313λ from the load.

The VSWR circle dashed (Fig. 2.22) is plotted which has a radius 2.4 and centre at the Smith chart's centre. A point *P* is located at a circum distance 0.313λ from the zero position (extreme left) in the direction "wavelengths towards load". The intersection of the line joining point *P* and the centre of the chart with the VSWR circle gives normalized load impedance which is (1.4+*j*). The load impedance is

$$Z_R(Z_L)=50(1.4+j)=70+50j\Omega.$$

**3. Single Stub Matching**

A transmission line is terminated in an unknown load. The standing wave measurement data yield the following information. VSWR $\rho=3$ and the

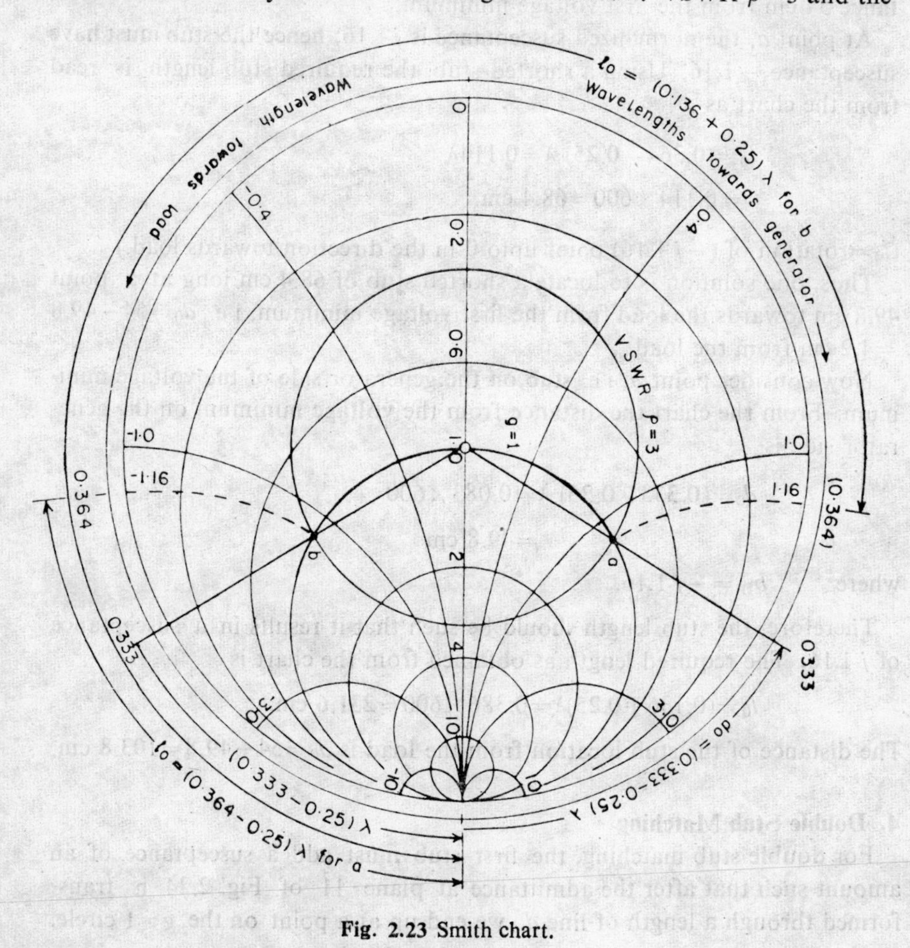


**Fig. 2.23** Smith chart.

first minimum is at a distance 54 cm from the termination, and the first maximum is 204 cm away from the termination. We have to design a single stub matching section so that the line is properly terminated (i.e. $\rho=1$).

To start with, draw VSWR=3 circle. Since admittance maximum coincides with a voltage minimum, locate $V_{min}$ on the chart as shown in Fig. 2.23. For matching, the normalized input admittance should be unity, i.e. $y_{in}=1$. Therefore, rotate on $\rho=3$ circle until the real part of $y_{in}=1$. This may be done in two ways, viz. rotation towards *load* (counter clockwise) to reach point $a$ or rotation towards *generator* (clockwise) to reach point $b$. Consider first the rotation towards *load*, i.e. point $a$. The distance from the position of voltage minimum where the stub is to be located, is

$$d_0'=(0.333 \text{ (position of } \boldsymbol{a})-0.25)\lambda$$

$$=0.083\lambda=0.083\times600=49.8 \text{ cm.}$$

It should be noted that this is a possible solution because load is at a distance 54 cm from the first voltage minimum.

At point $a$, the normalized susceptance is $j\,1.16$; hence the stub must have susceptance $-1.16$. Using a shorted stub, the required stub length is read from the chart as

$$l_0=(0.364-0.25)\ \lambda=0.114\lambda$$

$$=0.114\times600=68.4 \text{ cm.}$$

($l_0$=rotation of $(-j\ 1.16)$ point upto 0 in the direction towards load.)

Thus, one solution is to locate a shorted stub of 68.4 cm long at a point 49.8 cm towards the load from the first voltage minimum, i.e. $d_0=54-49.8=4.2$ cm from the load.

Now consider point $b$, i.e. stub on the generator side of the voltage minimum. From the chart the distance from the voltage minimum on the generator side is

$$d_0=(0.333-0.25)\ \lambda=0.083\times600$$

$$=49.8 \text{ cm}$$

where $\qquad b_{in}=-j\,1.16.$

Therefore, the stub length should be such that it results in a susceptance of $j\,1.16$. The required length as obtained from the chart is

$$l_0=(0.136+0.25)\lambda=0.386\times600=231.6 \text{ cm.}$$

The distance of the stub location from the load is $d_0=54+49.8=103.8$ cm.

**4. Double Stub Matching**

For double stub matching, the first stub must add a susceptance of an amount such that after the admittance at plane 11 of Fig. 2.24 is transformed through a length of line $d$, we end up at a point on the $g=1$ circle.

As a numerical example, let us take up the following values referring to Fig. (2.24): $l=0.1\ \lambda$ (distance of the first stub from the load), $d=0.4\ \lambda$ (separation of the stubs). We have to determine $l_1$ and $l_2$, i.e. stub lengths so that

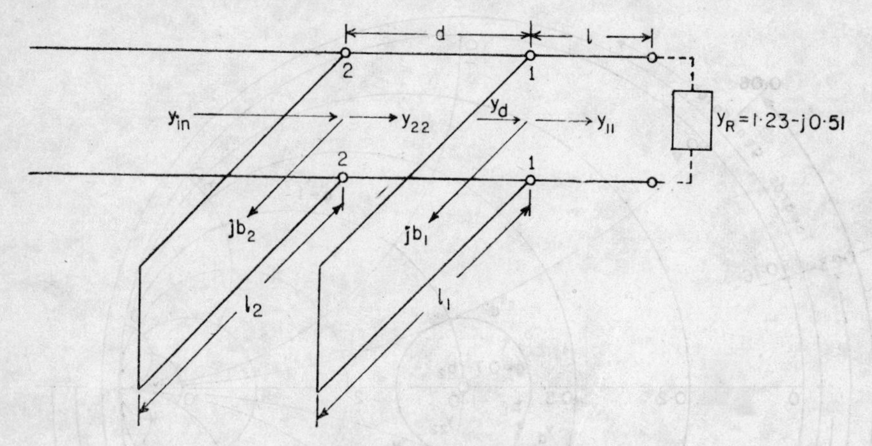


**Fig. 2.24** Double stub tunner.

the line is properly terminated when the normalized load is $1.23-j\,0.51$. First we locate $y_r$ as shown in Fig. 2.25. As the first stub is at a distance $0.1\ \lambda$, we rotate on the constant circle towards the generator by this distance and reach point $y_{11}$ (input impedance at 11 plane) where $y_{11}=(0.7-j\,0.3)$. Now to match, $y_d$ must lie on the locus at the intersection with the $g=0.7$ circle. Therefore, the stub must add a susceptance corresponding to $jb_1=j(0.3-0.14)=+j\,0.16$. Since stub 1 is shorted to find its length we enter independently a point $y_r=\infty$ and rotate this point towards load on the periphery a distance so as to have $b_1=+0.16$. We obtain

$$l_1=(0.50-0.25+0.025)\ \lambda=0.275\ \lambda$$

(length of the first stub).

Again we move point $y_a$ a distance $0.4\lambda$ on the constant $\rho$ circle to reach point $y_{22}$ which has a value $y_{22}=1-j\,0.4$. Hence stub 2 must add a susceptance of $j\,0.4$ so that finally we end at $g=1$ circle. The length of the second stub may be found by entering $y_r=\infty$ in the chart (stub short-circuited) and rotate it upto $b_2=0.4$ to reach the centre of the chart. We find,

$$l_2=(0.50-0.25+0.06)\ \lambda$$
$$=0.31\ \lambda.$$

It should be noted that for given values of $d$ and $l$ a match cannot always be obtained*, but double stub matching can easily be achieved by the hit and trial method moving two stubs one after the other in opposite directions.

*For details the reader is referred to R.E. Collin, *Foundations for Microwave Engineering New York: McGraw Hill Book Co. (1966).*

However, the tunning range is inversely proportional to the distance between two stubs.

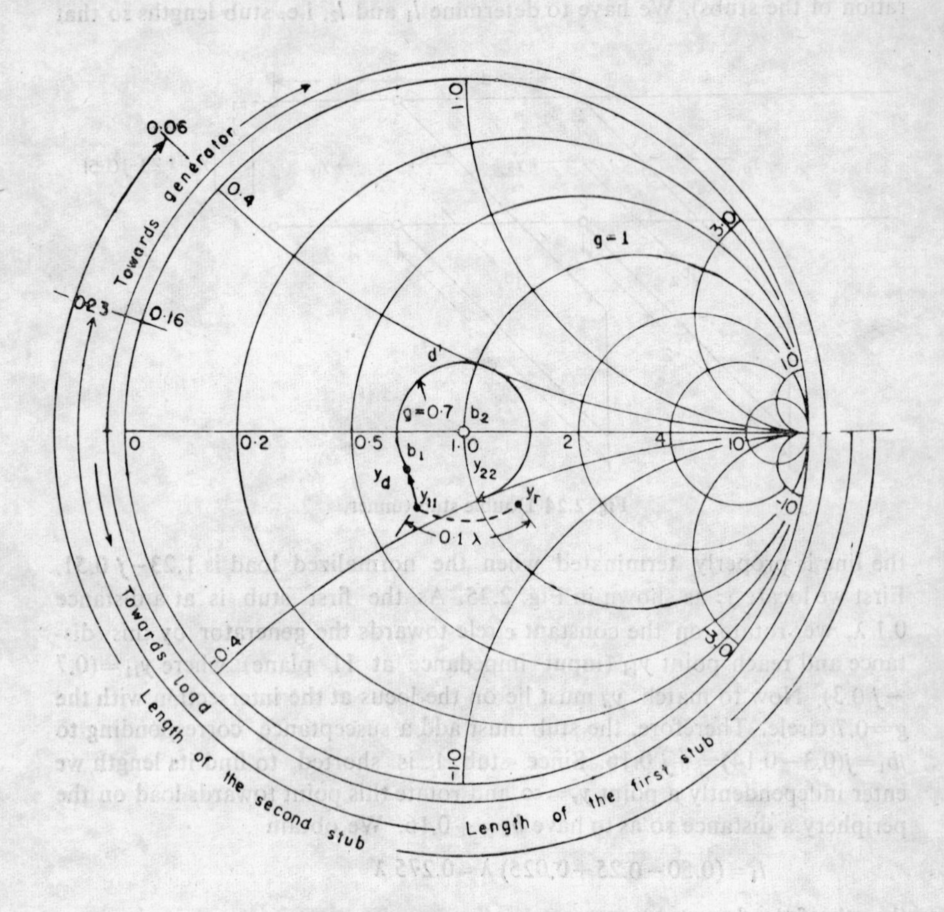


**Fig. 2.25** Smith chart.

### 5. Triple Stub Matching

Three stubs equally spaced are also used. The superiority of this system lies in the fact that the first stub transforms the impedance ($y_{\text{in}}$) in such a way that it reaches the limit of the tunning range of the double stub system which otherwise falls beyond its limit.

**Example 2.11**. A 50 Ω transmission line is terminated by a $(15-j\,20)$ Ω load. Find the following using the Smith chart.

(*a*) Reflection coefficient at the load.

(*b*) VSWR on the line.

(*c*) Distance of the first minimum of the standing wave pattern from the load.

(*d*) Line impedance at $l=0.05\ \lambda$.
(*e*) Line admittance at $l=0.05\ \lambda$.
(*f*) Location nearest to the load at which the real part of the line admittance equals the line characteristic admittance.

*Solution*: (*i*) The normalized load is

$$z_R\ (\text{or } z_L)=\frac{Z_R}{Z_0}=\frac{15-j\,20}{50}$$

$$=(0.3-j\,0.4).$$

(*ii*) Locate this load point on a normalized chart (A in Fig. Ex. 2.11). It is the point of intersection of the real circle of value 0.3 with imaginary circle of value $-0.4$.

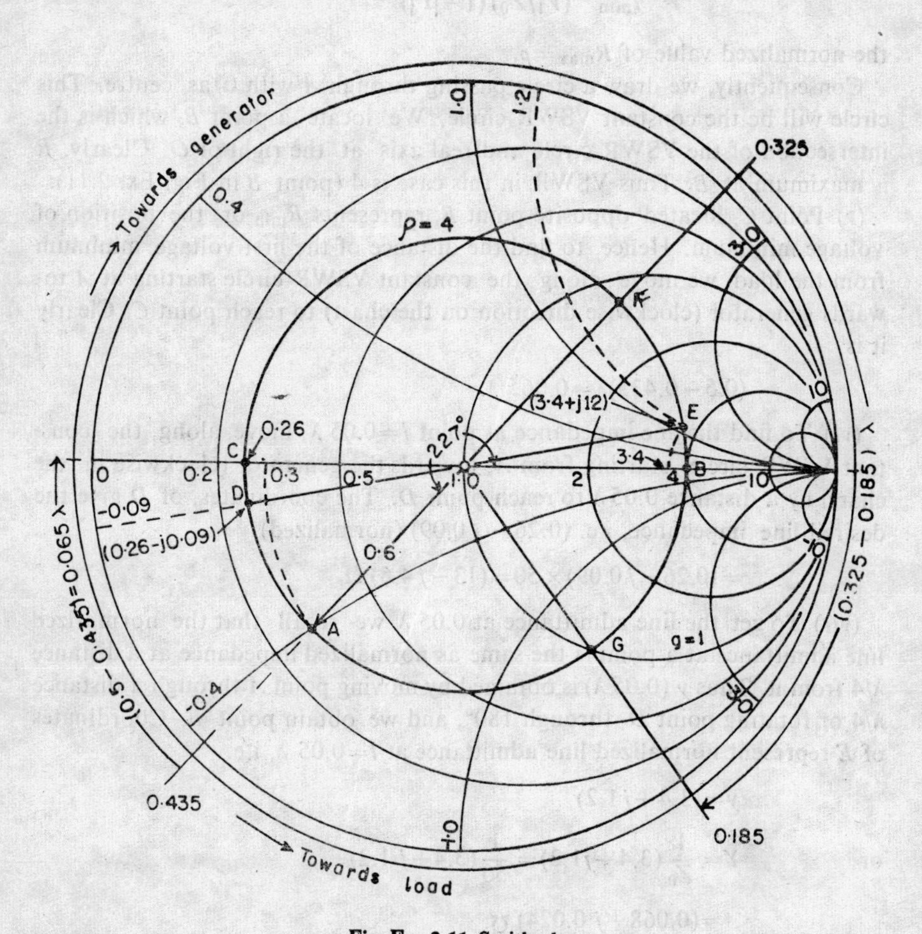


**Fig. Ex. 2.11** Smith chart.

(*iii*) The location of load is indirectly the location of the reflection coefficient. Since the Smith chart is the transformation in the $\Gamma$ plane, i.e. it has a value 0 at the centre and the locis of $\Gamma$ are circles with $O$ as the centre and radii proportional to $|\Gamma_1|$, the reflection coefficient becomes $OA=0.6$. The angle of the reflection coefficient is the angle made by $OA$ with reference to the real axis. That is

$$\underline{/\Gamma_{(0)}}=227^\circ$$

$$\Gamma_{(0)} \text{ (at load)}=0.6\, e^{j\,227^\circ}.$$

(*iv*) To find the VSWR, we must recall that the voltage maximum is located at a point where impedance is real and maximum. Denoting the maximum impedance by

$$R_{max}=\frac{V_{max}}{I_{min}}=\frac{V_1}{(V_1/Z_0)}\frac{(1+|\Gamma|)}{(1-|\Gamma|)}=Z_0$$

the normalized value of $R_{max}=\rho$.

Consequently, we draw a circle passing through $A$ with 0 as centre. This circle will be the constant VSWR circle. We locate a point $B$, which is the intersection of the VSWR circle and real axis at the right of $O$. Clearly, $R$ is maximum at $B$. Thus VSWR in this case is 4 (point $B$ in Fig. Ex. 2.11).

(*v*) Point $C$, located opposite point $B$, represents $R_{min}$ or the position of voltage minimum. Hence, to find the distance of the first voltage minimum from the load, we move along the constant VSWR circle starting at $A$ towards generator (clockwise direction on the chart) to reach point $C$. Clearly it is

$$(0.5-0.435)\lambda=0.065\ \lambda.$$

(*vi*) To find the line impedance at point $l=0.05\ \lambda$, move along the constant VSWR circle, starting from $A$, towards the generator (clockwise in the chart) by a distance $0.05\ \lambda$ to reach point $D$. The coordinates, of $D$ give the desired line impedance, i.e. $(0.26-j\,0.09)$ (normalized)

$$=(0.26-j\,0.09)\times 50=(13-j\,4.5)\ \Omega.$$

(*vii*) To get the line admittance at $0.05\ \lambda$ we recall that the normalized line admittance at a point is the same as normalized impedance at a distance $\lambda/4$ from it. Thus $y\ (0.05\ \lambda)$ is obtained by moving point $A$ through a distance $\lambda/4$ or rotating point $D$ through 180°, and we obtain point $E$. Coordinates of $E$ represent normalized line admittance at $l=0.05\ \lambda$, i.e.

$$y=(3.4+j\,1.2)$$

or

$$Y=\frac{1}{Z_0}(3.4+j1.2)=\frac{1}{50}(3.4+j\,1.2)$$

$$=(0.068+j\,0.024)\ \mho.$$

(*viii*) To find the location nearest to the load at which the real part of the line admittance equals the line characteristic admittance, we locate point $y_0$ (normalized load admittance which is diametrically opposite to $z_{(0)}$), point

*A*). We then move along the constant VSWR circle towards the generator until we intersect a circle passing through 1 on the real axis. This point is *G* and is on the "unit conductance circle". This distance from *F* to *G* as read on the chart is clearly $(0.325-0.185)\ \lambda=1.14\ \lambda$.

**Example 2.12.** Solve Exercise (2.10) using the Smith chart.

*Solution*: (*i*) The normalized load impedance is given by

$$z_R=\frac{Z_R}{Z_0}=\frac{30-j\,40}{50}=0.6-j\,0.8.$$

(*ii*) As above in problem (2.12), locate this load on the Smith chart. See point *A* in Fig. Ex. 2.12.

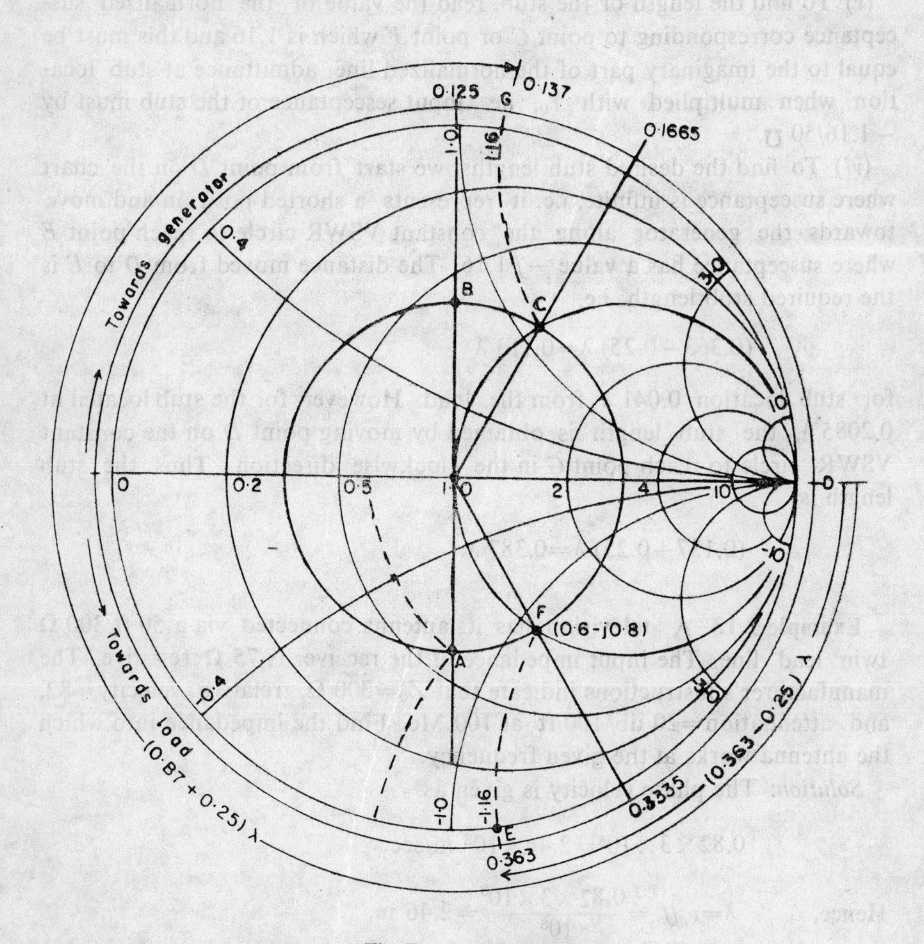


**Fig. Ex. 2.12** Smith chart.

(*iii*) Draw a circle with centre at 1 and passing through *A*. This will be the constant VSWR circle. Move point *A* diametrically opposite on the constant VSWR circle to *B* which corresponds to normalized load admittance.

(*iv*) Now, starting at point *B*, go around the constant VSWR circle towards the generator until point *C* or point *F* on the unit conductance circle is reached. Since point *C* and *F* are characterized by unit conductance, they represent the location of the stub (second transmission line). The distance moved from *B* to *C* or *B* to *F* represents the distance of the stub from the load, i.e.

$$(0.1665-0.125)\ \lambda=0.041\ \lambda. \qquad \text{(Pt. C)}$$

or

$$(0.3335-0.125)\ \lambda=0.2085\ \lambda \qquad \text{(Pt. F)}$$

(*v*) To find the length of the stub, read the value of the normalized susceptance corresponding to point *C* or point *F* which is 1.16 and this must be equal to the imaginary part of the normalized line admittance at stub location when multiplied with $Y_0$, i.e. input sesceptance of the stub must by $-1.16/50$ ℧.

(*vi*) To find the desired stub lengths, we start from point *D* on the chart where susceptance is infinite, i.e. it represents a shorted position and move towards the generator along the constant VSWR circle to reach point *E* where susceptance has a value $-j\,1.16$. The distance moved from *D* to *E* is the required stub length, i.e.

$$(0.363-0.25)\ \lambda=0.113\ \lambda$$

for stub location 0.041 λ from the load. However, for the stub located at 0.2085 λ, the stub length is obtained by moving point *D* on the constant VSWR circle to reach point *G* in the clockwise direction. Thus the stub length is

$$(0.137+0.25)\ \lambda=0.387\ \lambda.$$

**Example 2.13.** A television has its antenna connected via a 50 ft 300 Ω twin lead line. The input impedance of the receiver is 75 Ω resistive. The manufacturer's instructions indicate that $Z_0=300\ \Omega$, relative velocity=82, and attenuation=20 db/100 ft at 100 Mc. Find the impedance into which the antenna works at the given frequency.

*Solution*: The phase velocity is given as

$$0.82\times3\times10^8=2.46\times10^8 \text{ m/sec.}$$

Hence,
$$\lambda=v_p/f=\frac{0.82\times3\times10^8}{10^8}=2.46 \text{ m.}$$

Since
$$1\text{m}=3.281 \text{ ft}$$

length,
$$l=\frac{50}{3.281}\text{m}=\frac{50}{3.281}\,\frac{\lambda}{2.46}=6.19\ \lambda$$

and consequently,

$$\alpha l = 2 \times \frac{50}{100} = 1 \text{ db}.$$

The normalized load (terminating) impedance is

$$z_R = \frac{Z_R}{R_0} = \frac{75}{300} = 0.25.$$

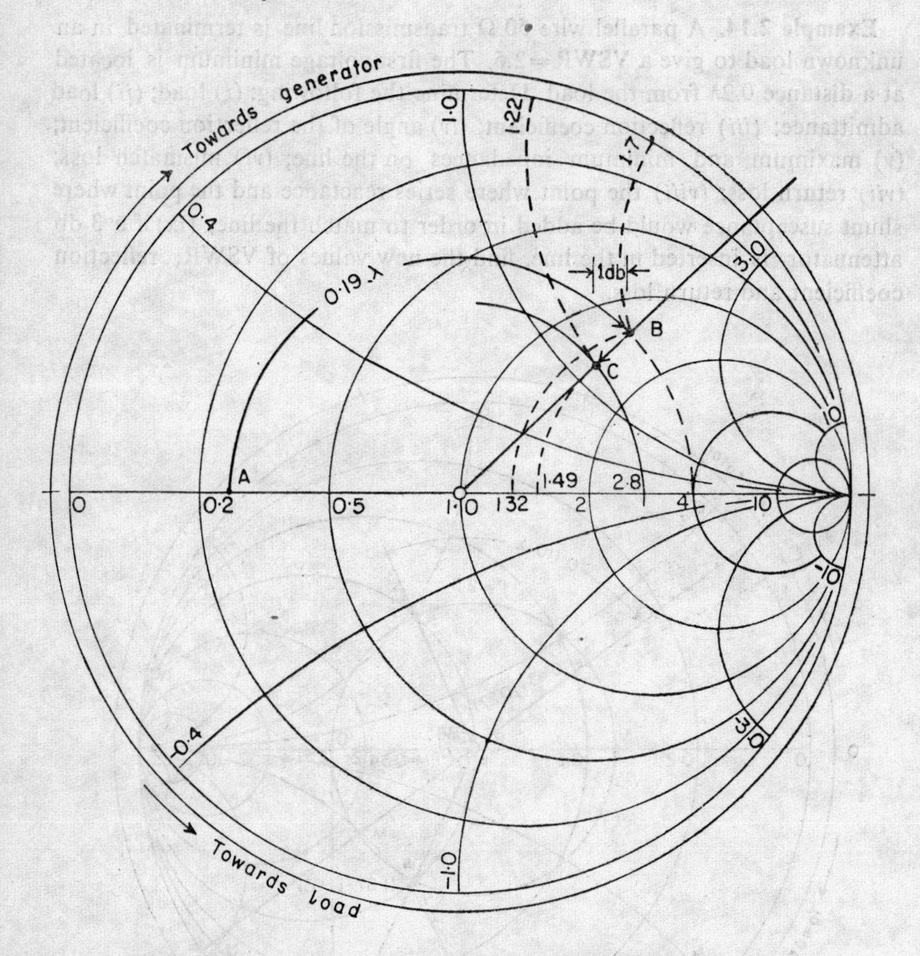


**Fig. Ex. 2.13 Smith chart.**

(*i*) Locate the impedance $z_R$ in the Smith chart at *A* in Fig. Ex. 2.13.

(*ii*) Rotate point *A* on the constant VSWR circle through a distance $6.19\lambda$, i.e. $0.19\lambda$ in the direction "wavelengths towards the generator". The new point *B* gives the normalized input impedance had the line been lossless.

(*iii*) The value of $z_{in}$ is $1.32 + j\,1.7$.

(*iv*) Clearly, this $z_{in}$ corresponds to a value of $\rho = \text{VSWR} = 4$.

($v$) To find a new value of VSWR for a lossy line, transfer (on the linear scale) $\alpha l$ on the db scale by 1 db (line being 50 ft) and locate the corresponding point on the VSWR scale. The new point has VSWR=2.8. Move radially inwards from point $B$ to $C$ to locate normalized input impedance for the given lossy line. Thus $z_{in}$ for the lossy line=$(1.49+j\ 1.22)300$

$$=447+j\ 366\ \Omega.$$

**Example 2.14.** A parallel wire 50 Ω transmission line is terminated in an unknown load to give a VSWR=2.5. The first voltage minimum is located at a distance 0.2λ from the load. Determine the following: ($i$) load; ($ii$) load admittance; ($iii$) reflection coefficient; ($iv$) angle of the reflection coefficient; ($v$) maximum and minimum inpedances on the line; ($vi$) mismatch loss; ($vii$) return loss; ($viii$) the point where series reactance and the point where shunt susceptance would be added in order to match the line; ($ix$) if a 3 db attenuator is inserted in the line, find the new values of VSWR, reflection coefficient and return loss.

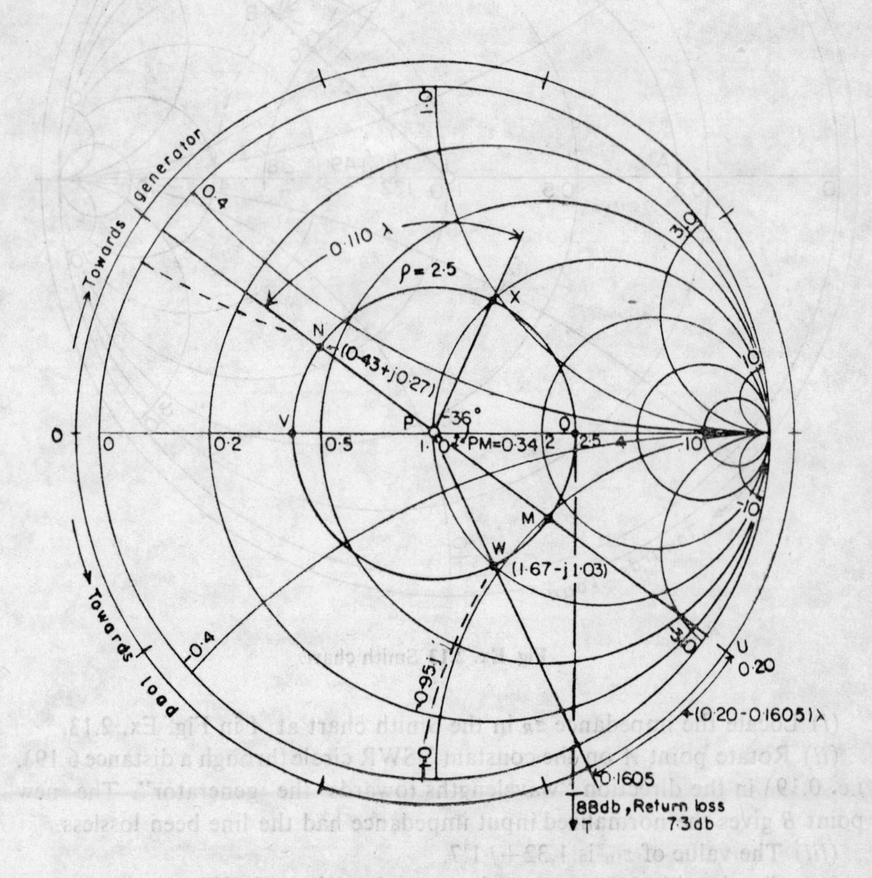


Fig. Ex. 2.14 Smith chart

*Solution*: The problem is solved using the Smith chart shown in Fig. Ex. 2.14. The wavelength of the propagating waves is

$$\lambda = \frac{C}{f} = \frac{3\times 10^8}{3\times 10^9} = 0.1 \text{ m} = 10 \text{ cm}.$$

Since VSWR=2.5, hence the VSWR circle is drawn with radius 2.5 and centre at $P$ (centre of the chart).

(*i*) The voltage minimum is located at a distance $0.2\lambda$ from the load moving a distance $0.2\lambda$ on the constant VSWR circle in the direction "wavelength towards the load" to reach point $M$. The normalized load is read to give $z_L = (1.67 - j\,1.03)$; the value of the load is thus $z_L = 50(1.67 - j\,1.03) = (83.5 - 51.5)\ \Omega$.

(*ii*) The load admittance is found by rotating point $M$ by 180° to $N$, to give normalized load admittance as

$$y_L = (0.43 + j\,0.27)$$

or
$$Y_L = \frac{1}{50}\,(0.43 + j\,0.27) = (0.0086 + j\,0.0054)\ \mho.$$

(*iii*) The reflection coefficient corresponding to the load corresponding to point $M$ is read on the linear scale to give $|\Gamma| = 0.34$.

(*iv*) Load point $M$ extended radially to $U$ gives the angle of the reflection coefficient as $-36°$.

(*v*) The maximum and minimum impedances are located at $O$ and $V$ are

$$(\text{VSWR})(Z_0) = 50 \times 2.5 = 125\ \Omega$$

and
$$\frac{Z_0}{\text{VSWR}} = \frac{50}{2.5} = 20\ \Omega \text{ respectively.}$$

(*vi*) Measure the distance $PO$, $O$ being the $Z_{max}$ point, and use that distance to read the mismatch loss on the right side radial scale. This value is 0.88 db.

(*vii*) Find the return loss at point $T$ on the radial scale. This is 7.3 db.

(*viii*) The series reactance required to match the line is located by travelling around the VSWR circle from the load at $M$ to a point where the real part of the normalized impedance reduces to unity, i.e. point $W$. The normalized impedance at this point is $1 - j\,0.95$. *A* series reactance of $+j\,0.95$ is required at this point to cancel the $-j\,0.95$ value. This reactance is located at a distance $0.20\lambda - 0.1605\lambda = 0.0395\lambda$ indicated on the circum wavelength scale.

The point at which the shunt admittance is to be added is located by travelling around the constant VSWR circle to the load admittance value of $1 + j\,0.095$ at $X$. A shunt susceptance of $-j\,0.95$ is placed at this point to obtain a match. The distance from the load is clearly $NX$ which on the circum wavelength scale comes out to be $(0.45 - 0.34)\lambda = 0.110\lambda$, from the load.

(*ix*) The VSWR 2.5 results in a return loss of 7.3 db indicated at point *T* (on the radial scale). The insertion of a 3 db attenuator corresponds to an additional return loss of 6 db. Thus total return loss becomes 6+7.3 =13.3 db. The corresponding values of VSWR and reflection coefficient as read on the radial scales are 1.55 and 0.215 indicated at *Y* and *Z* respectively.

## 2.19 NON-UNIFORM LINES

Lines which do not have $Z_0$ a constant but continuously varying according to certain low are called as *tapered or non-uniform lines.* These lines are generally used as impedance transformers to provide a match between the source and given load. The tapering of these lines may be continuous, such as the exponential taper shown in Fig. 2.26a, or discrete as in Fig. 2.26(b). In such lines the parameters *L, C, R* and *G* are all functions of positions.

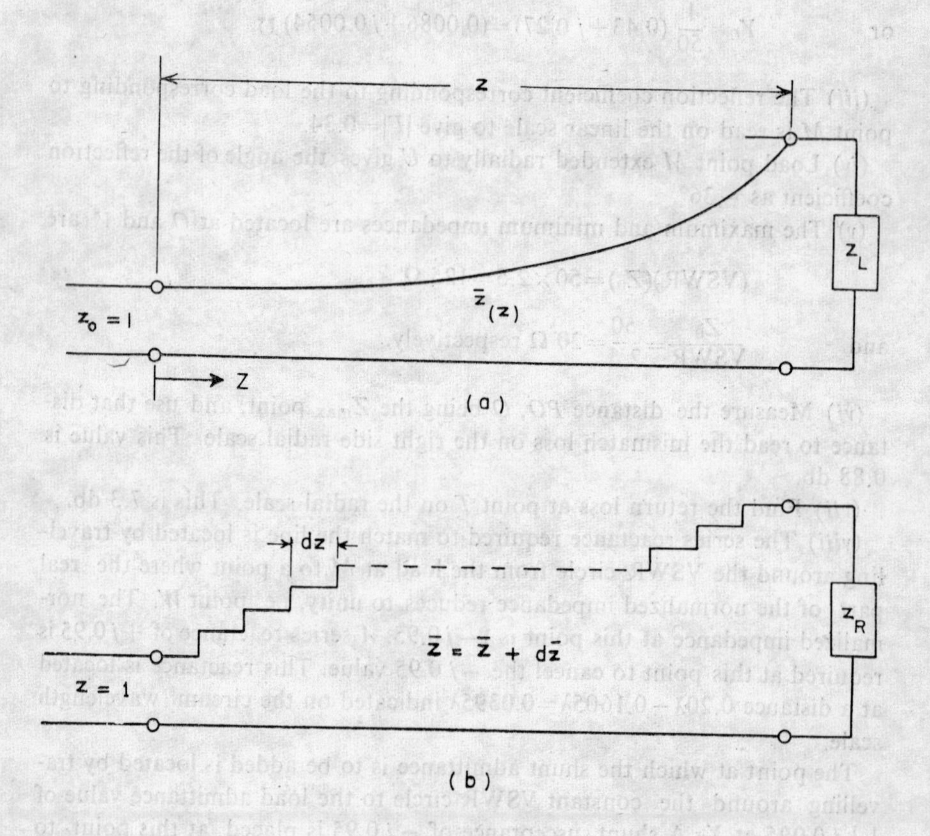


**Fig. 2.26** Tapered transmission line sections (*a*) Exponential line (*b*) dicrete taper line.

Our object in this section, is to provide an introductory view; however, detailed information may be found in the literature*.

Consider an infinitesimal line element of length $dz$ at a distance $z$ from the *generator end* or *sending end*. Then

$$I(z)-I(z+dz)=VY\,dz \qquad (2.135)$$
$$V(z)-V(z+dz)=IZ\,dz$$

where $Y$ and $Z$ are respectively the admittance and impedance per unit length of the line at a distance $z$, each of which is a function of $z$. Equations (2.135) can be rewritten as

$$\frac{dI}{dz}=-VY$$
$$\frac{dV}{dz}=-IZ. \qquad (2.136)$$

Separating variables $V$ and $I$ to have independent equations one for each, i.e. differentiating the first of Eqs. (2.136) and substituting the value of $dV/dz$ from the second.

$$\frac{d^2I}{dz^2}=-Y\frac{dV}{dz}-V\frac{dY}{dz}$$
$$=YZI-\left(-\frac{1}{Y}\frac{dI}{dz}\right)\frac{dY}{dz}=YZI+\left(\frac{1}{Y}\frac{dY}{dz}\right)\frac{dI}{dz}$$

or
$$\frac{d^2I}{dz^2}-\frac{d}{dz}(\ln Y)\frac{dI}{dz}-YZI=0 \qquad (2.137)$$

and similarly
$$\frac{d^2V}{dz^2}-\frac{d}{dz}(\ln Z)\frac{dV}{dz}-YZV=0. \qquad (2.138)$$

Equations (2.137) and (2.138) form the pair of fundamental wave equations required for the discussion of voltage and current existing on a tapered line.

It can be seen by obtaining the solution of these equations, that for an exponential taper

$$L(z)=L_s\,e^{-2\delta z}$$
$$C(z)=C_s\,e^{2\delta z} \qquad (2.139)$$
$$Z_0(z)=Z_{os}\,e^{-2\delta z}$$

*1. L. Solymar, "Some notes on the optimum design of stepped transmission line transformers", *IRE Trans.* Vol. M.T.T-6 (1958).

2. R.W. Klopfenstein, "Optimum tapered transmission line matching", Proc. IRE, Vol. 44 (April 1956).

3. R.E., Collin, *Foundations of Microwave Engineering*, New York: McGraw Hill Book Co. (1962).

where $\delta$ is the *transformation coefficient* of the taper and Eq. (2.139) is the *transformation equations.* By the use of these equations, parameters at any point on the line may be obtained. The subscripts represents sending end.

The phase constant for such a line is given by

$$\beta=\sqrt{\omega^2LC-\delta^2}. \tag{2.140}$$

Equation (2.140) indicates that $\beta$ may be real or imaginary depending on whether $\omega^2LC$ is greater than or less than $\delta^2$. When $\omega^2LC<\delta^2$, $\beta$ will be real and there will be flow of energy down the line. However, when $\omega^2LC>\delta^2$ $\beta$ will be imaginary and so $j\beta$ will be real; consequently, there will be no flow of energy. Thus a tapered line works as a high pass filter network with cut off frequency $\omega_c=\delta/\sqrt{LC}=v\delta$ where $v$ is the velocity of propagation.

The terminal or load impedance ($Z_R$), i.e. impedance at a distance $z$ from the source is given by

$$Z_R=\sqrt{L/C}\left[\sqrt{1-\left(\frac{\omega_c}{\omega}\right)^2}+j\frac{\omega_c}{\omega}\right]. \tag{2.141}^*$$

The fractional change in the reflection coefficient at any point $z$ of such a line can be written as*

$$d\Gamma_0=\frac{\bar{z}+d\bar{z}-\bar{z}}{\bar{z}+d\bar{z}+\bar{z}}\simeq\frac{d\bar{z}}{2\bar{z}}$$

or $$d\Gamma_0=\frac{1}{2}\frac{d}{dz}(\ln\bar{z})\,dz.$$

The input reflection coefficient of the tapered line of length $L$ is given by

$$\Gamma_{in}=\frac{1}{2}\int_0^L e^{-2j\beta z}\frac{d}{dz}(\ln\bar{z})\,dz \tag{2.142}$$

whose value for an exponential taper is

$$\underset{\text{(exponential)}}{\Gamma_{in}}=\tfrac{1}{2}\,e^{-j\beta L}\ln\bar{z}L\,\frac{\sin\beta L}{L}. \tag{2.143}$$

In case of a triangular taper it has a value

$$\underset{\text{(triangular)}}{\Gamma_{in}}=\tfrac{1}{2}\,e^{-j\beta L}\ln\bar{z}L\,\frac{\sin^2(\beta L/2)^2}{\beta L/2} \tag{2.144}$$

Equations (2.143) and (2.144) are of enormous importance in designing impedance transformers.

*H.A. Wheeler, "Transmission lines with exponential taper", *Proc. IRE*, Vol. 27, (Jan.1939).

**Bars have been used on the normalized impedances to distinguish them from the variation in $z$, i.e. $dz$.

## 2.20 MICROWAVE STRIP LINES

Now-a-days, microstrip transmission lines are extensively used in microwave circuits because they can be fabricated easily by employing printed circuit techniques. Three types of microstrips are most commonly used. There are parallel strip lines (Fig. 2.27a), a single strip over an infinite ground plane (Fig. 2.27b), and a shielded strip line (Fig. 2.27c).

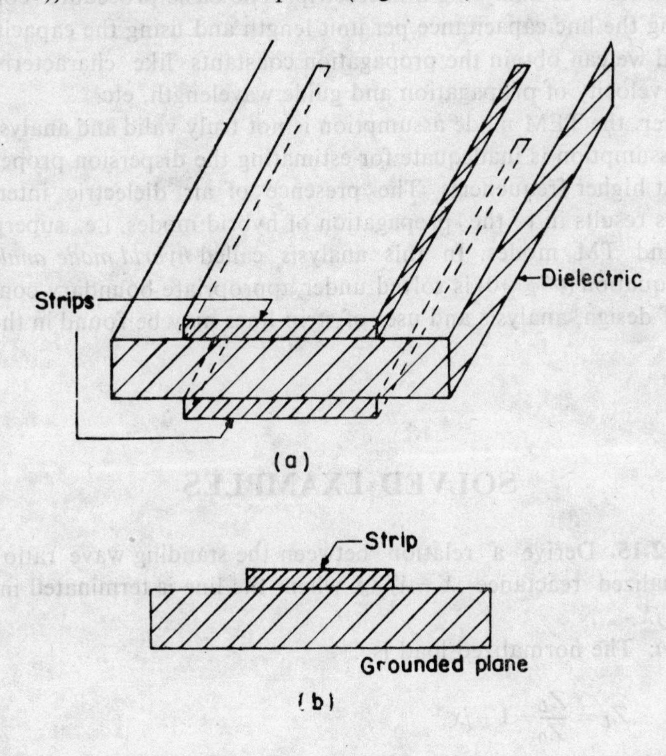


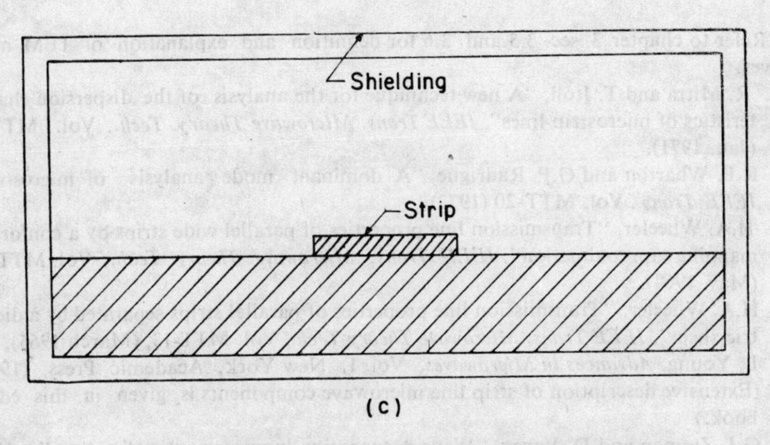


Fig. 2.27 (*a*) Parallel microstrip line (*b*) Microstrip on a grounded infinite plane (*c*) Shielded stripline

The dominant mode of propagation in these lines is the TEM* and the characteristic impedance of the line depends upon (in case of Fig. 2.27a) strip width, thickness of the dielectric constant and the dielectric constant of the substrate material. Most of the energy propagates in the dielectric just *below* the strip conductor and the remaining in a fringe field some of which extends in the air space above the dielectric. There are various methods of analysing a microstrip. The basic procedure consists of calculating the line capacitance per unit length and using the capacitance so calculated we can obtain the propagation constants like characteristic impedance, velocity of propagation and guide wavelength, etc.

However, the TEM mode assumption is not truly valid and analysis based on this assumption is inadequate for estimating the dispersion properties of the line at higher frequencies. The presence of air dielectric interface in these lines results in to the propagation of hybrid modes, i.e. superposition of TE and TM modes. In this analysis called *hybrid mode analysis* the Laplace equation ($\nabla^2\phi=0$) is solved under appropriate boundary conditions. Details of design, analysis and uses of strip lines may be found in the literature**

## SOLVED EXAMPLES

**Example 2.15.** Derive a relation between the standing wave ratio $\rho$ and the normalized reactance $X=X/Z_0$ when the line is terminated in a load $Z_L=Z_0+jX$.

*Solution*: The normalized load is

$$Z_L=\frac{Z_L}{Z_0}=1+jx$$

*Refer to chapter 3, sec. 3.5 and 3.6 for definition and explanation of TEM mode waves.

**1. R. Mitra and T. Itoh, "A new technique for the analysis of the dispersion characteristics of microstrip lines", *IEEE Trans. Microwave Theory Tech.*, Vol. MTT-19 (Jan. 1971).
2. R.P. Wharton and G.P. Radrigue "A dominant mode analysis of microstrip," *IEEE Trans.*, Vol. MTT-20 (1972).
3. H.A. Wheeler, "Transmission line properties of parallel wide strips by a conformed mapping approximation", *IEEE Trans. Microwave Theory Tech.*, Vol. MTT-12, (May 1964).
4. H.A. Wheeler, "Transmission line properties of parallel strips separated by a dielectric sheet", *IEEE Trans. Microwave Theory Tech.*, Vol. MTT-13, (March 1965).
5. L. Young, *Advances in Microwaves*, Vol. 1, New York, Academic Press, (1966). (Extensive description of strip line microwave components is given in this edited book.)
6. G.I. Zysman and D. Varon, "Wave propagation in microwave strip lines", *IEEE G-MTT Int. Microwave Symp. Dig.* (May 1969).

The reflection coefficient at the load

$$\Gamma_0 = \frac{z_L - 1}{z_L + 1} = \frac{jx}{2 + jx}$$

or

$$|\Gamma_o| = \frac{x}{\sqrt{4 + x^2}}. \qquad (a)$$

The standing wave ratio (VSWR) is given by

$$\rho = \frac{1 + |\Gamma_0|}{1 - |\Gamma_0|} \quad \text{from which} \quad |\Gamma_0| = \frac{\rho - 1}{\rho + 1}. \qquad (b)$$

Combining Eqs. (*a*) and (*b*)

$$\frac{|x|}{\sqrt{4 + x^2}} = \frac{\rho - 1}{\rho + 1}$$

or

$$x^2 = \frac{(\rho - 1)^2}{(\rho + 1)^2} (4 + x^2)$$

or

$$x = \frac{\rho - 1}{\sqrt{\rho}}.$$

**Example 2.16.** A 50 Ω lossless line is terminated in an unknown load. The distance between adjacent voltage minima is $d = 8$ cm, the VSWR is $= 2$, and the first voltage minimum is situated at $d_m = 1.5$ cm from the load. What is the value of $Z_L$.

*Solution*: $d = \lambda/2 = 8$ cm, thus $\lambda = 16$ cm.

At a voltage minimum, the impedance is given by $Z_m = Z_0/\rho$. Therefore,

$$Z_m = \frac{Z_0}{\rho} = Z_0 \frac{Z_L + jZ_0 \tan \beta d_m}{Z_0 + jZ_L \tan \beta d_m}.$$

Using $\beta = 2\pi/\lambda$, the normalized impedance can be calculated as

$$z_L = \frac{Z_L}{Z_0} = \frac{\frac{1}{\rho} - j \tan \beta d_m}{1 - j\frac{1}{\rho} \tan \beta d_m} = \frac{1 - j\rho \tan \beta d_m}{\rho - j \tan \beta d_m}.$$

Since $\beta d_m = \frac{2\pi}{16} \times 0.15 \simeq 0.59$, $\tan \beta d_m = 0.67$

and $z_L = \frac{1 - j\,1.34}{2 - j\,0.67} \simeq 0.65 - j\,0.45$

so that $Z_L = Z_0 z_L = (32.5 - j\,22.5)\ \Omega$.

**Example 2.17.** Let $v^+$ and $i^+$, $v^-$ and $i^-$ be the complex voltage and current of the incident and reflected waves respectively and $\Gamma_0 = |\Gamma_0|\, e^{j\theta_0}$ be the reflection coefficient from the end of the line terminated in a load. Determine the positions of voltage maxima and minima.

*Solution*:

$$v=v^+ + v^- = v_1\, e^{-j\beta x} + v_2\, e^{j\beta x}$$

$$i=i^+ + i^- = \frac{1}{Z_0}\,(v_1\, e^{-j\beta x} - v_2\, e^{j\beta x})$$

where $Z_0$ is the characteristic impedance.

Using Fig. Ex. 2.17 we have

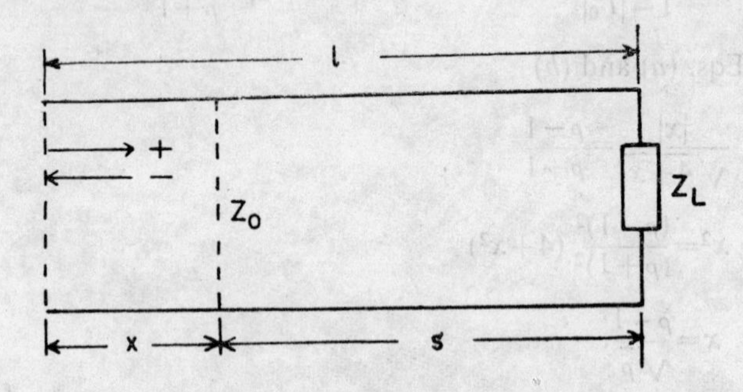


Fig. Ex. 2.17

$$v(s)=v_1\, e^{-j\beta l}\, e^{j\beta s} + v_2\, e^{j\beta l}\, e^{-j\beta s}$$

at the end of the line $s=0$

and $$v(0)=v_1\, e^{-j\beta l} + v_2\, e^{j\beta l}$$

$$= v^+(l)(1+\Gamma_o)$$

with $$\Gamma_0 = \frac{v_1}{v_2}\, e^{2j\beta l}.$$

Thus $$v(s)=v_1\, e^{-j\beta l}\,(e^{j\beta s} + \Gamma_o\, e^{-j\beta s})$$

$$v(s)v^*(s)=|v|^2=v_1 v_1^* \quad [e^{j\beta s} + \Gamma_0\, e^{-j\beta s}]\,[e^{-j\beta s} + \Gamma_0^*\, e^{j\beta s}]$$

$$v_1 v_1^* = v^+\,(v^+)^* = |v^+|^2$$

and

$$\left|\frac{v}{v^+}\right|^2 = 1+|\Gamma_o|^2 + |\Gamma_o|\,[e^{-j(\theta_o-2\beta s)} + e^{j(\theta_o-2\beta s)}]$$

$$=1+|\Gamma_o|^2+2\,|\Gamma_o|\cos(\theta_o-2\theta).$$

Thus $|v|$ is maximum when $\theta_o - 2\theta = \pm 2n\pi$,

$$v=|v^+|\,(1+|\Gamma_0|)$$

so that the positions of maximum are

$$s_{\max} = \left(\frac{\theta_0}{4\pi}\right)\lambda \pm n\,\frac{\lambda}{2}.$$

Similarly, positions of minimum are determined by the condition

$$\theta_0 - 2\theta = \pm(2n+1)\pi$$

which yields

$$s_{\min} = \left(\frac{\theta_0}{4\pi}\right) \pm \left(n + \frac{1}{2}\right)\frac{\lambda}{2}.$$

**Example 2.18.** A terminated (100 Ω) lossy line of length 24 cm ($\lambda = 10$ cm) has VSWR at the load and at the input respectively. $\rho_L = 4$ and $\rho_i = 3$. The first voltage minimum is at a distance 1 cm from the load. Find the attenuation constant of the line, input impedance and the load terminating the line.

*Solution*: We have reflection coefficients at input and load respectively as

$$|\Gamma_0| = \frac{\rho_L - 1}{\rho_L + 1} = \frac{3}{5} = 0.6$$

$$|\Gamma_i| = \frac{\rho_L - 1}{\rho_i + 1} = \frac{2}{4} = 0.5.$$

Now $\quad \Gamma_i = |\Gamma_0|\, e^{-2\gamma l}$

or $\quad |\Gamma_i| = |\Gamma_0|\, e^{-2\alpha l}$

which give

$$e^{-2\alpha l} = \frac{|\Gamma_i|}{|\Gamma_o|} = \frac{5}{6}$$

or $\quad \alpha_1 = \tfrac{1}{2} \ln 1.2 = 1.15 \log 1.2 = 0.09$ neper

$$\alpha = \frac{0.09}{2.4} = 0.0375 \text{ neper/wavelength}$$

$$= 0.375 \text{ neper/metre.}$$

Now $d_{\min} = 0.1\lambda$ and $\alpha\lambda_{\min} = 0.00375$ neper which may be neglected.

Then $\quad \theta_o - \dfrac{4\pi d_{\min}}{\lambda} = -\pi$

from which $\quad \theta_0 = -0.6\,\pi = -1.885$ radian

$$\Gamma_0 = 0.6\, e^{-j0\cdot 6\pi} = -0.6(0.309 + j\,0.95)$$

$$= -(0.185 + j\,0.57).$$

Also $\quad z_L = \dfrac{1+\Gamma_L}{1-\Gamma_L} = 0.37 - j\,0.66$

and $\quad Z_L = (37 - j\,66).$

Again $\quad \Gamma_i = |\Gamma_0|\, e^{j\theta_0}\, e^{-2\alpha l}\, e^{-2j\beta l} = |\Gamma_i|\, e^{j(\theta_0 - 2\beta l)}$

with $\quad 2\beta l = \dfrac{4\pi}{\lambda}\cdot 2.4\lambda = 9.6\pi$ and $\theta_o - 2\beta l = -10.2\,\pi.$

Then $\Gamma_i=|\Gamma_i|\, e^{-j0.2\pi}=0.404-j0.293\simeq 0.4-j\,0.3$

and $z_i=\dfrac{1.4-j\,0.3}{0.6+j\,0.3}=(1.67-j\,1.33)\ \Omega$

or $Z_i=(167-j\,133)\ \Omega.$

**Example 2.19.** Calculate the efficiency $\eta$ of a transmission line as a function of $\Gamma_0$, the reflection coefficient. Under what condition does $\eta$ have its maximum value $\eta_{max}$. Calculate $\eta$ for $\alpha=1.4$ dB/m, $l=2$m and $R_L=3\ z_0$.

*Solution*: $\Gamma_i=\Gamma_o\, e^{-2\gamma l}$.

The input power transmitted along the line

$$P_i=P_o(1-|\Gamma_i|^2)=P_o(1-|\Gamma_o|^2\, e^{-4\gamma l})$$

where $P_o$ is power supplied at the input. The power arriving at the load

$$P_o'=P_o\, e^{-2\alpha l}$$

and power delivered to the load

$$P_L=P_o'\,(1-|\Gamma_o|^2)$$

$\eta$ is maximum when $Z_L=Z_0$ so that $\Gamma_0=0_r$.

Then, $\eta=\eta_{max}=e^{-2\alpha l}$

or $$\eta=\eta_{max}\frac{1-|\Gamma_o|^2}{1-|\Gamma_o|^2\,\eta_{max}^2}.$$

If $Z_L=R_L,\ |\Gamma_0|^2=\left(\dfrac{R_L-Z_o}{R_L+Z_o}\right)^2=\left(\dfrac{1/\rho-1}{1/\rho+1}\right)^2.$

Thus $$\eta=\eta_{max}\frac{4\rho}{(\rho+1)^2-(\rho-1)^2\,\eta_{max}^2}.$$

If $l=2m$ and $\alpha=1.4$ dB/m

$$2.8=10\log\frac{1}{\eta_{max}}\quad\text{or}\quad \eta_{max}=0.527.$$

For $\dfrac{R_L}{Z_o}=3=\rho,\quad \eta=0.423$ or $\eta=42.3\%$.

**Example 2.20.** In a coaxial line the outer conductor is held coaxially to the inner conductor by means of two dielectric rings of thickness $t$ and relative permittivity $\epsilon_r$ and separated by a distance $d$. What relation must exist between $d$, $t$ and $\epsilon_r$ so that the line is matched.

*Solution*: Consider the coaxial line shown in Fig. Ex. 2.20. The line will be matched if there is no reflection at plane $E$ due to a matched load at $T$.

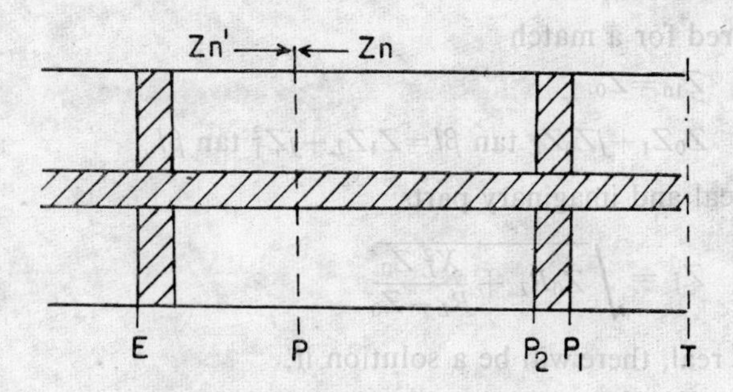


**Fig. Ex. 2.20**

Let $Z_0$ and $Z_1$ be the characteristic impedances of the line empty and filled with dielectric respectively and let $\beta_0$ and $\beta_1$ be respective propagation constants. Then

$$Z_1 = \frac{Z_0}{\sqrt{\epsilon_r}}, \quad \beta = \beta_0 \sqrt{\epsilon_r}.$$

Now impedance at $P_1$ is $Z_0$ which at $P_2$ becomes

$$Z(t) = Z_1 \frac{Z_0 + jZ_1 \tan \beta_1 t}{Z_1 + jZ_0 \tan \beta_1 t}$$

or at position $P$ impedance seen from the left

$$Z(P) = Z_n = Z_0 \frac{Z(t) + jZ_0 \tan \beta_0 d/2}{Z_0 + jZ(t) \tan \beta_0 d/2}.$$

Similarly, impedance at $P$ as seen from the right (obtained by replacing $t$ by $-t$ and $d$ by $-d$)

$$Z_n' = Z(P) = \frac{Z(-t) - jZ_0 \tan \beta_0 d/2}{Z_0 + jZ(-t) \tan \beta_0 d/2}.$$

For matching $Z_n' = Z_n$

which on substitution and simplification yields

$$\frac{2\sqrt{\epsilon_r}}{1+\epsilon_r} = \tan \beta_0 \sqrt{\epsilon_r} \cdot t \tan \beta_0 d.$$

**Example 2.21.** A given load $Z_L$ is matched by a line of characteristic impedance $Z_0$ by connecting a line of characteristic impedance $Z_1$ and length $l$. Calculate $Z_1$ and $l$. Is the match always possible?

*Solution*: The input impedance of the line of length $l$ terminated in load $Z_L$ is

$$Z_{in} = Z_1 \frac{Z_L + jZ_1 \tan \beta l}{Z_1 + jZ_L \tan \beta l}.$$

As required for a match

$$Z_{in}=Z_0.$$

Thus, $$Z_0Z_1+jZ_0Z_L \tan \beta l=Z_1Z_L+jZ_1^2 \tan \beta l.$$

Equating real and imaginary parts

$$Z_1 = \sqrt{Z_0R_L + \frac{X_L^2 Z_0}{R_L-Z_0}}.$$

Since $Z_1$ is real, there will be a solution if,

$$Z_0 R_L+\frac{X_L^2 Z_0}{R_L-Z_0}>0.$$

Note that (*i*) For $R_L>Z_0$ the match is always possible

(*ii*) For $R_L<Z_0$, $Z_0R_L>\frac{X_L^2 Z_0}{Z_0-R_L}$

or $$Z_oR_L>|Z_L|^2.$$

**Example 2.22.** A 50 Ω loss-less line is terminated in a matched load. If 25 percent of the incident power is reflected when the line is shunted by a

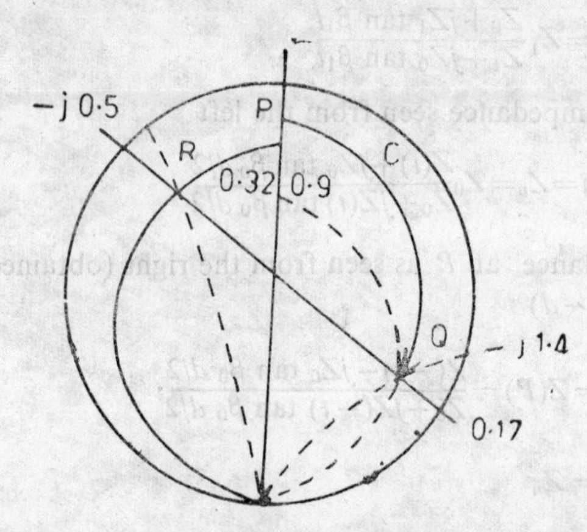


**Fig. Ex. 2.22**

conductance $G$, what is the value of $G$? Further, if the incident power is 40 mW, how much power is dissipated in the matched load? What is the attenuation of the system?

*Solution*: The reflection coefficient at $G$ is

$$\Gamma=\frac{Y_0-Y}{Y_0+Y}=\frac{-G}{G+2Y_0}.$$

Now $$\frac{P_r}{P_i}=|\Gamma|^2=\left|\frac{G}{G+2Y_o}\right|^2=\frac{1}{4} \text{ (given)}.$$

From which $G=2Y_o=2\times10^{-3}\ \Omega^{-1}$.

The power dissipated in the matched load

$$P_{Y_0}=\frac{Y_o}{G+Y_o}\,P_t=\frac{P_t}{3}=10 \text{ mW}.$$

The attenuation is

$$\alpha=10\log\frac{P_i}{P_{Y_0}}=10\log 4=6 \text{ dB}.$$

**Example 2.23**. A 50 Ω loss-less line of length 34 cm is terminated by a resistance 12.5 Ω. Calculate the parameters of the equivalent series circuit at the input of this line for 150 MHz and also for the parallel equivalent circuit.

*Solution*: $f=150$ MHz

$$\therefore \qquad \lambda=\frac{3\times10^8}{1.5\times10^8}=2 \text{ m}$$

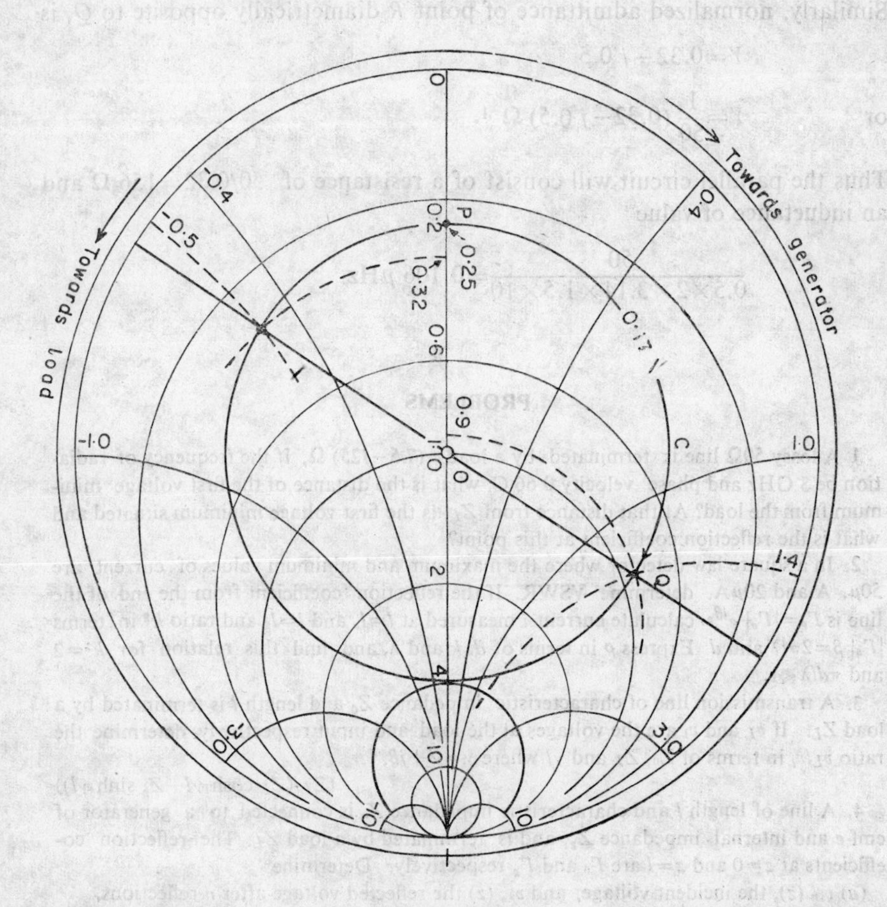


Fig. Ex. 2.23

so that $1/\lambda = 0.17.$

Since $r = R/Z_0 = \frac{12.5}{50} = 0.25$

We start at point $P$ on the chart corresponding to $r=0.25$ and draw about $O$ the circle $C$ of constant VSWR as shown in Fig. Ex. 2.23.

Following the circle $C$ through a circum distance $0.17\lambda$ on the outer circle we can find the normalized input impedance represented by point $Q$ as

$$z = 0.9 + j\,1.4$$

or $Z = (45 + j70)\ \Omega.$

The equivalent circuit will have a resistance of 45 Ω and an inductance of value

$$L = \frac{70}{\omega} = \frac{70}{2 \times 3.14 \times 1.5 \times 10^8} = 7.45 \times 10^{-8}\ \text{Hz.}$$

Similarly, normalized admittance of point $R$ diametrically opposite to $Q$, is

$$Y = 0.32 - j\,0.5$$

or $Y = \frac{1}{50}\,(0.32 - j\,0.5)\ \Omega^{-1}.$

Thus the parallel circuit will consist of a resistance of $50/0.32 = 156\ \Omega$ and an inductance of value

$$\frac{50}{0.5 \times 2 \times 3.14 \times 1.5 \times 10} = 0.106\ \mu\text{Hz.}$$

## PROBLEMS

1 A lossy 50Ω line is terminated by a load $=(7.5-j25)\ \Omega$. If the frequency of radiation be 3 GHz and phase velocity 0.66 C, what is the distance of the first voltage minimum from the load? At that distance from $Z_L$, is the first voltage minimum situated and what is the reflection coefficient at this point?

2. In a square law detector where the maximum and minimum values of current are 50μ. A and 20μA, determine VSWR. If the reflection coefficient from the end of the line is $\Gamma_o = |\Gamma_o|\, e^{i\theta_o}$, calculate current $i$ measured at $l = l_o$ and $l = l_2$ and ratio $k^2$ in terms $|\Gamma_o|$ $\beta = 2\pi/\lambda$ and $d$. Express $\rho$ in terms of $d$, $k$ and $\lambda$, and find this relation for $k^2 = 2$ and $\pi d/\lambda \ll 1$.

3. A transmission line of characteristic impedance $Z_0$ and length $l$ is terminated by a load $Z_L$. If $v_L$ and $v_i$ are the voltages at the load and input respectively. determine the ratio $v_L/v_i$ in terms of $Z_0$, $Z_L$ and $\gamma l$ where $\gamma = \alpha + j\beta$.

$(Z_L/(Z_L \cosh \gamma l + Z_0 \sinh \gamma l))$

4. A line of length $l$ and characteristic impedance $Z_0$ is connected to a generator of emf $e$ and internal impedance $Z_g$, and is terminated by a load $Z_L$. The reflection coefficients at $z=0$ and $z=l$ are $\Gamma_0$ and $\Gamma_g$ respectively. Determine

(a) $v_{in}(z)$, the incident voltage, and $v_{r_n}(z)$ the reflected voltage after $n$ reflections.

(b) Potential $u(x)$ in terms of $z_0$, $\Gamma_g$, $\Gamma_o$, $e_g$ and $\gamma l$.

5. A transmission line with $Z_0=100\ \Omega$ is terminated in an impedance $50+j\ 50\ \Omega$. Find the reflection coefficient, standing wave ratio, and the fraction of the incident power delivered to the load.

6. On a 50 Ω transmission line, the voltage at a distance $0.4\lambda_0$ from the load is $(4+j2)$. The corresponding current is —2 A. Determine normalized load impedance..

7. An air-filled coaxial line having $Z_0=100\ \Omega$ is terminated in an impedance. It is found that two adjacent voltage minima occur at 0.15 m and 0.45 m from the termination. Find (*a*) the frequency of the source, (*b*) the value of the terminating impedance, (*c*) the fraction of the incident power absorbed by the load.

What can be inferred about the value of $Z_r$ if, for the same value of $S$, minima occur at different distances from the termination? Evaluate $z_r$ for a particular case of the first minimum occurring 0.3 m from the load. *(London University)*

8. An air-space lossless transmission line is terminated in a load impedance that produces a voltage reflection coefficient of $0.5\ \angle 60°$. If the characteristic impedance of the line is 100 Ω, find the value of the load impedance, the position of voltage maximum nearest to the load, and the voltage standing wave ratio on the line. Also determine one position and one length of a short-circuited stub, having the same characteristic impedance which will render a matched condition between this stub and the source, if the frequency of the latter is 100 MHz. *(Bristol University, 1968)*

9. A coaxial transmission line is exactly one half-wavelength long. The input impedance is $(1+j0)\ \Omega$ with a short circuit applied, and $10^4\ (1+j0)\ \Omega$ with an open circuit line The characteristic impedance is $(100+j0)\ \Omega$ and the supply frequency is 150 MHz. Determine the attenuation constant $\alpha$. *(Southampton University)*

10. A loss-less open-wire feeder connects a transmitter to an antenna. The operating frequency is 90 MHz, the characteristic impedance of the line is 300 Ω and the input impedance of the antenna is resistive and equal to 75 Ω.

Determine the closest point to the load at which a single short-circuited stub may be connected in order to match the line from the source to that point and find the minimum length of the stub required.

If the antenna is replaced by an adjustable resistive load, estimate the maximum and minimum load resistances that can be tolerated if the voltage standing wave ratio on the main-feeder is not to exceed 1.4:1, with the stub connected as previously.

*(Bristol University, 1975)*

11. A line has the following primary constants per loop mile $R=40\ \Omega$, $L=1$ MH, $C=0.05\ F$, $G$ is negligible. Loading coils with inductance $L_d$ of 100 MHz and negligible resistance are added at intervals of one mile, giving rise to an attenuation coefficient which is nominally constant upto a frequency $f=[\pi^2(L_d+lL)\ lC]^{-1/2}$. Determine the minimum value of $l$ if $f=4$KHz and estimate the attenuation coefficient at 5 krad $s^{-1}$ before and after loading, assuming the additional inductance to be uniformly distributed. *(Bristol University, 1972)*

12. Show that for an air-filled coaxial line, minimum attenuation occurs when $x \ln x=1+x$, $x=b/a$. What is the corresponding characteristic impedance?

13. Evaluate $z_0$ for a lossy coaxial line and computed values of $R$, $G$, $L$ and $C$. Assume $b=3a=1$ cm, $f=1$ GHz, and $\epsilon=(2.56-j0.001)\ \epsilon_0$. Verify that $I_m(Z_0) \ll R_e(Z_0)$ and $Z_0 \simeq \sqrt{L/C}$.

14. In a coaxial line the outer and inner conductors are held coaxially by two dielectric rings of thickness $t$ and relative permittivity $\epsilon_r$ separated by a distance $d$. Show that the line will be matched when

$$\frac{2\sqrt{\epsilon_r}}{2+\epsilon_r}=\tan \beta_o \sqrt{\epsilon_r}\ t \tan \beta_o d$$

$\beta_o$ being the propagation constant of an air-filled coaxial line.

15. A load $Z_L=(80+j45)\ \Omega$ is matched to a 75 Ω line via transmission line of length $l$ and characteristic impedance $Z_0$. Determine $l$ and $Z_0$.

16. A 50 Ω loss-less line terminated in a load of impedance $Z_L$ is supplied an incident power $P_i=500$ mW by a generator. If VSWR is 1.4, what is the power reaching the load? What are the maximum and minimum values of impedance, voltage and current.

17. A loss-less transmission line has a characteristic impedance of 100 Ω and is terminated in a load of 75 Ω. The line is 0.75 λ long. Determine the sending end impedance and the reactance which, if connected across the sending end of the line, will make the input impedance a pure resistance.

18. A 75 Ω line is terminated by an impedance $Z_L$. With the aid of a Smith chart find the reflection coefficient and standing wave ratio due to the load when it takes the following values :

$$Z_1=225\ \Omega,\ Z_2=75\ \Omega,\ Z_3=15\ \Omega,\ Z_4=j45\ \Omega,$$
$$Z_5=-j\,225\ \Omega,\ Z_6=(45+j\,120)\ \Omega.$$

19. A $Z_0$ line is joined to a bifurcation of two lines of characteristic impedance $Z_0$ and $Z_0/2$. Evaluate the voltage reflection coefficient and VSWR for a wave travelling towards the junction on each of the three lines.

(viewd from $Z_0$, 1/2, −1/2 and viewed from $Z_0/2$, 1, ).

20. A 100 Ω line operating at 1 GHz is terminated in a load $(25-j\,75)\ \Omega$. Determine the lengths and positions (s) of 100 Ω stub

21. A line is terminated in a load $Z_L=(100+j\,100)\ \Omega$. The load is required to be matched by a double stub system of characteristic impedance 50 Ω. The first stub is placed at 0.40λ away from the load and the spacing of the stubs is (3/8)λ. Determine the length of the short-circuited stubs when a match is achieved. What terminations are forbidden for matching the line by the double stub device?

## *REFERENCES*

BADEN FULLES A.J., *Microwaves, Pergamon Press.*

BAHL, I.J. and BHARTIA, P., 'The Design of Broadside-coupled strip-line Circuits.' *IEEE Trans.*, MTT-29 (2) 1981.

BRONWELL, A.B and Beam, R.E. *Theory and Applications of Microwaves*, McGraw-Hill Book Co. (1947).

BROWN, R.G., LINES, *Waves and Antennas*, 2nd ed., Ronald Press, New York (1973).

COLLIN, R.E. *Foundation for Microwave Engineering*; McGraw-Hill Book Co. (1966).

COLLIN R.E., *Field Theory of Guided Waves*, McGraw-Hill New York (1960).

DAVIDSON C.W., '*Transmission Lines for Communication*, Macmillan Press, London (1978).

DELOGNE, P.P. AND LALOUX, A.A., 'Theory of Slotted Coaxial Cable, *IEEE Trans*, MTT-28 (10) 1980.

DENLUNGER, E.J. '*Losses of Microstrip lines*' *IEEE Trans. on Microwave Theory and Tech*. MTT-28 (6) 1980.

EVERITT, W.L. and ANNER, G.E., *Communication Engineering*, McGraw-Hill Book Co.

GESTSENGER, W.J., 'Microstrip Dispersion Model. *IEEE Trans.* MTT-21 (11) 1973.

GRIVET, P., *The Physics of Transmission Lines at High and Very High Frequencies*, Academic Press London (1969).

IEEE TRANS. Special Issue on 'Microwave Communication', *MTT-23* (4) 1975.

IEEE TRANS. Special Issue on 'Microwave and Millimeter Wave-Integrated Circuits, MTT-26 (10) 1978.

JACKSON, W., *High Frequency Transmission Lines,* Melhuen London (1951).

JANSEN, R.H., 'Hybrid Mode Analysis of End Effects of Planer Microwave and Millimeter-wave Transmission Lines' *IEEE Proc.* 128 H (2) 1981.

JORDAN E.C. and BALMAIN K.G. *Electromagnetic Waves and Radiating Systems*', Prentice-Hall Inc (1968).

KARBOWIAK and IRWING, D.H., 'Theory of Coupled Transmission Lines and Its Applications to Optical Fiber, *ATR*-**13** (1) 1979.

KING, R., *High Frequency Transmission Lines*, Dover, New York (1964).

KING, R., *Transmission Lines, Antennas and Waveguides,* Dover New York (1965).

KNORR, J.B., and Shayda, P.M. 'Millimeter Wave Fin-Line Characteristics', *IEEE Trans.* MTT-28 (7) 1980.

KRAKUS, J.D. and Carver K.R., *Electromagnetics,* McGraw-Hill Book Co. (1973).

KUESTER, E.F. and Chang, D.C. 'Theory of Disperson in Microstrip of Arbitrary Width' *IEEE Trans.* MTT-28 (3) 1980.

LAMONT V.B., *Transmission Lines and Waveguides*, Wiley New York (1969).

MAGNUSSON, P.C., *Transmission Lines and Wave Propagation,* 2nd ed., Allyn and Bacon, Boston Mars (1970).

MATSUMURA, K. AND TOMABECHI. Y., 'Mode Transformer for Connection between Rectangular and Circular Dielectric Waveguide', *Trans. IECE J* (Japan) J 62-B (3) 1979,

OGAWA, H. AND AIKAWA, M., 'Analysis of Coupled Microstrip Slot lines, *Trans IECEJ* J 62-B (4) 1979.

PARK, C. and SAMMUT, R.A., 'Developments in the Theory or Fiber Optics' (Review) *Proc. IREE* (Australia) **40** (3) 1979.

RAGAN, G.K., *Microwave Transmission Circuits,* McGraw-Hill, New York (1948).

RAMO, S., WHINNERY J.R. and T. VAN DAZER, *Fields and Waves in Communication Electronics*', John Wiley (1965).

RYDER J.D., *Networks, Lines and Fields,* Prentice-Hall of India.

SAMUEL Y.L. *Microwave Devices and Circuits,* Prentice-Hall In. N.J. (1980).

SANDER K.F., *Transmission and Propagation of Electromagnetic Waves,* Cambridge University Press (1978).

SESHADRI S.R., *Transmission Lines and Electromagnetic Fields,* Addison Wesley (1971).

SINNEMA W., *Electronic Transmission Technology Lines, Waves and Antennas,* Prentice-Hall Inc. N.J, (1979).

SLATER, J.C., *Microwave Transmission,* McGraw-Hill., New York (1942).

SOUTHWORTH G.C., '*Principles and Application of Waveguide Transmission*', D Van Nostrand Co. New York (1950).

STEER, M.B. and KAHN, P.J., 'Wide-band Equivalent Circuit for Radial Transmission Lines', *IEEE Proc.*, 128 H (2) 1981.

STEPHEN R., *Transmission Lines and Antennas,* Hollet Rinchat and Winston (1969).

STEWART, J.L., *Circuit Analysis of Transmission Lines,* Wiley, New York (1962).

VOGEL, R., 'Microstrip—Slotline Components for Microwave IC's. *Microwave J.,* 23 (5) 1980.

WHEELER, H.A., 'Transmission-Line Conductors of Various Cross-Sections', *IEEE Trans.* MTT-28 (2) 1980.

WHEELER, H.A., 'Transmission Line Properties of a Strip of a Dielectric Sheet on a Plane', *IEEE Trans.* MTT-25 1977.

ZEHENTNER, J., 'Analysis and Synthesis of Coupled Microstrip Lines by Polynomials', *Microwave J.* 23 (5) 1980.

# 3

# BASIC ELECTROMAGNETICS

## 3.1 INTRODUCTION

The transmission line theory, discussed so far using current and voltage wave concepts, is limited in its application to lines with two or more conductors such as two parallel telephone wires, coaxial lines, shielded two conductor lines, etc. Even these simple lines may propagate waves in which the distribution of currents and electric fields is more complicated than the assumptions that conventional theory permit. Therefore, a more comprehensive approach has to be made. This is possible by the application of the electromagnetic theory. Whereas conventional theory chiefly considers currents and voltages, electromagnetic theory is primarily concerned with electric and magnetic fields associated with these currents and voltages. With this in view we shall study basic electromagnetic theory in this chapter. This will include Maxwell's equations, their solutions and the wave equation along with the derivation and application of the appropriate electromagnetic boundary conditions. We shall also introduce and discuss the concepts of propagation constant, attenuation constant, phase constant, wave impedance and power flow in electromagnetic waves.

## 3.2 MAXWELL'S EQUATIONS

In order to study transmission lines (any guiding structure) electromagnetically, we need laws governing time varying electromagnetic fields. Instead of presenting different laws individually, we shall develop a set of Maxwell's equations which collectively express all the laws of electromagnetism. It should, however, be noted that our aim is not to obtain these equations in their basic mathematical abstract form but rather to gain some understanding of their physical significance and to learn how to obtain their solutions in practical situations of interest in microwave engineering.

Maxwell's equations, which basically represent four fundamental laws namely, Faraday's law of induced electromotive force (emf), Ampare's circuital law, Gauss's law of electric field and Gauss's law of magnetic field, deal with four vector fields defined as functions of time and space. These four vectors may be grouped in two pairs. The electric field $\vec{E}$ and the magnetic induction $\vec{B}$ being one pair and the electric displacement $\vec{D}$ and the magnetic field intensity $\vec{H}$ the other. The vectors of the first pair $\vec{E}$ and $\vec{B}$ determine force acting on a charge $q$ moving with velocity $\vec{v}$ that is

$$\vec{F} = q(\vec{E} + \vec{v} \times \vec{B}) \tag{3.1}$$

where $\vec{F}$ is the force in newton ($N$), $q$ is the charge in coulombs ($C$), $\vec{E}$ is the electric field measured in volts/metre (V/m), $\vec{B}$ is the magnetic induction measured in Weber/metre² (W/m²), and $\vec{v}$ is the velocity in metres/sec (m/sec). This force law is called the Lorentz force and the equation the Lorentz force equation. The second pair of vectors $\vec{D}$ and $\vec{H}$ are determined from charges and currents in the fields, as lines of force originate from the charges and currents. $\vec{D}$ and $\vec{B}$ are related to $\vec{E}$ and $\vec{H}$ as

$$\vec{B} = \mu\vec{H} \quad \text{and} \quad \vec{D} = \epsilon \vec{E} \tag{3.2}$$

where $\mu$ is the magnetic permeability and $\epsilon$ is the electric permittivity of the medium. In free space the above equations are reduced to

$$\vec{B} = \mu_0\vec{H} \quad \text{and} \quad \vec{D} = \epsilon_0 \vec{E}$$

where $\mu_0 = 4\pi \times 10^{-7}$ Henry/metre (H/m) and is the magnetic permeability of vacuum; and

$$\epsilon = \frac{10^{-9}}{36\,\pi} = 8.854 \times 10^{-12}$$

Farad/metre (F/m) and is known as the electric permittivity of vacuum.

### (*a*) Maxwell's First Curl Equation for Electromagnetic Fields

One of the basic laws of electromagnetic phenomena is Faraday's law, which states that a time varying magnetic field generates an electric field. Mathematically, this law can be expressed as

$$\oint_c \vec{E} \cdot d\vec{l} = -\frac{\partial}{\partial t}\int_s \vec{B} \cdot d\vec{s} \tag{3.3}$$

where $\frac{\partial}{\partial t}\int_s \vec{B} \cdot d\vec{s}$ is the time rate of change of total magnetic flux passing through a non-moving surface $s$ bounded by an arbitrary closed curve $C$ as shown in Fig. 3.1. This time rate of change of total magnetic flux is equal to the negative value of the total voltage measured around $C$ which is $-\oint_C \vec{E} \cdot d\vec{l}$, the line integral of the electric field $\vec{E}$ around $C$. Although Eq. (3.3) is in a form that can readily be interpreted physically, it is not in a form suitable for the analysis of physical problems where we require a differential equation which is equivalent to Eq. (3.3). This is obtained by using

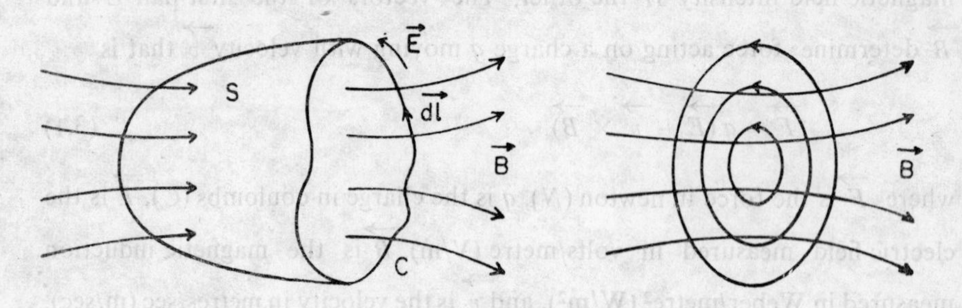


Fig. 3.1 Illustration of Faraday's laws.

Stokes' theorem of vectors which states that the line integral of a vector around a closed contour $C$ is equal to the normal component of the curl of the vector over any surface $s$, bounded by the contour $C$. In the present case Stokes' theorem can mathematically be expressed as

$$\oint_c \vec{E} \cdot d\vec{l} = \int_s (\nabla \times \vec{E}) \cdot d\vec{s}$$

where $\nabla \times \vec{E}$ is the curl of vector $\vec{E}$ and is a measure of the circulation of the vector field $\vec{E}$ at a point. If the surface referred to above is the same as in Eq. (3.3), then

$$\int_s (\nabla \times \vec{E}) \cdot d\vec{s} = -\frac{\partial}{\partial t} \int_s \vec{B} \cdot \vec{ds}$$

As the surface $s$ is stationary and is completely arbitrary, hence

$$\nabla \times \vec{E} = -\frac{\partial \vec{B}}{\partial t} \tag{M-1)*$$

Equation (M-1) is the desired differential equation describing Faraday's law and is the Maxwell's first curl equation for electromagnetic field.

If, however, the surfaces is considered above moves with a velocity $\vec{v}$ Eq. (M-1) takes the form**

$$\nabla \times \vec{E'} = -\frac{\partial}{\partial t} \vec{B} + \nabla \times (\vec{v} \times \vec{B}) \tag{3.4}$$

*We are numbering Maxwell's equations as M-1, M-2 . . . etc. because these shall be frequently referred to in our future discussions.

**N.N. Rao, *Basic Electromagnetics with Applications*, New Delhi: Prentice-Hall of India Private Ltd, 1974.

where $\vec{E'}$ is the electric field as measured by an observer moving with velocity $\vec{v}$ relative to the magnetic field $\vec{B}$.

**(b) Maxwell's Second Curl Equation for Electromagnetic Fields**

The second basic law of electromagnetic phenomena is Ampere's circuital law which states that the line integral of magnetic intensity around a closed path is equal to the current linking the path. The said line integral $\oint \vec{H} \cdot \vec{dl}$ is known as the magnetomotive force (mmf) in analogy with electromotive force (emf). Thus Ampere's circuital law can mathematically be written as

$$\oint_c \vec{H} \cdot \vec{dl} = I \text{ (total current)} = \int_s \vec{J} \cdot \vec{ds} \tag{3.5}$$

where $\vec{J}$ is surface current density.

The application of Stokes' theorem on the left hand side of Eq. (3.5) yields,

$$\int_s (\nabla \times \vec{H}) \cdot \vec{ds} = \int_s \vec{J} \cdot \vec{ds}.$$

The surface $s$ bounded by the contour $C$ in Eq. (3.5) is arbitrary, hence

$$\nabla \times \vec{H} = \vec{J}. \tag{3.6}$$

Diverging from Eq. (3.6) yields $\nabla \cdot \vec{J} = 0$, the equation of continuity for steady currents, i.e. currents that do not vary with time. Therefore, Eq. (3.6) is not consistent with the time varying equation of continuity $\nabla \cdot \vec{J} = -\partial\rho/\partial t$, where $-\partial\rho/\partial t$ is the rate of decrease in the free charge density. This inconsistency led Maxwell to seek a modification to Ampere's circuital law. This modification he found by substituting Gauss's law $\nabla \cdot \vec{D} = \rho$ (See Eq. M-37 into the equation of continuity for time varying fields, giving

$$\nabla \cdot \vec{J} = -\frac{\partial}{\partial t} (\nabla \cdot \vec{D}). \tag{3.7}$$

Interchanging the derivatives wlth respect to space and time (being independent of one another) we have,

$$\nabla \cdot \left( \frac{\partial \vec{D}}{\partial t} \times \vec{J} \right) = 0. \tag{3.8}$$

Equation (3.8) may be put into an integral form by integrating over a volume and then applying the divergence theorem in vectors.

$$\oint \left( \frac{\partial \vec{D}}{\partial t} + \vec{J} \right) \cdot \vec{ds} = 0 \tag{3.9}$$

The two equations (3.8) and (3.9) suggest that $(\partial \vec{D}/\partial t+\vec{J})$ may be regarded as the total current density for the time varying fields. Since $\vec{D}$ is the electric displacement flux density, $\partial \vec{D}/\partial t$ is known as the displacement current density.* Therefore, in Ampere's law the total current density $(\partial \vec{D}/\partial t+\vec{J})$ should replace $\vec{J}$. Hence Eq. (3.6) becomes

$$\nabla\times\vec{H}=\vec{J}+\frac{\partial\vec{D}}{\partial t}.$$

Since $\vec{D}=\epsilon\vec{E}$, $\nabla\times\vec{H}=\sigma\vec{E}+\epsilon\frac{\partial \vec{E}}{\partial t}$. (M-2)

This is Maxwell's second curl equation for electromagnetic field. $\sigma$ is conductivity (mho/metre) of the medium.

In integral form Eq. (3.5) is expressed as

$$\oint_c \vec{H}\cdot d\vec{l}=\int_s\left(\frac{\partial\vec{D}}{\partial t}+\vec{J}\right)\cdot d\vec{s}. \tag{3.10}$$

Equation (3.10) states that the magnetomotive force (mmf) around a closed path is equal to the conduction current plus the time derivative of electric displacement through the surface $s$ bounded by path $c$.

**(*c*) Maxwell's Divergence Equation for Electric Field**

As mentioned above the third basic law in electromagnetism is Gauss's law for electric field which states that the net electric flux emanating from a closed surface is equal to the net charge enclosed by the surface divided by $\epsilon_0$, the permittivity of the medium inside the closed surface. In equation form, Gauss's law is written as

$$\oint_{s\,(\text{closed})} \vec{E}\cdot d\vec{s}=q/\epsilon_o \tag{3.11}$$

where $q$ is the charge enclosed in surface $s$.

Using $\vec{D}=\epsilon\vec{E}$ in the above equation, we have

$$\oint \vec{D}\cdot d\vec{s}=q.$$

The total charge $q$, enclosed by the surface may be written as

$$q=\int_v \rho dv,$$

*For details about displacement current refer to

1. S.A. Schelkunoff,, *Electromagnetic Waves*, D. Van Nostrand Co. Inc. (1943).
2. N.N. Rao, *Basic Electromagnetics with Applications*, New Delhi: Prentice-Hall of India (1974).

where $v$ is the volume enclosed by the surface $s$ and $\rho$ is the free volume charge density. Thus we have

$$\oint_s \vec{D} \cdot d\vec{s} = \int_v \rho \, dv.$$

But, according to the Gauss's divergence theorem in vectors,

$$\oint_s \vec{D} \cdot d\vec{s} = \int_v (\nabla \cdot \vec{D}) \, dv.$$

Thus, we have

$$\int_v (\nabla \cdot \vec{D}) \, dv = \int_v \rho \, dv.$$

The volume under consideration is completely arbitrary so the above equation is reduced to

$$\nabla \cdot \vec{D} = \rho. \tag{M-3}$$

This is Maxwell's third equation for electromagnetic field which states that the divergence of electric displacement at any point is equal to the free volume charge density at that point.

### ($d$) Maxwell's Divergence Equation for Magnetic Field

The fourth fundamental law of electromagnetism is Gauss's law for magnetic field which states that the total magnetic flux emanating from any closed surface is zero. In equation form this law can be expressed as

$$\oint_s \vec{B} \cdot d\vec{s} = 0.$$

The application of the divergence theorem yields,

$$\nabla \cdot \vec{B} = 0. \tag{M-4}$$

This is Maxwell's fourth equation for magnetic field which states that the divergence of magnetic induction through any closed surface is zero.

To summarize we rewrite the four Maxwell equations as follows:

$$\nabla \times \vec{E} = \frac{\partial \vec{B}}{\partial t} \tag{M-1}$$

$$\nabla \times \vec{H} = \vec{J} + \frac{\partial \vec{D}}{\partial t} \tag{M-2}$$

$$\nabla \cdot \vec{D} = \rho \tag{M-3}$$

$$\nabla \cdot \vec{B} = 0 \tag{M-4}$$

where the equation of continuity

$$\nabla \cdot \vec{J} = -\partial \rho / \partial t \tag{3.12}$$

is contained in the above equations.

It should, however, be borne in mind that although $\frac{-\delta \vec{B}}{\delta t}$ may be regarded as a source for $\vec{E}$ and $\frac{\delta \vec{D}}{\delta t}$ as a source for $\vec{H}$, the ultimate sources for an electromagnetic field are the current $\vec{J}$ and charge $\rho$. For time varying fields, $\rho$ varies with time and is thus related to $\vec{J}$ by the equation of continuity. Hence in such cases, it is possible to derive time varying electromagnetic fields from the knowledge of current density $\vec{J}$ alone.

It is easy to visualize that Eqs. (M-1) and (M-2) lead to the propagation of electromagnetic disturbance in space, i.e. wave propagation.

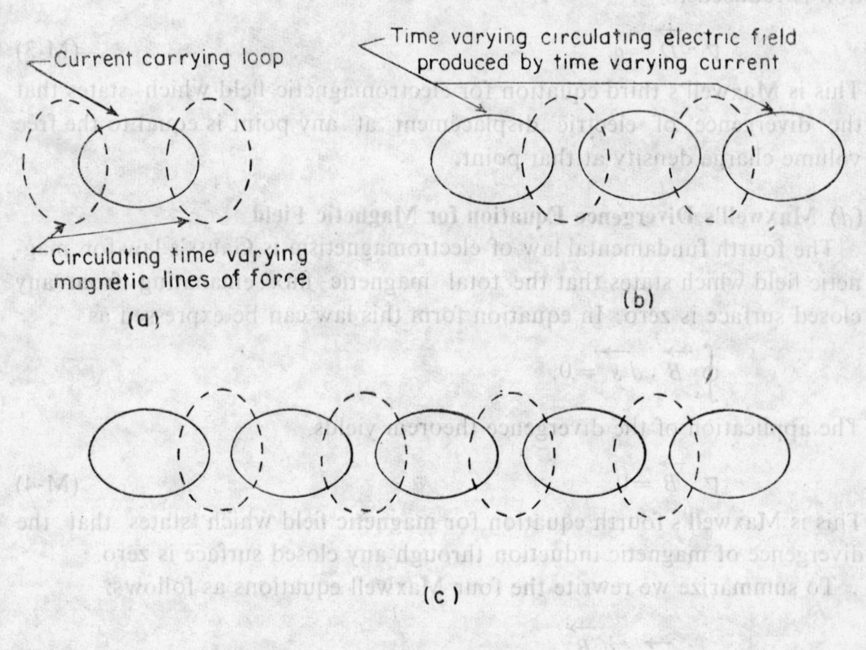


**Fig. 3.2** Generation of E.M. waves from current loop and propagation of E.M. Energy in Space.

Consider a loop of wire carrying time varying current. According to Eq. (M-2), this current will cause a circulating (curling) time varying magnetic field normal to the current as shown in Fig. 3.2 (a). (For clarity a few magnetic lines of force are shown.) This time varying magnetic field in turn causes a circulating (curling) time varying electric field as shown in Fig. 3.2(b) $[\partial \vec{B}/ \partial t=-(\nabla \times \vec{E})]$. These electric field lines are perpendicular to the encircling magnetic lines of force and encircle them. This circulating time varying electric field again results in a time varying circulating magnetic field in the space around the source wire loop and the process is repeated. As a result there is continual induction and spreading of the electromagnetic field into all space surrounding the current loop as shown in Fig. 3.2 (c). In other words, electromagnetic energy propagates in space with the velocity of light.

In microwave electronics we generally come across sinusoidally time varying fields of the form $e^{j\omega t}$. Consequently, using the phasor representation, the time derivative $\partial/\partial t$ may be replaced by the factor $(j\omega)$ since $\frac{\partial}{\partial t}(e^{j\omega t}) = j\omega\, e^{j\omega t}$. Therefore, Maxwell's equations with steady state sinusoidal time dependence become

$$\nabla \times \vec{E} = -j\omega \vec{B} \tag{3.13a}$$

$$\nabla \times \vec{H} = j\omega \vec{D} + \vec{J} \tag{3.13b}$$

$$\nabla \cdot \vec{D} = \rho \tag{3.13c}$$

$$\nabla \cdot \vec{B} = 0. \tag{3.13d}$$

## 3.3 THE MAXWELL-LORENTZ EQUATIONS

The set of Eqs. (M-1) through (M-4) were developed by Maxwell when the electrical origin of matter was unknown. As such these equations describe the electromagnetic phenomena macroscopically and are known as Maxwell's macroscopic equations.

Before we take up the derivation and solution of wave equations, it would be instructive at this point to inquire into the atomic origin of the field quantities $\vec{E}$, $\vec{H}$, etc. especially in regard to the study of the electric and magnetic properties of solids and gases at microwave frequencies. Knowing the electrical origin of matter, Lorentz* proposed a set of four equations by probing into subatomic levels interms of charges in vacuum. i.e. without the introduction of dielectric and magnetic media. These equations are:

$$\nabla \times \vec{e} = -\mu_0 \frac{\partial \vec{h}}{\partial t} \tag{3.14a}$$

$$\nabla \times \vec{h} = \rho' \vec{v} + \epsilon_0 \frac{\partial \vec{e}}{\partial t} \tag{3.14b}$$

$$\nabla \cdot \vec{e} = \rho'/\epsilon. \tag{3.14c}$$

$$\nabla \cdot \vec{h} = 0 \tag{3.14d}$$

where $\mu_0$ and $\epsilon_0$ are respectively the permeability and permittivity of vacu-

*H.A. Lorentz, *Theory of Electrons and its Applications to the Phenomena of Light and Radiant Heat*, New York; Dover Pub. Co. 1952

um. $\vec{J} = \rho' \vec{v}$ is the convection current density and $\rho'$ is the charge density with velocity $\vec{v}$.

These equations are the microscopic counterparts of Maxwell's equations (M-1) through (M-4) and are known as the Maxwell-Lorentz equations.

It is to be noted that the microscopic fields $\vec{e}$, $\vec{h}$ and the charge density $\rho'$ are not the same as their counterparts $\vec{E}$, $\vec{H}$ and $\rho$ in the macroscopic equations. It is obvious that Eqs. (3.14a) through Eq. (3.14d) are atomic or molecular in nature whereas the macroscopic equations are statistical. Hence $\vec{D}$, $\vec{E}$, $\vec{B}$, $\vec{H}$ and $\rho$ must be correlated to the average of the microscopic fields and charges over a large number of molecules. Van Vleck* obtained the following correlation:

$$\vec{E} = \langle \vec{e} \rangle = \frac{1}{n_0} \sum_{i=1}^{n_0} e_i$$

$$\vec{B} = \mu_0 \langle \vec{h} \rangle = \frac{\mu_0}{n_0} \sum_{i=1}^{n_0} \vec{h}_i$$

where $\vec{e}_i$ and $\vec{h}_i$ are the corresponding fields on the $i^{\text{th}}$ molecule. $\langle \vec{e} \rangle$ and $\langle \vec{h} \rangle$ are average fields over a very small physical volume inaccessible ordinary methods of physical measurement but still contains a large number of molecules $n_0$.

## 3.4 THE WAVE EQUATIONS

Two curl equations (M-1) and (M-2) are first order coupled equations which may be decoupled by solving them as simultaneous equations, to yield two explicit equations one each for electric and magnetic fields. These are known as wave equations. To accomplish this we take curl on both sides of Eq. (M-1) which yieles,

$$\nabla \times (\nabla \times \vec{E}) = -\nabla \times \frac{\partial \vec{B}}{\partial t}.$$

Because time and space derivatives are independent and so interchangeable, hence,

$$\nabla \times (\nabla \times \vec{E}) = -\frac{\partial}{\partial t}(\nabla \times \vec{B}).$$

*J.H. Van Vleck, *Electric and Magnetic Susceptibilities*, New York: Oxford University Press, (1962).

Using the vector identity

$$\nabla\times(\nabla\times\vec{A})=\nabla(\nabla\cdot\vec{A})-\nabla^2 A$$

we have

$$\nabla(\nabla\cdot\vec{E})-\nabla^2\vec{E}=-\frac{\partial}{\partial t}(\nabla\times\vec{B}). \tag{3.15}$$

For an isotropic medium we can write $\nabla\times\vec{B}=\mu(\nabla\times\vec{H})$. Then using Eq. (M-2) in Eq. (3.15), we have

$$\nabla(\nabla\cdot\vec{E})-\nabla^2\vec{E}=-\mu\frac{\partial\vec{J}}{\partial t}-\mu\frac{\partial^2\vec{D}}{\partial t^2}. \tag{3.16}$$

Similarly, taking curl on Eq. (M-2) and substituting the value of $\nabla\times\vec{E}$ from Eq. (M-1) we have,

$$\nabla(\nabla\cdot\vec{H})-\nabla^2\vec{H}=\nabla\times\vec{J}-\epsilon\frac{\partial^2\vec{B}}{\partial t^2}. \tag{3.17}$$

For the region where $\vec{J}=\rho=0$, as in the case of free space, Eqs. (3.16) and (3.17) respectively become

$$\nabla^2\vec{E}=\mu\epsilon\frac{\partial^2\vec{E}}{\partial t^2} \tag{3.18}$$

$$\nabla^2\vec{H}=\mu\epsilon\frac{\partial^2\vec{H}}{\partial t^2}. \tag{3.19}$$

These are wave equations in free space.

Now returning to the general case, Eqs. (3.16) and (3.17) can be solved when $\vec{J}$ is expressed in terms of $\vec{E}$. In a conductor $\vec{J}$ and $\vec{E}$ are related by Ohm's law

$$\vec{J}=\sigma\vec{E} \tag{3.20}$$

$\sigma$ is the conductivity of the material. Then using Eq. (3.20) in Eq. (3.17) and substituting the value of $(\nabla\times\vec{E})$ from Eq. (M-1), Eq. (3.17) is reduced to

$$\nabla(\nabla\cdot\vec{H})-\nabla^2\vec{H}=-\mu_0\frac{\partial\vec{H}}{\partial t}-\mu\epsilon\frac{\partial^2\vec{H}}{\partial t^2} \tag{3.21}$$

Similarly Eq. (3.16) can be reduced to

$$\nabla(\nabla\cdot\vec{E})-\nabla^2 E=-\mu\sigma\frac{\partial\vec{E}}{\partial t}-\mu\epsilon\frac{\partial^2\vec{E}}{\partial t^2}. \tag{3.22}$$

Note that although $\vec{E}$ and $\vec{H}$ obey a similar equation, $\vec{E}$ is not equal to $\vec{H}$. Moreover, Eqs. (3.21) and (3.22) would be identical in form if $\nabla \cdot \vec{E} = \nabla \cdot \vec{H} = 0$. No doubt $\nabla \cdot \vec{H}$ is always zero because of the non-existence of free magnetic mono poles but unlike $\nabla \cdot \vec{H}$, $\nabla \cdot \vec{E}$ has a value $\rho/\epsilon$ which is in general non-zero because free electric charges do exist. However, any charge density that exists in a conductor decays in extremely rapidly to zero. This can be shown by taking the divergence on both the sides of Eq. (M-2), that is

$$\nabla \cdot (\nabla \times \vec{H}) = 0 = \sigma \left( \frac{\nabla \cdot \vec{D}}{\epsilon} \right) + \frac{\partial}{\partial t} (\nabla \cdot \vec{D}).$$

Using Eqs. (M-3) and (3.20) the above equation reduces to

$$\frac{\sigma}{\epsilon} + \frac{\partial \rho}{\partial_t} = 0.$$

Its solution is

$$\rho = \rho_0 \, e^{-(\sigma/\epsilon)t}. \tag{3.23}$$

That is, the charge density $\rho$ decays with the characteristic decay time constant $\epsilon/\sigma$. For a typical metal, this time constant is of the order $10^{-18}$ sec, an extremely short interval compared to the microwave period of the order $10^{-10}$ sec. Therefore, $\nabla \cdot \vec{E}$ and $\nabla \cdot \vec{B}$ can be set equal to zero at microwave frequencies. Hence $\vec{E}$ and $\vec{H}$ obey identical equations, which are

$$\nabla^2 \vec{E} = \mu \left( \sigma \frac{\partial \vec{E}}{\partial t} + \epsilon \frac{\partial^2 \vec{E}}{\partial t^2} \right) \tag{3.24}$$

$$\nabla^2 \vec{H} = \mu \left( \sigma \frac{\partial \vec{H}}{\partial t} + \epsilon \frac{\partial^2 \vec{H}}{\partial t^2} \right). \tag{3.25}$$

Taking the time dependence of the fields of the form $e^{j\omega t}$, Eq. (3.24) is reduced to

$$\nabla^2 \vec{E} = j\omega\mu(\sigma + j\omega\epsilon)\, \vec{E}$$

or

$$\nabla^2 \vec{E} = \gamma^2 \vec{E}. \tag{3.26}$$

Similarly Eq. (3.25) reduces to

$$\nabla^2 \vec{H} = \gamma^2 \vec{H} \tag{3.27}$$

where

$$\gamma = \pm \sqrt{j\omega\mu(\sigma + j\omega\epsilon)} = \alpha + j\beta \tag{3.28}$$

is a complex quantity depending upon the properties of the medium. In one dimension, Eqs. (3.26) and (3.27) are analogous to the current voltage transmission line equations (Section 2.2) and $\gamma$ is analogous to the propagation constant of the transmission line. It is, therefore, known as the intrinsic propagation constant, the real part of $\gamma$ being the attenuation constant, $\alpha$, measured in neper/metre and the imaginary part, $\beta$, the phase shift constant measured in radian/metre.

In case of perfect dielectrics such as free space $\sigma=0$; thus Eq. (3.28) gives

$$\gamma=\alpha+j\beta=\pm\sqrt{j\omega\mu\cdot j\omega\epsilon}=\pm j\omega\sqrt{\mu\epsilon}.$$

Thus

$$\alpha=0 \quad \text{and} \quad \beta=\pm j\omega\sqrt{\mu\epsilon}=\pm j\,\frac{\omega}{v}. \tag{3.29}$$

This implies that waves pass unattenuated with a phase shift of $\omega/v$ rad/m.

For a poor conductor $\sigma$ is small so that Eq. (3.28) may be written as

$$\gamma=\alpha+j\beta=\pm\sqrt{-\omega^2\mu\epsilon(\epsilon-j\,\sigma/\omega\epsilon)}.$$

The square root may be expanded using the binomial series and retaining first order terms in $\sigma$. We obtain

$$\gamma=\alpha+j\beta\simeq\pm j\omega\sqrt{\mu\epsilon}\left(1-j\frac{\sigma}{2\omega\epsilon}\right)$$

$$\simeq\pm\left(\frac{\sigma}{2}\sqrt{\mu/\epsilon}+j\omega\sqrt{\mu\epsilon}\right) \tag{3.30}$$

giving us an attenuation of $\alpha=\sigma/2\sqrt{\mu/\epsilon}$ neper/metre in addition to the phase shift $\beta=\omega\sqrt{\mu\epsilon}$ rad/m.

In the case of good conductors $\sigma$ is large, so we can neglect term $\epsilon$ as compared to $\sigma/\omega$ so that Eq. (3.30) simplifies to

$$\gamma=\alpha+j\beta=\pm\sqrt{j\omega\mu\sigma}=\pm(1+j)\sqrt{\frac{\omega\mu\sigma}{2}} \tag{3.31}$$

$$=\sqrt{\omega\mu\sigma}\,\underline{/45^\circ}.$$

It may be noted that since $\vec{E}$ and $\vec{H}$ obey identical equations, either Eq. (3.26) or Eq. (3.27) may be solved without loss of generality to obtain the desired propagating fields. Once $\vec{E}$ or $\vec{H}$ is obtained by solving these equations the other may easily be obtained from the curl Eqs. (M-1) or (M-2) as the case may be.

Before obtaining the solution of these equations for a general case, it is instructive to consider a simple but important case of plane waves in free space.

## 3.5 PLANE WAVES IN FREE SPACE

In free space, the wave equation for the electric field is

$$\nabla^2 \vec{E} = \mu\epsilon \frac{\partial^2 \vec{E}}{\partial t^2}. \tag{3.32}$$

This vector equation is equivalent to three scalar wave equations one for each of the scalar components of the electric field. Our aim is to show mathematically that time varying electric and magnetic fields lead to electromagnetic wave propagation. To simplify the problem, let us consider the case of a plane wave in which $\vec{E}$ is independent of $x$ and $y$ but depends on $z$ and $t$ only. Then Eq. (3.32) may be written as

$$\frac{\partial^2 \vec{E}}{\partial z^2} = \mu\epsilon \frac{\partial^2 \vec{E}}{\partial t^2}. \tag{3.33}$$

Comparing Eq. (3.33) with the standard wave equation

$$\left(\frac{\partial^2 y}{\partial z^2} = \frac{1}{v^2}\frac{\partial^2 y}{\partial t^2}\right)$$

leads to the conclusion that this equation represents propagating electric field.

Similarly,

$$\frac{\partial^2 \vec{H}}{\partial z^2} = \mu\epsilon \frac{\partial^2 \vec{H}}{\partial t^2}$$

where $\vec{H}$ is a function of $z$ and $t$ only. This equation represents propagating magnetic field. The velocity of these propagating electromagnetic fields is $v = \frac{1}{\sqrt{\mu\epsilon}}$. For free space its value is

$$v = \frac{1}{\sqrt{\mu\epsilon}} = \frac{1}{\sqrt{4\pi \times 10^{-7} \times \frac{1}{36\pi \times 10^9}}} = 3 \times 10^8 \text{ m/sec.} \tag{3.34}$$

Let us now discuss some characteristics of these plane elecrtomagnetic waves. Equation (M-3) when applied to these waves in free space yields

$$\frac{\partial E_z}{\partial z} = 0. \tag{3.35}$$

Therefore, there is no variation of $E_z$ in the direction of propagation. Moreover, from Eq. (3.33) it follows that $\partial^2 \vec{E}/\partial t^2 = 0$. This implies that $E_z$ is either zero, constant in time, or increases uniformly with time. Such a field $E_z$ cannot be a part of wave motion and so $E_z$ can be put equal to zero.

A similar analysis carried out for magnetic field shows that $H_z=0$. It is, therefore, concluded that uniform plane electromagnetic waves are transverse and have components $\vec{E}$ and $\vec{H}$ only in the direction perpendicular to the direction of propagation and are therefore called *transverse electromagnetic* waves or TEM waves.

To evaluate the orientation of $\vec{E}$ and $\vec{H}$ in the plane perpendicular to the direction of motion and the interrelation between these ($\vec{E}$ and $\vec{H}$), we assume the solution of Eq. (3.27), representing TEM wave propagation, to be of the form ($\gamma=j\beta$: see Eq. 3.29)

$$E_{(x \text{ or } y \text{ component})}=\vec{E_0}\, e^{j(\omega t-\beta z)} \tag{3.36}$$

where $\omega$ is the angular frequency and $\beta$ is the phase constant.

Now, substituting Eq. (3.36) in Eq. (M-1), and bearing in mind that $\partial/\partial y=\partial/\partial x=0$ along with $E_z=0$ for TEM waves, we have

$$\hat{x}j\beta E_y-\hat{y}j\beta E_x=-j\omega\mu(H_x\hat{x}+H_y\hat{y})$$

which is equivalent to the following two equations

$$j\beta E_y=-j\omega\mu H_x \tag{3.37}$$

$$j\beta E_x=+j\omega\mu H_y. \tag{3.38}$$

These equations yield

$$\frac{E_x}{H_y}=-\frac{E_y}{H_x}=\frac{|\vec{E}|}{|\vec{H}|}=\frac{\omega\mu}{\omega\sqrt{\mu\epsilon}}=\sqrt{\frac{\mu}{\epsilon}}=\eta. \tag{3.39}$$

The last step follows from Eq. (3.29). Equation (3.39) indicates that in TEM waves travelling in free space there is a definite ratio between the amplitudes of $\vec{E}$ and $\vec{H}$ and that ratio is equal to the square root of the ratio of permeability to permittivity (dielectric constant) of the medium. Since $E$ is measured in volts/metre and $H$ is measured in amperes/metre, the ratio will have dimensions of impedance or ohms. It is therefore customary to refer to this ratio, $\sqrt{\mu/\epsilon}$ as the characteristic impedance or intrinsic impedance of non-conducting or loss-less medium which is generally denoted by $\eta$. For free space it has a value

$$\eta=\sqrt{4\pi\times10^{-7}\times36\pi\times10^{9}}=120\pi\ \Omega.$$

It may be noted that for TEM waves travelling in free space, the expressions for the propagation constant, ($\omega\sqrt{\mu\epsilon}$), and characteristic impedance, ($\sqrt{\mu/\epsilon}$), are analogous to the corresponding expressions for a lossless transmission line $\omega\sqrt{LC}$ and $\sqrt{L/C}$ respectively.

Taking the scalar product of $\vec{E}$ with $\vec{H}$ and substituting the values of $E_x$ and $E_y$ from Eq. (3.39), we have

$$\vec{E} \cdot \vec{H} = E_x H_x + E_y H_y = \eta H_y H_x - \eta H_x H_y = 0. \tag{3.40}$$

Thus, in a TEM wave $\vec{E}$ and $\vec{H}$ are at right angles to each other. Further information about field vectors may be obtained from their *cross product*.

$$\vec{E} \times \vec{H} = \hat{z}(E_x H_y - E_y H_x) = \frac{\hat{z}}{\eta} H^2. \tag{3.41}$$

Thus, $\vec{E} \times \vec{H}$ gives the direction in which the wave travels. Therefore, for a plane polarized* wave travelling in positive $z$-direction, $\vec{E}$ and $\vec{H}$ are in the directions of $x$ and $y$ respectively.

Having evaluated the structure of TEM waves we now pass on to determine some of its important properties. The general solution of the wave Eq. (3.33) may now be written as

$$E_x = E_1\, e^{j(\omega t - \beta z)} + E_2\, e^{j(\omega t + \beta z)} \tag{3.42}$$

where $\beta = \omega\sqrt{\mu\epsilon}$.

Equation (3.42) may be recognized to represent two waves. The velocity of each wave may be obtained by identifying some point on the wave form and observing its velocity. Thus for the equiphase point given by $\omega t - \beta z = \text{const}$,

$$\frac{dz}{dt} = v_p = \omega/\beta$$

is the phase velocity of the wave. Thus Eq. (3.42) represents the sum of two travelling waves, one each in positive and negative $z$-directions with phase velocity.

$$v_p = \omega/\beta. \tag{3.43}$$

Another important quantity which has yet to be defined for plane waves is the wavelength denoted by $\lambda$. It is defined as the distance over which the sinusoidal waveform (as represented by Eq. 3.39) passes over a full cycle of $2\pi$ radians. Then by definition

$$\beta \cdot \lambda = 2\pi$$

or

$$\lambda = \frac{2\pi}{\beta}. \tag{3.44}$$

*Polarization of an electromagnetic wave refers to the time rate of variation of the electric field at a fixed point in space.

Combining Eqs. (3.44) and (3.45) we have

$$v_p=\omega/\beta=\frac{2\pi f}{2\pi/\lambda}=f\lambda. \tag{3.45}$$

Equation (3.45) establishes the famous relation in wave motion, stated as the phase velocity of a wave in metre/sec. which equals the product of the frequency ($f$) in hertz and the wavelength in metres.

For TEM waves,

$$v_p=\frac{\omega}{\beta}=\frac{\omega}{\omega\sqrt{\mu\epsilon}}=\frac{1}{\sqrt{\mu\epsilon}}$$

i.e. phase velocity of the TEM waves is equal to the velocity of a plane e.m. wave. It will be $c$ in free space for $\mu=\mu_0$ and $\epsilon=\epsilon_0$.

## 3.6 PLANE TEM WAVES IN A CONDUCTING MEDIUM

Let us now consider the propagation of plane TEM waves in a medium of conductivity $\sigma$. The wave equation for the electric field in a conducting medium is

$$\frac{\partial^2 E_x}{\partial z^2}=\gamma^2 E_x. \tag{3.46}$$

(Electric field is taken along the axis of $x$ when the waves are propagating in a $z$-direction) $\gamma$ is the propagation constant given by Eq. (3.28). The general solution of Eq. (3.46) may be written as

$$E_x=E_1\,e^{-\gamma z}+E_2\,e^{\gamma z}. \tag{3.47}$$

The corresponding magnetic field is obtained from the curl equation (M-1)

$$\nabla\times\vec{E}=-\frac{\partial\vec{B}}{\partial t}=-j\omega\mu\,\vec{H}.$$

Thus

$$H_y=-\frac{1}{j\omega\mu}\frac{\partial}{\partial z}[E_1\,e^{-\gamma z}+E_2\,e^{\gamma z}]$$

$$=\frac{\gamma}{j\omega\mu}[E_1\,e^{-\gamma z}-E_2\,e^{\gamma z}].$$

Also $H_y$ can be written as

$$H_y=H_1\,e^{-\gamma z}+H_2\,e^{\gamma z}$$

where $H_1=\frac{\gamma E_1}{j\omega\mu}$ and $H_2=-\frac{\gamma E_2}{j\omega\mu}$.

It therefore follows that

$$\frac{E_1}{H_1}=\frac{E_2}{H_2}=\frac{j\omega\mu}{\gamma}=\frac{\gamma}{\sigma+j\omega\epsilon}=\eta_m \tag{3.48}$$

The last step follows from Eq. (3.28).

As in transmission line theory the ratio of incident electric field to the incident magnetic field is the *characteristic impedance* of the given medium denoted by $\eta_m$. The expression for the attenuation and phase constants for a conducting medium are given by Eq. (3.30). It may be noted that these are analogous to the corresponding transmission line expressions. For a good conducting media $\sigma \gg \omega\epsilon$ even at the upper limit of microwave frequencies. Therefore, Eq. (3.48) yields

$$\eta_m=\frac{\gamma}{\sigma}$$

which on using Eq. (3.31) reduces to

$$\eta_m=\pm\frac{(1+j)\sqrt{\frac{\omega\mu\sigma}{2}}}{\sigma}=\pm(1+j)\sqrt{\frac{\omega\mu}{2\sigma}}=\sqrt{\frac{\omega\mu}{\sigma}}\angle 45^\circ. \tag{3.49}$$

It is noted from Eq. (3.49) that the intrinsic impedance of the conducting medium is very small and has a reactive component. This implies that there is a low ratio of the electric to the magnetic field and the fields are 45° in phase [Eq. (3.49)].

From Eq. (3.31), we see that both $\alpha$ and $\beta$ are large. This means that the wave is attenuated greatly as it progresses through the conductor and the phase shift per unit length is also great. The velocity of the wave being inversely proportional to $\beta$, is very small in good conductors.

Table 3.1 shows the calculated wave propagation properties of free space (good dielectric) relative to that of a good conductor, silver, at 100 MHz.

**TABLE 3.1 Comparison of the properties of conducting and non-conducting media at 100 MHz**

| *S. No.* | *Property* | *Dielectric* (free space) | *Metal silver* (conducting medium) |
|---|---|---|---|
| 1. | $\eta$ | $(120\,\pi)$ 376.6 Ω | $(0.0025+j\,0.0025)$ Ω |
| 2. | $\lambda$ | 3 m | 0.00004 m |
| 3. | $v_p$ | $3\times10^8$ m/sec. | $4\times10^3$ m/sec. |
| 4. | $\alpha$ | Almost zero | $15.7\times10^4$ neper/m |
| 5. | $\beta$ | 2.094 rad/m | $1.57\times10^5$ rad/m |
| 6. | $\gamma$ | $j$ (2.094) | $(15.7+j\,15.7)\ 10^5$ |

## 3.7 PLANE TEM WAVES IN LOW-LOSS DIELECTRICS

So far we have discussed two extreme cases of propagation media where $\sigma=0$ or it is very large. Now we come to the practical case of the propagation of electromagnetic waves in *good dielectrics*. In this case no doubt $\sigma/\omega\epsilon$ is much smaller than unity but it is not zero.

To evaluate the propagation properties, we may recall Eq. (3.28) and separate $\gamma$ into real and imaginary parts to yield $\alpha$ and $\beta$ as

$$\alpha=\omega\sqrt{\frac{\mu\epsilon}{2}\left(\sqrt{1+\sigma^2/\omega^2\epsilon^2}-1\right)}$$

$$\beta=\omega\sqrt{\frac{\mu\epsilon}{2}\left(\sqrt{1+\sigma^2/\omega^2\epsilon^2}+1\right)}.$$

For lowloss dielectrics, to a very good approximation, we may write

$$\sqrt{1+\frac{\sigma^2}{\omega^2\epsilon^2}}\cong 1+\frac{\sigma^2}{2\omega^2\epsilon^2}$$

where only the first two terms of the binomial expression have been retained. The expressions for $\alpha$ and $\beta$ become

$$\alpha=\frac{\sigma}{2}\sqrt{\mu/\epsilon}=\frac{\sigma}{2}\eta \tag{3.50}$$

$$\beta=\omega\sqrt{\mu\epsilon}\left(1+\frac{\sigma^2}{8\omega^2\epsilon^2}\right). \tag{3.51}$$

Equations (3.50) and (3.51) may be compared with the corresponding expressions $\alpha=\frac{G}{2}Z_0$ and $\beta=\omega\sqrt{LC}\left\{1+\frac{G^2}{8\,\omega^2\epsilon^2}\right\}$ for a low-loss transmission line having zero series resistance. It can be seen that due to a small amount of loss in the dielectric an additional phase shift $\sigma^2/8\,\omega^2\epsilon^2$ occurs as the perturbation or a correction factor.

The phase velocity of the propagating wave in such a dielectric is given by

$$v'_{ph}=\frac{\omega}{\beta}=\frac{1}{\sqrt{\mu\epsilon}\,(1+\sigma^2/8\,\omega^2\epsilon^2)}$$

$$=v'_{ph}\left(1-\frac{\sigma^2}{8\omega^2\epsilon^2}\right) \tag{3.52}$$

where $v_{ph}=1/\sqrt{\mu\epsilon}$ is the velocity of the wave in the dielectric when the conductivity is zero. The effect of a small amount of loss is to reduce slightly the velocity of propagation of the wave.

The intrinsic impedance of the medium having finite conductivity becomes

$$\eta'_m=\sqrt{\frac{\mu}{\epsilon}\,\frac{1}{1+\sigma/j\omega\epsilon}}\cong\eta\left(1+j\frac{\sigma}{2\omega\epsilon}\right). \tag{3.53}$$

with the same approximation as above, where $\eta$ is the intrinsic impedance of the dielectric when $\sigma=0$. It is seen that the chief effect of a small amount of loss is to add a small reactive component to the intrinsic impedance.

The expression for $\eta_m'$ in Eq. (3.53) is also analogous to the corresponding expression for low-loss transmission lines with zero series resistance [Eq. 2.39 with $R=0$].

The losses in a dielectric can be interpreted in yet an another way by rewriting Eq. (3.26) as

$$\nabla^2 \vec{E} = -\omega^2\mu\epsilon\,(1-j\sigma/\omega\epsilon)\,\vec{E}$$

$$= -\omega^2\mu\epsilon^*\,\vec{E}. \tag{3.54}$$

Thus a low-loss medium is characterized by a *complex permittivity*, $\epsilon^*$, whose real part is the permittivity of the loss-less medium and imaginary part represents losses. The complex permittivity $\epsilon^*$ is sometimes written as

$$\epsilon^*=\epsilon(1-j\tan\delta) \tag{3.55}$$

then, $\tan\delta=\sigma/\omega\epsilon$ represents the losses in the dielectric and is, therefore, called the *loss tangent*.

The concept of complex permeability is very useful in the analysis of cavity resonators and other problems.

## 3.8 POWER FLOW IN ELECTROMAGNETIC FIELD: THE POYNTING VECTOR

When electromagnetic waves propagate through space from their source to distinct receiving points, there is a transfer of energy from the source to the receivers. The rate of this propagating energy bears a simple and direct relation to the magnitudes, distributions and phases of the electric and magnetic fields of the wave. This relation can be obtained from Maxwell's equations as follows:

Consider the following vector identity

$$\nabla\cdot(\vec{E}\times\vec{H})=\vec{H}\cdot(\nabla\times\vec{E})-\vec{E}\cdot(\nabla\times\vec{H}). \tag{3.56}$$

Substituting Eqs. (M-1) and (M-2) in Eq. (3.56), yields

$$\nabla\cdot(\vec{E}\times\vec{H})=\mu\vec{H}\cdot\frac{\partial\vec{H}}{\partial t}-\vec{E}\cdot\vec{J}-\epsilon\,\vec{E}\frac{\partial\vec{E}}{\partial t}.$$

Integrating the above equation over volume V, bounded by surface $s$, and applying the divergence theorem, we get

$$\int_v \nabla\cdot(\vec{E}\times\vec{H})\,dV=\int_s(\vec{E}\times\vec{H})\cdot\vec{ds}$$

$$=-\int_v \vec{E}\cdot\vec{J}\;dV-\frac{\partial}{\partial t}\int_v\left(\frac{\epsilon E^2}{2}+\frac{\mu H^2}{2}\right)dV \tag{3.57}$$

This equation represents the conservation of electromagnetic energy in volume V. The first term on the RHS represents the generalized joule dissipation in volume V. As the electric field, $\vec{E}$, measured in volts/metre, may be recognized as the source producing electric current density $\vec{J}$, measured amp. per square metre, $\vec{E} \cdot \vec{J}$ represents the work done by the impressed forces (currents) per unit volume. Hence the first term $\int_v \vec{E} \cdot \vec{J} \, dV = \int_v E^2/\sigma \, dv$ from Ohm's law $\vec{J} = \sigma E$) represents the instantaneous power dissipated in the given volume V.

The second term on the RHS is the negative time derivative of the sum of two quantities $\epsilon E^2/2$ and $\mu H^2/2$. These quantities may be identified* to represent the time average energy stored per unit volume in the electric and magnetic fields respectively. Thus this term represents the rate of decrease of electromagnetic energy stored in volume V.

It therefore follows from the conservation of energy principle as expressed in Eq. (3.57), that the term on the LHS can be interpreted as the flow of energy through the surface $s$ enclosing volume V. From the definition of divergence of a vector, we can interpret energy as the flux of a vector $\vec{P} = \vec{E} \times \vec{H}$ such that the surface integral of the component $\vec{P}$ normal to the surface gives the energy flowing out of the given volume. It is seen that the vector

$$\vec{P} = \vec{E} \times \vec{H} \tag{3.58}$$

has the dimensions watt/metre$^2$ and is called the *poynting vector*. The poynting theorem states that the poynting vector at any point is a measure of the rate of energy flow per unit area at that point. The direction of flow is perpendicular to $\vec{E}$ and $H$ in the direction of vector $\vec{E} \times \vec{H}$.

So far we have considered $\vec{E}$ and $\vec{H}$ to be real. However, if $\vec{E}$ and $\vec{H}$ are complex as in the case of time varying fields, the time average power transmitted across a closed surface $s$ is given by the integral of the real part of one-half of the normal component of the complex poynting vector $\vec{E} \times \vec{H}^*$, i.e.

$$\text{Power} \quad = \text{Re} \frac{1}{2} \oint_s (\vec{E} \times \vec{H}^*) \cdot \vec{ds}. \tag{3.59}$$

This result may be derived by considering the complex vector identity $\nabla \cdot (\vec{E} \times H^*) = (\nabla \times \vec{E}) \cdot \vec{H}^* - (\nabla \times \vec{H}^*) \cdot \vec{E}$ and proceeding as before (in arriving at Eq. 3.57 and interpreting it).

*Refer J.A. Starton, *Electromagnetic Theory*, New York: McGraw-Hill Book Co. (1941).

The reactive part of the power can be written as

$$P_{react}=I_m\frac{1}{2}\oint_s (\vec{E}\times\vec{H}^*)\cdot d\vec{s}. \tag{3.60}$$

## 3.9 ELECTROMAGNETIC BOUNDARY CONDITIONS

Maxwell's equations in differential form basically express the relationship that exists between the four field vectors $\vec{E}$, $\vec{D}$, $\vec{H}$ and $\vec{B}$ at any point in the neighbourhood of which the physical properties of the medium vary continuously, i.e. within a continuous medium. However, in practical situations which generally involve material bodies with boundaries, a knowledge of the behaviour of the electromagnetic fields at the boundary (or surface) separating one medium from another, is required. As there are sharp changes in $\mu$, $\epsilon$ and $\sigma$ at the interface, one can therefore expect a corresponding occurrence of discontinuities in the electric and magnetic fields at the separating boundary. The conditions which electromagnetic fields obey at the boundary separating two material media of different electrical properties, are known as boundary conditions. These conditions are essential to obtain the proper and unique solution of Maxwell's equations in a given situation. From the mathematical point of view, also, the solution of a partial differential equation such as a wave equation in region V is not unique unless the boundary conditions are specified, i.e. when the behaviour of the fields at the boundary of V are known. In this section we shall develop the appropriate electromagnetic boundary conditions.

Consider a geometrical surface separating two media of dielectric constants $\epsilon_1$ and $\epsilon_2$. If a unit charge is carried a short distance $\vec{dl}$ parallel to the boundary on the surface of medium 1 and an equal distance in medium 2 to bring it to the starting point as shown in Fig. 3.3. From Faraday's law (or Maxwell's first equation in integral form) we have

$$\oint_c \vec{E}\cdot d\vec{l}=(E_{t2}-E_{t1})\cdot\Delta\vec{l}=\int_s \frac{\partial\vec{B}}{\partial t}\cdot d\vec{s}. \tag{3.61}$$

If for the contour $c$ in the above equation, width $h$ is made to approach zero, the magnetic flux flowing through this contour vanishes giving

$$E_{t1}=E_{t2}. \tag{3.62}$$

Furthermore, as $h\to O$ for the same contour $c$, the total displacement current directed through the contour vanishes so that,

$$\lim_{h\to 0}\oint_c \vec{H}\cdot d\vec{l}=(\vec{H}_{t2}-\vec{H}_{t1})\cdot\Delta\vec{l}=\lim_{h\to 0}\left(\int_s \frac{\partial\vec{D}}{\partial t}\cdot d\vec{s}\right)=0 \tag{3.63}$$

giving

$$H_{t2} = H_{t1} \tag{3.64}$$

where $t$ denotes the components tangential to the boundary surface. Equations (3.62) and (3.64) indicate that the tangential components of $\vec{E}$ and $\vec{H}$ are continuous across the boundary.

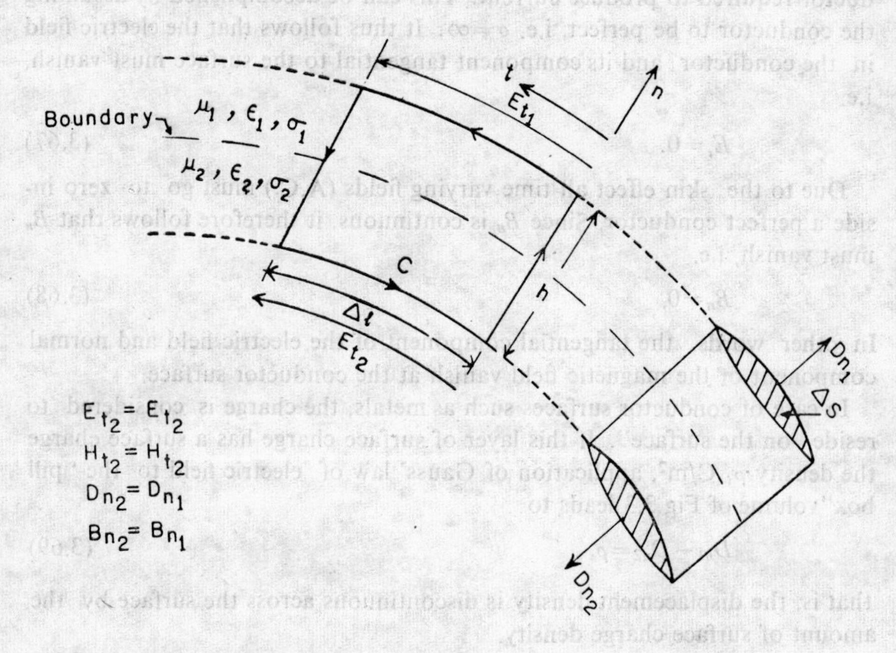


Fig. 3.3 Electromagnetic boundary conditions.

To obtain the behaviour of normal components we consider two very small elements of area $\Delta S$, one on either side of the boundary between any two material media with a surface charge density $\rho_s$ existing on the boundary. Applying Gauss' theorem on this "pill box" volume of Fig. 3.3, we obtain

$$(D_{n1} - D_{n2})\, \Delta S = \rho_s \Delta S$$

or $\quad D_{n1} - D_{n2} = \rho_s.$

For a chargefree boundary

$$D_{n1} = D_{n2} \tag{3.65}$$

that is, if there is no surface charge, the normal component of $\vec{D}$ is continuous across the surface. Similarly, the application of the Gauss', law of magnetic field to the said "pill box" volume leads to

$$B_{n1} = B_{n2}. \tag{3.66}$$

Because there are no isolated magnetic charges, Eq. (3.66) is always valid. Hence, the normal component of magnetic flux density is always continuous across the boundary.

Equations (3.62) and (3.65) are modified when applied at the conducting boundary. Though no conductor is perfect, in many practical situations it becomes necessary to neglect the small but finite electric field along the conductor required to produce current. This can be accomplished by assuming the conductor to be perfect, i.e. $\sigma=\infty$. It thus follows that the electric field in the conductor and its component tangential to the surface must vanish, i.e.

$$E_t=0. \tag{3.67}$$

Due to the skin effect all time varying fields (A.C.) must go to zero inside a perfect conductor. Since $B_n$ is continuous, it therefore follows that $B_n$ must vanish, i.e.

$$B_n=0. \tag{3.68}$$

In other words, the tangential component of the electric field and normal component of the magnetic field vanish at the conductor surface.

In case of conductor surfaces such as metals, the charge is considered to reside "on the surface". If this layer of surface charge has a surface charge the density $\rho_s$ C/m$^2$, application of Gauss' law of electric field to the "pill box"volume of Fig.3.3 leads to

$$D_{n1}-D_{n2}=\rho_s \tag{3.69}$$

that is, the displacement density is discontinuous across the surface by the amount of surface charge density.

For metallic conductors the displacement density $D$ within the conductor is a negligibly small quantity (zero in a static case or perfect conductor $\sigma=\infty$). Therefore, $D_{n2}=0$. Hence

$$D_{n1}=\rho_s \tag{3.70}$$

that is, the normal component of displacement density in the dielectric is equal to the surface charge density on the conductor.

In a good conductor a high frequency current will flow in a thin sheet near the surface, the depth of this sheet approaching zero as the conductivity approaches infinity. Thus, a metal boundary may be considered to be a *thin current sheet*. The application of Ampere's law to the contour of Fig. 3.3 leads to

$$H_{t2}-H_{t1}=J_s \tag{3.71}$$

where $J_s$ is the current per unit width along the surface of a perfect conductor. Now the electric field is zero within a perfect conductor (for A.C. fields), therefore, $H_{t2}$ must vanish. So

$$H_{t1}=-J_s. \tag{3.72}$$

In other words, the current per unit width along the surface of a perfect conductor is equal to the magnetic field strength $H$ just outside the surface.

The magnetic field and surface current will be parallel to the surface but perpendicular to each other.

Thus, the boundary conditions at the metal boundary (Eqs. 3.67, 3.68, 3.70 and 3.72) may be written in the following vector form:

$$\begin{aligned} \hat{n} \times \vec{E} &= 0 \\ \hat{n} \cdot \vec{B} &= 0 \\ \hat{n} \cdot \vec{D} &= \rho_s \\ \hat{n} \times \vec{H} &= J_s. \end{aligned} \tag{3.73}$$

## 3.10 REFLECTION OF ELECTROMAGNETIC WAVES

When a travelling electromagnetic wave impinges on the interface of two media it undergoes partial reflection. The reflected part of the wave combines with the incident wave to produce *standing waves* in the incident (first medium). The ratio of the maximum to the minimum electric field of such *a*. standing wave is a measure of mismatch and is called the *standing wave ratio* When the intrinsic impedances of two media $\eta_1$ and $\eta_2$ are approximately equal, most of the electromagnetic energy is transmitted into the second medium. On the other hand, if the intrinsic impedance of the first medium $\eta_1$ differs greatly from the intrinsic impedance of the second medium, most of the energy is reflected back in the first medium.

Consider a plane TEM wave incident at the interface of two media, extending infinitely in all directions except at the interface. Taking the interface on the $y=0$ plane as shown in Fig. 3.4, the field intensities in medium 1 can be written as

$$\vec{E}_x = \vec{E}_1\, e^{-\gamma_1 z} + \vec{E}_2\, e^{\gamma_1 z} \tag{3.74}$$

$$\vec{H}_y = \frac{\vec{E}_1}{\eta_1}\, e^{-\gamma_1 z} - \frac{\vec{E}_2}{\eta_2}\, e^{\gamma_1 z} \tag{3.75}$$

where $\gamma_1$ and $\eta_1$ are the propagation constant and intrinsic impedance of medium 1 respectively. $\vec{E}_1$ and $\vec{E}_2$ are the incident and reflected wave electric field intensities at the surface $z=0$, and similarly the magnetic field $\vec{H}_1$ and $\vec{H}_2$.

Now the boundary conditions (Sec. 3.8) require that the tangential components of $\vec{E}$ and $\vec{H}$ must be continuous at the interface of the two media.

This implies that the ratio $|\vec{E}|/|\vec{H}|$ must be the same on either side of the interface. The wave in medium 2 is only an outgoing wave (medium 2 extends infinitely), hence

$$\frac{|\vec{E_3}|}{|\vec{H_3}|}=\eta_2$$

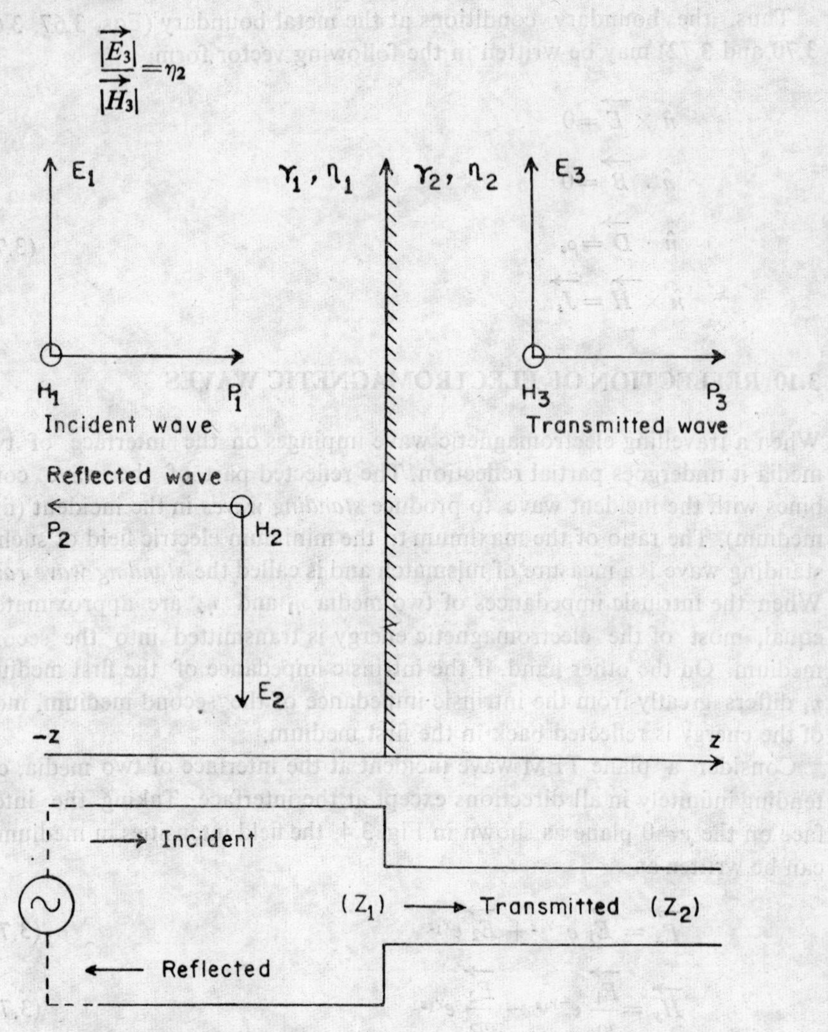


**Fig. 3.4 Reflection of TEM waves and transmission line analogy.**

where $\vec{E_3}$ and $\vec{H_3}$ are the transmitted field intensities. Thus, if $\vec{E}$ and $\vec{H}$ are the resultant field intensities on the surface in medium 1, the boundary conditions require

$$\frac{|\vec{E}|}{|\vec{H}|}=\frac{|\vec{E_3}|}{|\vec{H_3}|}=\eta_2. \tag{3.76}$$

We, therefore, conclude that the impedance terminating medium 1 is the intrinsic impedance of medium 2. This result is in striking contract to the junction of two transmission lines with different characteristic impedances where the impedance terminating line 1 is the characteristic impedance of line 2.

Thus, at $y=0$ plane we have

$$\vec{E} = \vec{E_1} + \vec{E_2}$$

$$\vec{H} = \frac{1}{\eta_1}(\vec{E_1} - \vec{E_2}).$$

Substituting $\vec{H} = \vec{E}/\eta_2$ into the second of the above equations and solving them for $\vec{E_1}$ and $\vec{E_2}$, we obtain

$$\vec{E_1} = \frac{\vec{E}}{2}\left(1+\frac{\eta_1}{\eta_2}\right),$$

$$\vec{E_2} = \frac{\vec{E}}{2}\left(1-\frac{\eta_1}{\eta_2}\right). \tag{3.77}$$

Substituting Eq. (3.77) in Eq. (3.74) we get

$$\vec{E_x} = \frac{\vec{E}}{2}\left(1+\frac{\eta_1}{\eta_2}\right) e^{-\gamma_1 z} + \frac{\vec{E}}{2}\left(1-\frac{\eta_1}{\eta_2}\right) e^{\gamma_1 z} \tag{3.78}$$

$$\vec{H_y} = \frac{\vec{E}}{2\eta_1}\left(1+\frac{\eta_1}{\eta_2}\right) e^{-\gamma_1 z} - \frac{\vec{E}}{2\eta_1}\left(1-\frac{\eta_1}{\eta_2}\right) e^{\gamma_1 z} \tag{3.78a}$$

or when expressed in hyperbolic form

$$\vec{E_x} = \vec{E_1}\left(\cosh \gamma_1 z - \frac{\eta_1}{\eta_2} \sinh \gamma_1 z\right) \tag{3.78b}$$

$$\vec{H_y} = \frac{\vec{E_1}}{\eta_2}\left(-\sinh \gamma_1 z \, \frac{\eta_1}{\eta_2} \cosh \gamma_1 z\right). \tag{3.78c}$$

Hence the wave impedance is given by

$$Z = \frac{|\vec{E_x}|}{|\vec{H_y}|} = \frac{\eta_2 - \eta_1 \tanh \gamma_1 z}{\eta_1 - \eta_2 \tanh \gamma_1 z} \tag{3.79}$$

(the wave is travelling in positive $z$-direction).

Equation (3.79) is analogous to the input impedance of a line that has terminated into another line. (The difference in sign is due to the fact that the transmission line is taken along the negative $z$-axis.)

Since the tangential electric field intensities are equal at the interface, the field intensities at the boundary in the second medium are given by

$$\vec{E_3} = \vec{E}\, e^{-2\gamma z}$$

$$\vec{H_3} = \frac{\vec{E}}{\eta}\, e^{-2\gamma z}. \tag{3.80}$$

The reflection coefficient $\Gamma$ defined as the ratio of reflected to incident field intensity is given by

$$|\Gamma| = \left|\frac{\vec{E_2}}{\vec{E_1}}\right| = \frac{\eta_2 - \eta_1}{\eta_2 + \eta_1} \tag{3.81}$$

(last step follows from Eq. (3.59)).

Similarly, the transmission coefficient $T$ is given by

$$|T| = \left|\frac{\vec{E_3}}{\vec{E_1}}\right| = \left|\frac{\vec{E}}{\vec{E_1}}\right| = \frac{2\eta_2}{\eta_2 + \eta_1} \tag{3.82}$$

(the last step follows from Eqs. (3.80) and (3.77)).

It is noted that Eq. (3.79) through Eq. (3.82) are analogous to their respective transmission line equations.

## 3.11 REFLECTION OF ELECTROMAGNETIC WAVES FROM A CONDUCTING SURFACE

Consider a TEM wave incident normally on a conducting surface. If the surface is a perfect conductor its intrinsic impedance will be zero. Consequently, from Eqs. (3.81) and (3.82), $|\Gamma| = -1$ and $|T| = 0$, that is, there is total reflection of the incident wave. Further, the boundary conditions require $\vec{E}$ to be zero at the surface. Hence the fields in medium 1 become [Eq. (3.78b) and (3.78c)].

$$\vec{E_x} = \vec{H}\, \eta_1 \sinh \gamma_1 z$$

$$\vec{H_y} = \vec{H} \cosh \gamma_1 z$$

and

$$Z = \frac{|E_x|}{|H_y|} = \eta_1 \tanh \gamma_1 z$$

where $E_x$ has been written in terms of $\vec{H}$ to avoid indeterminate forms. These equations are similar to those obtained when a transmission line is terminated in a short circuit.

In practice no conductor is perfect, i.e. it has a finite conductivity $\sigma$. This imperfectness of the conductor results in a very small portion of energy entering the conductor. i.e. the second medium. However, due to the high value of $\sigma$, the waves are highly attenuated by a factor $e^{-\alpha z}$, where $\alpha$ is the real part of $\gamma = \eta\sigma_m$ in Eq. (3.49). Since $\alpha$ is a function of $\sigma$, therefore, the attenuation depends upon the nature of the metal. The property of the metal to attenuate the fields is expressed in terms of a quantity defined as the *depth of penetration or skin depth*. It is defined as the distance within the metal in which the amplitude of the travelling wave is attenuated to $1/e$ or 36.8 percent of its value just above the metal surface. Then by definition, if $\delta$ is the skin depth

$$\alpha\delta = 1 \text{ or } \delta = \frac{1}{\alpha}$$

which on using Eq. (3.31) reduces to

$$\delta = \sqrt{\frac{2}{\omega\mu\sigma}}. \tag{3.83}$$

For copper and silver $\delta$ has values 0.00667 mm and 0.00637 mm, respectively, at 100 MHz frequency. Thus, in good conductors such as copper, silver, etc. at microwave frequencies, the wave is attenuated to a negligibly small value within few thousandths of a centimetre.

In this case $|\Gamma|$ is not equal to $-1$ but is very near to it because $\eta_1 \gg \eta_2$. The transmission coefficient $|T|$ is also different from zero but small and has a value

$$|T| = 2\frac{\eta_2}{\eta_1}\cdot \tag{3.84}$$

(since $\eta_1 \gg \eta_2$)

The time average power density entering the surface of the conductor may by obtained from the Poynting theorem (Eq. 3.59) and is given by

$$P_{av} = \frac{|H|^2}{2}\sqrt{\frac{\omega\mu_2}{2\sigma}} = \frac{|H|^2}{2}\underset{\text{(say)}}{R_s}. \tag{3.85}$$

In an A.C. circuit the time average power consumed in a resistance $R$ is given by $P_{av} = I_m^2 R/2$ where $I_m$ is the peak value of current. Since the peak value of current flowing through a cross-section of the conductor parallel to the $y-z$ plane with unit length in the $y$-direction and infinitely long in the $z$-direction equals the surface value of the magnetic field. Hence, $|H|^2$ in our present case equals the square of the peak current. It therefore follows from Eq. (3.85) that $Rs$ is the *skin effect resistance or surface resistance* of the plane conducting surface and is given by

$$R_s = \sqrt{\frac{\omega\mu_2}{2\sigma}} = \frac{1}{\sigma\delta}. \tag{3.86}$$

(the last step follows from Eq. (3.87)).

Using Eq. (3.86), Eq. (3.49) may be written as

$$\eta_m = \pm \frac{1+j}{R_s} \cdot \tag{3.87}$$

So $\eta_m$ may also be called surface impedance.

## SOLVED EXAMPLES

**Example 3.1.** A perfect dielectric medium $x<0$ is bounded by a perfect conductor ($x=0$). If the electric field intensity in the dielectric medium is given by

$$\vec{E}_{(x,y,z,t)} = \hat{y}\,[E_1 \cos(\omega t - \beta x \cos\theta - \beta z \sin\theta) + E_2 \cos(\omega t + \beta x \cos\theta - \beta z \sin\theta)]$$

where $E_1$, $E_2$, $\omega$, $\beta$ and $\theta$ are constants, determine the relationship between $E_1$ and $E_2$, and the surface current density on the conducting surface $z=0$.

*Solution*: The boundary conditions require that the tangential component of the electric field intensity at the surface of the perfect conductor must vanish. Thus

$$[E_y]_{x=0} = (E_1+E_2)\cos(\omega t - \beta z \sin\theta) = 0. \tag{i}$$

For this equation to be true for all values of $z$ and $t$, $E_1+E_2$ must vanish. Therefore,

$$E_2 = -E_1. \tag{ii}$$

Hence the electric field intensity in the dielectric region is given by

$$\vec{E} = \hat{y}\,[E_1 \cos(\omega t - \beta x \cos\theta - \beta z \sin\theta) - E_1 \cos(\omega t + \beta x \cos\theta - \beta z \sin\theta)]$$

$$= 2E_1 \hat{y} \sin(\beta x \cos\theta) \sin(\omega t - \beta z \sin\theta)]. \tag{iii}$$

The corresponding magnetic flux density $\vec{B}$ can be obtained from Maxwell's curl equation for $\vec{E}$ (Eq. M-1). That is

$$-\frac{\partial \vec{B}}{\partial t} = \nabla x \vec{E} = -\frac{\partial E_y}{\partial z}\hat{x} + \frac{\partial E_y}{\partial x}\hat{z} \tag{iv}$$

Substituting the value of $E_y$ from Eq. (*iii*) into Eq. (*iv*) and integrating with respect to $t$ we have.

$$\vec{B} = -\frac{2E_1\beta}{\omega}[\sin\theta \sin(\beta x \cos\theta) \sin(\omega t - \beta z \sin\theta)\,\hat{x} - \cos\theta \cos(\beta x \cos\theta)\cos(\omega t - \beta z \sin\theta)\,\hat{z}].$$

The magnetic flux density at the surface of the perfect conductor is

$$\vec{[B]}_{x=0} = \frac{2E_1\beta}{\omega} \cos\theta \cos(\omega t - \beta z \sin\theta)\,\hat{z}.$$

The surface current density at the conducting surface is given by

$$\vec{J_s} = -\,\hat{x} \times \vec{[H]}_{x=0}$$

$$= \hat{y}\frac{2E_1\beta}{\omega\mu} \cos\theta \cos(\omega t - \beta z \sin\theta).$$

**Example 3.2.** Show that the wavelength of an electromagnetic wave in the direction of propagation is related to apparent wavelengths $\lambda_x$, $\lambda_y$ and $\lambda_z$ along the coordinate axes $x$, $y$ and $z$ respectively as

$$\frac{1}{\lambda^2} = \frac{1}{\lambda_x^2} + \frac{1}{\lambda_y^2} + \frac{1}{\lambda_z^2}$$

and the phase velocity $v_p$ as

$$\frac{1}{v_p^2} = \frac{1}{v_{px}^2} + \frac{1}{v_{py}^2} + \frac{1}{v_{pz}^2}.$$

*Solution*: The wavelength along the direction of propagation is given by

$$\lambda = \frac{2\pi}{\beta}$$

where $\beta$ is the propagation or phase constant. We know that the constant phase plane of a propagating wave is given by

$$\omega t - \vec{\beta} \cdot \vec{\gamma} = \text{constant}$$

or $\qquad \omega t - \beta_x x - \beta_y y - \beta_z z = \text{constant}$

or at $\qquad t = 0,\ \beta_x x - \beta_y y - \beta_z z = \text{constant}.$

Now the direction of the gradient of this scalar function $\beta_x x + \beta_y y + \beta_z z$ $=$constant gives the direction of propagation, whereas the magnitude of the gradient gives the rate of change of phase with respect to distance. Therefore,

$$\nabla(\beta_x x + \beta_y y + \beta_z z) = \beta_x \hat{x} + \beta_y \hat{y} + \beta_z \hat{z}.$$

Hence the phase constant along the direction of propagation is

$$\beta = (\beta_x^2 + \beta_y^2 + \beta_z^2)^{1/2}. \qquad (i)$$

Substituting

$$\beta = \frac{2\pi}{\lambda},\ \beta_x = \frac{2\pi}{\lambda_x},\ \beta_y = \frac{2\pi}{\lambda_y} \quad \text{and} \quad \beta_z = \frac{2\pi}{\lambda_z}$$

we have

$$\frac{1}{\lambda^2}=\frac{1}{\lambda_x^2}+\frac{1}{\lambda_y^2}+\frac{1}{\lambda_z^2}.$$

The phase velocity along the direction of propagation is

$$v_p=\frac{\omega}{\beta} \qquad \text{or} \qquad \beta=\frac{\omega}{v_p}.$$

Substituting the above result in Eq. ($i$) yields

$$\frac{1}{v_p^2}=\frac{1}{v_{px}^2}+\frac{1}{v_{py}^2}+\frac{1}{v_{pz}^2}.$$

**Example 3.3.** The propagation vector $\beta$ (phase constant) of a 12 MHz plane wave propagation in free space makes an angle 30° upwards with the horizontal $x$-$y$ plane and its projection on the $x$-$y$ plane makes an angle 60° with the $x$-axis. The electric field has no z-component and its magnitude as a function of time at $x=0$, $y=0$ and $z=0$ is 10 cos ($\omega t-30°$) V/m, where $\omega$ is the angular frequency.

Find the complex field vectors $\vec{E}$ and $\vec{H}$.

*Solution*: $\beta=\frac{\omega}{v_p}=\frac{2\pi f}{v_p}=\frac{2\pi\times 12\times 10^6}{3\times 10^8}=0.08\ \pi$

From the problem it is clear that

$$\begin{aligned}\beta&=0.08\pi\ (\cos 30°\cos 60°\ \hat{x}+\cos 30°\sin 60°\ \hat{y}+\sin 30°\ \hat{z})\\&=0.08\pi(\sqrt{3}\,\hat{x}+3\hat{y}+2\hat{z}).\end{aligned}$$

Since $\vec{E}$ has no $z$ component it can be written as

$$\vec{E_0}=E_{x0}\,\hat{x}+E_{y0}\,\hat{y}.$$

Further, $\vec{\beta}\cdot\vec{E_0}$ must vanish so

$$\beta_x\,E_{x0}+\beta_y\,E_{y0}=0.$$

Since $\beta_x$ and $\beta_y$ are real, either $E_{x0}$ and $E_{y0}$ are in phase or out of phase. Hence, let

$$E_{x0}=|E_{x0}|\ e^{j\alpha},\ E_{y0}=|E_{y0}|\ e^{j\alpha}$$

so that

$$\vec{E_0}=\{|E_{x0}|\hat{x}+|E_{y0}|\hat{y}\}\ e^{j\alpha}.$$

Moreover,

$$\begin{aligned}\beta_x\,|E_{x0}|+\beta_y\,|E_{y0}|&=0.02\pi\ (\sqrt{3}\,|E_{x0}|+3\ |E_{y0}|)=0\\|E_{x0}|&=-\sqrt{3}\,|E_{y0}|.\end{aligned}$$

It is given that for $x=0$, $y=0$ and $z=0$, i.e. $\vec{r}=0$ we have

$$|\{|E_{x0}|\hat{x}+|E_{y0}|\hat{y}\}|\ e^{j\alpha}=10\ e^{-j30°}$$

or

$$\{|E_{x0}|^2+|E_{y0}|^2\}^{1/2}=10$$

and $\alpha=-\pi/6.$

Substituting $|E_{x0}|=-\sqrt{3}\ |E_{y0}|$ in the above equation yields

$$4\ |E_{y0}|^2=100$$

or $|E_{y0}|=5$

and $|E_{x0}|=-5\sqrt{3}.$

Hence the required expression for $\vec{E}$ is

$$\vec{E}=5(-\sqrt{3}\,\hat{x}+\hat{y})\ e^{-j\pi/6}\ e^{-j0.02\pi(\sqrt{3}x+3y+2z)}.$$

The corresponding expression for $\vec{H}$ follows from the expression

$$\vec{H}=\frac{1}{\omega\mu}\ \vec{\beta}\times\vec{E}$$

$$=\frac{1}{96\pi}\begin{vmatrix}\hat{x} & \hat{y} & \hat{z}\\ \sqrt{3} & 3 & 2\\ -\sqrt{3} & 1 & 0\end{vmatrix}$$

$$=\frac{1}{48\pi}\ (-\hat{x}-\sqrt{3}\,\hat{y}+2\sqrt{3}\,\hat{z})\ e^{-j\pi 6}\ e^{-j0.02\pi(\sqrt{3}x+3y+2z)}.$$

**Example 3.4.** If the electric field radiated by an antenna located at the origin of a spherical coordinate system is given by

$$\vec{E}=\frac{E_0\sin\theta}{r}\cos(\omega t-\beta\gamma)\,\hat{\theta}$$

where $E_0$, $\omega$ and $\beta=\omega\sqrt{\mu\epsilon}$ are constants, determine the magnetic induction $\vec{B}$ associated with the electric field and then find the power radiated by the antenna within a spherical sphere of radius $\gamma$ centred at the antenna.

*Solution*: We know that

$$-\frac{\partial\vec{B}}{\partial t}=\nabla\times\vec{E}$$

$$\text{or} \qquad \frac{\partial \vec{B}}{\partial t} = - \begin{vmatrix} \frac{\hat{\gamma}}{r^2 \sin\theta} & \frac{\hat{\theta}}{r \sin\theta} & \frac{\hat{\phi}}{r} \\ \frac{\partial}{\partial r} & \frac{\partial}{\partial \theta} & \frac{\partial}{\partial \phi} \\ 0 & E_0 \sin\theta \times \cos(\omega t - \beta\gamma) & 0 \end{vmatrix}$$

$$\text{or} \qquad \vec{B} = \frac{\beta E_0}{\omega r} \sin\theta \cos(\omega t - \beta r)\, \hat{\phi}.$$

The power radiated over the given volume is

$$\int \vec{E} \times \vec{H} \cdot ds \quad (s = \text{spherical surface of radius } r)$$

$$\text{Power} = \int_{\theta=0}^{\pi} \int_{\phi=0}^{2\pi} \frac{\beta E_0^2 \sin^2\theta}{\mu_0\, \omega\, r^2} \cos^2(\omega t - \beta r)\, \hat{r} \cdot r^2 \sin\theta\, d\theta\, d\phi\, \hat{r}$$

$$= \frac{2\pi\beta E_0^2 \cos^2(\omega t - \beta r)}{\mu_0 \omega} \int_{\theta=0}^{\pi} \sin^3\theta\, d\theta$$

$$= \frac{8\pi\beta E_0^2 \cos^2(\omega t - \beta r)}{3\, \mu_0 \omega}.$$

**Example 3.5.** For uniform plane waves in sea water find the value of $\alpha$, $\beta$, $\eta$ and $\lambda$ at $10^5$ MHz given that

$$\sigma = 4\ \mho/\text{m}, \quad \epsilon = 80\, \epsilon_0 \quad \text{and} \quad \mu = \mu_0.$$

*Solution*: At $10^5$ MHz frequency we note that

$$\omega\epsilon = 2\pi \times 10^{11} \times 80 \times \frac{1}{4\pi \times 9 \times 10^9} = 444.44 \gg \sigma.$$

Therefore sea water is a good dielectric at this frequency. Hence

$$\alpha = \tfrac{1}{2}\sigma\sqrt{\mu/\epsilon}\left(1 - \frac{\sigma^2}{8\omega^2\epsilon^2}\right) \simeq \tfrac{1}{2}\sigma \sqrt{\frac{\mu}{\epsilon}} = \frac{\sigma}{2}\sqrt{\frac{\mu_0}{80\, \epsilon_0}}$$

$$= 2 \times \frac{377}{\sqrt{80}} = 84.3 \text{ neper/metre}$$

$$\beta = \omega\sqrt{\mu\epsilon}\left(1 + \frac{\sigma^2}{8\, \omega^2 \epsilon^2}\right) \simeq \omega\sqrt{\mu\epsilon}$$

$$= \frac{2\pi \times 10 \times 10^9 \sqrt{80}}{3 \times 10^8} = 1873 \text{ rad/m}$$

$$= \sqrt{\frac{\mu}{\epsilon}}\left[\left(1 - \frac{3}{8}\frac{\sigma^2}{\omega^2\epsilon^2}\right) + j\frac{\sigma}{\omega\epsilon}\right] \simeq \sqrt{\frac{\mu}{\epsilon}}$$

$$=\frac{377}{\sqrt{80}}=42.15\ \Omega$$

$$\lambda=\frac{2\pi}{\beta}=\frac{2\pi}{1837}=3.353\times10^{-3}\ \text{m}.$$

**Example 3.6**. Find the attenuation constant and phase velocity of plane waves of frequency 10 GHz propagating in polythylene having permittivity six times that of air and a loss-tangent $2\times10^{-4}$.

*Solution*: The phase velocity is given by Eq. (3.34)

$$v_p=\frac{1}{\sqrt{\mu\epsilon}}=\frac{\sqrt{\epsilon_0/\epsilon}}{\sqrt{\mu_0/\mu}}=\frac{3\times10^8}{2\cdot3}=1.98\times10^8\ \text{m/sec.}$$

The attenuation constant is

$$\alpha=\tfrac{1}{2}\,\sigma(\mu/\epsilon)=\frac{\sigma\omega\sqrt{\mu\epsilon}}{2\ \omega\epsilon}$$

$$=\pi f\tan\delta/v_p$$

$$=\frac{10^{10}\times2\times10^{-4}}{1.98\times10^8}=3.17\times10^{-2}$$

$$=0.28\ \text{db/m}.$$

## PROBLEMS

1. Show that the velocity of an electromagnetic wave in a non-ferous medium is proportional to the square root of its relative permittivity.

2. Define polarization of an electromagnetic wave and obtain an expression for the electric and magnetic fields of a circularly polarized electromagnetic wave.

3. A uniform plane wave in space has $Y$-polarization given by

$$\overrightarrow{E_y}=E_0\ e^{\gamma(l_x+nz)}$$

Determine magnetic intensity and the Poynting vector.

4. Find the average Poynting vector for a plane wave of amplitude $H_0=1$ A/m in a homogeneous isotropic medium for which $\mu_r=1$ and $\epsilon_r=4$.

5. A plane electromagnetic wave is incident on sea water, $\sigma=4$ ℧/m rms $\epsilon=80\ \epsilon_0$ $\mu=u_0$. What is the electric intensity of the reflected wave on the surface if the rms electric intensity of the incident wave is 1 V/m? At what depth does the rms electric field equal $\mu\gamma$/m?

6. Considering oblique incidence of a plane polarized electromagnetic ray, establish Snell's and Brewster's laws.

7. Show that if the boundary conditions are satisfied for either electric intensity or magnetic intensity, and the fields satisfy Maxwell's equations, then the boundary conditions for the other field intensity are automatically satisfied.

8. A sinusoidally time varying plane electromagnetic wave in free space is normally incident on a plane dielectric slab of thickness $a$. The slab has a perfectly con-

ducting sheet coated on its back. Determine the electromagnetic fields in regions 1 and 2.

9. Determine the depth of penetration of brass, $\sigma = 1.45 \times 10^7$ ℧/m at frequencies 1 MHz, 1 GHz and 10 GHz.

10. The half space $Z>0$ is filled with a material with permeability $\mu$ and permittivity $\epsilon$. When a plane wave is incident normally on this material, show that the reflection and transmission coefficients are

$$\Gamma = \frac{Z - Z_0}{Z + Z_0}, \quad T = 1 + \Gamma = \frac{2Z}{Z + Z_0}$$

where $\quad Z = \sqrt{\frac{\mu}{\epsilon}} \quad$ and $\quad Z_0 = \sqrt{\frac{\mu_0}{\epsilon_0}}$. Choose $\vec{E} = E_x \hat{x}$.

11. A plane electromagnetic wave is normally incident on a semi-infinite dielectric. The standing-wave ratio in the free space (medium 1) is 1.73 and there is an average rate of flow of energy in the $z$-direction of 1 W/m². Determine the relative permittivity of the dielectric, the maximum and minimum electric field intensities in medium 1 and the electric field intensity in medium 2.

12. The reflectivity of a metal is defined as the ratio of reflected power to incident power for a plane wave normally incident onto a plane surface of the metal. At a wavelength 500 nm the reflectivities for aluminium, silver and steel are found to be 89% 95% and 56% respectively. Estimate the conductivity of each metal at this wavelength.

13. The conductivity of copper measured at 24 GHz is $3.05 \times 10^7$ ℧/m. Calculate the attenuation constant, phase constant, phase velocity and wavelength of a uniform plane wave propagating in copper at this frequency.

14. A charge $Q_0$ is placed on the surface of a conductor. How this charge distribute on the conductor surface with time? Calculate the relaxation time and justify the utility of copper in designing a waveguide at 24 GHz.

15. A slab of dielectric (permittivity) is covered with a plane layer of another dielectric and a plane wave is made to fall normally from air onto the covered slab. Show that no reflection takes place if the thickness of the layer is one-quarter of the wavelength in the layer.

## *REFERENCES*

BLEANEY, B.I. AND BLEANY, B., '*Electricity and Magnetism*'. Oxford University Press (1978).

BRONWELL, A.B. AND BEAM, R.E., '*Theory and Applications of Microwaves.*' McGraw Hill Book Co. (1966).

COLLIN, R.E., '*Foundations for Microwave Engineering*'. McGraw-Hill Book Co. (1960).

HAYT, W.H., '*Engineering Electromagnetics.*' McGraw-Hill Book Co. (1974).

HAMMAND P., '*Electromagnetism for Engineers*—An Introductory Course' Oxford Pregaman (1978).

JAKSON J.D., '*Classical Electrodynamics*'. John Wiley New York (1975).

JORDAN, E.C. AND BALMAIN, K.G., '*Electromagnetic Waves and Radiating Systems*'. Printice Hall of India (1971).

PANOFSKY W.K.H., AND PHILLIPS M., '*Classical Electricity and Magnetism*' Addision-Wesley Inc. 1962.

RAO, N.N., '*Basic Electromagnetics with Applications*'. Printice-Hall of India (1974).

SANDER K.F. AND REED, G.A.L., '*Transmission and Propagation of Electromagnetic Waves*'. Cambridge University Press. London (1978).

SCHELKUNOFF, S.A., '*Electromagnetic Waves*'. D. Van Nostrand Co. Inc. (1943).

SEELY, S., *Electrodynamics Classical and Modern Theory and Applications*'. Marcel Dekker New York (1979).

SLATER J.C.. '*Microwave Electronics*'. D Van Nostrand (1950).

STARTON, J.A., *Electromagnetic Theory*'. McGraw Hill Book Co. (1941).

# 4

# WAVEGUIDE THEORY

## 4.1 INTRODUCTION

In this chapter we shall study waveguides of commonly used cross-section, viz. rectangular and circular, However, a brief reference to various other types of waveguides will also be made.

## 4.2 WAVEGUIDE PROPAGATION AS A REFLECTION PHENOMENA

The fields in waveguides comprise a series of plane waves, each travelling with a velocity characteristic of the medium inside and all of which multiply when reflected from wall to wall in a guide in such a way that they travel in a zig-zag path down the guide. The analysis of the guide can therefore, be built up on the basis of the oblique reflections of plane waves. Such analysis not only throws light on the nature of guided waves* but also enables us to deduce some important relations pertaining to waveguides, that otherwise require rather complicated mathematical analysis through the solution of Maxwell's equations under given boundary conditions. Let us first consider waveguides from the viewpoint of a plane wave reflection.

Consider a plane wave of arbitrary polarization incident on a plane conductor as shown in Fig. 4.1. With reference to this figure, the electric field of the incident wave of the incident wave can be written as (following Eq. 3.36.)

$$\vec{E_i} = E_0 \left(e^{-j\vec{\beta_i} \cdot r}\right)\hat{y}$$

$$= E_0 \, e^{-(\beta\cos\theta_i \hat{x} + \beta\sin\theta_i \hat{z}) \cdot \vec{r}} \, \hat{y}$$

$$= E_0 \, e^{-j(\beta x\cos\theta_i + \beta z\sin\theta_i)}\hat{y} \tag{4.1}$$

where $\vec{\beta_i}$ $(|\vec{\beta_i}|) = \omega\sqrt{\mu\epsilon}$ is the phase constant of the incident wave, $\theta_i$ is the angle of incidence and $E_0$ is the complex amplitude of the incident electric field. The incident magnetic field can be obtained from

$$\vec{H_i} = \frac{1}{\omega\mu} \vec{\beta_i} \times \vec{E_i}$$

*Wave that propagate longitudinally along the guiding structure such as hollow metal pipes.

$$=\sqrt{\epsilon/\mu}\left\{-E_0 \sin\theta_i\,\hat{x}+E_0\cos\theta_i\,\hat{z}\right\} e^{-j(\beta x\cos\theta_i+\beta z\sin\theta_i)}. \quad (4.2)$$

The boundary condition of zero tangential electric field at the metal boundary dictates the presence of the reflected wave which completely cancels

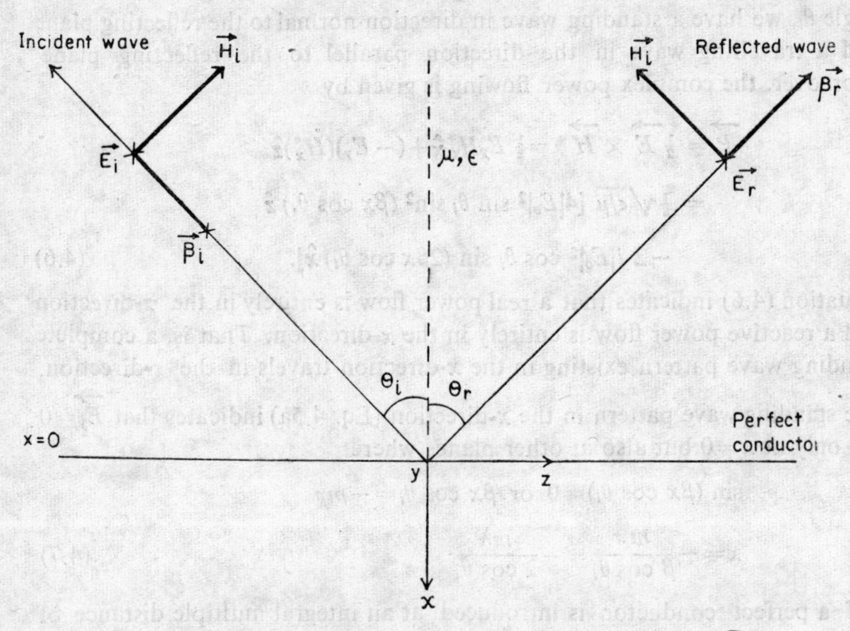


**Fig. 4.1** Oblique incidence of a uniform plane wave on a perfect conductor.

the tangential component of the electric field vector of the incident wave. This implies that the electric field of the reflected wave at the surface of the conductor in the $y$-direction. Hence the reflected wave will have the electric field

$$\vec{E_y}=E_o'\,(e^{j\vec{\beta_r}\cdot\vec{r}})\hat{y}$$

$$=E_o'\,e^{j(\beta x\cos\theta_r-\beta z\sin\theta_r)}\,\hat{} \quad (4.3)$$

and $$\vec{H_r}=\sqrt{\epsilon/\mu}\,\,(-E_o'\sin\theta_r\hat{x}-E_o'\cos\theta_r\hat{z})\,e^{j(\beta x\cos\theta_r-\beta z\sin\theta_r)}. \quad (4.4)$$

Combining incident and reflected electric and magnetic fields and applying the boundary condition that tangential electric field vanishes at the metal surface $[\vec{E_y}]_{x=0}=0$, we find that $\theta_r=\theta_i$ and $E_o'=-E_o'$.

The fields are then given by

$$\vec{E_y}=-2j\,\vec{E_o}\sin(\beta x\cos\theta_i)\,e^{-j\beta z\sin\theta_i} \quad (4.5a)$$

$$\vec{H_x}=\sqrt{\epsilon/\mu}\,2j\,\vec{E_o}\sin\theta_i\sin(\beta x\cos\theta_i)\,e^{-j\beta z\sin\theta_i} \quad (4.5b)$$

$$\vec{H_z} = \sqrt{\epsilon/\mu}\ 2\ \vec{E_o} \cos\theta_i \cos(\beta_x \cos\theta_i)\ e^{-j\beta z \sin\theta_i}. \tag{4.5c}$$

The exponential factors in Eqs. (4.5a to 4.5c) impart progressive character to the fields while sine and cosine factors impart a standing wave character. Thus, if a uniform plane wave impinges upon a conducting surface at an angle $\theta_i$, we have a standing wave in direction normal to the reflecting plane and a travelling wave in the direction parallel to the reflecting plane. Moreover, the complex power flowing is given by

$$\vec{P} = \tfrac{1}{2}\vec{E} \times \vec{H}^* = \tfrac{1}{2} E_y H_z^* \hat{x} + (-E_y)(H_x^*)\hat{z}$$

$$= \tfrac{1}{2}\sqrt{\epsilon/\mu}\ [4|E_o|^2 \sin\theta_i \sin^2(\beta x \cos\theta_i)\,\hat{z}$$

$$-2\,j|E_o|^2 \cos\theta_i \sin(2\beta x \cos\theta_i)\,\hat{x}]. \tag{4.6}$$

Equation (4.6) indicates that a real power flow is entirely in the $z$-direction and a reactive power flow is entirely in the $x$-direction. That is, a complete standing wave pattern existing in the $x$-direction travels in the $z$-direction.

The standing wave pattern in the $x$-direction (Eq. 4.5a) indicates that $\vec{E_y}=0$ not only at $x=0$ but also at other planes where

$$\sin(\beta x \cos\theta_i)=0 \quad \text{or} \quad \beta x \cos\theta_i = -m\pi$$

or

$$x = -\frac{m\pi}{\beta\cos\theta_i} = -\frac{m\lambda}{2\cos\theta_i}. \tag{4.7}$$

If a perfect conductor is introduced at an integral multiple distance of $\lambda/2\cos\theta_i$ from the reflecting plane under discussion, it does not alter in any way the field configuration, once it is established. The new conductor configuration is a parallel plane waveguide. Thus, the wave propagation in a waveguide* is nothing but a reflection phenomenon. It comprises recurring oblique bouncing of plane electromagnetic waves between two conducting planes as shown in Fig. 4.2. The total magnetic field has a component

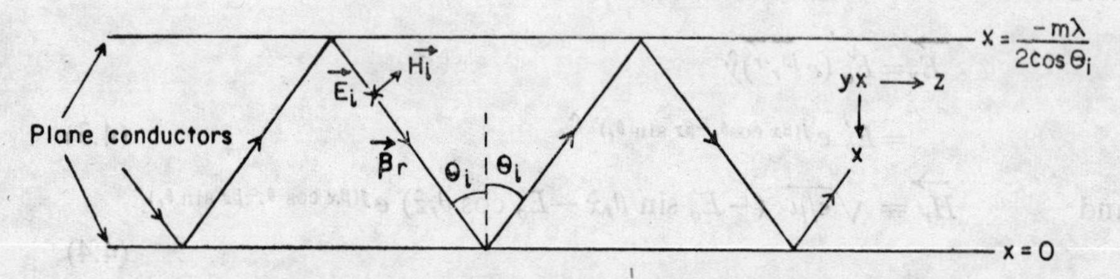


**Fig. 4.2** Oblique bouncing of plane electromagnetic waves between two conducting plane.

in the $z$-direction, which is the direction of time-average power flow whereas the electric field is entirely transverse to the $z$-direction. For this reason, the

*This discussion is for a parallel plane waveguide but will be extended later to a rectangular waveguide.

waves are known as Transverse Electric or TE waves. If the magnetic field becomes totally transverse to the direction of propagation we have Transverse Magnetic or TM waves; and, if both electrics and magnetic fields are perpendicular to the direction of motion we have Transverse Electromagnetic or TEM waves. Further, for any given spacing between the conductors there are various discrete values of $m$ and $\lambda$. Which means that the TE and TM wave have various modes of propagation.

## 4.3 GENERAL CHARACTERISTICS OF WAVEGUIDE PROPAGATION

In this section we use a phenomenological discussion of waveguide propagation as a reflection phenomena to study many important characteristics of waveguide propagation.

### (*i*) Velocities of Propagation and the Cut-off Phenomenon

The condition in Eq. (4.7) may be interpreted in another way. If two conducting surfaces are spaced by a distance $a$, then the condition for wave propagation will be

$$\cos\theta_i = \frac{m\lambda}{2a} \tag{4.8}$$

where $\lambda$ is the free space wavelength.

From the above relation we note that for very high frequencies, $\lambda \simeq 0$, $\cos\theta_i \simeq 0$, and $\theta_i \simeq \pi/2$, the waves slide between the plates almost like in a TEM wave (Fig. 4.3a). As the frequency is decreased, $\lambda$ increases, $\cos\theta_i$

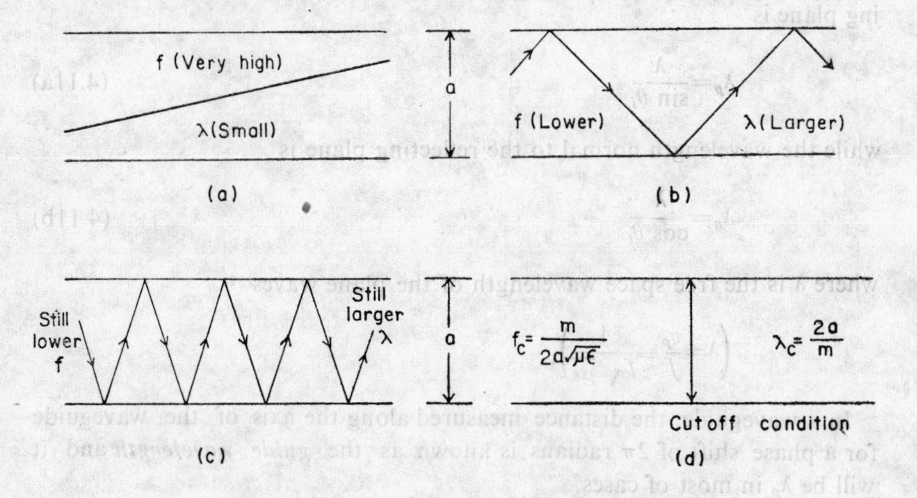


**Fig. 4.3** Cut-off phenomenon; Bouncing of uniform plane waves of different frequencies between parallel plane conductors of fixed spacing.

increases and $\theta_i$ decreases, and the waves bounce obliquely between the plates progressing in a $z$-direction (Fig. 4.3,b c). However, at a frequency for which $\lambda=2a/m$, $\cos\theta_i=1$ and $\theta_i=0°$, the waves bounce back and forth between the plates and normal to them so that there is no progress in the $z$-direction as shown in Fig. 4.3(d). For $\lambda>2a/m$, $\cos\theta_i>1$ and $\sin\theta_i=\sqrt{1-\cos^2\theta_i}$ become imaginary and the exponents in the expressions for the field configuration Eq. (4.5a) through (4.5c) become imaginary. This situation does not correspond to any wave propagation. The fields diminish in magnitude along $z$.

Thus, there is a wavelength below which propagation occurs and above which there is attenuation or no propagation. This wavelength is known as the *cut-off wavelength* and is denoted by $\lambda_c$. Here,

$$\lambda_c=\frac{2a}{m}, \quad m=1, 2, 3 \ldots \tag{4.9}$$

the corresponding cut-off frequency is

$$f_c=\frac{v_p}{\lambda_c}=\frac{m}{2a\sqrt{\mu\epsilon}} \tag{4.10}$$

where $v_p=1/\sqrt{\mu\epsilon}$ is the phase velocity along the direction of incidence. The propagation occurs for $f>f_c$ or $\lambda\leqslant\lambda_c$ only.

Figure 4.4 depicts a plane wave incident obliquely on a plane conductor sheet (reflected wave is omitted in the figure). When we apply the general definition of wavelength, i.e. the distance between two successive equiphase points at any instant of time, we find that there are many different ways in which the wavelength can be taken. The wavelength parallel to the reflecting plane is

$$\lambda_p=\frac{\lambda}{\sin\theta_i} \tag{4.11a}$$

while the wavelength normal to the reflecting plane is

$$\lambda_n=\frac{\lambda}{\cos\theta_i} \tag{4.11b}$$

where $\lambda$ is the free space wavelength of the plane waves

$$\left(\lambda=\frac{v_p}{f}=\frac{1}{f\sqrt{\mu\epsilon}}\right).$$

In a waveguide, the distance measured along the axis of the waveguide for a phase shift of $2\pi$ radians is known as the *guide wavelength* and it will be $\lambda_p$ in most of cases.

$$\lambda_g=\frac{\lambda}{\sin\theta_i}. \tag{4.12}$$

Substituting Eq. (4.9) in Eq. (4.8), we have

$$\cos\theta_i = \lambda/\lambda_c, \quad \sin\theta_i = \sqrt{1-(\lambda/\lambda_c)^2}.$$

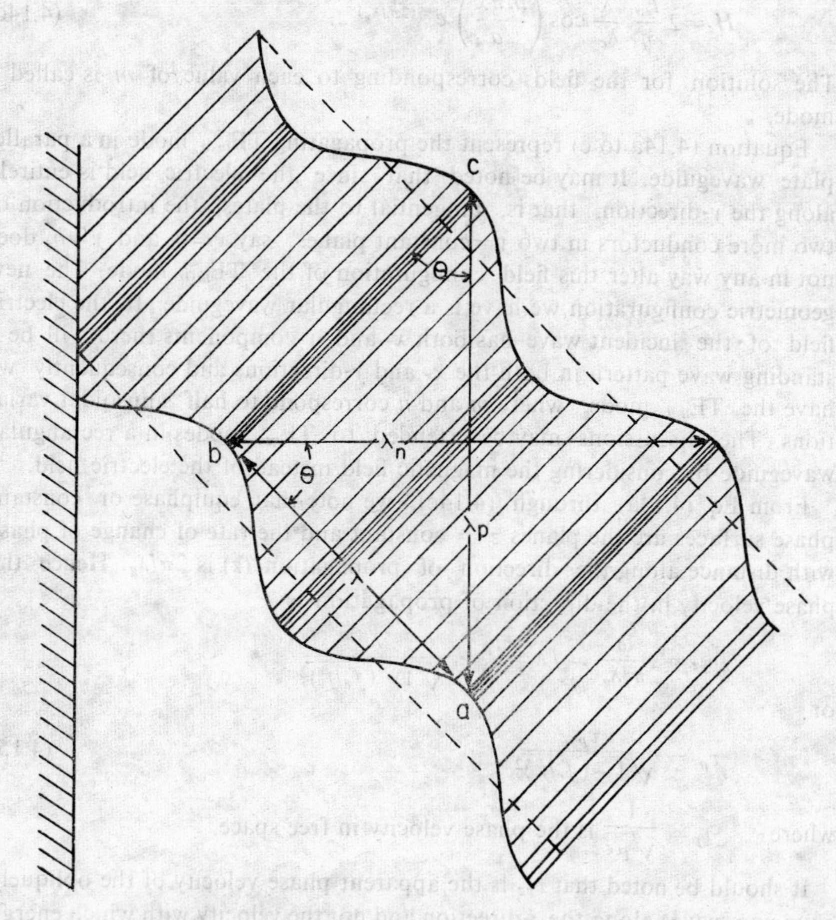


**Fig. 4.4** Incident wave showing wavelengths parallel to and normal to the reflecting surface.

Hence, the guide wavelength is given by

$$\lambda_g = \frac{\lambda}{\sin\theta_i} = \frac{\lambda}{\sqrt{1-(\lambda/\lambda_c)^2}} = \frac{\lambda}{\sqrt{1-(f_c/f)^2}}. \tag{4.13}$$

Substituting the values of $\cos\theta_i$ and $\sin\theta_i$ in Eqs. (4.5a through 4.5c) we obtain*

$$E_y = -2\,jE_o \sin\left(\frac{m\pi x}{a}\right) e^{-j(2\pi/\lambda_g)Z} \tag{4.14a}$$

*These equations may be directly obtained from Maxwell's equations by solving them for the TE mode of propagation.

$$H_x = 2j \frac{E_o}{\eta} \frac{\lambda}{\lambda_g} \sin\left(\frac{m\pi x}{a}\right) e^{-j(2\pi/\lambda_g)z} \tag{4.14b}$$

$$H_z = 2 \frac{E_o}{\eta} \frac{\lambda}{\lambda_c} \cos\left(\frac{m\pi x}{a}\right) e^{-j(2\pi/\lambda_g)z}. \tag{4.14c}$$

The solution for the fields corresponding to each value of $m$ is called a mode.

Equation (4.14a to c) represent the propagating $TE_{m,o}$ mode in a parallel plate waveguide. It may be noted that since the electric field is entirely along the $y$-direction, that is, tangential to the plates, the introduction of two more conductors in two $y$=constant planes, say $y=0$ and $y=b$, does not in any way alter this field configuration of the $TE_{m,o}$ mode. The new geometric configuration we have is a rectangular waveguide. If the electric field of the incident wave has both $x$- and -$y$ components there will be a standing wave pattern in both the $x$- and $y$-directions and consequently we have the $TE_{mn}$ mode, where $m$ and $n$ correspond to half sinusoidal variations. The discussions may be extended to $TE_{mn}$ modes in a rectangular waveguide by considering the magnetic field instead of the electric field.

From Eq. (4.14a) through (4.14c), we note that equiphase or constant phase surfaces are the planes $z$ = constant and the rate of change of phase with distance along the direction of propagation ($z$) is $2\pi/\lambda_g$. Hence the phase velocity in the direction of propagation is

$$v_{pz} = \frac{\omega}{2\pi/\lambda_g} = f\lambda_g = \frac{v_p}{\lambda} \cdot \frac{\lambda}{\sqrt{1-(f_c/f)^2}}$$

or

$$v_{pz} = \frac{v_p}{\sqrt{1-(f_c/f)^2}} \tag{4.15}$$

where $v_p = \frac{1}{\sqrt{\mu\epsilon}}$ is the phase velocity in free space.

It should be noted that $v_{pz}$ is the apparent phase velocity of the obliquely bouncing waves along the $z$-direction and not the velocity with which energy propagates.

### (*ii*) Group Velocity and Dispersion

The concept of phase velocity introduced above applies only to fields that are periodic in space and consequently represent wave train of infinite duration. Since a wave train of finite duration cannot be represented in simple harmonic form, the concept of phase velocity loses its significance. To illustrate this point let us consider a group of two waves each with amplitude $E_o$ and closely spaced frequencies $\omega$ and $\omega+\delta\omega$ and phase constants $\beta$ and $\beta+\delta\beta$. The superposition results in a wave

$$\vec{E} = \vec{E_o} \cos(\omega t - \beta z) + E_o \cos[(\omega+\delta\omega)t + (\beta+\delta\beta)z]$$

$$= 2\vec{E_o} \cos\left(\frac{\delta\omega t - \delta\beta z}{2}\right) \cos\left[\left(\omega + \frac{\delta\omega}{2}\right)t - \left(\beta - \frac{\delta\beta}{2}\right)z\right] \tag{4.16}$$

Since $\delta\omega$ is very small the resultant field $\vec{E}$ of Eq. (4.16) still has a phase constant $\beta$ and oscillates at a frequency that differs negligibly from $\omega$. Hence the velocity of a plane of constant phase or *phase velocity* is the same, $v_p=\omega/\beta$. On the other hand, it is clear from Eq. (4.16) that the amplitude of the resultant wave is

$$2\,E_o \cos\left(\frac{\delta\omega t-\delta\beta z}{2}\right)$$

which varies slowly from 0 to $2E_o$. Consequently, the interference of two waves of slightly different frequencies and phase constants, results in a field distribution in space and time that has an envelope with a series of periodically repeated beats. In this case, the velocity with which the constant amplitude of the wave travels (envelope of Fig. 4.5) is defined as *group velocity* denoted by $v_g$, and is given by

$$\delta\omega t-\delta\beta z=\text{constant}$$

or

$$v_g=\frac{dz}{dt}=\frac{d\omega}{d\beta}. \tag{4.17}$$

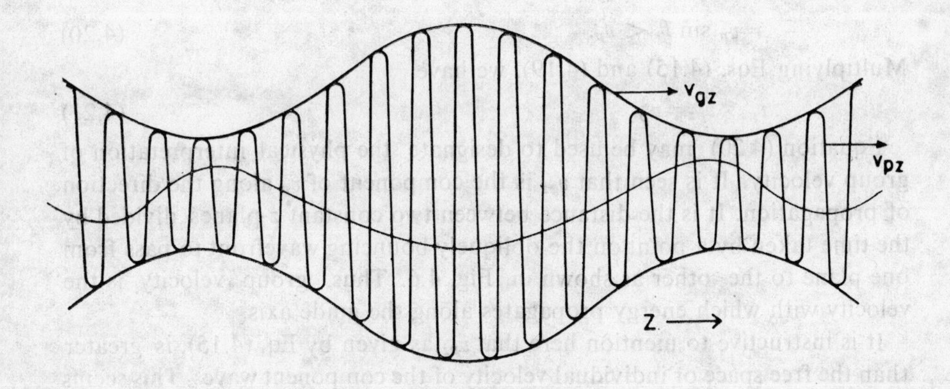


**Fig. 4.5 Groups and phase velocities of a wave.**

It should be noted that the concept of group velocity developed for two constituent waves is applicable to a band of frequencies provided we can find group velocity for each pair (two frequencies taken together) and it is the same for all pairs. This condition can be satisfied for a narrow band of frequencies. Hence the concept of group velocity is useless when we talk of a broad band of frequencies. Moreover, referring to Fig. 4.5 we may say that group velocity is the velocity of envelope (of the band of frequencies) while phase velocity is the velocity of the component wave. For $v_p>v_g$, we would expect the high frequency wave to slip through the envelope with the passage of time and as such group velocity is the velocity with which energy propagates. To establish a relation between group and phase velocities we may recall that the phase constant in the $z$-direction is given by

$$\beta_z = \frac{\omega}{v_{pz}} = \frac{\omega}{v_p}\sqrt{1-(f_c/f)^2}$$

$$= \beta\sqrt{1-(f_c/f)^2} \tag{4.18}$$

Differentiating Eq. (4.18) with respect to $\omega$ we have,

$$\frac{d\beta_z}{d\omega} = \beta\left[1-\left(\frac{\omega_c}{\omega}\right)^2\right]^{-1/2}\frac{\omega_c^2}{\omega^3} + \frac{d\beta}{d\omega}\left[1-\left(\frac{\omega_c}{\omega}\right)^2\right]^{1/2}$$

$\left[\frac{d\omega}{d\beta}\right.$ is free space group velocity which is same as $\left.v_p\right]$

$$= \frac{\omega_c^2}{v_p\omega^2}\left[1-\left(\frac{\omega_c}{\omega}\right)^2\right]^{-1/2} + \frac{1}{v_p}\left[1-\left(\frac{\omega_c}{\omega}\right)^2\right]^{1/2}$$

$$= \frac{1}{v_p}\left[1-\left(\frac{\omega_c}{\omega}\right)^2\right]^{-1/2}$$

or
$$v_{gz} = \frac{d\omega}{d\beta_z} = v_p\sqrt{1-\left(\frac{\omega_c}{\omega}\right)^2} = v_p\sqrt{1-\left(\frac{f_c}{f}\right)^2}$$

$$= v_p\sqrt{1-\left(\frac{\lambda}{\lambda_c}\right)^2} \tag{4.19}$$

$$= v_p \sin\theta < v_p. \tag{4.20}$$

Multiplying Eqs. (4.15) and (4.19), we have

$$v_{gz} \times v_{pz} = v_p^2. \tag{4.21}$$

Equation (4.20) may be used to designate the physical interpretation of group velocity. It is seen that $v_{gz}$ is the component of $v_p$ along the direction of propagation. It is the distance between two constant $z$-planes divided by the time taken by a point on the obliquely bouncing wavefront to pass from one plane to the other as shown in Fig. 4.6. Thus, group velocity is the velocity with which energy propagates along the guide axis.

It is instructive to mention here that $v_{pz}$ as given by Eq. (4.15), is greater than the free space or individual velocity of the component waves. This seems to contradict the fundamental principle of relativity that the velocity of light is the maximum achieveable velocity. But since energy propagates at a velocity $v_{gz}$ which is always less than $v_p$, this contradiction is removed. From Eq. (4.2) we see that the product of $v_{pz}$ and $v_{gz}$ is the square of the phase velocity of an individual wave; hence, it may be, concluded that as phase velocity increases group velocity or the envelope of Fig. 4.5 regresses.

Further, when $v_{pz}$ and $v_{gz}$ coincide with $v_p$ if the cut-off frequency is infinite as in the case of two conductor lines, such lines are known as *non-dispersive lines.* Hollow conductor waveguides, where $v_{pz} > v_{gz}$, are known as *dispersive lines.* The term dispersion used here signifies that the energy of various frequency components travels with different velocities and that the $\omega-\beta$ curve is not a linear. However, the dispersion discussed above is related to the geometry of the line and is known as *geometric dispersion.* There are

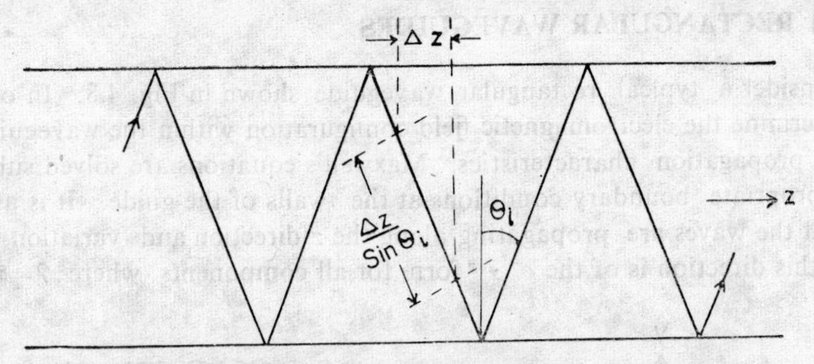


**Fig. 4.6** Illustration that the group velocity is the velocity with which energy propagates along the axis of the waveguide.

other types of dispersion for which the reader is referred to the literature*.

Figure 4.7 depicts the group velocity of a dispersive line as a function of frequency.

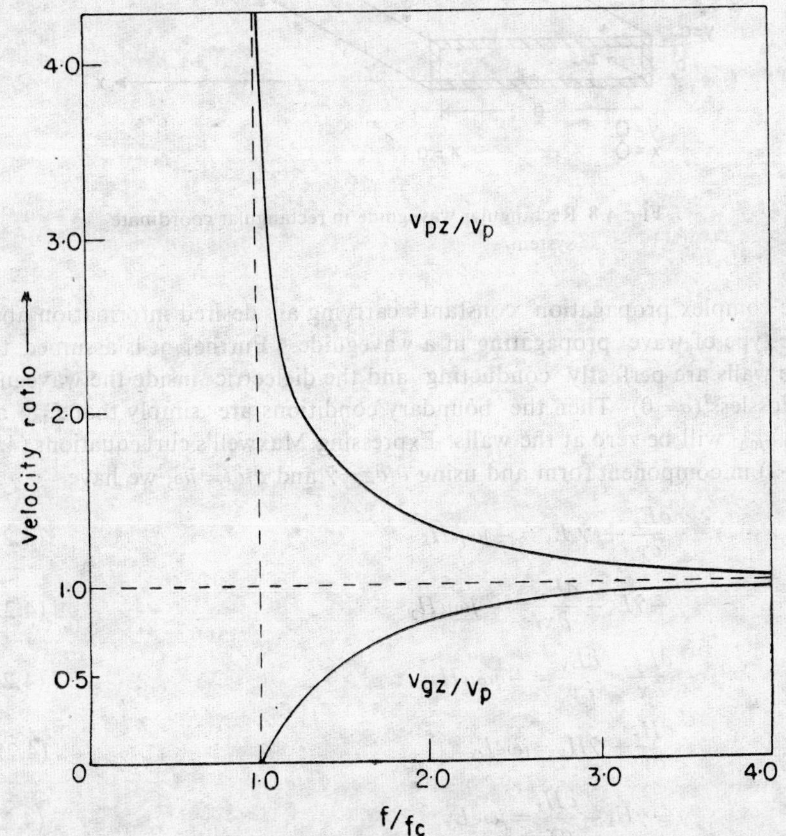


**Fig. 4.7** Group Velocity in a dispersive transmission line as a function of frequency.

*J.A. Stratton, *Electromagnetic Theory*, New York: Mc Graw-Hill Book Co., (1941). N.N. Rao, *Basic Electromagnetics with Applications* New Delhi: Prentice-Hall of India (1974).

## 4.4 RECTANGULAR WAVEGUIDES

Consider a typical rectangular waveguide shown in Fig. 4.8. In order to determine the electromagnetic field configuration within the waveguide and the propagation characteristics, Maxwell's equations are solved subject to appropriate boundary conditions at the walls of the guide. It is assumed that the waves are propagating along the $z$-direction and variation of field in this direction is of the $e^{-\bar{\gamma} z}$ form for all components where $\bar{\gamma}=\bar{\alpha}+j\bar{\beta}$ is

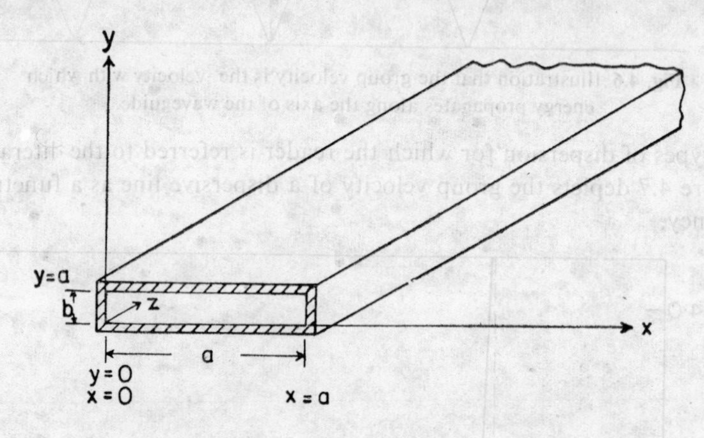


**Fig. 4.8** Rectangular waveguide in rectangular coordinate system.

the complex propagation constant* carrying all desired information about the type of wave propagating in a waveguide. Further, it is assumed that the walls are perfectly conducting and the dielectric inside the waveguide is lossless ($\sigma=0$). Then the boundary conditions are simply that $E_{tan}$ and $H_{normal}$ will be zero at the walls. Expressing Maxwell's curl equations ($M$-4, $M$-2) in component form and using $\partial/\partial z=\bar{\gamma}$ and $\partial/\partial t=j\omega$, we have

$$\frac{\partial E_z}{\partial y}+\bar{\gamma}\, E_y=-j\omega\mu H_x \tag{4.22a}$$

$$-\bar{\gamma}E_x-\frac{\partial E_z}{\partial x}=-j\omega\mu H_y \tag{4.22b}$$

$$\frac{\partial E_y}{\partial x}-\frac{\partial E_x}{\partial y}=-j\omega\mu H_z \tag{4.22c}$$

$$\frac{\partial H_z}{\partial y}+\bar{\gamma}H_y=j\omega\epsilon E_x \tag{4.22d}$$

$$-\gamma H_x-\frac{\partial H_z}{\partial x}=j\omega\epsilon E_y \tag{4.22e}$$

*In general, $\bar{\gamma}$ is different from $\gamma$ but reduces to $\gamma$ in case of uniform plane waves in free space. The $\gamma$ of transmission lines discussed in Chapter 2 also differs in many ways and will be made clear in future discussions.

$$\frac{\partial H_y}{\partial x}-\frac{\partial H_x}{\partial y}=j\omega\epsilon E_z. \tag{4.22f}$$

The wave Eqs. (3.26 and 3.27) for $E_z$ and $H_z$ become

$$\frac{\partial^2 E_z}{\partial x^2}+\frac{\partial^2 E_z}{\partial y^2}+\bar{\gamma}^2 E_z=-\omega^2\mu\epsilon E_z \tag{4.23a}$$

$$\frac{\partial^2 H_z}{\partial x^2}+\frac{\partial^2 H_z}{\partial y^2}+\bar{\gamma}^2 H_z=-\omega^2\mu\epsilon H_z. \tag{4.23b}$$

Here $E_x$, $E_y$, $H_x$ . . . are all functions of $x$ and $y$ only.

Substituting the value of $E_y$ from Eq. (4.22e) in Eq. (4.22a), we have

$$-j\omega\mu H_x=\bar{\gamma}\left\{-\frac{\bar{\gamma}H_x+\partial H_z/\partial x}{j\omega\epsilon}\right\}+\frac{\partial E_z}{\partial y}$$

or $$H_x=\frac{1}{\omega^2\mu\epsilon+\bar{\gamma}^2}\left\{j\omega\epsilon\,\frac{\partial E_z}{\partial y}-\bar{\gamma}\,\frac{\partial H_z}{\partial x}\right\}$$

or $$H_x=\frac{1}{k_c^2}\left\{j\omega\epsilon\,\frac{\partial E_z}{\partial y}-\bar{\gamma}\frac{\partial H_z}{\partial x}\right\}. \tag{4.24a}$$

Similarly, we can express other field components in terms of $E_z$ and $H_z$, i.e. the substitution of $E_x$ from Eq. (4.22d) in Eq. (4.22b) gives $H_y$, the substitution of $H_y$ from Eq. (4.22d) in Eq. (4.22b) gives $E_x$, and the substitution of $H_x$ from Eq. (4.22a) in Eq. (4.22e) gives. Thus

$$H_y=\frac{1}{k_c^2}\left\{j\omega\epsilon\,\frac{\partial E_z}{\partial x}+\bar{\gamma}\,\frac{\partial H_z}{\partial y}\right\} \tag{4.24b}$$

$$E_x=-\frac{1}{k_c^2}\left\{j\omega\mu\,\frac{\partial H_z}{\partial y}+\bar{\gamma}\,\frac{\partial E_z}{\partial x}\right\} \tag{4.24c}$$

$$E_y=\frac{1}{k_c^2}\left\{j\omega\mu\,\frac{\partial H_z}{\partial x}-\bar{\gamma}\,\frac{\partial E_z}{\partial y}\right\} \tag{4.24d}$$

where we have introduced a new constant $k_c$

$$k_c^2=k^2+\bar{\gamma}^2=\omega^2\mu\epsilon+\bar{\gamma}^2;\quad k^2=\omega^2\mu\epsilon \tag{4.25}$$

whose significance will be made clear in due course. It is noted that if $E_z=H_z=0$ all field components are zero. Hence the TEM mode cannot exist in a waveguide.

Though Eq. (4.24) are for a loss-less dielectric within the guide, the same equations are valid for a low-less dielectric but the permittivity $\epsilon$ has to be replaced by a complex permittivity $\epsilon^*$ such that

$$\epsilon^*=\epsilon\,(1+\sigma/j\omega\epsilon)=\epsilon\left(1-\frac{j\sigma}{\omega\epsilon}\right). \tag{4.26}$$

Now we turn our attention on the wave equations. We note that both electric and magnetic fields satisfy a similar wave equation of the form

$$\nabla^2\,\vec{A}\,=k^2\,\vec{A}$$

where $A$ is $E_z$ or $H_z$.

In order to solve the wave equation we decompose the Laplacian operator into transverse and axial components

$$\nabla^2 = \nabla_T^2 (x, y) + \nabla_z^2 = \nabla_T^2 + \frac{\partial^2}{\partial z^2}.$$

The wave equations (4.23a and 4.23b) can be written as

$$\nabla_T^2 E_z = -(\bar{\gamma}^2 + K^2) E_z = k_c^2 E_z \tag{4.27a}$$

$$\nabla_T^2 H_z = -(\bar{\gamma}^2 + k^2) H_z = -k_c^2 H_z. \tag{4.27b}$$

Form Eq. (4.25), we have

$$\bar{\gamma} = \sqrt{k_c^2 - k^2} = \sqrt{k_c^2 - \omega^2 \mu \epsilon}. \tag{4.28}$$

For $k_c^2 > \omega^2 \mu \epsilon$, $\bar{\gamma}$ is real and since fields vary as $e^{-\bar{\gamma} z}$, all the fields are exponentially attenuated without any propagation. On the other hand, for $k_c^2 < \omega^2 \mu \epsilon$, $\bar{\gamma}$ is imaginary and so fields propagate down the waveguide without any attenuation. In this way, a waveguide is like a high pass filter with a transition frequency $\omega_c$ from high loss to low-loss equal to $k_c / \sqrt{\mu \epsilon}$. However, we shall see later that unlike the lumped parameter, high pass filter $k_c$, may take on a number of discrete values, each for a particular waveguide mode. This constant, $k_c$, is known as the *cut-off wave number* associated with *cut-off frequency* $\omega_c$. The corresponding cut-off wavelength is given by

$$\lambda_c = \frac{2\pi}{k_c} = \frac{2\pi}{\omega_c \sqrt{\mu \epsilon}}. \tag{4.29}$$

The cut-off frequency and cut-off wavelengths are geometrical parameters depending on waveguide cross-sectional configurations. The phase constant or propagation factor for frequencies greater than cut-off is given by

$$\beta_z = \sqrt{\omega^2 \mu \epsilon - k_c^2} = \sqrt{\omega^2 \mu \epsilon - \omega_c^2 \mu \epsilon} \tag{4.30}$$

The guide wavelength is readily seen to be given by

$$\lambda_g = \frac{2\pi}{\beta_z} = \frac{\lambda}{\sqrt{1 - (\lambda_o/\lambda_c)^2}} = \frac{\lambda}{\sqrt{1 - (f_c/f)^2}}.$$

This expression is identical with Eq. (4.13).

Moreover Eq. (4.30) may be rearranged as

$$\omega \sqrt{\mu \epsilon} = \sqrt{k_c^2 + \beta_z^2}. \tag{4.31}$$

Equation (4.31) suggests that the $\omega - \beta$ plot for a waveguide is non-linear. For a waveguide operating well above cut-off $\beta_z \gg k_c$, we have

$$\frac{\omega}{\beta_z} \simeq \frac{1}{\sqrt{\mu \epsilon}} = v_p.$$

Hence the $\omega - \beta$ diagram has a linear character. But as we move near cut-off it becomes a non-linear hyperbolic relationship and the waveguide becomes a dispersive device. Figure 4.9 shows the $\omega-\beta$ diagram for a waveguide and is known as the dispersion diagram.

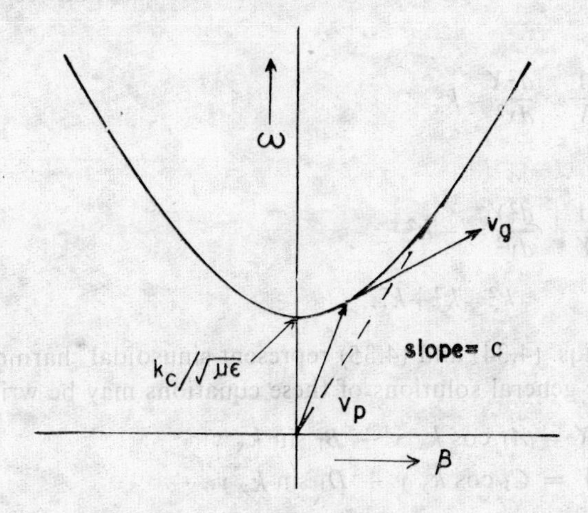


**Fig. 4.9** $\omega-\beta$ diagram for a waveguide.

## Transverse Magnetic Waves

Equations (4.24) express all the field components in terms of the axial components of electric and magnetic fields. For transverse magnetic waves $H_z=0$, hence $E_z$ has to be evaluated from Eq. (4.23a) under the boundary conditions

$$E_x=E_z=0 \quad \text{at } y=0 \quad \text{and} \quad y=b$$

$$E_y=E_z=0 \quad \text{at } x=0 \quad \text{and} \quad x=a.$$

Assuming the solution of the partial differential equation (4.23a) to be

$$E_z\,(x, y, z) = E_z^{o*}\,(x, y)\; e^{-\bar{\gamma}z}$$

to separate variables, let

$$E_z^o = X(x)\, Y(y). \tag{4.32}$$

Inserting Eq. (4.32) in Eq. (4.23), we have

$$Y\,\frac{d^2X}{dx^2}+X\,\frac{d^2Y}{dy^2}+\bar{\gamma}^2\,XY=-\,\omega^2\mu\epsilon XY.$$

Using Eq. (4.25) and dividing both sides of the above equation by $XY$ and separating variables, we have

$$\frac{1}{X}\;\frac{d^2X}{dx^2}+k_c^2=-\;\frac{1}{Y^2}\;\frac{d^2Y}{dy^2}. \tag{4.33}$$

*Suffix zero overhead implies the presence of a multiplicative factor $e^{(j\omega t-\bar{\gamma}z)}$

In Eq. (4.33) the L.H.S. in a function of $x$ alone while the R.H.S. is a function of $y$ alone. Thus, equating each of these functions to a constant we have,

$$\frac{1}{X} \frac{d^2X}{dx^2} = k_x^2 \tag{4.34}$$

and

$$\frac{1}{Y} \frac{d^2Y}{dy^2} = -k_y^2 \tag{4.35}$$

with
$$k_c^2 = k_x^2 + k_y^2. \tag{4.36}$$

Each of the Eqs. (4.34) and (4.35) represent sinusoidal harmonic variation and hence the general solutions of these equations may be written as

$$X = A_1 \cos k_x x + B_1 \sin k_x x \tag{4.37}$$

$$Y = C_1 \cos k_y y + D_1 \sin k_y y. \tag{4.38}$$

Boundary conditions require $E_y^o$ to vanish at $x=0$ for all values of $y$. This is possible when $A_1=0$. Similarly, $E_z^o$ should vanish at $y=0$ for all values $x$; this yields $C_1=0$. Hence the solution for $E_z^o$ is

$$\begin{aligned} E_z^o &= B_1 D_1 \sin k_x x \sin k_y y \\ &= A \sin k_x x \sin k_y y \end{aligned} \tag{4.39}$$

where $A=B_1D_1$ is a new constant.

Furthermore, $E_y^o$ must vanish at $x=a$ for all values of $y$. This requires

$$\sin k_x a = 0$$

or
$$k_x a = m\pi \text{ or } k_x = \frac{m\pi}{a} \tag{4.40}$$

$$m = 1, 2, 3 \ldots$$

Finally, the vanishing of $E_z^o$ at $y=b$ for all values of $x$ requires

$$\sin k_y b = 0$$

$$k_y b = n\pi \text{ or } k_y = \frac{n\pi}{b} \tag{4.41}$$

where
$$n = 1, 2, 3 \ldots$$

Therefore, the final expression for the axial component of the electric field is

$$E_z^o = A \sin\left(\frac{m\pi}{a} x\right) \sin\left(\frac{n\pi}{b} y\right) = A \sin k_x x \sin k_y y. \tag{4.42}$$

Using Eq. (4.42) and $H_z=0$ in Eq. (4.24) we obtain the following field configuration for a loss-less rectangular waveguide operating $TM_{mn}$ mode or $E_{mn}$ mode.

$$E_z^o = A \sin k_x x \sin k_y y$$

$$H_x^o = \frac{1}{k_c^2}\left(j\omega\epsilon \frac{\partial E_z^o}{\partial y}\right) = \frac{j\omega\epsilon}{k_c^2} A k_y \sin k_x x \cos k_y y$$

$$H_y^o = \frac{1}{k_c^2}\left(j\omega\epsilon \frac{\partial E_z^o}{\partial x}\right) = \frac{j\omega\epsilon}{k_c^2} A k_y \cos k_x x \sin k_y y$$

$$E_x^o = -\frac{j\beta_z}{k_c^2} \frac{\partial E_z}{\partial_x} = -\frac{j\beta_z}{k_c^2} A k_x \cos k_x x \sin k_y y \tag{4.43}$$

$$E_y^o = -\frac{j\beta_z}{k_c^2} \frac{\partial E_z}{\partial y} = -\frac{j\beta_z}{k_c^2} A k_y \sin k_x x \cos k_y y$$

where

$$k_x = \frac{m\pi}{a}, \quad k_y = \frac{n\pi}{a} \text{ and } k_c^2 = k_x^2 + k_y^2 \tag{4.44}$$

and $m$ and $n$ are integers.

The field configuration in Eq. (4.43) satisfies Maxwell's equations and the boundary conditions and hence represents the possible field configuration of a TM mode in a rectangular waveguide. However, the dimensional factors $m$ and $n$ can take infinite integral values and hence are doubly infinite modes. The integers $m$ and $n$ pertain to the number of standing wave interference maxima occurring in the field solutions that describe the variation of the fields along the two transverse coordinates. It can be inferred from Eq. (4.28) by substituting $k_c^2 = k_x^2 + k_y^2$ that each mode has an associated cut-off wavenumber, cut-off frequency and cut-off wavelength depending on the dimensions of the waveguide.

If $m=n=0$, the fields in Eq. (4.43) vanish; hence the $TM_{oo}$ mode is impossible. Further, $m=0$ or $n=0$, causing all the field to vanish and so the $TM_{on}$ and $TM_{mo}$ modes are also impossible. Moreover, integers $m$ and $n$ may have positive as well as negative integral values but the negative values do not correspond to any new solutions so these are not considered.

Equations (4.43) indicate that fields vary periodically and show standing wave patterns in the $x$- and $y$-directions. To obtain the variation of field components in the $z$-direction and with time, multiply each field component by a factor $e^{j\omega t - \bar{\gamma} z}$ and take out the real part. This will represent fields at any instant of time.

**Field Distribution of TM Waves**

Consider a $TM_{11}$ (or $E_{11}$) mode. The magnetic field is in $(x,y)$ plane and has components

$$H_x \propto \sin(\pi/a\, x) \cos(\pi/b\, y)$$

$$H_y \propto \cos(\pi/a\, x) \sin(\pi/b\, y). \tag{4.45}$$

The plot the lines of force arising from these fields draw two lines $AA$ and $BB$ at right angles passing through the centre of the waveguide cross-section as shown in Fig. 4.10(a). Since at $AA$, $y=b/2$, $H_x$ vanishes irrespective of the values of $x$ and similarly at $BB=a/2$, $H_y$ vanishes irrespective of the values of $y$. Thus, magnetic lines of force must cross $AA$ and $BB$ normally. In the rest of the waveguide cross-section both $H_x$ and $H_y$ exist and lines of force take the shape of distorted ellipses tending to become rectangular in the vicinity of the waveguide walls [ $\overrightarrow{H}_{\text{normal}} = 0$]. Furthermore, $H_x=0$ at $x=0$ and $x=a$ but is maximum at $x=a/2$; this implies half wave variation of $H_x$ along the $x$-axis. Similarly, $H_y$ has a half wave variation along the $y$-axis in the $TM_{11}$ mode. The variation of $H_x$ and $H_y$ has been shown in Fig. 4.10b.

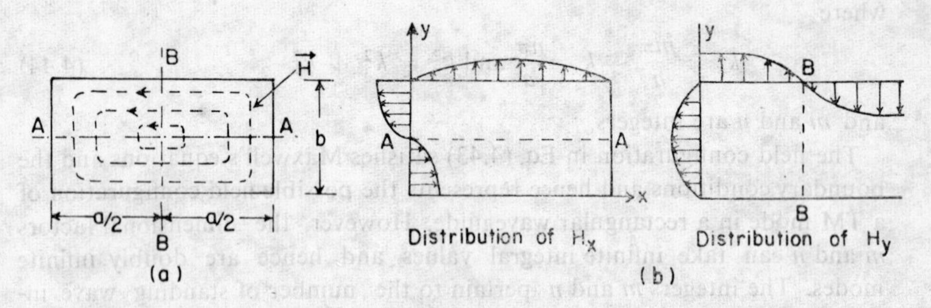


**Fig. 4.10** Diagrams of magnetic field distribution and pattern of magnetic lines of force in waveguide cross-section at $TM_{11}$ mode.

To obtain the variation in the magnetic field with $z$ at any instant of time, multiply each $H_x$ and $H_y$ with $e^{j\omega t - \bar{\gamma} z}$ and take the real part. This results in a multiplicative factor $\cos \beta_z z$. The field components periodically vary with $z$ with the distance between equiphase planes equal to $\lambda_g/2$.

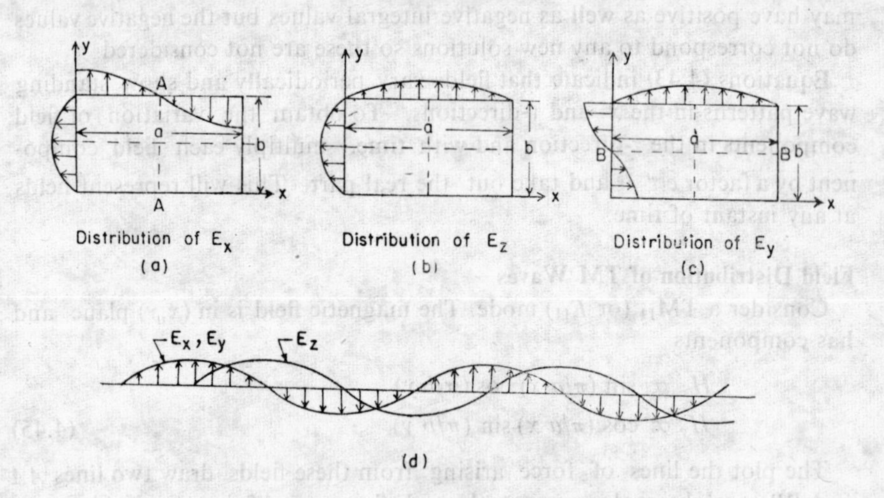


**Fig. 4.11** Diagrams of electric field distribution in transverse and longitudinal cross-section of a rectangular waveguide at $TM_{11}$ mode,

Since the electric field has all the three components its distribution has to be examined in two planes. The field components* are

$$E_x \propto \cos(\pi/a\ x)\sin(\pi/b\ y)$$
$$E_y \propto \sin(\pi/a\ x)\cos(\pi/b\ y) \quad (4.46)$$
$$E_z \propto j\sin(\pi/a\ x)\sin(\pi/b\ y)$$

The field distribution corresponding to the first two equations is shown in Fig. 4.11. (a, b, c). The presence of factor $j$ in the equation for $E_z$ indicates that the maximum of $E_z$, as distinct from the maximum of the transverse field components $E_x$ and $E_y$, is displaced along the $z$-axis by $\pi/2$, i.e. by a quarter waveguide wavelength as shown in Fig. 4.11(d). The electric lines of force arising from these fields are shown in Fig. 4.12. As desired by

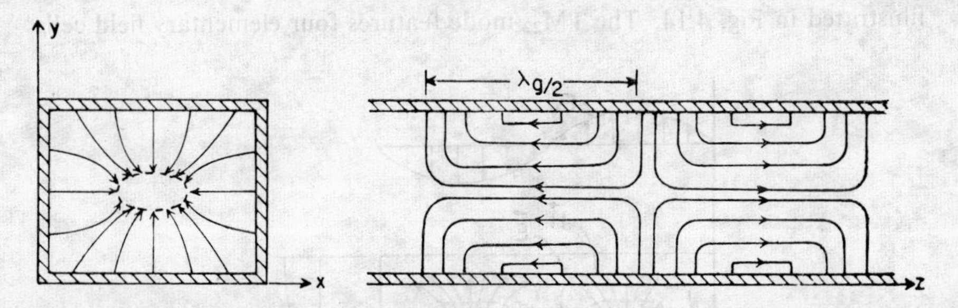


**Fig. 4.12** Electric field pattern in a rectangular waveguide at $TM_{11}$ mode.

the boundary conditions, the flux of the lines of force is almost zero near the walls and increases as the distance from the walls increases. Also the electric lines of force in sections *AA* and *BB* lie entirely in planes passing through these axis and are parallel to the *z*-axis.

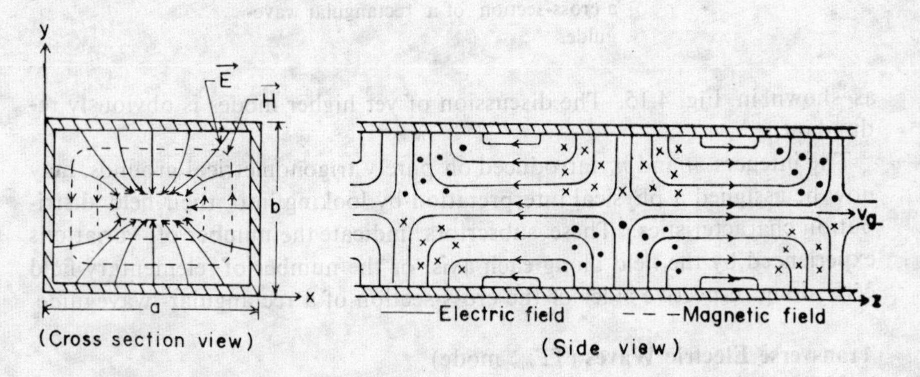


**Fig. 4.13** The composite field pattern of a rectangular waveguide at $TM_{11}$ mode.

*$j$ has been taken as common throughout.

When plotting a composite field pattern ( $\vec{E}$ and $\vec{H}$ together), it should be remembered that, as indicated in the equations for $E_x$, $H_x$, $E_y$ and $H_y$, all the transverse field components are in phase. Factor $j$, indicating a phase shift by $\pi/2$, is present only in the equation describing $E_z$. Therefore, the maximum of the transverse electric field coincides with the maximum of the transverse magnetic field. This is to be expected because the energy (poynting vector) propagates along the $z$-direction. The composite field pattern corresponding to the $TM_{11}$ mode is shown in Fig. 4.13. This field pattern moves along the waveguide axis with phase velocity $v_{pz}$ while energy propagates with group velocity.

The $TM_{21}$ mode differs from the $TM_{11}$ mode only in the presence of a second half wave in the field distribution along the $x$-axis. This means that a second elementary cell of the field is developed along dimension $a$, as illustrated in Fig. 4.14. The $TM_{22}$ mode features four elementary field cells

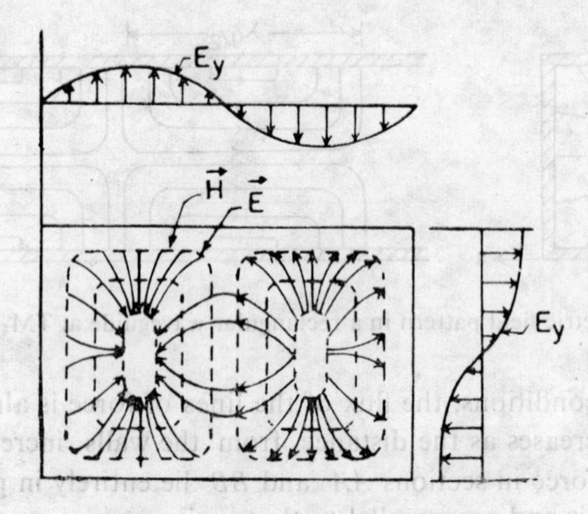


**Fig. 4.14** Field pattern of $TM_{21}$ mode in transverse a cross-section of a rectangular waveguide.

as shown in Fig. 4.15. The discussion of yet higher modes is obviously redundant.

The integers $m$ and $n$ introduced on purely trigonometrical grounds, may now be assigned a physical interpretation by looking into their field distribution characteristics. These subscripts indicate the number of variations experienced by the field along each axis, or the number of elementary field cells along respective sides of the cross-section of a rectangular waveguide.

**Transverse Electric Waves** ($TE_{mn}$ mode)

In this case the boundary conditions are that the derivative of $H_z$ vanishes at $x=0$, $x=a$, $y=0$ and $y=b$ (at metal boundaries). This leads to the values of constants $B_1$ and $D_1$ in Eqs. (4.37) and (4.38) to be zero. The general solution for the wave equation (4.27b) for $H_z$ becomes

$$H_z^o = A_1C_1 \cos k_x x \cos k_y y$$
$$= B \cos k_x x \cos k_y y. \quad (4.47)$$

$A_1C_1 = B$ is a new constant.

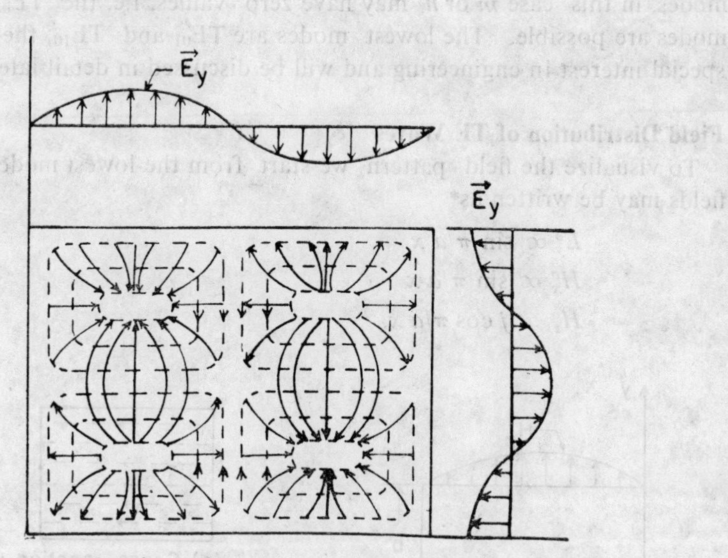


**Fig. 4.15** Fixed pattern of $TM_{22}$ mode in transverse a cross-section of a rectangular waveguide.

The vanishing of the derivative of $H_z^o$ at $x=a$ for all values of $y$ and at $y=b$ for all values of $x$, requires

$$k_x a = m\pi \text{ or } k_x = \frac{m\pi}{a}$$

$$k_y b = n\pi \text{ or } k_y = \frac{n\pi}{b}.$$

We see that the values of $k_x$ and $k_y$ are the same as that for the $TM_{mn}$ mode. Substitution of Eq. (4.47) with $E_z=0$ in Eqs. (4.24) yields the complete field configuration

$$H_z^o = B \cos k_x x \cos k_y y$$

$$H_x^o = \frac{j\beta_z B k_x}{k_c^2} \sin k_x x \cos k_y y$$

$$H_y^o = \frac{j\beta_z B}{k_c^2} k_y \cos k_x x \sin k_y y \quad (4.48)$$

$$E_x^o = \frac{j\omega\mu}{k_c^2} B k_y \cos k_x x \sin k_y y$$

$$E_y^o = \frac{j\omega\mu}{k_c^2} B k_x \sin k_x x \cos k_y y$$

with $k_x = m\pi/a$, $k_y = n\pi/b$.

Equation (4.48) reveals that in this case too there are doubly infinite modes. A particular mode may be designated as $TE_{mn}$ or $H_{mn}$. Moreover, it is easy to verify that the $TE_{oo}$ mode is impossible. However, unlike TM modes, in this case $m$ or $n$ may have zero values, i.e. the $TE_{on}$ and $TE_{mo}$ modes are possible. The lowest modes are $TE_{01}$ and $TE_{10}$; the latter is of special interest in engineering and will be discussed in detail later.

**Field Distribution of TE Waves**

To visualize the field pattern we start from the lowest mode $TE_{10}$. The fields may be written as*

$$
\begin{aligned}
E_y^o &\propto \sin \pi/a \, x \\
H_x^o &\propto \sin \pi/a \, x \\
H_z^o &\propto j \cos \pi/a \, x.
\end{aligned}
\tag{4.49}
$$

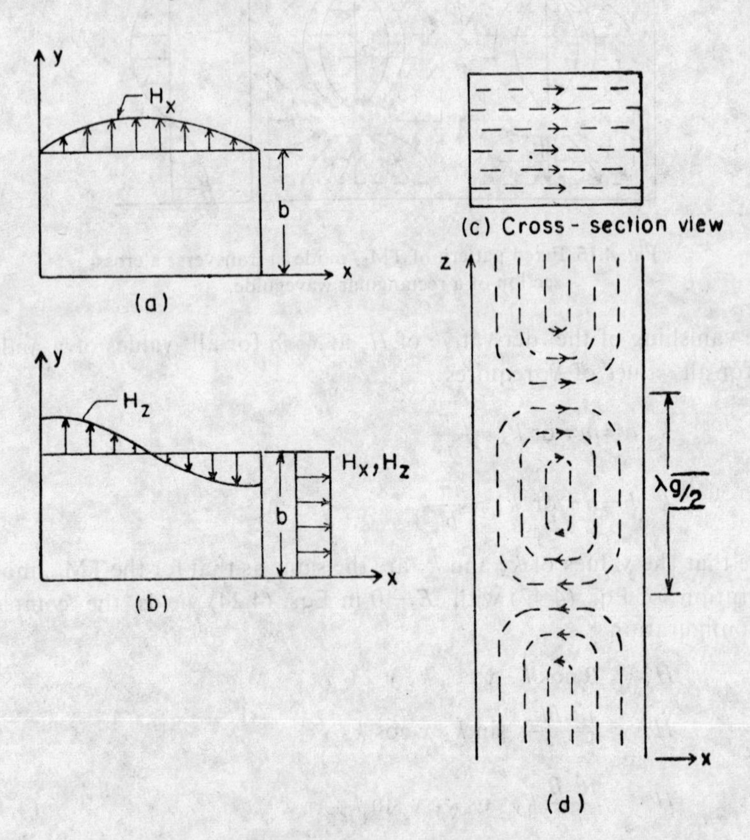


**Fig. 4.16** Magnetic field distribution of a rectangular waveguide operating on $TE_{10}$ mode.

*Constants have not been considered to avoid unnecessary complications and all equations have been multiplied by $j$ which does nothing but shift the point of observatoin in space by $\lambda_g/4$.

It is seen from Eqs. (4.49) that $E_y^o$ is independent of $y$, i.e. there is no variation in $E_y^o$ with dimension $b$ while it varies sinusoidally with dimension $a$. The field is maximum at $x=a/2$, i.e. electric lines of force have maximum density at $x=a/2$ and almost zero density at $x=0$ and $x=a$, as shown in Fig. 4.16a. Magnetic lines of force are confined to the $x-z$ plane. The locus of the magnetic field in the $xz$ plane is an ellipse.

$$\left(\frac{H_z^o}{B}\right)^2+\left(\frac{H_x^o}{\frac{jB\pi\beta_z}{k_c^2a^2}}\right)^2=1.$$

Due to the presence of $j$ in $H_z^o$, the maximum of the longitudinal component $H_z^o$ is displaced along the $z$-axis by $\lambda_g/4$ relative to the maximum of the transverse component $H_x^o$. Both $H_x^o$ and $H_z^o$ are independent of $y$. The field pattern in various planes is shown in Figs. (4.16b, c and d), and the composite field pattern is illustrated in Figs. (4.17a, b and c).

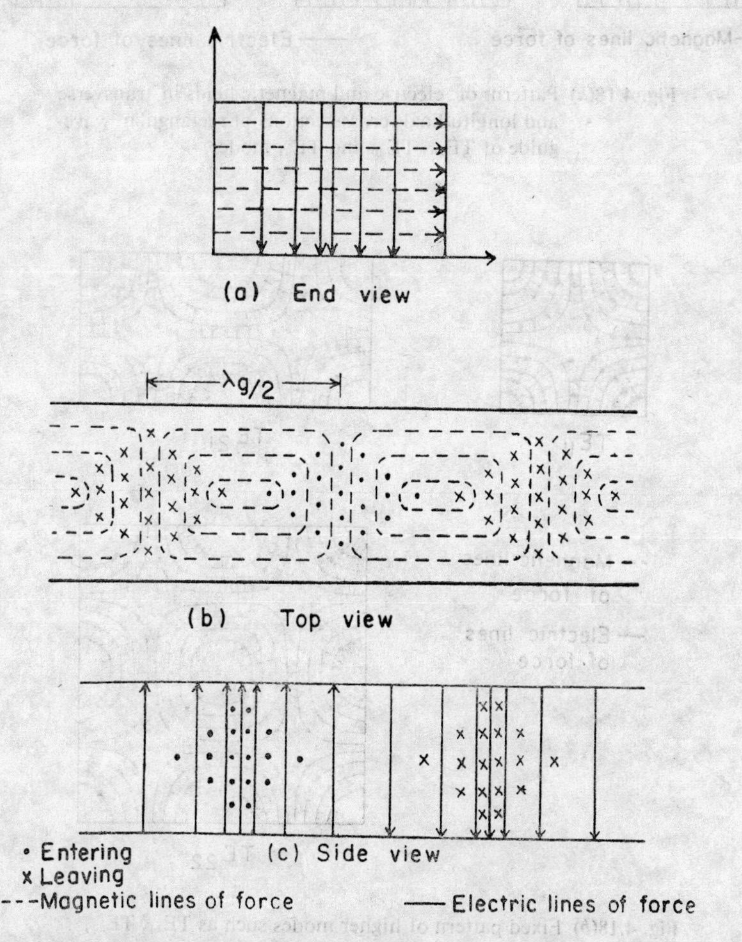


Fig. 4.17 Field pattern of $TE_{10}$ mode on a rectangular waveguide.

Higher modes such as $TE_{20}$, $TE_{30}$, etc. feature, two, three and more variations respectively, of the field along dimension $a$ of the waveguide as shown in Fig. (4.18). The $TE_{01}$ mode differs from the $TE_{10}$ mode only in the position of its plane of polarization; consequently, the order in which subscripts $m$ and $n$ occur carries a physical meaning only when the ratios of cross-sectional dimensions $a$ and $b$ is known a priori or when the system of coordinates is assumed to be positioned relative to the cross-section of the waveguide under examination.

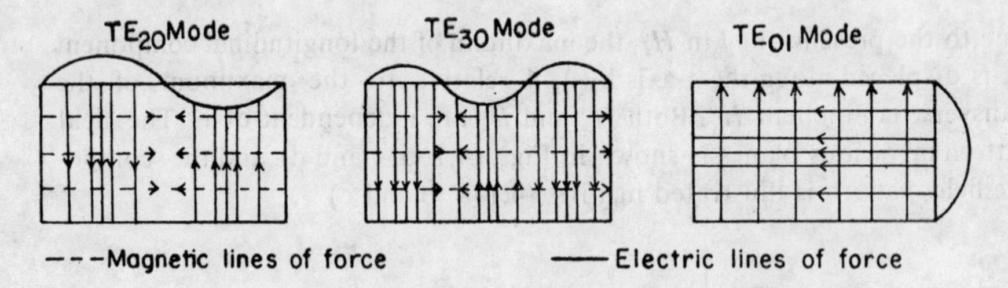


**Fig. 4.18(*a*)** Pattern of electric and magnetic fields in transverse and longitudinal cross-sections of rectangular waveguide of $TE_{20}$, $TE_{30}$ and $TE_{01}$ modes.

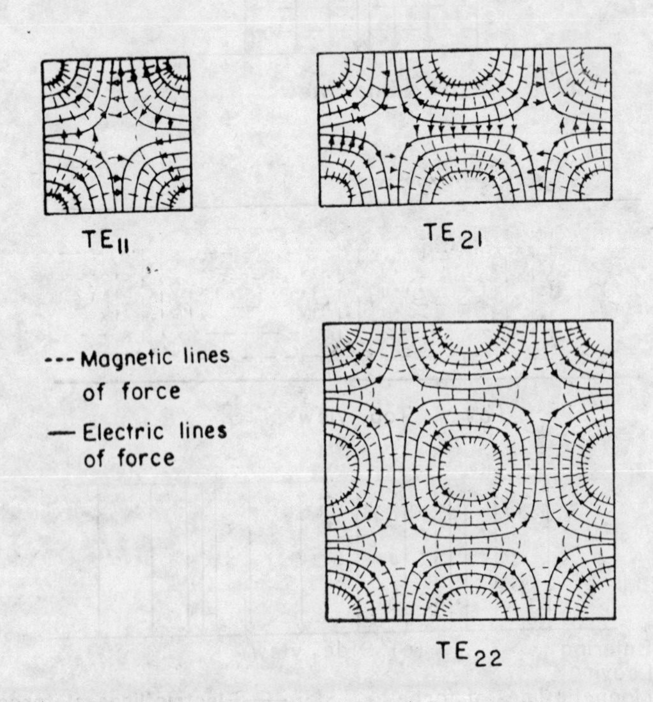


**Fig. 4.18(*b*)** Fixed pattern of higher modes such as $TE_{11}$, $TE_{21}$, and $TE_{22}$.

Higher modes $TE_{21}$, $TE_{22}$, etc. are more complex as shown in Fig. 4.19. However, these may be drawn bearing in mind the physical meaning of the subscripts $m$ and $n$. Higher modes are combinations of the simplest mode patterns.

## 4.5 CHARACTERISTICS OF TE AND TM WAVES IN RECTANGULAR WAVEGUIDES

### (i) Cut-off Frequencies

The cut-off wave number for $TE_{mn}$ and $TM_{mn}$ modes is the same and is given by Eq. (4.36)

$$k_{c,mn}^2 = k_x^2 + k_y^2 = \left(\frac{m\pi}{a}\right)^2 + \left(\frac{n\pi}{b}\right)^2.$$

The propagation constant for the $mn$th mode is

$$\bar{\gamma}_{mn} = j\bar{\beta}_{z,mn} = j\,(k^2 - k_{c,mn}^2)^{1/2}$$
$$= j\left[\left(\frac{2\pi}{\lambda}\right)^2 - \left(\frac{m\pi}{a}\right)^2 - \left(\frac{n\pi}{b}\right)^2\right]^{1/2}. \tag{4.50}$$

It may be noted from Eq. (4.50) that when $k > k_{c,mn}$, $\bar{\beta}_{z,mn}$ is pure, real and mode propagates; when $k < k_{c,mn}$, then $\bar{\gamma}_{mn}$ is real but $\bar{\beta}_{z,mn}$ is imaginary indicating thereby that mode decays rapidly with distance from the point where it is excited. It should be noted that this decay is not associated with energy loss, but is a characteristic feature of the solution and may be attributed to the local diffraction or fringing of the fields that exists in the vicinity of the coupling probes and obstacles in waveguides. The cut-off frequency separating the propagation and non-propagation bands of frequencies is given by

$$f_{c,mn} = \frac{v_p}{2\pi}\, k_{c,mn} = \frac{v_p}{2\pi}\left[\left(\frac{m\pi}{a}\right)^2 + \left(\frac{n\pi}{b}\right)^2\right]^{1/2}$$

or
$$f_{c,mn} = \frac{1}{2\sqrt{\mu\epsilon}}\left[\left(\frac{m}{a}\right)^2 + \left(\frac{n}{b}\right)^2\right]^{1/2}. \tag{4.51}$$

The correspodinng cut-off wavelength is

$$\lambda_{c,mn} = \frac{2\pi}{k_{c,mn}} = \frac{2}{\sqrt{(m/a)^2 + (n/b)^2}}$$

or
$$\lambda_{c,mn} = \frac{2\,ab}{\sqrt{(mb)^2 + (na)^2}}. \tag{4.52}$$

The typical cut-off wavelengths and frequencies for some lower modes are

$$\lambda_{c,10} = 2a, \quad f_{c,10} = \frac{1}{\sqrt{\mu\epsilon}\;2a}$$

$$\lambda_{c,01} = 2b, \quad f_{c,01} = \frac{1}{\sqrt{\mu\epsilon}\;2b}.$$

It is noted that, in general, for a waveguide $a>b$, the $TE_{10}$ mode has the largest cut-off wavelength and the lowest cut-off frequency. $TE_{10}$ mode is the *dominant mode*. For a waveguide having dimensions

$$a=2b, \quad \lambda_{c,mn}=\frac{2a}{\sqrt{m^2+4n^2}},$$

we have a band of wavelengths from $a$ to $2a$, i.e. frequency band

$$\frac{v_p}{2a}<f<v_p/a$$

($v_p$ is the characteristic velocity of waves in the medium.)

Figure 4.19 illustrates the cut-off frequency spectrum for various modes for the dimension ratios $b/a=1$ and $b/a=1/2$.

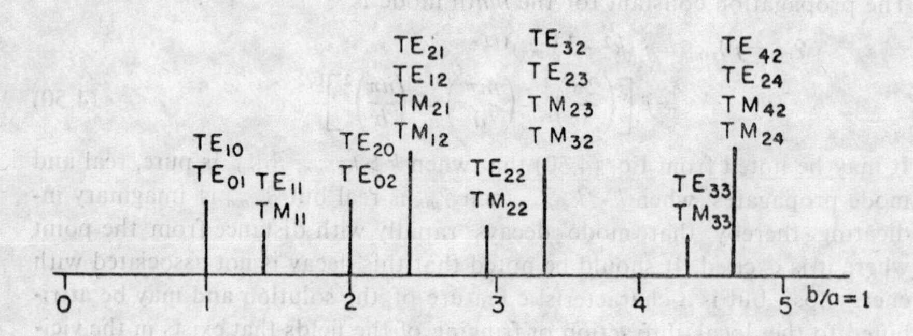


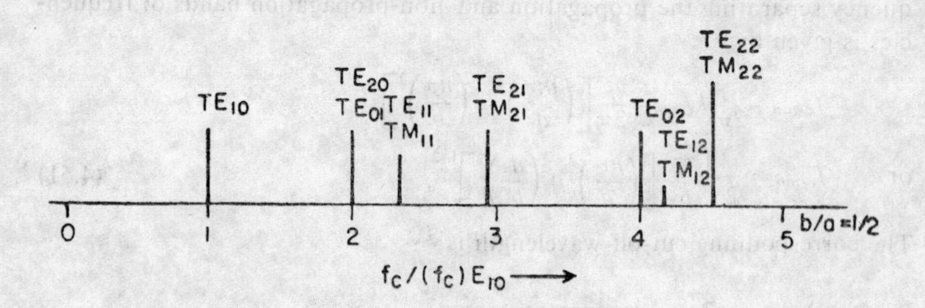


Fig. 4.19 Cut-off frequency spectrum of waves in rectangular waveguide.

***(ii)* Guide Wavelength**

The waveguide wavelength for the $TE_{mn}$ or $TM_{mn}$ mode is obtained by substituting the value of $\lambda_{c,mn}$ in Eq. (4.13)

$$\lambda_{g,mn}=\frac{2\pi}{\beta_{z,mn}}=\frac{\lambda}{\sqrt{1-\dfrac{\lambda^2(m^2b^2+n^2a^2)^{1/2}}{4\,a^2b^2}}}=\frac{\lambda}{\sqrt{1-\dfrac{\lambda^2}{\lambda_c^2}}}. \tag{4.53}$$

It may be noted that when

$$\lambda > \lambda_c = \left( \frac{2ab}{\sqrt{m^2b^2+n^2a^2}} \right)$$

propagation cannot occur because $\lambda_{g,mn}$ becomes imaginary. The variation of $\lambda_g/\lambda$ with $\lambda/\lambda_c$ is shown in Fig. 4.20 for TE and TM modes. For the TEM mode (which does not exist in waveguides) $\lambda_g=\lambda$ is also shown.

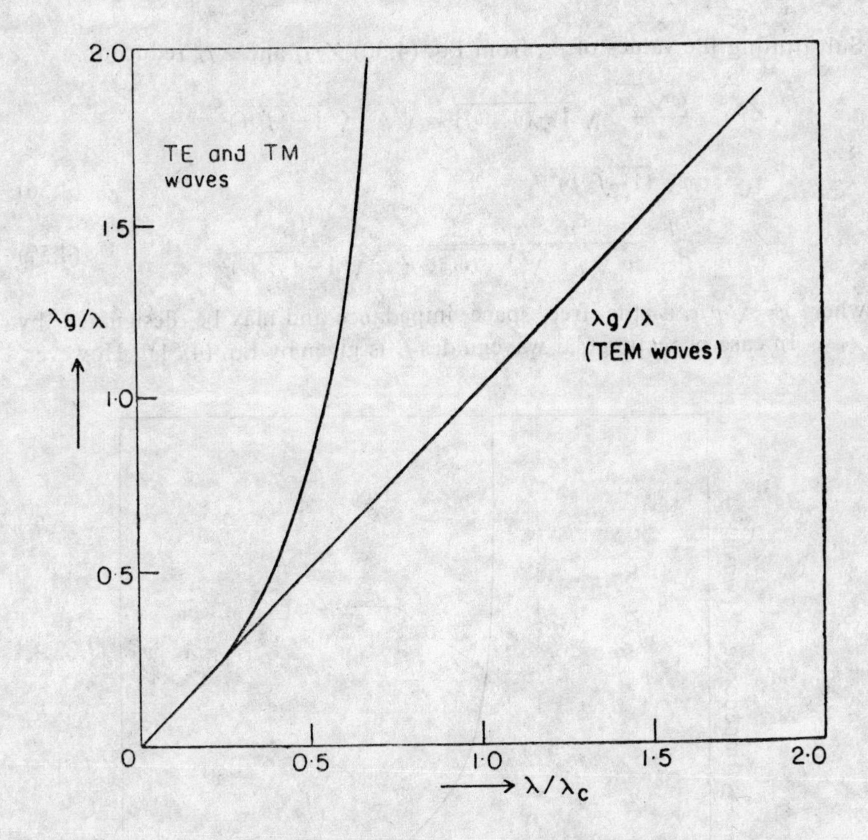


**Fig. 4.20** $\lambda_g/\lambda$ as a function of $\lambda/\lambda_c$ for a waveguide.

The *wave impedance* or characteristic impedance of a waveguide is the ratio of the transverse electric field to the transverse magnetic field. For TM waves, $Z_{TM}$ is given by

$$Z_{TM} = \frac{E_x}{H_y} = -\frac{E_y}{H_x} = \frac{\sqrt{E_x^{o2}+E_y^{o2}}}{\sqrt{H_y^{o2}+H_x^{o2}}} \tag{4.53}$$

which on substituting values from Eq. (4.43) reduces to

$$Z_{TM} = \frac{\overline{\beta}_z}{\omega\epsilon}. \tag{4.54}$$

Similarly, wave impedance for the TE mode is given by

$$Z_{\mathrm{TE}}=\frac{E_x^o}{H_y^o}=-\frac{E_y^o}{H_x^o}=\frac{\sqrt{E_x^{o2}+E_y^{o2}}}{\sqrt{H_x^{o2}+H_y^{o2}}}$$

which on using Eqs. (4.48) gives

$$Z_{\mathrm{TE}}=\frac{\omega\mu}{\beta_2}. \tag{4.55}$$

Substituting the values of $\bar{\beta}_z$, from Eq. (4.30) $Z_{TM}$ and $Z_{TE}$ reduce to

$$Z_{\mathrm{TM}}=\frac{\omega\sqrt{\mu\epsilon}}{\omega\epsilon}\sqrt{1-(\omega_c/\omega)^2}=\sqrt{\mu/\epsilon}\sqrt{1-(f_c/f)^2}$$

$$=\eta\sqrt{(1-f_c/f)^2}. \tag{4.56}$$

$$Z_{\mathrm{TE}}=\frac{\omega\mu}{\beta_z}=\frac{\omega\mu}{\omega\sqrt{\mu\epsilon}\sqrt{1-(\omega_c/\omega)^2}}=\frac{\eta}{\sqrt{1-(f_c/f)^2}} \tag{4.57}$$

where $\eta=\sqrt{\mu/\epsilon}$ is the free space impedance and may be designated by $Z_{\mathrm{TEM}}$. In case of rectangular waveguides $f_c$ is given by Eq. (4.51). However,

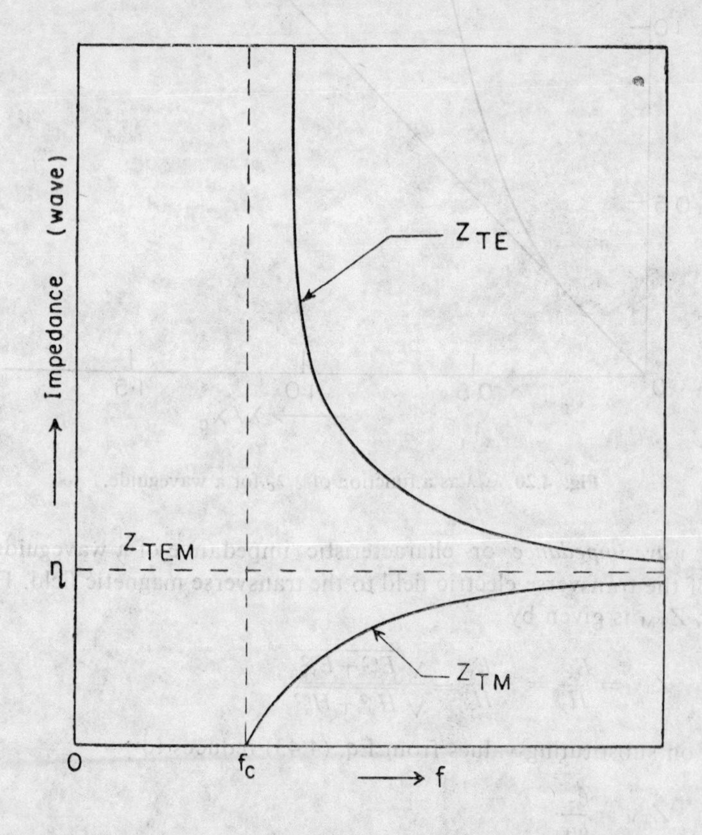


Fig 4.21 Wave impedance $Z_{\mathrm{TE}}$ and $Z_{\mathrm{TM}}$ as a function of frequency.

$Z_{TE}$ and $Z_{TM}$ expressed in Eqs. (4.56) and (4.57) are most general and applicable to any of the waveguides. Figure 4.21 illustrates variation of $Z_{TE}$ and $Z_{TM}$ with the frequency ratio $f_c/f$. It is noted that

$$\sqrt{Z_{TE}\,Z_{TM}}=\eta=(Z_{TEM}). \tag{4.58}$$

## 4.6 CURRENTS IN WAVEGUIDE WALLS

From the general engineering concept it can easily be inferred that the inner surfaces of waveguide walls carry radio frequency currents whose pattern is closely related to the field distribution inside the waveguide. It can be shown by applying Ampere's m.m.f. equation on a small rectangle *abcd* shown in Fig. 4.22, whose width in the metal exceeds the skin depth, that the density of the surface current at every point of the inner surface of a waveguide wall is numerically equal to the intensity of the tangential magnetic field at this point.

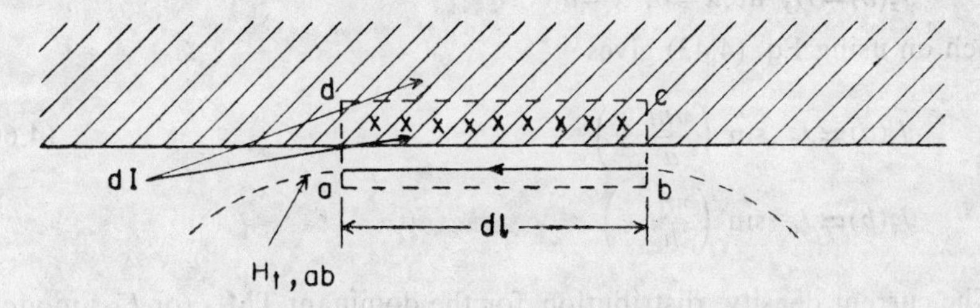


**Fig. 4.22** Determination of surface current in a waveguide wall.

$$dI=|H_{t,ab}-H_{t,cd}|\,dl=|H_{t,ab}|\,dl \text{ (because } |H_{t,cd}|=0\text{; depth exceeds skin depth)}$$

or

$$J=\frac{dI}{dl}=|H_{t,ab}|=|H_{\tan}|. \tag{4.59}$$

From the right-hand screw rule, the longitudinal current density $J_z$ is due to the transverse components of magnetic fields $H_x$ and $H_y$, and vice versa i.e. the transverse currents $J_x$ and $J_y$ are developed by longitudinal components of $H_z$ of the magnetic field.

The knowledge of the direction and magnitude of surface currents in waveguide walls is important for solving many practical problems in microwave engineering. Firstly, the knowledge of wall currents is very useful for calculating the active losses arising from the finite conductivity of waveguide walls. Secondly, in many microwave circuit components such as directional couplers, slotted sections, coupling networks and filters, etc. slots and holes need to be drilled. These slots and holes cause energy radiation into the surrounding space if they disturb the current paths in the waveguide walls.

In certain cases such as directional couplers, these effects can be useful to couple the power. Slots and holes in the case of directional couplers are drilled so as to cut *maximum* current paths. However, in other components such as slotted sections this radiation is highly undesirable since it can disturb the surface currents in the waveguide walls and produce reflections in the wave being propagated. Thus, the slot in the slotted section should be made so as to disrupt *minimum* surface current.

**Current Distribution in the $TM_{mn}$ Mode**

Since in this case, there is no longitudinal (or axial) component of the magnetic field, there will be no transverse currents in the waveguide walls, i.e.

$$J_x = J_y = 0.$$

The longitudinal currents in broad ($a$) and narrow walls respectively are given by Eq. (4.59)

$$J_z(a) = H_x \text{ at } y=0,\ y=b$$

$$J_z(b) = H_y \text{ at } x=0,\ x=a$$

which on using Eq. (4.43) gives

$$J_z(a) \simeq J_o \sin\left(\frac{\pi m}{a} x\right) \tag{4.60}$$

$$J_z(b) \simeq J_o \sin\left(\frac{\pi n}{b} x\right).$$

The current density distribution for the dominant $TM_{11}$ (or $E_{11}$) mode is illustrated in Figs. (4.23a, b and c). The direction of current is the same in every cross-section of the waveguide. The line lengths indicate the magnitude of current density and are drown at any instant of time.

The ribs of the waveguide operating on the $TM_{11}$ mode do not carry any current and hence a good electrical contact is not needed. It can be seen from the current distribution that the propagation of $TM_{mn}$ modes remain undisturbed if a narrow slot is cut lengthwise. On the other hand, any transverse slot will produce significant distortions of the propagating fields.

The current distributions for the higher modes may be determined in a similar way.

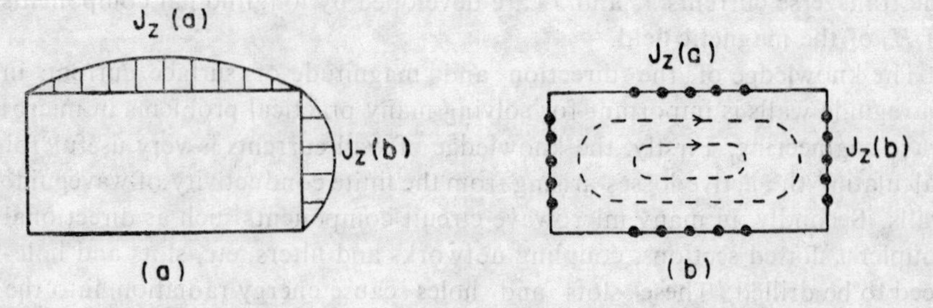


**Fig. 4.23**

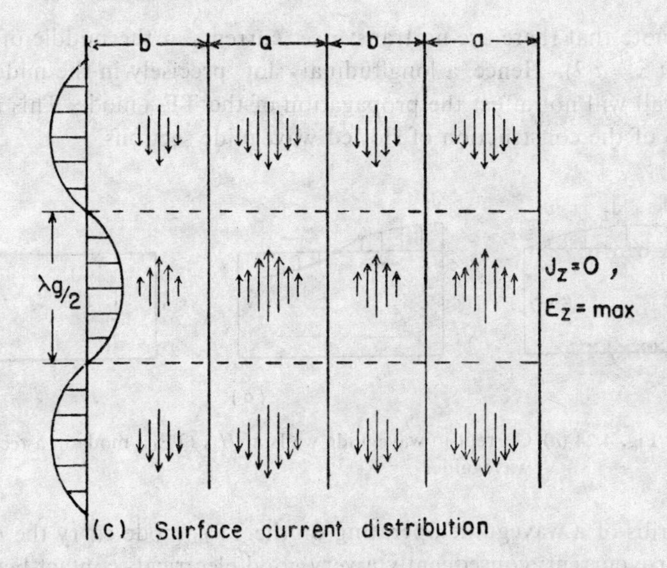


**Fig. 4.23** *(a, b & c)* Current distribution in waveguide walls at $TE_{11}$ mode *(a)* and *(b)* current in walls.

## Current Distribution in the $TE_{mn}$ Mode

In the $TE_{mn}$ mode, we note that since, in general, all the three components of the magnetic field are present, there must be transverse as well as longitudinal currents in the waveguide walls. Following Eq. (4.59), the transverse current density is given by

$$J_x(a)=H_z \text{ at } y=0,\ y=b$$
$$J_y(b)=H_z \text{ at } x=0\ x=a$$

which on using Eq. (4.48) and the right-hand screw rule reduces to the following form for the dominant $TE_{10}$ mode.

$$J_{zx}=J_z(a)=\pm\frac{j\beta_z{}^{\beta}}{k_c^2}\frac{\pi}{a}\sin\left(\frac{\pi x}{a}\right)$$
$$J_{zy}=J_z(b)=0$$
$$J_{xx}=J_x(a)=\pm B\cos\left(\frac{\pi x}{a}\right) \tag{4.61}$$
$$J_{yy}=J_y(b)=B.$$

Figure 4.24(a) illustrates the current distribution in the waveguide cross-section. The narrow waveguide wall does not carry any longitudinal current and the transverse current in this wall is constant in any fixed cross-section of the waveguide. Any longitudinal slot in the narrow wall will therefore inevitably be a source of considerable distortion of field at the $TE_{10}$ mode. On contrary, a narrow transverse slot in this wall should not cause any disturbance to the propagating mode.

We note that there are no transverse currents at the middle of the wide wall (at $x=a/2$). Hence, a longitudinal slot precisely in the middle of the wide wall will not affect the propagation of the $TE_{10}$ mode. This is an illustration of the construction of slotted waveguide sections.

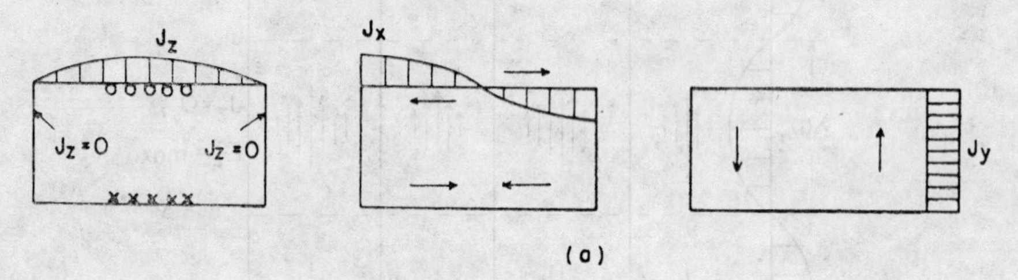


**Fig. 4.24** (*a*) Current in waveguide walls at $H_{10}$ ($TE_{10}$) mode in a rectangular waveguide.

The ribs of a waveguide operating on the $TE_{10}$ mode carry the maximum transverse current; consequently a very good electrical contact between the waveguide walls at the ribs has to be ensured, if the waveguide is fabricated from separate strips of metal.

The instantaneous current density distribution in a rectangular waveguide operating on the $TE_{10}$ mode is shown in Fig. 4.24(b) where waveguide walls are mapped into planes.

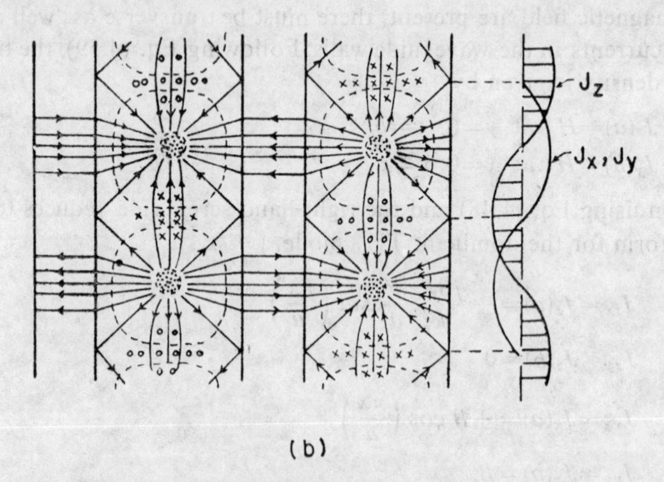


**Fig. 4.24**(*b*) Current distribution in rectangular waveguide for a travelling wave.

For higher $TE_{mn}$ modes, the current distribution may similarly be obained.

The maximum current density which depends upon various design factors determines the power handling capacity of the waveguide operating on a given mode.

## 4.7 DOMINANT $TE_{10}$ MODE IN RECTANGULAR WAVEGUIDES

Since $TE_{10}$ is the dominant and most commonly used mode we summarize below some of its important characteristics.

The field configuration for this mode is obtained by substituting $m=1$ and $n=0$ in Eq. (4.48) (propagation factor $e^{j(\omega t-\bar{\beta}_z z)}$ is also written

$$E_x=H_y=E_z=0$$

$$H_z=B\cos\left(\frac{\pi}{a}x\right)e^{j(\omega t-\bar{\beta}_z z)}$$

$$H_x=\frac{j\bar{\beta}_z B}{k_c^2}\left(\frac{\pi}{a}\right)\sin\left(\frac{\pi}{a}x\right)e^{j(\omega t-\bar{\beta}_z z)} \tag{4.62}$$

$$E_y=-\frac{j\omega\mu}{k_c^2}B\left(\frac{\pi}{a}\right)\sin\left(\frac{\pi}{a}x\right)e^{j(\omega t-\bar{\beta}_z z)}$$

where the cut-off wave number is $k_c=\pi/a$.

The cut-off wavelength and cut-off frequency respectively are given by

$$\lambda_c=\frac{2\pi}{k_c}=\frac{2\pi a}{\pi}=2a \tag{4.63}$$

$$f_c=\frac{v_o}{\lambda_c}=\frac{1}{2a\sqrt{\mu\epsilon}}.$$

The electric field is the gradient of potential so the peak voltage $V_o$ is $b$ times the peak electric field which is at $x=a/2$. Hence, barring the time part,

$$V_o=b\,[E_y]_{x=a/2}=\frac{jb\omega\mu Ba}{\pi}e^{j\bar{\beta}_z z}. \tag{4.64}$$

The total longitudinal current in either the top or bottom face of the wave-guide as shown in Fig. (4.25) is given by

$$I=-\int_o^a H_x dx$$

(minus sign appears because waves assumed to be propagating in the $-z$-direction)

$$=-\frac{1}{Z_{TE}}\int_o^a E_y dx$$

$$=-\frac{1}{Z_{TE}}\int_o^a \frac{-j\omega\mu B}{(\pi/a)^2}(\pi/a)\sin\frac{\pi}{a}x\,e^{j\bar{\beta}_z z}$$

$$=\frac{1}{Z_{TE}}\left(\frac{j\omega\mu B}{\pi^2/a^2}\right)2a^2\,e^{j\bar{\beta}_z}$$

or

$$I=\frac{2\,j\omega\mu Ba^2}{Z_{TE}\,\pi^2}e^{j\bar{\beta}z} \tag{4.65}$$

We note that the current is a function of the broader dimension $a$ only.

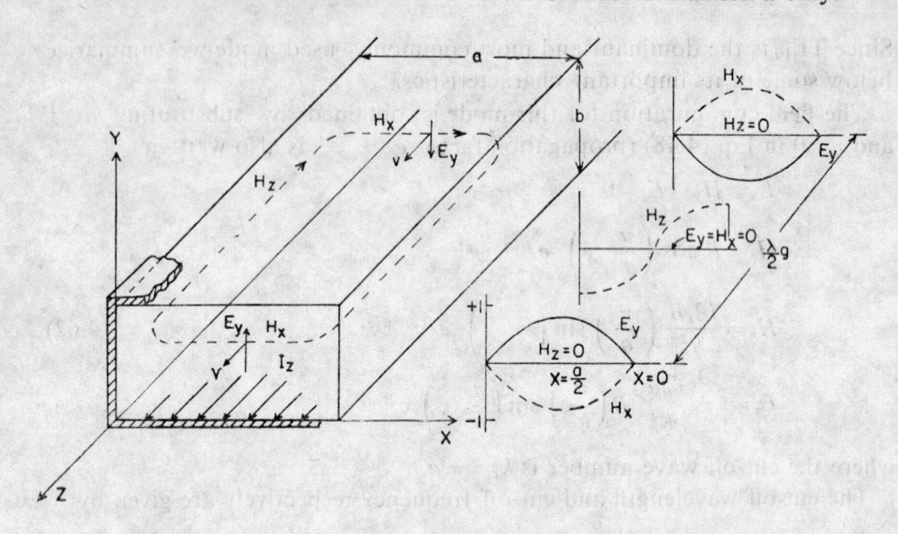


**Fig. 4.25** Longitudinal currents and relative magnitude directions of intensity components $\vec{E}$ and $\vec{H}$ in a hollow rectangular waveguide.

The power flowing through the waveguide in the $TE_{10}$ mode in a rectangular waveguide is given by

$$P_T = \frac{1}{2} \operatorname{Re} \oint_S (\vec{E} \times \vec{H}^*) \cdot \vec{ds}$$

$$= \frac{1}{2} \int_0^a \int_0^b H_x E_y \, dx dy.$$

Using $Z_{TE} = E_y / H_x$,

$$P_T = \frac{ab\ \omega^2 \mu^2 B^2}{4\ Z_{TE} (\pi/a)^2}. \tag{4.66}$$

We have defined wave impedances $Z_{TE}$ and $Z_{TM}$ using transverse field components. Schelkunoff* gives three different ways in which characteristic impedances for the waveguides may be defined. For the $TE_{10}$ mode

$$Z_{V,I} = \frac{V_0}{I} = \frac{\pi}{2} \frac{b}{a} \frac{E_y}{H_x}$$

$$Z_{P,V} = \frac{V_0^2}{2P} = 2 \frac{b}{a} \frac{E_y}{H_x} \tag{4.67}$$

$$Z_{P,I} = \frac{2P}{I^2} = \frac{\pi^2}{8} \frac{b}{a} \frac{E_y}{H_x}.$$

*S.A. Schel Kunoff, *Electromagnetic Waves*, New York; D. Van Nostrand Co. (1943).

where $V_0$ is the maximum of peak voltage, $I$ is the total longitudinal current and $P$ is power. It is noted that all the three definitions for characteristic impedance reveal that it is proportional to $E_y/H_x$ or the wave impedance. Hence $Z_{V,I}$, $Z_{P,V}$ and $Z_{P,I}$ differ only in multiplicative constants.

The $Z_{P,V}$ characteristic impedance is very close to the experimental results and hence may be used to represent characteristic impedance of the rectangular waveguide operating in the $TE_{10}$ mode.

Other important parameters are

$$v_{pz}=\frac{1}{\sqrt{\mu\epsilon}}\frac{1}{\sqrt{1-(\lambda/\lambda_c)^2}}=\frac{1}{\sqrt{\mu\epsilon}}\frac{1}{\sqrt{1-(\lambda/2a)^2}}$$

$$v_g=\frac{1}{\sqrt{\mu\epsilon}}\sqrt{1-\left(\frac{\lambda}{\lambda_c}\right)^2}=\frac{1}{\sqrt{\mu\epsilon}}\sqrt{1-(\lambda/2a)^2} \tag{4.68}$$

$$Z_{TE}=\frac{\eta}{\sqrt{1-(\lambda/\lambda_c)^2}}=\frac{\eta}{\sqrt{1-(\lambda/2a)^2}}.$$

It will be seen in Section (4.11) that the $TE_{10}$ mode besides having simple and useful field distribution, has minimum attenuation arising from various factors.

## 4.8 CYLINDRICAL WAVEGUIDES OF CIRCULAR CROSS-SECTION

We now consider the cylindrical waveguide of a circular cross-section arranged in a coordinate system as shown in Fig. 4.26. Assuming the variation of

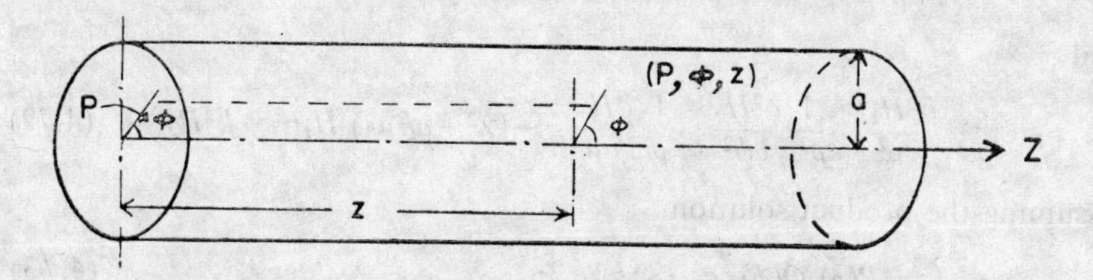


**Fig. 4.26** Cylindrical waveguide and cylindrical coordinate system.

fields as $e^{-\bar{\gamma}z}$, Maxwell's equations in cylindrical coordinates $(\rho, \phi z)$ may be written as

$$\frac{\partial H_z^o}{\rho\partial\phi}+\bar{\gamma}H_\phi^o=j\omega\epsilon E_\rho^o \tag{4.69a}$$

$$\frac{\partial E_z^o}{\rho\partial\phi}+\bar{\gamma}E_\phi^o=-j\omega\mu H_\rho^o \tag{4.69b}$$

$$-\bar{\gamma}H_\rho^o-\frac{\partial H_z^o}{\partial\rho}=j\omega\epsilon E_\phi^o \tag{4.69c}$$

$$-\bar{\gamma}E_\rho^o-\frac{\partial E_z^o}{\partial \rho}=-j\omega\mu E_\phi^o \tag{4.69d}$$

$$\frac{1}{\rho}\left\{\frac{\partial\,(\rho H_\phi^o)}{\partial \rho}-\frac{\partial H_\rho^o}{\partial \phi}\right\}=j\omega\epsilon H_z^o \tag{4.69e}$$

$$\frac{1}{\rho}\left\{\frac{\partial\,(\rho E_\phi^o)}{\partial \rho}-\frac{\partial E_\rho^o}{\partial \phi}\right\}=-\,j\omega\mu H_z^o. \tag{4.69f}$$

Solving these Eq. (4.69) for transverse components in terms of axial components we obtain,

$$H_\rho^o=\frac{1}{k_c^2}\left(\frac{j\omega\epsilon}{\rho}\frac{\partial E_z^o}{\partial \phi}-\bar{\gamma}\frac{\partial H_z^o}{\partial \rho}\right) \tag{4.70a}$$

$$H_\phi^o=-\frac{1}{k_c^2}\left(j\omega\epsilon\frac{\partial E_z^o}{\partial \rho}+\frac{\bar{\gamma}}{\rho}\frac{\partial H_z^o}{\partial \phi}\right) \tag{4.70b}$$

$$E_\rho^o=-\frac{1}{k_c^2}\left(\bar{\gamma}\frac{\partial E_z^o}{\partial \rho}+\frac{j\omega\mu}{\rho}\frac{\partial H_z^o}{\partial \phi}\right) \tag{4.70c}$$

$$E_\phi^o=\frac{1}{k_c^2}\left(j\omega\mu\frac{\partial H_z^o}{\partial \rho}-\frac{\bar{\gamma}}{\rho}\frac{\partial E_z^o}{\partial \phi}\right) \tag{4.70d}$$

where $\quad k_c^2=\bar{\gamma}^2+\omega^2\mu\epsilon=\bar{\gamma}^2+k^2.$ (4.70e)

The wave equations for $E_z$ and $H_z$ components are

$$\nabla_T^2\,(\rho,\phi)\,(E_z \text{ or } H_z)+\frac{\partial^2}{\partial z^2}(E_z \text{ or } H_z)=-\,\omega^2\mu\epsilon\,(E_z \text{ or } H_z)$$

that is,

$$\frac{\partial^2 E_z}{\partial \rho^2}+\frac{1}{\rho^2}\frac{\partial^2 E_z}{\partial \phi^2}+\frac{1}{\rho}\frac{\partial E_z}{\partial \rho}=-\,(\bar{\gamma}^2+\omega^2\mu\epsilon)\,E_z=-\,k_c^2E_z \tag{4.71}$$

and

$$\frac{\partial^2 H_z}{\partial \rho^2}+\frac{1}{\rho^2}\frac{\partial^2 H_z}{\partial \phi^2}+\frac{1}{\rho}\frac{\partial H_z}{\partial \rho}=-\,(\bar{\gamma}^2+\omega^2\mu\epsilon)\,H_z=-\,k_c^2H_z. \tag{4.72}$$

Assuming the product solution

$$E_z=P(\rho)\,Q(\phi),\,e^{-\bar{\gamma}z}=E_z^o\,e^{-\bar{\gamma}z}. \tag{4.73}$$

Substituting Eq. (4.73) in Eq. (4.71) and dividing by $E_z=P(\rho)\;Q(\phi)$ we obtain an equation in which variables $P(\rho)$ and $Q(\phi)$ are separated

$$\frac{1}{\rho}\frac{d^2P}{d\rho^2}+\frac{1}{\rho P}\frac{dP}{d\rho}+\frac{1}{Q\rho^2}\frac{d^2Q}{d\phi^2}+k_c^2=0. \tag{4.74}$$

As before, Eq. (4.74) may be broken into two ordinary differential equations

$$\frac{d^2Q}{d\phi^2}=-\,n^2Q \tag{4.75}$$

$$\frac{d^2P}{d\rho^2}+\frac{1}{\rho}\frac{dP}{d\rho}+\left(k_c^2-\frac{n^2}{\rho^2}\right)P=0 \tag{4.76}$$

where $n$ is a constant. The general solution of the harmonic equation (4.75) is

$$Q = A_n \cos n\phi + B_n \sin n\phi. \tag{4.77}$$

Dividing Eq. (4.76) throughout by $\rho^2$ reduces it to the standard form of Bessel's equation in $\rho k_c$

$$\frac{d^2P}{d(\rho k_c)^2} + \frac{1}{(\rho k_c)}\frac{dP}{d(\rho k_c)} + \left\{1 - \frac{n^2}{\rho^2 k_c^2}\right\} P = 0. \tag{4.78}$$

The solution of Eq. (4.78) is expressed in terms of the Bessel functions of the first and second kind*

$$P = A_2 J_n(\rho k_c) + B_2 N_n(\rho k_c) \tag{4.79}$$

where $J_n(\rho k_c)$ is the Bessel function of the first kind and $n^{th}$ order and $N_n$ is the Bessel function of the second kind and $n^{th}$ order. Since Bessel and Neumann functions are the real and imaginary parts of Hankel functions

$$H_n^{(1)}(\rho k_c) = J_n(\rho k_c) + jN_n(\rho k_c)$$

$$H_n^{(2)}(\rho k_c) = J_n(\rho k_c) - jN_n(\rho k_e)$$

where $H_n^{(1)}(\rho k_c)$ and $H_n^{(2)}(\rho k_c)$ are the Hankle functions of the first and second kind and order $n$. The general solution of the Eq. (4.78) may be written as

$$P = A_2 J_n(\rho k_c) + B_2 H_n^{(1)}(\rho k_c).$$

Substitution of Eqs. (4.77) and (4.80) in Eq. (4.73) yields

$$H_z^o \text{ or } E_z^o = \{A_2 J_n(\rho k_c) + B_2 H_n^{(1)}(\rho k_c)\} \times \{A_n \cos n\phi + B_n \sin n\phi\}. \tag{4.80}$$

Since $E_z^o$ and $H_z^o$ obey similar wave equations so Eq. (4.80) is also the solution for $H_z^o$.

All the functions used here have been tabulated by Jahnke and Emde**

## 4.9 CHARACTERESTICS OF TM AND TE WAVES IN CIRCULAR GUIDES

As in the case of rectangular waveguides, we divide the possible solutions in TE and TM modes.

*Bessel functions of the second kind are also known as Neumann functions denoted by $N_n$.

**E. Jahnke and F. Emide, *Tables of Functions*, New York: Dover Publications, (1943).

For TE waves, $E_z$ is *identically zero* so $H_z^o$ has to be considered in Eq. (4.80).

The value $\rho=0$ implies a point on the axis of the guide, i.e. in the dielectric region where there is a wave propagation, hence the Hankle function cannot be the solution because $N_n(o)=\infty$. So the solution for $H_z^o$ in this case is

$$H_z^o = J_n(\rho k_c)\{A_n \cos n\phi + B_n \sin n\phi\}\, e^{-\bar{\gamma} z}.$$

Moreover, amplitude constants $A_n$ and $B_n$ only determine the orientation of the fields in the guide and so any one of them may be put to zero. In wave guides with $\phi=0$ as axis the use of factors cos $n\phi$ or sin $n\phi$ becomes a matter of choice. Thus,

$$H_z^o = A_n J_n(k_c \rho) \cdot {}^{\cos}_{\sin}\, n\phi\, e^{-\bar{\gamma} z} \tag{4.81a}$$

Other field components may easily be obtained by substituting Eq. (4.81a) in Eq. (4.70) bearing in mind that $E_z^o=0$ for the TE mode.

$$H_\rho^o = \frac{1}{k_c^2}\left[-\bar{\gamma}\frac{\partial}{\partial \rho}\left\{A_n J_n(\rho k_c)\cdot {}^{\cos}_{\sin}\, n\phi\right\}\right]$$

or

$$H_\rho^o = -\frac{j\beta_z}{k_c} A_n J_n'(\rho k_c) \cdot {}^{\cos}_{\sin}\, n\phi \tag{4.81b}$$

$$H_\phi^o = -\frac{j\beta_z n A_n}{\rho k_c^2} J_n(\rho k_c) \cdot {}^{-\sin}_{\cos}\, n\phi \tag{4.81c}$$

$$E_\rho^o = -\frac{1}{k_c^2}\frac{j\omega\mu}{\rho}\frac{\partial}{\partial \phi}\left\{A_n J_n(\rho k_c) {}^{\cos}_{\sin}\, n\phi\right\}$$

$$= \frac{\omega\mu}{\beta_z} H_\phi^o = Z_{TE} H_\phi^o \tag{4.81d}$$

and

$$E_\phi^o = \frac{1}{k_c^2}\left(-\frac{j\beta_z}{\rho}\frac{\partial E_z^o}{\partial \phi}\right) = -\frac{\omega\mu}{\beta_z} H_\rho^o = -Z_{TE}\, H_\rho^o \tag{4.81e}$$

where

$$Z_{TE} = \frac{E_\rho^o}{H_\phi^o} = -\frac{E_\phi^o}{H_\rho^o} = \frac{\omega\mu}{\beta_z} \tag{4.82}$$

is the *transverse wave impedance* of the circular waveguide operating on the TE mode. This expression is the same as that for rectangular waveguides. However, the value of $\beta_z$ is different in the two cases. The qualitative behaviour of $Z_{TE}$ in the two cases will be the same.

For TM waves, $H_z^o=0$, so, we have

$$E_z^o = B_n J_n(\rho k_c) \cdot {}^{\cos}_{\sin}\, n\phi \tag{4.83a}$$

Substitution of Eq. (4.83a) in Eq (4.70) with $H_z=0$ gives us

$$H_\rho^o = \frac{jB_n\omega\epsilon n}{\rho k_c^2} J_n(\rho k_c) \cdot {}^{-\sin}_{\cos} n\phi \tag{4.83b}$$

$$H_\phi^o = -\frac{jB_n\omega\epsilon}{k_c} J_n'(\rho k_c) \cdot {}^{\cos}_{\sin} n\phi \tag{4.83c}$$

$$E_\rho^o = \frac{\overline{\beta}_z}{\omega\epsilon} H_\phi^o = Z_{TM} H_\phi^o \tag{4.83d}$$

$$E_\phi^o = -\frac{\overline{\beta}_z}{\omega\epsilon} H_\rho^o = -Z_{TM} H_\phi^o \tag{4.83e}$$

where

$$Z_{TM} = \frac{\overline{\beta}_z}{\omega\epsilon} = \frac{E_\rho^o}{H_\phi^o} = -\frac{E_\varphi^o}{H_\rho^o} \tag{4.84}$$

is the transverse wave impedance of the cylindrical waveguide operating on the TM mode. It is noted that

$$Z_{TM} \times Z_{TE} = \frac{\mu}{\epsilon} = \eta^2$$

or

$$\eta = \sqrt{Z_{TM} \times Z_{TE}}.$$

**Cut-off Properties**

Consider the case of TM waves. From boundary conditions, $E_z^o$ must vanish at the conductor, surface, i.e. at $\rho=a$. This yields

$$J_n(k_c a) = 0. \tag{4.85}$$

$J_n(k_c a)$ behaves like a damped sinusoidal function and passes through zero in a quasi-periodic fashion. Thus Eq. (4.85) may be satisfied for infinite number of values of $k_c$ for which $J_n(k_c a)$ vanishes and consequently we have infinite propagating modes. Let the $m^{th}$ root of Eq. (4.85) be designated by $p_{nm}$; the allowed values (eigen values) of $k_c$ are

$$k_{c,nm} = \frac{p_{nm}}{a}. \tag{4.86}$$

Each choice of $n$ and $m$ specifies a particular $TM_{nm}$ mode and differs from others in respect of field configuration, cut-off and propagation properties. The propagation constant for the $nm^{th}$ mode is given by

$$\overline{\beta}_{g,nm} = \sqrt{\omega^2\mu\epsilon - \left(\frac{p_{nm}}{a}\right)^2}. \tag{4.87}$$

For TE waves, the derivative of $H_z^o$ must vanish at $\rho=a$. This requires

$$J_n'(k_c a) = 0 \tag{4.88}$$

$J_n'(k_c a)$ also has a quasi-periodic character and hence in this case too we have infinite propagating modes.

If $p'_{nm}$ is the $m^{\text{th}}$ root of Eq. (4.88) then the cut-off wave number for the $\text{TF}_{nm}$ mode is given by

$$k_{c,nm}=\frac{p'_{nm}}{a}. \tag{4.89}$$

The propagation constant for the $\text{TE}_{nm}$ mode is

$$\beta_{z,nm}=\sqrt{\omega^2\mu\epsilon-\left(\frac{p'_{nm}}{a}\right)^2} \tag{4.90}$$

Though there are doubly infinite modes of propagation in each case, for wave propagation $\omega^2\mu\epsilon>p_{nm}/a$ or $\omega^2\mu\epsilon>p'_{nm}/a$, i.e. either frequency of operation must be very large or there will be wave propagation corresponding to the first few roots $p_{nm}$ or $p'_{nm}$. Table 4.1 summarizes the first few important roots for TE and TM modes.

**TABLE 4.1. First few roots of bessel functions $J_n(k_c a)$ and $J_n'(k_c a)$**

| *Values of n* | $p_{nm}$ (TM *Waves*) | | | $p'_{nm}$ (TE *Waves*) | | |
|---|---|---|---|---|---|---|
| | $p_{n1}$ | $p_{n2}$ | $p_{n3}$ | $p'_{n1}$ | $p'_{n2}$ | $p'_{n3}$ |
| 0 | 2.405 | 5.520 | 8.654 | 3.832 | 7.016 | 10.174 |
| 1 | 3.832 | 7.016 | 10.174 | 1.841 | 5.331 | 8.536 |
| 2 | 5.135 | 8.417 | 11.602 | 3.054 | 6.706 | 9.970 |

The cut-off wavelengths and corresponding cut-off frequencies are given by

$$(\lambda_c)_{\text{TM}nm}=\frac{2\pi}{k_{c,nm}}=\frac{2\pi a}{p_{nm}}$$

$$(f_c)_{\text{TM}nm}=\frac{k_{c,nm}}{2\pi\sqrt{\mu\epsilon}}=\frac{p_{nm}}{2\pi a\sqrt{\mu\epsilon}} \tag{4.91}$$

$$(\lambda_c)_{\text{TE}nm}=\frac{2\pi}{k_{c,nm}}=\frac{2\pi a}{p'_{nm}}$$

$$(f_c)_{\text{TE}nm}=\frac{k_{c,nm}}{2\pi\sqrt{\mu\epsilon}}=\frac{p'_{nm}}{2\pi a\sqrt{\mu\epsilon}}.$$

Table 4.2 shows the cut-off wavelengths of some lower modes.

**TABLE 4.2. Cut-off wavelengths of various modes**

| $TM_{nm}$ *Modes* | | $TE_{nm}$ *Modes* | |
|---|---|---|---|
| *Mode type* | *Cut-off wavelength* | *Mode type* | *Cut-off wavelength* |
| $TM_{01}$ | $2.62a$ | $TE_{01}$ | $1.64a$ |
| $TM_{02}$ | $1.14a$ | $TE_{02}$ | $0.90a$ |
| $TM_{03}$ | $0.72a$ | $TE_{03}$ | $0.62a$ |
| $TM_{11}$ | $1.64a$ | $TE_{11}$ | $3.41a$ |
| $TM_{12}$ | $0.90a$ | $TE_{12}$ | $1.18a$ |
| $TM_{13}$ | $0.62a$ | $TE_{13}$ | $0.74a$ |
| $TM_{21}$ | $1.22a$ | $TE_{21}$ | $2.06a$ |
| $TM_{22}$ | $0.75a$ | $TE_{22}$ | $0.94a$ |
| $TM_{23}$ | $0.99a$ | $TE_{31}$ | $1.49a$ |
| | | $TE_{41}$ | $1.18a$ |
| | | $TE_{51}$ | $0.98a$ |

Examination of Table 4.2 reveals that the longest critical wavelength does not feature in the mode with the lowest values of subscripts $m$ and $n$, but it is for $TE_{11}$, which is therefore the dominant mode. This peculiarity is another evidence of the difference between the meaning of the subscripts for modes in circular and rectangular waveguides. However, the $TE_{01}$ mode of propagation is of specific importance because of low-loss (see Sec. 4.11) in waveguide walls, peculiar field configuration and high $Q$. No doubt energy transmission through this mode involves difficulties as it is not a dominant mode. Figure 4.27 illustrates the line spectrum of the cut-off frequency as compared to the cut-off frequency of the dominant mode $(f_c)_{TE11}$.

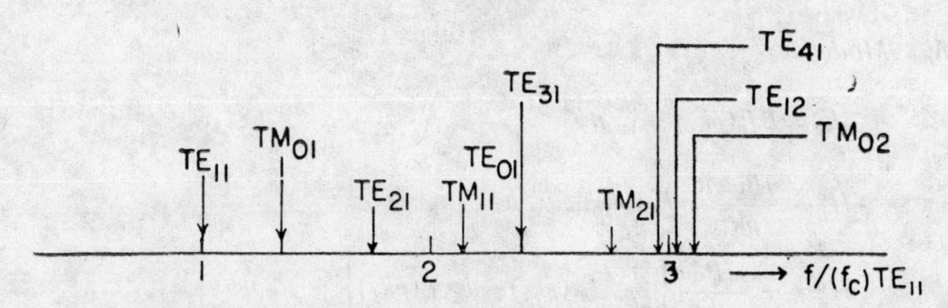


Fig. 4.27 Relative cut-off frequencies of waves in a circular guide

### (*ii*) Field Configuration

We see from Eqs. (4.81) and (4.83) that each of the $TE_{nm}$ and $TM_{nm}$ modes is doubly degenerate. Indeed, the presence of sine and cosine terms in all the equations referred to above testifies the existence of the mode differing in an odd or even variation of the field relative to an arbitrary origin of azimuthal angles. These mode pairs feature identical propagation constants provided the waveguide cross-section is absolutely symmetrical. Only $TM_{on}$ and $TE_{on}$ modes are noted for azimuthal symmetry, i.e. those having no field variations as a function of $\phi_o$ and are not degenerate. Further, following the field configuration (Eqs. 4.81 and 4.83), we can assign the physical interpretation to the subscripts $n$ and $m$. Evidently, $n$ represents the number of full cycles of variation of $E_z$ or $H_z$ as $\phi$ varies through $2\pi$ and $m$ represents the number of times $E_\phi$ is zero along the radial direction of the waveguide, extending from the axis to the inner surface of the guide, excluding zero on the axis.

To get a clear idea of the field pattern existing in a circular waveguide, we transform the derivative of the Bessel function in Eqs. (4.81) and (4.83) to higher order Bessel functions using the following formula:

$$J'_n(\rho k_c) = \frac{n}{\rho k_c} J_n(\rho k_c) - J_{n+1}(\rho k_c). \tag{4.92}$$

The field configuration for the $TM_{nm}$ and $TE_{nm}$ modes is then given by

$TE_{nm}$ *Mode*

$$H_z^o = A J_n(\rho k_c) \begin{matrix}\cos\\ \sin\end{matrix} n\phi$$

$$H_\rho^o = -\frac{j\beta_z A}{k_c}\left\{\frac{n}{\rho k_c} J_n(\rho k_c) - J_{n+1}(\rho k_c)\right\} \cdot \begin{matrix}-\sin\\ \cos\end{matrix} n\phi$$

$$H_\phi^o = -\frac{j\beta_z n A}{\rho k_c^2} J_n(\rho k_c) \cdot \begin{matrix}-\sin\\ \cos\end{matrix} n\phi$$

$$E_\rho^o = \frac{\omega\mu}{\beta_z} H_\phi^o = Z_{TE} H_\phi^o$$

$$E_\phi^o = -\frac{\omega\mu}{\beta_z} H_\rho^o = -Z_{TE} H_\rho^o$$

$TM_{nm}$ *Mode*

$$E_z^o = B J_n(\rho k_c) \cdot \begin{matrix}\cos\\ \sin\end{matrix} n\phi$$

$$H_\rho^o = \frac{jB\omega\epsilon n}{\rho k_c^2} J_n(\rho k_c) \cdot \begin{matrix}\sin\\ \cos\end{matrix} n\phi$$

$$H_\phi^o = -\frac{jB\omega\epsilon}{k_c}\left\{\frac{n}{\rho k_c} J_n(\rho k_c) - J_{n+1}(\rho k_c)\right\} \cdot \begin{matrix}\cos\\ \sin\end{matrix} n\phi$$

$$E_\rho^o = \frac{\beta_z}{\omega\epsilon} H_\phi^o = Z_{TM} H_\phi^o$$

$$E_\phi^o = \frac{\beta_z}{\omega\epsilon} H_\rho^o = -Z_{TM} H_\rho^o. \tag{4.93}$$

For $TM_{01}$ Mode, $n=0$, $m=1$, we have

$$H_z^o = H_\rho^o = E_\phi^o = 0$$

$$E_z^o = BJ_o\left(\frac{2.405}{a}\ \rho\right)$$

$$H_\phi^o = \frac{jB\omega a}{2.405\ \rho}\ J_1\left(\frac{2.405}{a}\ \rho\right) \tag{4.94}$$

$$E_\rho^o = \frac{\overline{\beta}_z}{\omega\epsilon}\ H_\phi^o.$$

The variation of the field components at any instant of time is obtained by multiplying each equation above with a factor $e^{j(\omega t-\overline{\beta}_z z)}$ and taking the real part. This is shown in Fig. 4.28. We see that there is no field variation along the azimuthal direction and the field variation in radial direction is governed by the Bessel function rather than the trigonometrical function (as in the case of rectangular waveguide). No doubt, the field variation along the $z$-axis is still sinusoidal with a phase shift of $\pi/2$ between the $E_z$ component and $E_\rho$ and $H_\phi$ components. As required by TM waves in general, the currents in the waveguide walls are purely longitudinal. The maximum current density coincides with the maximum intensity of $E_\rho$ and $H_\phi$ components. This field distribution resembles the field distribution of the $TM_{11}$ mode in rectangular waveguides (See Fig. 4.13).

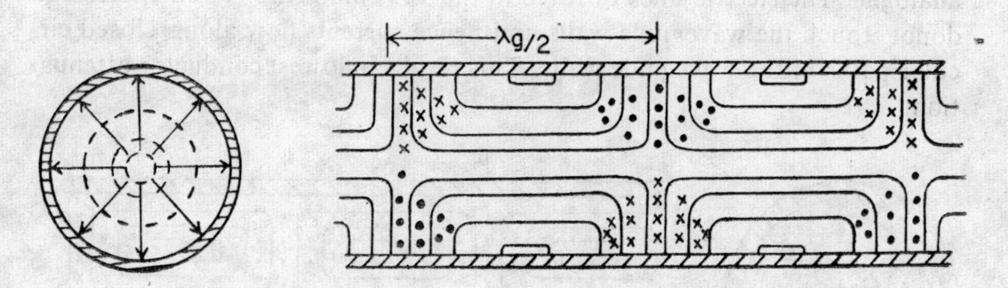


**Fig. 4.28** Field pattern in $TM_{01}$ mode in a cylindrical waveguide.

If $m=0$, all the field components in Eq. (4.93) are identically zero hence $TM_{no}$ modes cannot exist in a circular waveguide. Figures 4.29 and 4.30

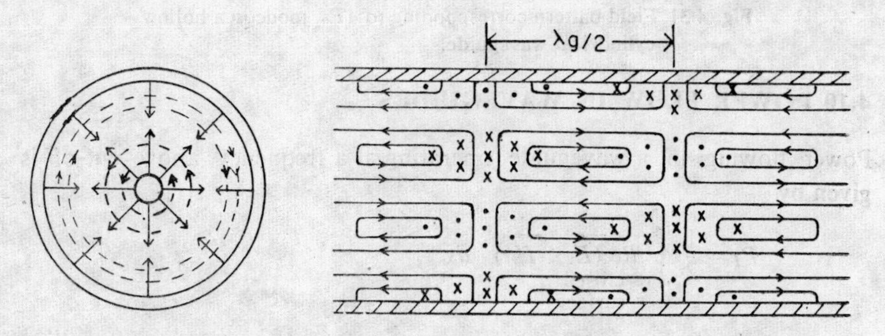


**Fig. 4.29** Field distribution corresponding to $TM_{02}$ mode in a holl ow cylindrical waveguide.

depict field distribution corresponding to $TM_{02}$ and $TM_{11}$ modes. The field configuration for higher modes may be anticipated bearing in mind the physical significance of $n$ and $m$. Note that the $TM_{11}$ mode does not possess any azimuthal symmetry.

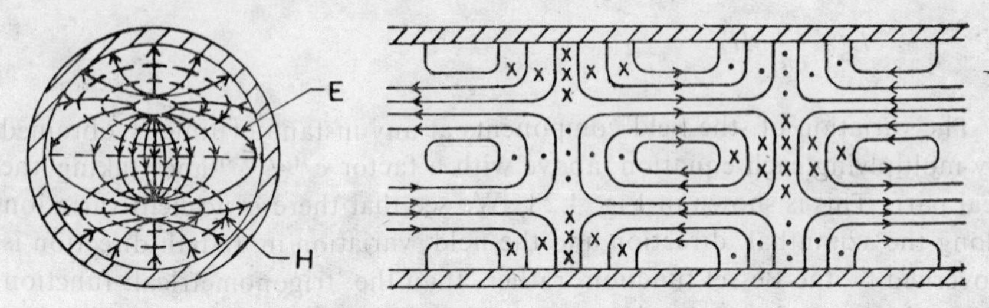


**Fig. 4.30** Field distribution corresponding to $TM_{11}$ mode in a hollow cylindrical waveguide.

The field distribution corresponding to the $TE_{01}$ and $TE_{11}$ modes is shown in Figs. 4.31 and 4.32 respectively. We see that the $TE_{11}$ mode in cylindrical waveguides resembles the $TE_{10}$ mode in rectangular waveguides, but the $TE_{01}$ mode in circular waveguides does not have any rectangular waveguide analogue. The electric lines of force in the $TE_{11}$ mode are closed circles and donot touch the waveguide walls and hence currents flow along closed circular paths in the waveguide walls. This results in lowest conductor attenuation.

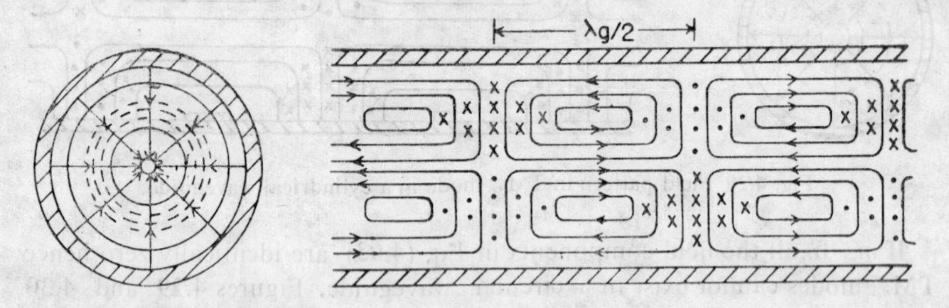


**Fig. 4.31** Field pattern corresponding to $TE_{01}$ mode in a hollow cylindrical waveguide.

## 4.10 POWER FLOW IN WAVEGUIDES

Power flowing in a waveguide operating at a frequency above cut-off is given by

$$P_T = \frac{1}{2} \oint_{s=\text{closed surface}} Re\,(\vec{E} \times \vec{H}^*) \cdot ds$$

$$= \frac{1}{2} \oint_s |E_T|\,|H_T|\, ds \qquad (4.95)$$

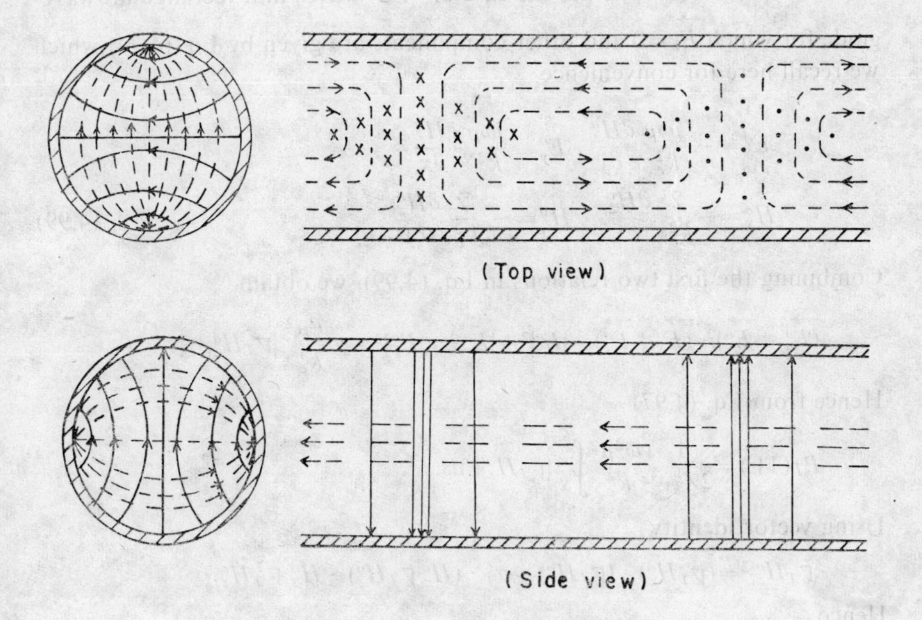


**Fig. 4.32** Field distribution corresponding to $TE_{11}$ mode in a hollow cylindrical waveguide.

where $|E_T|$ and $|H_T|$ are transverse components of the electric and magnetic fields. Using the general definition of wave impedance

$$Z_{TE} \text{ or } Z_{TM} = \frac{|E_T|}{|H_T|} = Z_o$$

$$P_T = \frac{Z_o}{2} \oint |H_T|^2 \, ds = \frac{1}{2Z_o} \oint_S |E_T|^2 \, ds. \quad (4.96)$$

For rectangular waveguides Eq. (4.96) becomes

$$P_T = \frac{1}{2Z_o} \int_o^a \int_o^b \{|E_x|^2 + |E_y|^2\} \, dx \, dy \quad (4.97)$$

and for circular waveguides

$$P_T = \frac{1}{2Z_o} \int_o^{2\pi} \int_o^b \{|E_\rho|^2 + |E_\phi|^2\} \, \rho d\rho \, d\phi. \quad (4.98)$$

Equations (4.97) and (4.98) may be used to evaluate power transfer in the desired mode of rectangular and circular waveguides respectively. However, axial components may also be used to evaluate the power flowing through a waveguide.

As an illustration consider the case of TE waves in a rectangular waveguide for which $\vec{E}_z=0$ and other components are given by Eq. (4.24) which we recall here for convenience

$$E_x^o=-\frac{j\omega\mu}{k_c^2}\frac{\partial H_z^o}{\partial y},\ E_y^o=\frac{j\omega\mu}{k_c^2}\frac{\partial H_z^o}{\partial x}$$

$$H_x^o=-\frac{\bar{\gamma}}{k_c^2}\frac{\partial H_z^o}{\partial x},\ H_y^o=-\frac{\bar{\gamma}}{k_c^2}\frac{\partial H_z^o}{\partial y}. \tag{4.99}$$

Combining the first two relations in Eq. (4.99), we obtain

$$(\vec{E}_x^o+\vec{E}_y^o)\cdot(\vec{E}_x^o+\vec{E}_y^o)=|E_x^o|^2+|E_y^o|^2=|E_T^o|^2=\frac{\omega^2\mu^2}{k_c^4}|\nabla_T H_z^o|^2.$$

Hence from Eq. (4.97)

$$P_T(\text{TE})=\frac{1}{2Z_{\text{TE}}}\frac{\omega^2\mu^2}{k_c^4}\oint_S|\nabla_T H_z^o|^2\,ds.$$

Using vector identity

$$|\nabla_T H_z^o|^2=(\nabla_T H_z^o)\cdot(\nabla_T H_z^o)=\nabla_T\cdot(H_z^o\,\nabla_T H_z^o)-H_z^o\,\nabla_T^2 H_z^o.$$

Hence

$$\oint_S|\nabla_T H_z^o|^2\,ds=\oint_S\nabla_T\cdot(H_z^o\nabla_T H_z^o)\,ds-\oint_S H_z^o\nabla_T^2 H_z^o\,ds.$$

Now, since

$$\oint_S\nabla_T\,(H_z^o\nabla_T H_z^o)\,ds=\oint H_z^o\frac{\partial H_z^o}{\partial n}\,dl=0$$

because $\partial H_z^o/\partial n=0$ at the boundary, and from the wave equation

$$\nabla_T^2 H_z^o=-k_c^2 H_z^o$$

we have

$$\oint_S(\nabla_T H_z^o)^2\,ds=k_c^2\oint_S|H_z^o|^2\,ds$$

$$P_T(\text{TE})=\frac{1}{2Z_{\text{TE}}}\frac{\omega^2\mu^2}{k_c^2}\oint_S|H_z^o|^2\,ds$$

or

$$P_T(\text{TE})=\frac{1}{2Z_{\text{TE}}}\eta^2\,(f/f_c)^2\oint_S|H_z^o|^2\,ds. \tag{4.100}$$

Similarly, for TM waves

$$P_T(\text{TM})=\frac{Z_{\text{TM}}}{2}\,(f/f_c)^2\,\frac{1}{\eta^2}\oint_S|E_z^o|^2\,ds. \tag{4.101}$$

The use of longitudinal components greatly simplifies the analytical part involved in the calculation of $P_T$. Moreover, the expressions for $P_T$ in Eqs. (4.100) and (4.101) are common for cylindrical and rectangular waveguides.

The use of these expressions shall be made when we discuss attenuation in waveguides in the next section. However, as an example, we consider the case of a rectangular waveguide operating on the $TE_{mn}$ mode. Using Eqs. (4.48) for the value of $|H_x^o|$ and $|H_y^o|$ in Eq. (4.97) we have

$$P_T(TE_{mn}) = \frac{Z_{TEnm}}{2} \int_o^a \int_o^b \{|H_x^o|^2 + |H_y^o|^2\}\, dxdy$$

$$= \frac{Z_{TEnm}}{2} \int_o^a \int_o^b \frac{\beta_z^2 B^2}{k_c^4} \{k_x^2 \sin^2 k_x x \cos^2 k_y y$$

$$+ k_y^2 \cos^2 k_x\, x \sin^2 k_y y\}\, dxdy.$$

Taking into consideration the fact that ($k_x = m\pi/a$, $k_y = n\pi/b$) and

$$\int_o^a \int_o^b \sin^2 k_x x \cos^2 k_y y\, dxdy = \int_o^a \int_o^b \cos^2 k_x x \sin^2 k_y y\, dxdy$$

$$= \begin{cases} ab/4 & m \neq 1,\ n \neq 0 \\ ab/2 & m \text{ or } n = 0 \end{cases}$$

we find

$$P_T(TE_{mn}) = \frac{Z_{TEnm}}{2} \frac{\beta_z^2 B^2}{k_c^4} (k_x^2 + k_y^2) \cdot \frac{ab}{2} \frac{1}{\epsilon_{om}\, \epsilon_{on}}$$

$$P_T(TE_{nm}) = \frac{B^2 ab}{2\epsilon_{om}\, \epsilon_{on}} \left(\frac{\beta_z}{k_c}\right)^2 Z_{TEmn} \tag{4.102}$$

where $\epsilon_{on}$ is the Neumann factor having value 1 for $n=0$ and 2 for $n>0$.

Equation (4.102) gives the power flowing in a rectangular waveguide in the $TE_{mn}$ mode. However, if many modes propagate in a waveguide then the total power is the sum due to individual modes, i.e.

$$P_T = p_{Tmn} + P_{Trs} + P_{T,pq} + \ldots$$

This is the general property of all waveguides and for details the reader may refer to the literature*.

## 4.11 ATTENUATION IN WAVEGUIDES

In our discussions so far we assumed waveguide walls to be perfectly conducting and the dielectric to be perfect. Both these assumptions are impractical as an electromagnetic wave propagating through a waveguide suffers attenuation arising from one or more of the following causes:

(*i*) Operating frequency is less than the cut-off frequency (cut-off phenomena);

(*ii*) Losses in the dielectrics;

(*iii*) Losses in the guide walls due to finite conductivity.

*R.E. Collin, *Foundations for Microwave Engineering*, New York: McGraw-Hill Book Co. (1986).

## Waveguide Under Cut-off

From Eq. (4.28) it is noted that the propagation constant $\bar{\gamma}$ becomes a real number when $f<f_c$ and consequently there cannot be any wave propagation. All the fields decay exponentially. Moreover, it is seen that both

$$Z_{TE}=\frac{\eta}{\sqrt{1-(f/f_c)^2}} \quad \text{and} \quad Z_{TE}=\eta\sqrt{1-\left(\frac{f}{f_c}\right)^2}$$

become imaginary, i.e. the waveguide offers a pure reactive load to the source causing perfect reflection of the waves. In this way the waveguide below cut-off resembles a low frequency filter consisting of reactive components ($L$ and $C$) and operating in the cut-off region. It should be mentioned that cut-off attenuation is a *reactive attenuation* arising from the reflection of electromagnetic energy and does not result in any dissipation of power. The attenuation constant arising from cut-off is given by

$$\bar{\alpha}_{\text{cut-off}}=\frac{2\pi}{\lambda_c}\sqrt{1-\left(\frac{\lambda_c}{\lambda}\right)^2}. \tag{4.103}$$

It is seen from Eq. (4.103) that at a frequency far below cut-off ($f\ll f_c$ or $\lambda\gg\lambda_c$), the attenuation approaches a constant value $\bar{\alpha}_{\text{cut-off}}$ $2\pi/\lambda_c$, independent of operating frequency. Since this attenuation depends only on transverse dimension and not on any shape, it is the property of every waveguide and is used to design waveguide attenuator*. Cut-off attenuators find many applications in microwave circuitry such as power control and discontinuity studies in microwave transmission systems.

It can be seen from the field configuration equations for various modes that for $f\ll f_c$ all the fields vanish quickly. However, just below cut-off $f/f_c\to 0$; in the case of TM waves, magnetic fields are reduced to zero but transverse components of electric field approach a constant value; on the other hand, in case of TE waves electric fields are reduced to zero while transverse components of the magnetic field approach a constant value.

Thus, a waveguide below cut-off in the TM mode provides simple electric coupling between the source and receiver with the waveguide acting as a shield. Similarly, a waveguide operating below cut-off in the TE mode provides simple magnetic coupling between source and receiver with the waveguide as a shield. Moreover, when we operate just below cut-off the effect of damping etectric field on magnetic field and vice-versa has to be considered. These effects in combination with finite conductivity losses modify the cut-off characteristics of a waveguide operating just below cut-off and this has been shown in Fig. 4.33.

*Constructional details are discussed in Ch. 6.

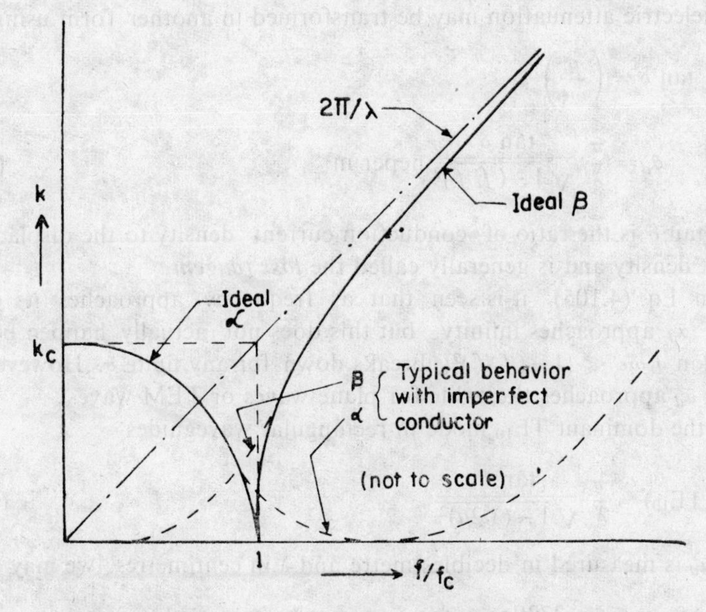


Fig. 4.33 Modification of propagation characteristics due to losses in a waveguide.

(*ii*) **Attenuation Due to Dielectric Losses**

As mentioned earlier, dielectric losses may be anticipated when we take the permittivity to be complex $\epsilon^* = \epsilon\ (1 - j\sigma/\omega\epsilon)$. The attenuation constant due to dielectric losses may be obtained by separating $\bar{\gamma}$ into real and imaginary parts.

$$\bar{\gamma} = \bar{\alpha}_d + \bar{\beta}_z = \sqrt{k_c^2 - \omega^2 \mu \epsilon^*}$$

$$= \sqrt{k_c^2 - \omega^2 \mu\epsilon\ (1 - j\sigma/\omega\epsilon)}$$

$$= \omega\sqrt{\mu\epsilon}\left\{\frac{k_c^2}{\omega^2\mu\epsilon} - 1 + j\frac{\sigma}{\omega\epsilon}\right\}^{1/2}$$

$$= j\omega\ \sqrt{\mu\epsilon}\ [\{1 - (f_c/f)^2\} - j\sigma/\omega\epsilon]^{1/2}.$$

At microwave frequencies $\sigma/\omega\epsilon \ll \{1 - f_c/f)^2\}$, the bracket above may be expanded using the binomial theorem

$$\bar{\gamma} = j\omega\ \sqrt{\mu\epsilon}\{1 - (f_c/f)^2\}\left\{1 - \frac{1}{2}\ \frac{j\sigma/\omega\epsilon}{\sqrt{1 - (f_c/f)^2}}\right\}$$

$$= \frac{\omega\sqrt{\mu\epsilon}}{2\omega\epsilon}\ \frac{\sigma}{\sqrt{1 - (f_c/f)^2}} + j\omega\ \sqrt{\mu\epsilon}\sqrt{1 - (f_c/f)^2}$$

$$= \bar{\alpha}_d + j\bar{\beta}_z.$$

We see that the value of $\bar{\beta}_z$ remains the same but the wave is attenuated with factor $e^{-\bar{\alpha} dz}$ as $\bar{\alpha}_d$ acts as perturbation to $j\bar{\beta}_z$.

$$\bar{\alpha}_d = \frac{\omega\sqrt{\mu\epsilon}}{2\omega\epsilon}\ \frac{\sigma}{\sqrt{1 - (f_c/f)^2}} = \frac{\eta\sigma}{2\sqrt{1 - (f_c/f)^2}}. \qquad (4.104)$$

The dielectric attenuation may be transformed in another form using

$$\tan \delta = \left(\frac{\sigma}{\omega\epsilon}\right)$$

$$\bar{\alpha}_d = \frac{\pi}{\lambda} \frac{\tan \delta}{\sqrt{1-(f_c/f)^2}} \text{ neper/m} \tag{4.105}$$

where tan δ is the ratio of conduction current density to the displacement current density and is generally called the *loss tangent*.

From Eq. (4.105), it is seen that as frequency approaches its cut-off values, $\bar{\alpha}_d$ approaches infinity, but this does not actually happen because condition $\sigma/\omega\epsilon \ll \{1-(f_c/f)^2\}$ breaks down for any finite $\sigma$. However, for $f \gg f_c$, $\bar{\alpha}_d$ approaches the value for plane waves or TEM waves.

For the dominant $TE_{10}$ mode in rectangular waveguides

$$\bar{\alpha}_d(TE_{10}) = \frac{\pi}{\lambda} \frac{\tan \delta}{\sqrt{1-(\lambda/2a)^2}} \tag{4.106}$$

When $\bar{\alpha}_d$ is measured in decibles/metre and $\lambda$ in centimetres, we may write

$$\bar{\alpha}_d(TE_{10}) = \frac{2730\sqrt{\epsilon_r} \tan \delta}{\lambda_o [1-(\lambda_o/2a\sqrt{\epsilon_r})^2]^{1/2}}$$

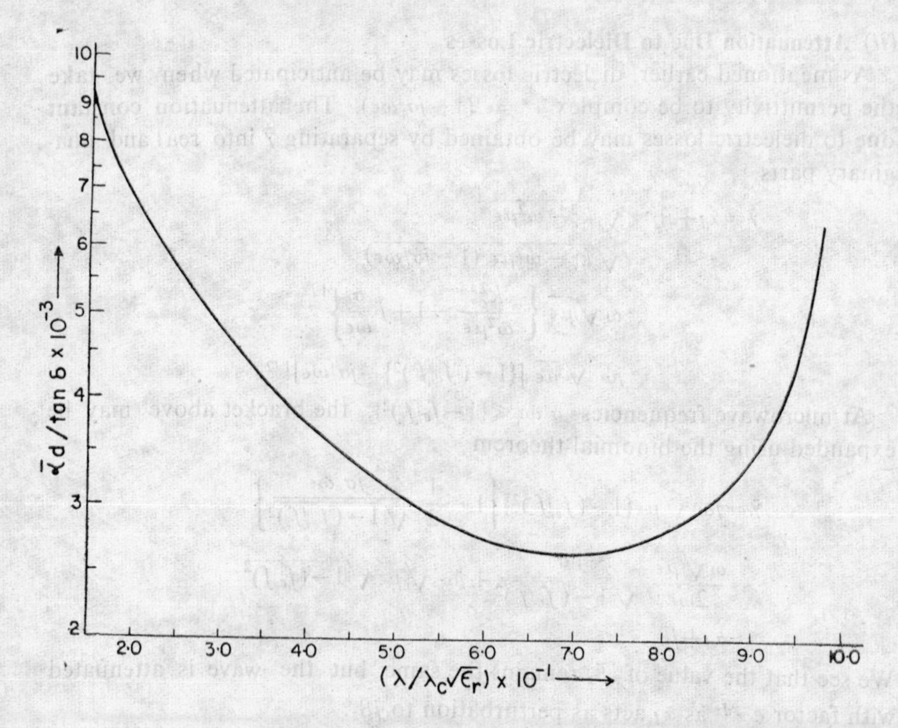


**Fig. 4.34** Variation of dielectric attention ($\bar{\alpha}d$/tan δ) with the wavelength ($\lambda/\lambda_c\sqrt{\epsilon_r}$) for $TE_{10}$ mode in a rectangular waveguide.

where $\epsilon_r$ is the relative permittivity of the dielectric medium. Figure 4.34 shows the variation of $\bar{\alpha}_d/\tan\delta$ versus $\lambda/\lambda_c\sqrt{\epsilon_r}$. It is observed that the dielectric attenuation is minimum for

$$\lambda \simeq a\sqrt{\epsilon_r} \times 1.4 \quad \text{or} \quad f = \frac{v_p}{1.4\, a\sqrt{\epsilon_r}}.$$

(*iii*) **Attenuation Due to Conductor Losses**

If the waveguide walls are not perfectly conducting, there exists a tangential component of electric field and normal component magnetic field at the walls. This results in average power delivered to the waveguide walls and dissipated in it and hence continuous attenuation of propagating electromagnetic waves.

As shown in Fig. (4.35), if $\vec{H}_t$ and $\vec{E}_t$ are the tangential components of electric and magnetic fields then the power loss per unit guide length is given by

$$P_L = \frac{1}{\Delta z}\frac{1}{2}\,\mathrm{Re}\oint_{S_1} (\vec{E}_t \times \vec{H}_t^*)\, d\vec{S}_1$$

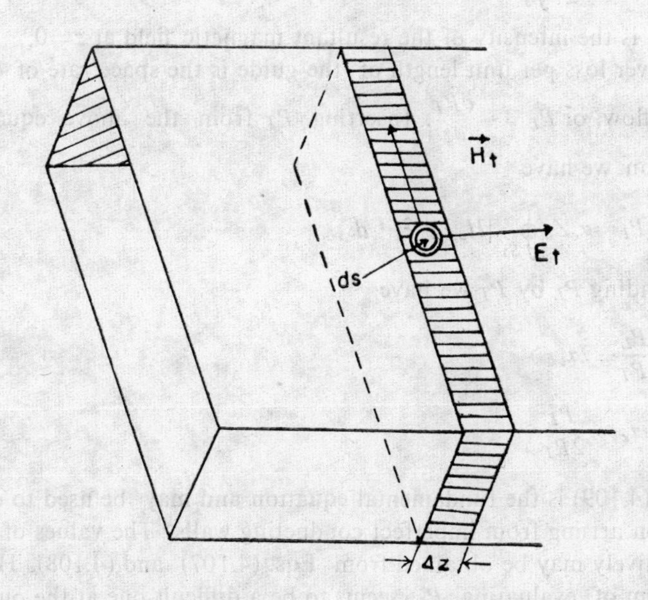


Fig. 4.35 Tangential components of electric and magnetic fields in a waveguide wall.

$\Delta z$ is the length element in Fig. (4.35).

Since $\vec{E}_t$ results from the flow of current due to $\vec{H}_t$ and the direction of current flow is normal to $\vec{H}_t$, hence $\vec{E}_t$ and $\vec{H}_t$ must be at right angles. The ratio of tangential electric field to the tangential magnetic field may be defined as high frequency conductor impedance

$$\eta_m = \frac{E_t}{H_t}.$$

Hence, $$P_L = \frac{1}{\Delta z} \operatorname{Re} \oint_{s1} (\eta_m |H_t|^2)\, ds_1$$

where $\eta_m$ is given by Eq. (3.45). Using

$$\operatorname{Re}(\eta_m) = R_s = \sqrt{\frac{\omega \mu_m}{2\sigma_m}}, \tag{4.107}$$

where $\mu_m$ and $\sigma_m$ are the permeability and conductivity of the metal, in Eq. (4.107), we have

$$P_L = \frac{R_s}{2\,\Delta z} \oint_{s1} |H_t|^2\, ds_1.$$

Due to the continuous loss in power the fields are attenuated by a constant $e^{-\alpha_c \Delta z}$ and consequently the power coming out of the length $\Delta z$ is given by

$$P_T = \frac{Z_o}{2} \oint_s |H_o|^2\, e^{-2\alpha_c z}\, ds \tag{4.108}$$

where $|H_o|$ is the intensity of the resultant magnetic field at $z=0$.

The power loss per unit length of the guide is the space rate of decrease of power flow, or $P_L = -\frac{\partial P_T}{\partial z}$. Inserting $P_T$ from the above equation in this relation we have

$$P_L = \alpha_c Z_o \oint_{S_1} |H_o|^2\, e^{-2\alpha_c z}\, ds_1.$$

Hence dividing $P_L$ by $P_T$ we have

$$\frac{P_L}{P_T} = 2\alpha_c$$

or $$\alpha_c = \frac{P_L}{2P_T}. \tag{4.109}$$

Equation (4.109) is the fundamental equation and may be used to evaluate attenuation arising from imperfect conducting walls. The values of $P_L$ and $P_T$ respectively may be obtained from Eqs. (4.107) and (4.108). However, the problem of evaluating $P_L$ seems to be a difficult one at the outset because the loss depends upon the field configuration within the guide, and the field configuration in turn, depends to some extent, upon the losses. The attack on this problem is one that is used quite often in engineering. We first write the equations for the electric and magnetic fields as though the guide were loss-less. Since the walls of the guide are assumed to be perfectly conducting, the tangential component of the electric field at the guide walls in zero. However, there will be a tangential component of magnetic intensity which is terminated by a current flowing on the surface of the conductor. If we now assume that the guide walls are imperfectly conducting, the tangential component of magnetic intensity will be substantially the same if the walls were perfectly conducting and hence $P_L$ may be evaluated.

Inserting Eqs. (4.108) and (4.96) in (4.109) we find that the attenuation constant arising from guide walls is given by

$$\alpha_c = \frac{R_s \oint_{s1} |H_t|^2 \, ds_1}{2Z_o \oint_s |H_T|^2 \, ds}. \tag{4.110}$$

The numberator of Eq. (4.110) is evaluated at each conductor surface whereas in the denominator it is over cross-section of the waveguide.

We now proceed to evaluate $\alpha_c$ for few waveguide modes of interest.

**Conductor Attenuation in Rectangular Waveguides**

$TM_{mn}$ *Mode.* Since $H_z = 0$ in this case, the power loss per unit length in the lower and upper walls of the waveguide is

$$P_{L1} = 2 \times \frac{R_s}{2} \oint_{s1} |H_t|^2 \, ds_1 = R_s \int_o^a |H_x^o|^2_{\substack{y=0 \\ y=b}}.$$

Using value $H_x^o$ from Eq. (4.43)

$$P_{L1} = \frac{R_s \omega^2 \epsilon^2 A^2 k_y^2}{k_c^4} \int_o^a \sin^2 k_x \, x dx = \frac{R_s \omega^2 \epsilon^2 A^2}{k_c^4} k_y^2 \frac{a}{2}.$$

Similarly, power loss per unit length in the side walls is

$$P_{L2} = 2 \times \frac{R_s}{2} \int_o^b |H_y^o|^2_{\substack{x=0 \\ x=a}} dy = R_s \frac{\omega^2 \epsilon^2 A^2}{k_c^4} k_x^2 \int_o^b \sin^2 k_y \, y dy$$

or

$$P_{L2} = R_s \frac{\omega^2 \epsilon^2 A^2}{k_c^4} k_x^2 \frac{b}{2}.$$

The power loss per unit length is

$$P_L = P_{L1} + P_{L2} = R_s \frac{\omega^2 \epsilon^2 A^2}{k_c^4} \left( \frac{k_y^2 \, a}{2} + \frac{k_x^2 \, b}{2} \right)$$

$$= \frac{R_s \, \omega^2 \epsilon^2 A^2 \pi^2}{2k_c^4} \left( \frac{n^2 a}{b^2} + \frac{m^2 b}{a^2} \right). \tag{4.111}$$

The power transmitted through the waveguide is given by

$$P_T = \frac{1}{2} R_e \oint_s (\vec{E} \times \vec{H}^*) \, ds = \frac{1}{2} Z_{TM} \oint_s |H_T|^2 \, ds$$

$$= \frac{Z_{TM}}{2} \int_o^a \int_o^b \{|H_x^o|^2 + |H_y^o|^2\} \, dxdy$$

which on using values of $|H_x^o|$ and $|H_y^o|$ from Eqs. (4.43) reduces to

$$P_T=\frac{Z_{\text{TM}}}{2}\left\{\int_o^a\int_o^b \frac{\omega^2\epsilon^2A^2k_y^2}{k_c^4}\sin^2 k_x x\cos^2 k_y y\,dxdy\right.$$

$$\left.+\int_o^a\int_o^b \frac{\omega^2\epsilon^2A^2}{k_c^4}k_x^2\cos^2 k_x x\sin^2 k_y y\,dxdy\right\}$$

$$=\frac{Z_{\text{TM}}}{2}\frac{\omega^2\epsilon^2A^2}{k_c^4}\left\{k_y^2\frac{b}{2}\frac{a}{2}+k_x^2\frac{b}{a}\frac{a}{b}\right\}$$

$$=\frac{Z_{\text{TM}}\ \omega^2\epsilon^2A^2\ ab}{8k_c^2}. \tag{4.112}$$

Substituting the values of Eqs. (4.111) and (4.112) in Eq. (4.109), yields

$$\alpha^{\square *}_{c(\text{TM})}=\frac{P_L}{2P_T}=\frac{2R_s\,\pi^2\left(\dfrac{n^2a}{b}+\dfrac{m^2b}{a}\right)}{Z_{\text{TM}}\ k_c^2}. \tag{4.113}$$

Using $Z_{\text{TM}}=\dfrac{\overline{\beta}_z}{\omega\epsilon}=\eta\sqrt{1-(f_c/f)^2}$ and $k_c^2=\left(\dfrac{2\pi}{\lambda_c}\right)^2=\dfrac{4\pi^2}{k_c^2}$.

Equation (4.113) reduces to a simple form

$$\alpha^{\square}_{c(\text{TM})}=\frac{R_s\,\lambda_c^2\left(\dfrac{n^2a}{b^2}+\dfrac{m^2b}{a^2}\right)}{2\eta\ ab\sqrt{1-(\lambda/\lambda_c)^2}}. \tag{4.114a}$$

Or, if we use

$$\lambda_c=\frac{2ab}{\sqrt{n^2a^2+m^2b^2}}$$

$$\alpha^{\square}_{c(\text{TM})}=\frac{2R_s}{b\eta\sqrt{1-(f_c/f)^2}}\ \frac{m^2\,(b/a)^3-n^2}{m^2\,(b/a)^2+n^2}. \tag{4.114b}$$

Similarly, the expression for $\alpha^{\square}_{c\,(\text{TE})}$ may be obtained as

$$\alpha^{\square}_{c(\text{TE}_{mn})}=\frac{2R_s}{b\eta\sqrt{1-(f_c/f)^2}}\left[\left(1+\frac{b}{a}\right)\left(\frac{f_c}{f}\right)^2\right.$$

$$\left.+\left\{1-\left(\frac{f_c}{f}\right)^2\right\}\left\{\frac{\dfrac{b}{a}\left(\dfrac{b}{a}m^2+n^2\right)}{(b^2m^2/a^2)+n^2}\right\}\right]. \tag{4.115}$$

Figure 4.36 depicts the variation of conductor attenuation with frequency for a few but important modes. The minimum attenuation is for the $\text{TE}_{10}$ mode for $b/a = 1$ but due to obvious reasons (to be discussed later) these dimensions are not suitable. However, for best suited dimensions $b/a = 2$,

*Suffixes □ and O indicate the expressions for rectangular and circular waveguides respectively.

the TE$_{10}$ mode carries minimum attenuation and is given by (substitute $m=0$ and $n=1$ in Eq. 4.115).

$$\alpha_{c(\mathrm{TE}_{10})}^{\square}=\frac{R_s\left[1+\frac{2b}{a}\left(\frac{f_c}{f}\right)^2\right]}{\eta b\sqrt{1-(f_c/f)^2}}. \tag{4.116}$$

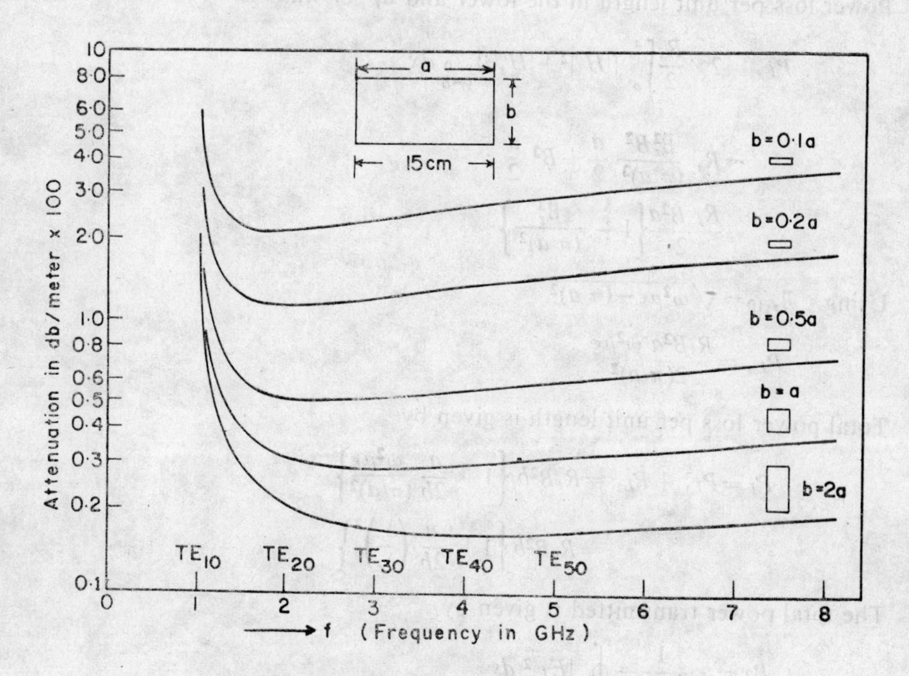


**Fig. 4.36** Calculated conductor attenuation of hollow rectangular, copper waveguides of various proportion.

**Example 4.1.** The field pattern of the TE$_{10}$ mode in a rectangular waveguide is

$$E_x=H_y=E_z=0$$

$$H_z^o=B\cos(\pi x/a)$$

$$H_x^o=\frac{j\bar{\beta}_z B}{\pi/a}\sin(\pi x/a)$$

$$E_y^o=-j\frac{\omega\mu}{\pi/a}B\sin(\pi x/a).$$

Show that the expression for conductor attenuation is

$$\alpha_c=\frac{R_s\left\{1+\frac{2b}{a}\left(\frac{f_c}{f}\right)^2\right\}}{\eta b\sqrt{1-(f_c/f)^2}}.$$

*Solution*: Power loss per unit length of two side walls is

$$P_{L1} = 2\times\frac{R_s}{2}\int_o^b H_z^o|^2_{\substack{x=0\\x=a}}\,dy$$

$$=R_s\,\beta^2\,b.$$

Power loss per unit length in the lower and upper walls is

$$P_{L_2} = 2\times\frac{R_s}{2}\int_o^a \{|H_x^o|^2+|H_z^o|^2\}_{\substack{y=0\\y=b}}\,dx$$

$$=R_s\,\frac{\beta_z^2 B^2}{(\pi/a)^2}\,\frac{a}{2}+B^2\,\frac{a}{2}$$

$$=\frac{R_s\,B^2a}{2}\left\{1+\frac{\beta_z^2}{(\pi/a)^2}\right\}.$$

Using $\beta_{z,10}=\sqrt{\omega^2\mu\epsilon-(\pi/a)^2}$

$$P_{L_2}=\frac{R_sB^2a\;\omega^2\mu\epsilon}{2(\pi/a)^2}.$$

Total power loss per unit length is given by

$$P_L=P_{L_1}+P_{L_2}=R_sB^2b\left\{1+\frac{a}{2b}\,\frac{\omega^2\mu\epsilon}{(\pi/a)^2}\right\}$$

$$=R_sB^2b\left\{1+\frac{a}{2b}\left(\frac{\lambda}{\lambda_c}\right)^2\right\}.$$

The total power transmitted is given by

$$P_T=\frac{1}{2Z_{\mathrm{TE}_{10}}}\oint_s |E_T|^2\,ds$$

or

$$P_T=\frac{1}{2Z_{\mathrm{TE}_{10}}}\int_o^a\int_o^b \frac{\omega^2\mu^2}{(\pi/a)^2}\,B^2\sin^2(\pi a/x)\,dxdy$$

$$=\frac{\omega^2\mu^2B^2a^2}{2Z_{\mathrm{TE}_{10}}\,\pi^2}\cdot\frac{a}{2}\cdot b.$$

Using $Z_{\mathrm{TE}_{10}}=\dfrac{\eta}{\sqrt{1-(\lambda/\lambda_c)^2}}$, $\omega^2=\dfrac{4\pi^2}{\lambda^2\,\mu\epsilon}$

we have

$$P_T=\frac{4\;\mu^2B^2a^3b\,\sqrt{1-(\lambda/\lambda_c)^2}}{\lambda^2\,\mu\epsilon\;4\eta}=\frac{\eta\;ab\;B^2}{4}\left(\frac{\lambda_c}{\lambda}\right)^2\sqrt{1-\left(\frac{\lambda}{\lambda_c}\right)^2}.$$

Hence

$$\alpha_{c\mathrm{TE}_{10}}=\frac{P_L}{2P_T}=\frac{R_sB^2b\left\{1+\dfrac{a}{2b}\left(\dfrac{\lambda_c}{\lambda}\right)^2\right\}}{2\eta\,\dfrac{ab}{4}\left(\dfrac{\lambda_c}{\lambda}\right)^2B^2\sqrt{1-\left(\dfrac{\lambda}{\lambda_c}\right)^2}}$$

$$= \frac{R_s\left\{1+\frac{2b}{a}\left(\frac{f_c}{f}\right)^2\right\}}{\eta b\sqrt{1-(f_c/f)^2}}. \qquad Q.E.D.$$

**Conductor Attenuation in Cylindrical Waveguides**

The field configuration for the $TE_{nm}$ mode is given by Eqs. (4.83a to e). Power loss per unit guide length is given by

$$P_L = \frac{R_s}{2}\oint_{s_1 = \text{conductor surface}} |H_t|^2\, ds_1$$

$$= \frac{R_s}{2}\int_o^{2\pi}\left\{|H_\rho^o|^2 + |H_\phi^o|^2\right\}_{\rho=a} a\,d\phi$$

which on using Eqs. (4.83b) and (4.83c), gives

$$P_L = \frac{R_s}{2}\int_o^{2\pi}\left\{\frac{B_n^2\,\omega^2\epsilon^2 n^2}{a^2 k_c^4}\, J_n^2\,(ak_c)\sin^2 n\phi + \frac{B_n^2\,\omega^2\epsilon^2}{k_c^2}\, J_n'^2\,(k_c a)\cos^2 n\phi\right\} a\,d\phi.$$

Now from the boundary condition $J_n\,(ak_c)=0$ the first term in the integral vanishes

$$P_{L(\text{TM})} = \frac{R_s}{2}\,\frac{B_n^2\,\omega^2\epsilon^2}{k_c^2}\, J_n'^2\,(ak_c)\, a\int_o^{2\pi}\cos^2 n\phi\, d\phi$$

$$= \frac{\pi a B_n^2}{\eta^2}\left(\frac{f}{f_c}\right)^2\frac{R_s}{2}\, J_n'^2\,(k_c\, a). \qquad (4.117)$$

Power transmitted through the waveguide is given by Eq. (4.101). Substituting the value of $E_z^o$ from Eq. (4.83a), we have

$$P_{T(\text{TM})} = \frac{Z_{\text{TM}}}{2}\left(\frac{f}{f_c}\right)^2\frac{1}{\eta^2}\oint_{s=\text{cross-section}} B_n^2\, J_n^2\,(\rho k_c)\cos^2 n\phi\cdot\rho d\rho\, d\phi$$

$$= \frac{Z_{\text{TM}}}{2}\left(\frac{f}{f_c}\right)^2\frac{1}{\eta^2}\int_o^{2\pi}\int_o^a B_n^2\, J_n^2\,(\rho k_c)\cos^2 n\phi\,\rho d\rho\, d\phi$$

$$= \frac{Z_{\text{TM}}}{2}\left(\frac{f}{f_c}\right)^2\frac{1}{\eta^2}\, B_n^2\,\pi\int_o^a \rho J_n\,(k_c\rho)\, d\rho.$$

Now, using the property of the Bessel function

$$\int \rho J_n^2\,(\rho k_c) = \frac{\rho^2}{2}\left\{J_n'^2\,(\rho k_c) + \left(1-\frac{n^2}{\rho^2 k_c^2}\right) J_n^2\,(\rho k_c)\right\}. \qquad (4.118)$$

We may write

$$\int_o^a \rho\, J_n^2\,(\rho k_c)\, d\rho = \frac{a^2}{2}\left\{J_n'\,(ak_c) + \left(1-\frac{n^2}{k_c^2 a^2}\right) J_n^2\,(k_c a)\right\}$$

$$= \frac{a^2}{2}\, J_n'^2\,(ak_c).$$

The last step follows from the boundary condition $J_n(k_c a)=0$. Hence,

$$P_T = \frac{Z_{TM}}{2}\left(\frac{f}{f_c}\right)^2 \frac{1}{\eta^2} B_n^2 \frac{\pi a^2}{2} J_n'^2(k_c a). \tag{4.119}$$

Substituting Eqs. (4.117) and (4.119) in Eq. (4.109), we obtain,

$$\alpha^o_{c(TM)} = \frac{\dfrac{\pi a B_n^2}{\eta^2}\left(\dfrac{f}{f_c}\right)^2 \dfrac{R_s}{2} J_n'^2(k_c a)}{2\times\dfrac{Z_{TM}}{2}\left(\dfrac{f}{f_c}\right)^2 \dfrac{1}{\eta^2} B_n^2 \dfrac{\pi a^2}{2} J_n'^2(k_c a)}$$

$$= \frac{R_s}{a\, Z_{TM}} = \frac{R_s}{a\eta\sqrt{1-(f_c/f)^2}} \text{ nep/m.} \tag{4.120}$$

For TE waves, using the field configuration of Eq. (4.81a to 4.81e), we find that power loss per unit guide length is given by,

$$P_L = \frac{R_s}{2}\oint_{s_1=\text{conductor surface}} |H_t|^2\, ds_1$$

$$= \frac{R_s}{2}\int_o^{2\pi}\left\{|H^o_\phi|^2 + |H^o_z|^2\right\}_{\rho=a} a\,d\phi\cdot dz.$$

Substituting values of $|H^o_\phi|$ and $|H^o_z|$ from Eqs. (4.81c) and (4.81a), in the above equation we obtain,

$$P_L = \frac{R_s}{2}\int_o^{2\pi}\left\{\frac{\beta_z^2 n^2 A_n^2}{a^2 k_c^4} J_n^2(k_c a)\sin^2 n\phi + A_n^2 J_n^2(k_c a)\cos^2 n\phi\right\} a\,d\phi$$

$$(dz=1)$$

or

$$P_L = \frac{R_s}{2}\, a\, A_n^2\, \pi\left\{\left(\frac{\beta_z^2 n^2}{a^2 k_c^4} + 1\right)\right\} J_n^2(k_c a). \tag{4.121}$$

Power transmitted is given by Eq. (4.100)

$$P_T = \frac{1}{2 Z_{TM}}\eta^2\left(\frac{f}{f_c}\right)^2 \oint_{s=\text{cross-section}} |H^o_z|^2\, ds.$$

Substituting the value of $|H^o_z|$ from Eq. (4.81a) in above equation, we have

$$P_T = \frac{1}{2 Z_{TE}}\eta^2\left(\frac{f}{f_c}\right)^2 \int_o^a \int_o^{2\pi} A_n^2 J_n^2(k_c\rho)\cos^2 n\phi\, d\rho\cdot\rho d\phi$$

$$= \frac{\eta^2 (f/f_c)^2}{2 Z_{TE}} A_n^2\, \pi \int_o^a \rho J_n^2(k_c\rho)\, d\rho.$$

Using Eq. (4.118) to evaluate the above integral we have

$$P_T = \frac{\eta^2 (f/f_c)^2}{2 Z_{TE}} A_n^2\, \pi\, \frac{a^2}{2}\left\{J_n'^2(ak_c) + \left(1 - \frac{n^2}{(k_c a)^2}\right) J_n^2(k_c a)\right\}$$

$J_n'(ak_c)=0$ from the boundary condition, hence

$$P_T = \frac{a^2\eta^2 (f/f_c)^2}{4 Z_{TE}}\, \pi\, A_n^2 J_n^2(k_c a)\left(1 - \frac{n^2}{(k_c a)^2}\right). \tag{4.122}$$

Therefore, the conductor attenuation for the $TE_{nm}$ mode in cylindrical waveguides is obtained by substituting Eqs. (4.121) and (4.122) in Eq. (4.109)

$$\alpha^o_{c(\mathrm{TE}nm)}=\frac{\frac{R_s}{2}\, a\, A_n^2\, \pi\left(\frac{\beta_z^2 n^2}{a^2 k_c^4}+1\right) J_n^2\,(k_c a)}{2\times\frac{a^2\eta^2}{4\,Z_{\mathrm{TE}}}\,\pi A_n^2\left(1-\frac{n^2}{(k_c a)^2}\right) J_n^2(k_c a)}.$$

Using $Z_{\mathrm{TE}}=\frac{\eta}{\sqrt{1-(f_c/f)^2}}$ and $\beta_z=\sqrt{\omega^2\mu\epsilon-k_c^2}$

the above expression reduces to

$$\alpha^o_{c(\mathrm{TE}nm)}=\frac{R_s}{\eta a\sqrt{1-(f_c/f)^2}}\left\{\left(\frac{f_c}{f}\right)^2+\frac{n^2}{(k_c a)^2-n^2}\right\}$$

or

$$\alpha^o_{c(\mathrm{TE}nm)}=\frac{R_s}{\eta a\sqrt{1-(f_c/f)^2}}\left\{\left(\frac{f_c}{f}\right)^2+\frac{n^2}{(p'_{nm})^2-n^2}\right\} \tag{4.123}$$

where $p'_{nm}$ is the $m^{\text{th}}$ root of the Bessel function $J'_x\,(k_c\rho)=0$.

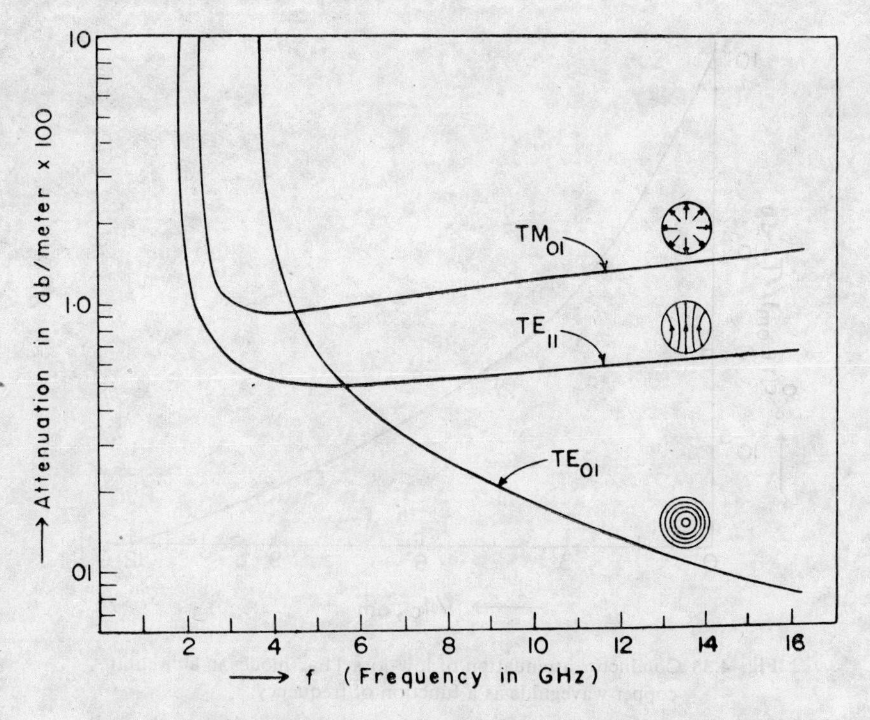


**Fig. 4.37** Conductor attenuation offered by a cylindrical waveguide (copper) of radius 5 cm for each of three important modes i.e. $TM_{01}$, $TE_{11}$ and $TE_{01}$.

An analysis of Eqs. (4.120) and (4.123) indicates that conductor attenuation in cylindrical waveguides is not only a function of frequency but also decreases as the diameter is increased. However, the diameter cannot be increased at random because of other complications such as multimode propagation, etc. Therefore, an optimum diameter has to be selected for a particular band of operation. Figure 4.37 shows the plot of attenuation constants for some lower TE modes versus waveguide diameter. It is noted that the $TE_{11}$ mode carries the minimum attenuation though it is not a dominant mode. The qualitative character of TM modes is similar. Moreover, $TE_{om}$ carries some interesting properties.

Substituting $m=0$ in Eq. (4.123) we find ($p'_{nm}=\infty$ for $m=0$)

$$\alpha^o_{c(TEom)}=\frac{R_s}{a\eta}\frac{(f_c/f)^2}{\sqrt{1-(f_c/f)^2}}.$$

Since $R_s$ increases as $f^{1/2}$ the attenuation $\alpha^o_{c(TEom)}$ falls as $f^{-3/2}$ at high frequencies. This rapid decrease in attenuation with frequency is the unique property of $TE_{om}$ modes that makes cylindrical waveguides very useful for low-loss communication links. Figure (4.38) depicts the decrease in $\alpha^o_{c(TEom)}$ with increase in $f/f_{c,\ om}$.

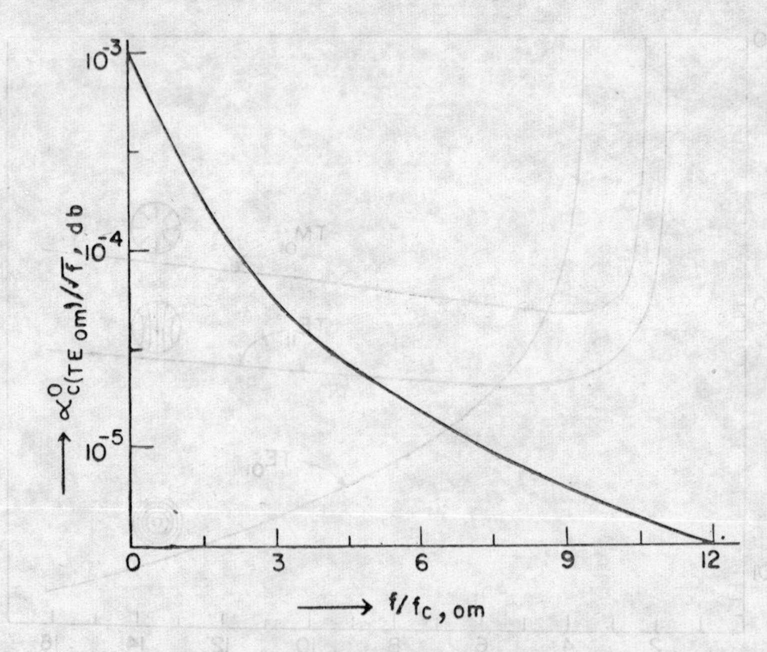


**Fig. 4.38** Conductor attenuation of low-loss $TE_{om}$ modes in a circular copper waveguide as a function of frequency.

Moreover, wall currents associated with $TE_{om}$ modes are in circumferential direction only. This property is utilized in mode filtration. Special cylindrical waveguides having high conductivity in circumferal direction and low conductivity in axial direction are used as mode supressors.*

## 4.12 '$Q$' OF WAVEGUIDES

In the case of waveguides too, we can define a quality factor '$Q$' following the same general definition of Eq. (2.102). The quality factor for walls may be written as

$$Q_w = \omega \frac{\text{Energy stored per unit length}}{\text{Energy lost per unit length per second}}.$$

In waveguides energy propagates at group velocity $v_{gz}$. Hence,
Energy transmitted per second $= v_{gz} \times$ energy stored per unit length
or
Energy stored per unit length $= 1/v_{gz} \times$ power transmitted

$$= P_T/v_{gz}$$

Thus

$$Q_W = \frac{\omega}{v_{gz}} \left( \frac{\text{Power transmitted}}{\text{Power lost per unit length}} \right)$$

$$= \frac{\omega}{2\alpha_c \, v_{gz}} = \frac{\omega}{2\alpha_c \, v_p \sqrt{1-(f_c/f)^2}} \qquad (4.124)$$

($\alpha_c$ is conductor attenuation per unit length.)

Waveguides are characterized by low values of $\alpha_c$, hence waveguide sections may have very high $Q$ values. Such sections find extensive applications in microwave resonators and filters.

Equation (4.124) may also be transformed into conventional form

$$Q_W = \frac{\omega/v_{gz}}{2\alpha} = \frac{\overline{\beta}_z}{2\alpha}. \qquad (4.125)$$

To anticipate losses in the dielectric waveguides, we sometimes make use of the concept of complex frequency $\omega = \omega_1 + j\omega_2$ and define '$Q_d$' for dielectric loss. $Q_d$ is defined as the ratio of the imaginary component of frequency to the real component, i.e.

$$Q_d = \frac{\omega_2}{\omega_1}. \qquad (4.126)$$

The propagation constant $\bar{\gamma}$ for a waveguide must be imaginary. That is,

$$\bar{\gamma} = \sqrt{[\{k_c^2 - (\omega_1^2 - \omega_2^2)\mu\epsilon - \sigma\omega_2\mu\} + j\{\omega_1\mu(\sigma - 2\,\omega_2\epsilon)\}]}$$

*For details the reader may refer to

1. T. Hosono and S. Kohno, "The transmission of $TE_{01}$ waves in Halix waveguides", *IRE Trans.*, Vol. MTT-7, (July 1959).
2. G.W. Luderer and H.G. Unger, "Circular electrical wave propagation in periodic structure" *Bell System Tech. J.*, Vol. 43 (March 1964).

must be imaginary or the first bracket must be negative and the second bracket must vanish. This yields

$$\omega_2 = \frac{\sigma}{2\epsilon}. \qquad (6.127)$$

Making use of the relation

$$\omega^2 = \omega_1^2 + \omega_2^2$$

and Eq. (4.127), Eq. (4.126) may be reduced to

$$Q_d = \frac{\omega_2}{\omega_1} = \sqrt{\left(\frac{\omega_1}{2\omega_2}\right)^2 + \frac{1}{4}} = \frac{\epsilon\omega_1}{\sigma}. \qquad (4.128)$$

Thus dielectric $Q_d$ turns out to be equal to the ratio of dielectric current to the conduction current in a waveguide.

In general, total $Q$ value, $Q_T$, is given by

$$\frac{1}{Q_T} = \frac{1}{Q_W} + \frac{1}{Q_d}$$

or $$Q_T = \frac{Q_W Q_d}{Q_W + Q_d}. \qquad (4.129)$$

## 4.13 WAVEGUIDE EXCITATION

The basic approach adopted for waveguide excitation is to orient an exciting element such as a probe, hole or loop in a waveguide to produce roughly the same field pattern as that of the mode to be excited. The following are the commonly used techniques.

### Electric Excitation

If the output from a microwave generator is in the form of a coaxial line then a part of the inner conductor is usually used as a probe by stripping off the dielectric coating and the outer conductor. The probe so formed serves as an electric dipole. This dipole is oriented in the direction of the electric field, at a place where the electric field is maximum. Consequently, the electric lines of force arising from the probe coincide with the lines of force of the desired mode both in magnitude (relative line density) and direction as shown in Fig. 4.39(a). The exact probe location and probe size (insertion) are determined by impedance matching and desired coupling characteristics.

### Magnetic Excitation

In this case the output coaxial line from the microwave generator is bent into a loop by connecting the inner conductor to the outer conductor as shown in Fig. 4.39(b). The loop forms a magnetic dipole and is placed at or near a point where magnetic field is maximum and oriented in such a way

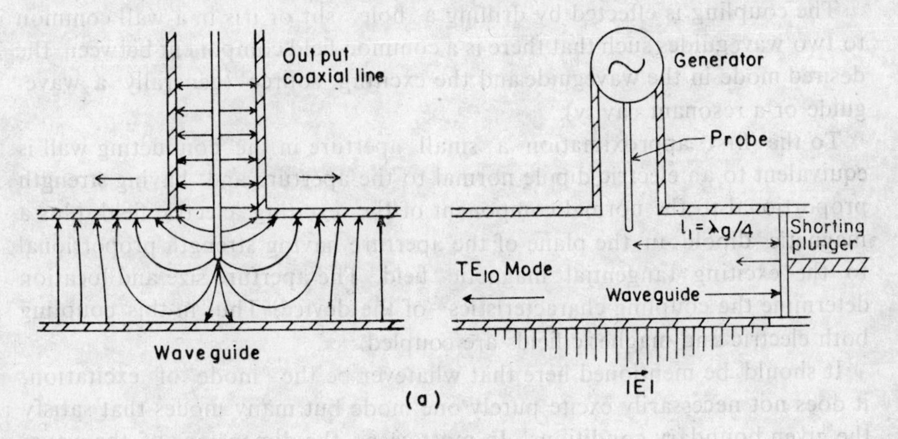


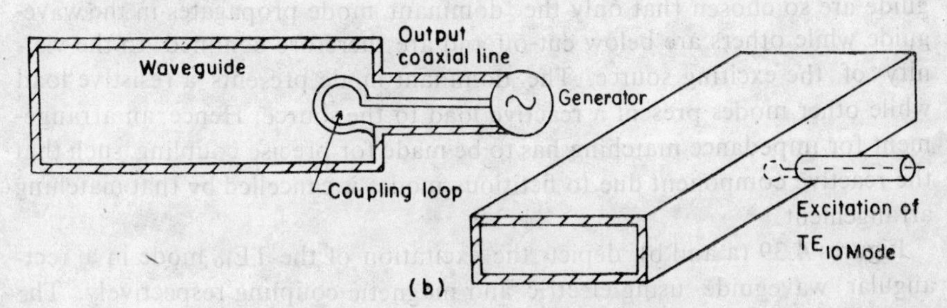


**Fig. 4.39** *(a)* Electric excitation of the dominant-$TE_{10}$ mode in a rectangular waveguide. *(b)* Magnetic coupling of a rectangular waveguide.

that its plane is normal to the magnetic flux lines of the desired mode. Consequently, the magnetic lines emanating from the loop coincide in magnitude and the direction with those of the desired mode. The loop may be mounted on the end wall of the shorted waveguide or on the top or bottom wall as an integral multiple of half wavelength distance from the shorted end. The location and orientation of the loop in this case too is determined by degree of coupling and impedance matching considerations. The coupling increases as the area of the loop is increased for the definite orientation of the loop within the guide. However, if the length of the loop conductor is not small as compared to wavelength, there is an appreciable gradient potential along the conductor and magnetic coupling may be accompanied by appreciable electric coupling.

**Aperture Excitation**

This type of excitation is commonly used to excite a waveguide from another already excited waveguide such as in directional couplers and waveguide junctions. The procedure is sometimes called coupling of wave-guides.

The coupling is effected by drilling a hole, slit or iris in a wall common to two waveguides such that there is a common field component between the desired mode in the waveguide and the exciting source (generally a wave-guide or a resonant cavity).

To the first approximation a small aperture in the conducting wall is equivalent to an electric dipole normal to the aperture and having strength proportional to the normal component of the exciting electric field, plus a magnetic dipole in the plane of the aperture having strength proportional to the exciting tangential magnetic field. The aperture size and location determine the coupling characteristics* of the device. Thus in this coupling both electric and magnetic fields are coupled.

It should be mentioned here that whatever be the mode of excitation, it does not necessarily excite purely one mode but many modes that satisfy the given boundary conditions. In most cases, the dimensions of the wave-guide are so chosen that only the dominant mode propagates in the wave-guide while others are below cut-off and are therefore confined to the vicinity of the exciting source. The dominant mode presents a resistive load while other modes present a reactive load to the source. Hence, an arrangement for impedance matching has to be made for precise coupling, such that the reactive component due to fictitious modes is cancelled by that matching arrangement.**

Figures 4.39 (a and b) depict the excitation of the $TE_{10}$ mode in a rectangular waveguide using electric and magnetic coupling respectively. The probe is passed through a hole in the wide wall of the waveguide and

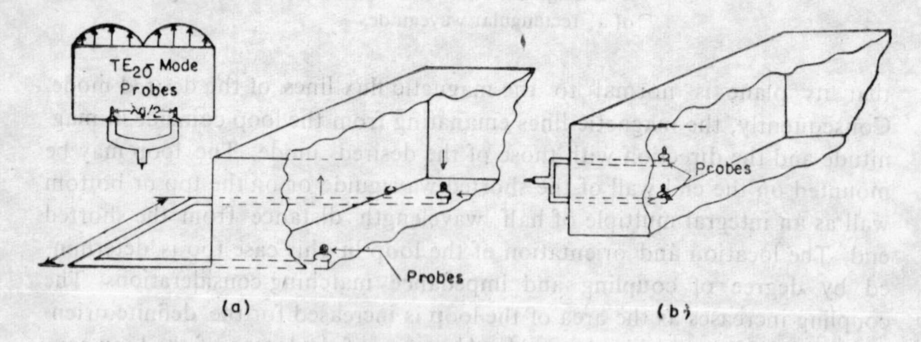


**Fig. 4.40** (*a*) Excitation of $TE_{20}$ mode in a rectangular waveguide.
(*b*) Excitation of $TE_{11}$ mode in a rectangular wave-guide.

*R.E. Collin, *Foundations for Microwave Engineering*, New York: McGraw-Hill Book Co. (1966).

**N. Marcwitz *Waveguide Hand Book*, New York: 1979 McGraw-Hill Book Co. (1951), Vol. 10, Rad. Lab. Series.

mounted parallel to the narrow wall, as desired by field configuration to be obtained. As the radiated fields may propagate in both the directions, a shorting plunger is used to reflect the fields propagating in an undesirable direction. Distances $l_1$ and $l_2$ are adjusted for maximum power coupling, i.e. good impedance match.

Similarly, magnetic coupling may be explained.

To excite the $TE_{20}$ mode, we need two probes spaced at a distance half the guide wavelength, as shown in Fig. 4.40(a) and parallel to the broader dimension of the waveguide. If the probes are made parallel to the narrow dimension, the $TE_{11}$ mode is excited (see Fig. 4.40b). A probe placed

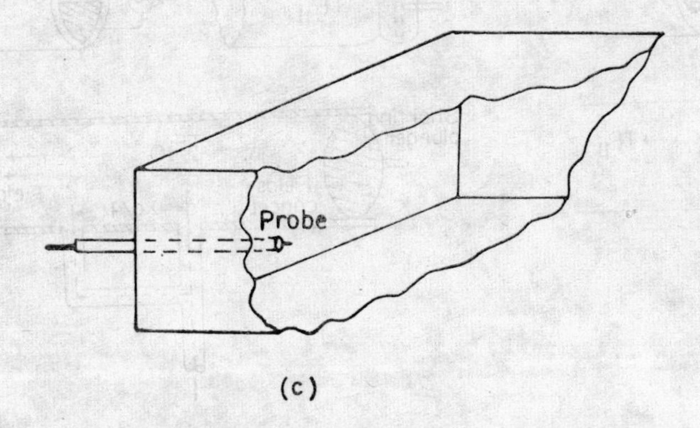


**Fig. 4.40** (*c*) Excitation of $TM_{11}$ mode in a rectangular waveguide.

parallel to the guide axis excites $TM_{11}$ (Fig. 4.40c). Excitation of other higher modes is shown in Fig. 4.40d.

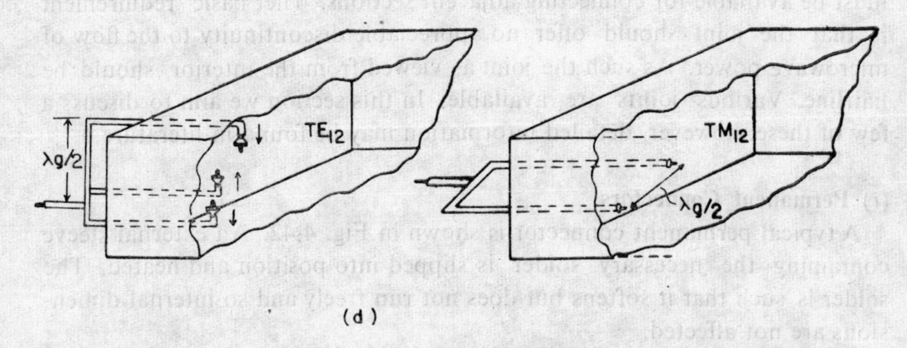


**Fig. 4.40** (*d*) Excitation of $TE_{12}$ and $TM_{12}$ modes in a rectangular waveguides.

Figure 4.41 illustrates the excitation of cylindrical waveguides.

The receiving problem is the reverse of the exciting problem, and in general, any method which works well for exciting will also work well for receiving as desired by the reciprocity theorem.

The mathematical details of various excitation problems may be found in literatures.*

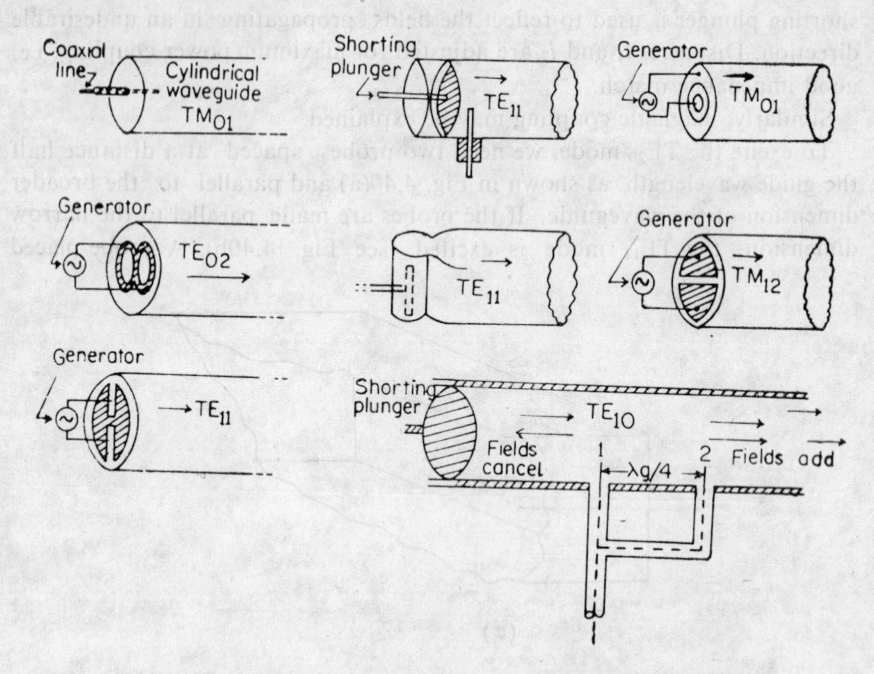


Fig. 4.41 Excitation of various modes in a cylindrical waveguides

## 4.14 WAVEGUIDE CONNECTORS AND FITTINGS

Regardless of the purpose for which waveguides may be used, some means must be available for connecting adjacent sections. The basic requirement is that the joint should offer no appreciable discontinuity to the flow of microwave power. As such the joint as viewed from the interior should be hairline. Various joints are available. In this section we aim to discuss a few of these; however, detailed information may be found in literature.*

### (*i*) Permanent Connectors

A typical permanent connector is shown in Fig. 4.42. An external sleeve containing the necessary solder is slipped into position and heated. The solder is such that it softens but does not run freely and so internal dimensions are not affected.

*J.C. Statter, *Microwave Electronics*, New York: D. Van Nostrand Co. Inc. (1950).

*1. G.C. Southworth, *Principles and Applications of Waveguide Transmission*, New York.
D. Van Nostrand Co. (1950).

*2. G. L. Ragan, *Microwave Transmission Circuits*, New York: McGraw-Hill Book Co. (1948), Vol. 9, Rad. Lab. Series.

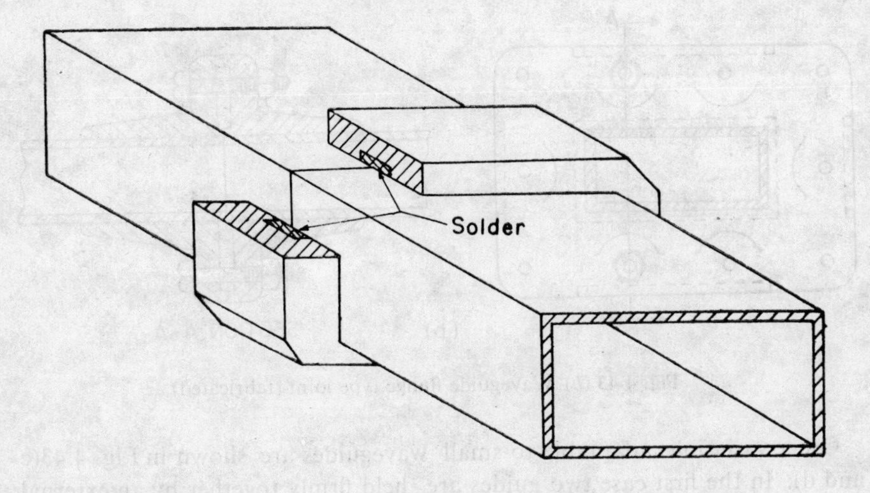


Fig. 4.42 A soldered-sleeve-type waveguide joint.

(*ii*) **Simple Flange Connectors**

In a laboratory permanent connectors do not serve the purpose and so waveguides are joined by means of simple flanges soldered to the ends of the pipes and latter screwed together. A typical simple and rather inexpensive flange is shown in Fig. 4.43(a). The banks are punched initially from brass sheet, soldered into position, and then bored with the help of an accu-

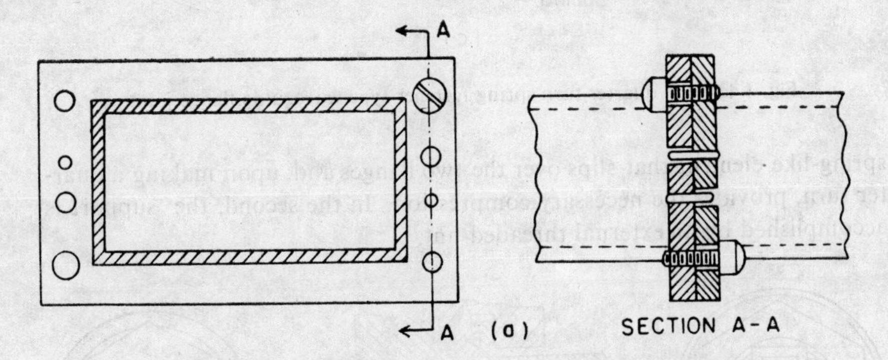


Fig 4.43 (*a*) A tailor-mode type waveguide flange joint with allignment dowels.

rate jig so that the various holes are in proper position relative to the inner walls. Two dowels provide the necessary alignment while four screws provide the necessary longitudinal strength. Two ends of the guides are machined or lapped. Sometimes pinch-type spring clips replace the screws as shown in Fig. 4.43(b) (fabricated).

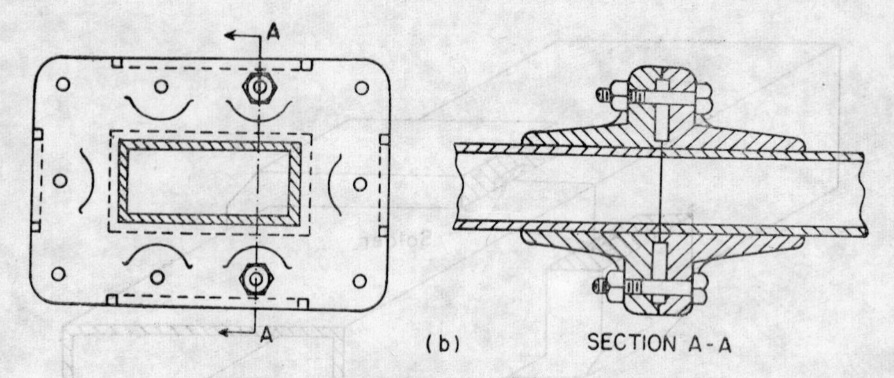


Fig. 4.43 (*b*) Waveguide flange type joint (fabricated).

Other variations adaptable to small waveguides are shown in Fig. 4.43(c and d). In the first case two guides are held firmly together by an external

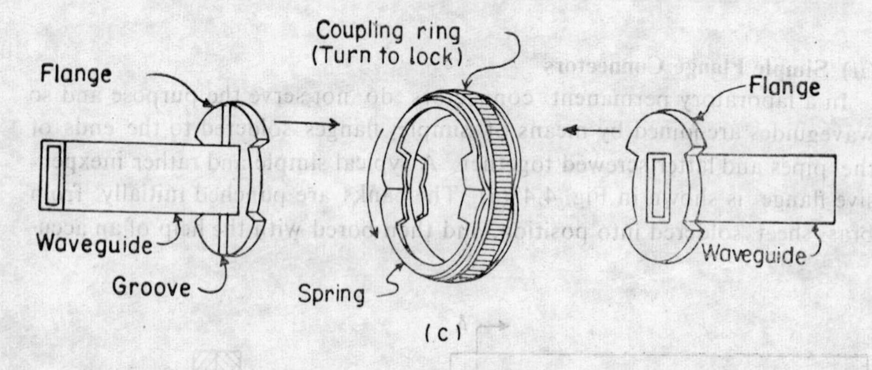


Fig. 4.43 (*c*) A quarter turn spring fastener type waveguide flange joint.

spring-like element that slips over the two flanges and, upon making a quarter turn, provides the necessary compression. In the second, the support is accomplished by an external threaded nut.

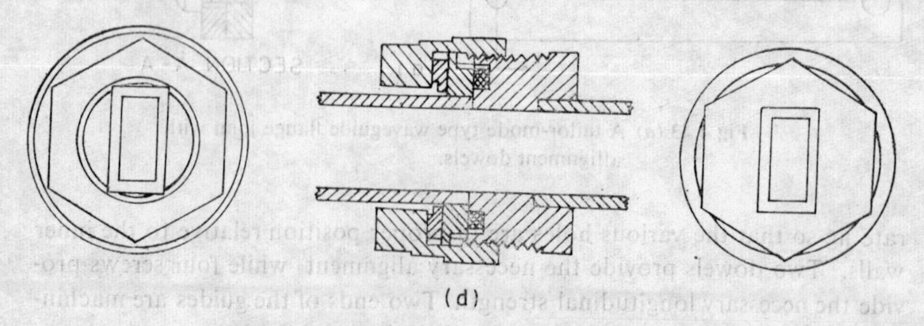


Fig. 4.43 (*d*) Threaded sleeve type waveguide flange.

### (*iii*) Choke-type Connectors

Figure 4.44 depicts the choke-type waveguide connectors. It utilizes the transmission line theory in a way that largely obviates the discontinuity due

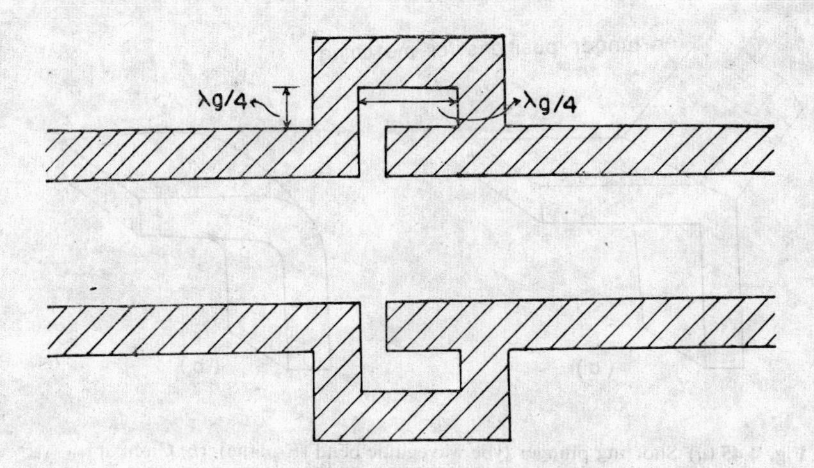


**Fig. 4.44** Choke type waveguide connector.

to a break in the conductivity of a waveguide wall. Two quarter-wave sections present a short circuit at the joint and electrical continuity is thus preserved.

### (*iv*) Other Waveguide Fittings

The use of waveguides in microwave transmission lines, necessarily calls for numerous associated fittings such as bends, rotators and transitions. Each of these entail discontinuities which are obviated by simple expedients. Two methods are commonly employed. In one, a second discontinuity is introduced that produces a reflection equal in magnitude and opposite in phase to the first, whereas, in the second method, the characteristics of one position of the line may be tapered gradually into the other.

*Bends*: Bends may be desired in the *E*-plane as well as in the *H*-plane. The bends provide a reactance that may be cancelled by providing a shorting plunger as shown in Fig. 4.45(a). There may many be positions in which the shorting piston can cancel reflections.

For tapering, the bends may be designed to form circular tapers of desired radii as shown in Fig. 4.45(b). A very satisfactory form of construction is illustrated in Fig. 4.45(c). A circular slot of the correct proportions is first turned in a metal disc. The disc is then cut by means of a thin saw into four quadrants. Two of these quadrants are sweated together to form a single 90° bend.

Figure 4.45(d) illustrates circular bends of both compensated and tapered type.

Bends and also other more complicated waveguide structures may be electroformed, i.e. constructed by depositing electrolytically the required thickness of metal on a carefully made matrix.

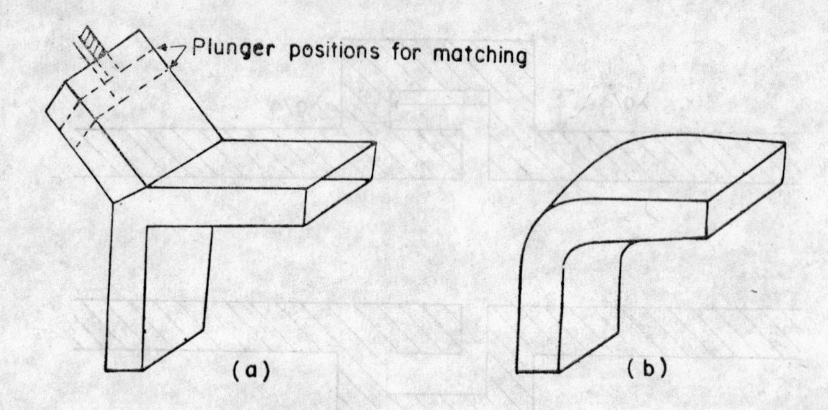


**Fig. 4.45** (*a*) Shorting plunger type waveguide bend (E-plane), (*b*) Circular taper type waveguide bend (E-plane).

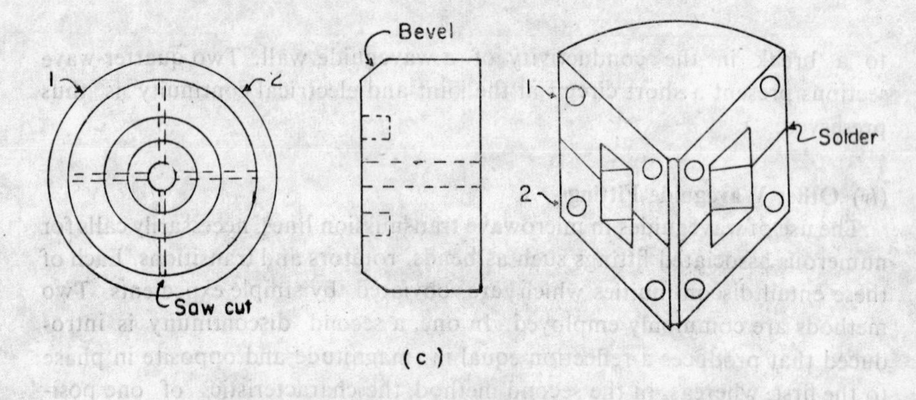


**Fig. 4.45** (*c*) Construction of an E-plane circular taper bend.

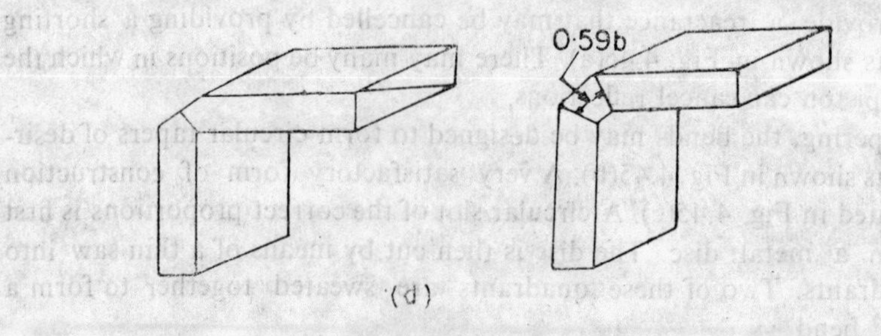


**Fig. 4.45** (*d*) Tapered bends (E-plane).

*Rotators*. Waveguide rotators are used to rotate the plane of polarization in a waveguide. In case of rectangular waveguides this may be done by filling the interior with a low melting point material and twisting the whole waveguide to the desired angle. The low melting point material is then removed. The circular waveguide rotator is made by placing a diametral septum with a progressive twist in the guide. The twists are shown in Fig. 4.46. These may also be electroformed.

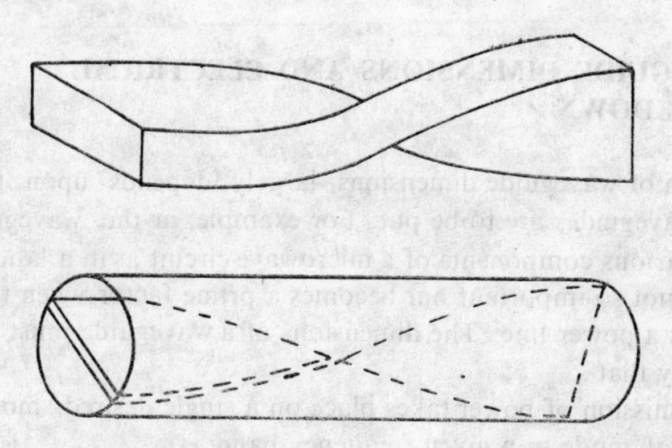

**Fig. 4.46** Waveguide twists or rotators (used to rotate plane of polarization).

*Transitions*. The transformation from one form or shape of a waveguide to another may be achieved by waveguide transformers shown in Fig. 4.47(a)

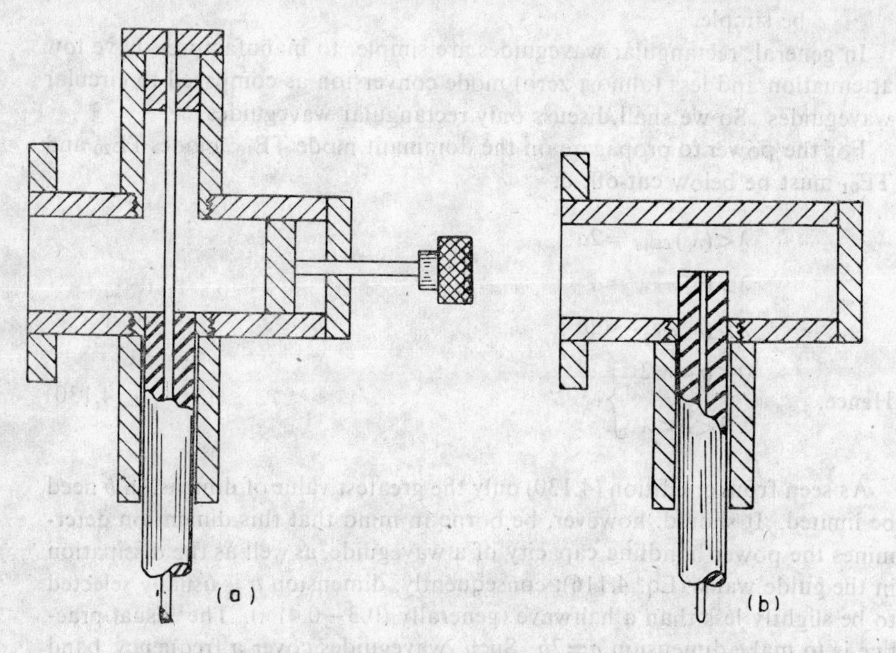


Fig. 4.47 Coaxial to waveguide transitions (*a*) turned induction type. (*b*) turned capacitance type.

(details of which may be found in chapter 6). However, it is mostly desired to have a transition from a coaxial line to a waveguide and vice versa. Figure 4.47(b) illustrates such transitions. Discontinuities resulting from sudden termination of the dielectric or transformation of the radial electric wave into a coaxial to the dominant wave in the waveguide are turned out by varying either coaxial tuner settings or the length of the central conductor protruding into the guide.

## 4.15 WAVEGUIDE DIMENSIONS AND ELECTRICAL BREAKDOWN

The selection of waveguide dimensions largely depends upon the use to which the waveguides are to be put. For example, in the waveguides used to connect various components of a microwave circuit as in a laboratory, attenuation is not so important but becomes a prime factor when the guide is to be used as a power line. The dimensions of a waveguide must be chosen in such a way that

(*i*) transmission of power takes place on a single desired, mostly fundamental, mode in a given frequency band;

(*ii*) the losses, i.e. attenuation, reflection, mode conversion and radiation losses are minimum; and

(*iii*) the power handling capacity is maximum for desired specifications such as weight and size. The process of manufacturing should also be simple.

In general, rectangular waveguides are simple to manufacture, have low attenuation and less (almost zero) mode conversion as compared to circular waveguides. So we shall discuss only rectangular waveguides.

For the power to propagate on the dominant mode $TE_{10}$, modes $TE_{20}$ and $TE_{01}$ must be below cut-off, or

$$\lambda < (\lambda_c)_{TE_{10}} = 2a$$

$$\lambda > (\lambda_c)_{TE_{20}} = a$$

$$\lambda > (\lambda_c)_{TE_{01}} = 2b$$

Hence,
$$\left.\begin{array}{l} 0 < b < \lambda/2 \\ \lambda/2 < a < \lambda \end{array}\right\}. \tag{4.130}$$

As seen from condition (4.130) only the greatest value of dimension $b$ need be limited. It should. however, be borne in mind that this dimension determines the power handling capacity of a waveguide, as well as the dissipation in the guide walls (Eq. 4.116); consequently, dimension $b$ is usually selected to be slightly less than a halfwave (generally $(0.3-0.4)\ \lambda$). The usual practice is to make dimension $a \simeq 2b$. Such waveguides cover a frequency band ranging from 6520 MHz to 13040 MHz for a $10 \times 23$ mm waveguide.

Coming to requirement (*iii*), in high power transmission waveguides there may be electric breakdown as in the case of low frequency lines, though this seldom happens. The breakdown occurs at the high or peak field region and appears in the form of a spark accompanied by a loud sound at atmospheric pressure. This results in very low conductivity and the reflections arising due to this fall in conductivity result in damage to the oscillator or the source which is highly undesirable. The power breakdown for $TE_{10}$ mode in a rectangular waveguide is given by

$$(P_{b \cdot d})_{TE_{10}} = \frac{ab}{4} \frac{E_{b \cdot d}^2}{\eta} \sqrt{1-(\lambda/2a)^2}. \quad (4.131)$$

$E_{b \cdot d}$ is the peak electric field when breakdown occurs. It may be noted that the larger the dimensions $a$ and $b$ the higher will be the breakdown power. But $a$ and $b$ are restricted by Eq. (4.130). Hence, optimum guide dimensions have to be selected. Conversely power ratings of a waveguide should be checked before use. Table 4.3 summarizes waveguide dimensions and power rating.

## 4.16 OTHER TYPES OF WAVEGUIDES

### (*i*) Coaxial Lines

While examining the possibilities of TEM modes in waveguides it was necessary to have a central conductor in which axial current can flow. Thus a circular conductor placed coaxially in a cylindrical waveguide will result in a guide that can propagate the TEM mode. This guide is known as the coaxial line and the dominant mode is shown in Fig. 4.48. (A discussion of the

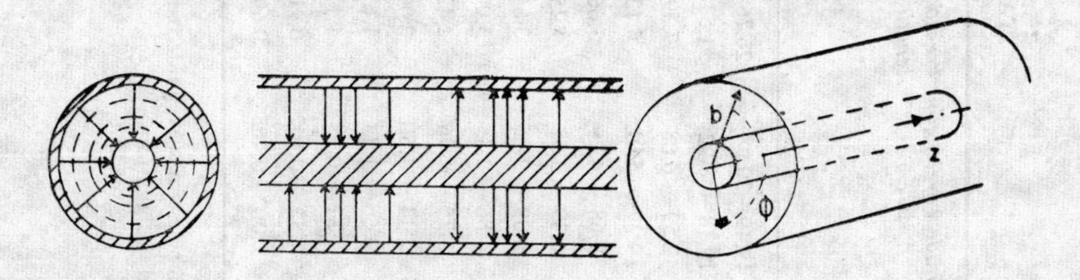


**Fig. 4.48** Dominal TEM mode in a coaxial line.

TEM mode has already been done in Chapter 2.) The field configuration may be obtained by substituting $H_z = E_z = 0$ in Eqs. 4.69 (a) and (b), that is

$$\frac{\partial E_\rho}{\partial z} = j\omega\mu H_\phi$$

$$\frac{\partial H_\phi}{\partial_z} = -j\omega\epsilon E_\phi$$

**TABLE 4.3 Waveguide Dimensions**

| S. No. | Band* | Frequency in GHz for $TE_{10}$ mode. | Guide wavelength in cm | Freespace wave-length in cm | Inner Dimensions in inches | | | Cutt-off for $TE_{10}$ Mode | | Theoretical CW power rating (mw) lowest to highest frequency |
|---|---|---|---|---|---|---|---|---|---|---|
| | | | | | (Broad) a | (Narrow) b | Tolerance | Frequency in GHz | Wave-length in cm | |
| 1. | L | 1.12– 1.70 | 45.49–20.98 | 26.76–17.63 | 6.500 | 3.250 | ±0.005 | 0.908 | 33.020 | 11.90–17.20 |
| 2. | S | 2.60– 3.95 | 19.18– 8.92 | 11.53– 7.09 | 2.840 | 0.340 | ±0.005 | 2.078 | 14.430 | 2.20– 3.20 |
| 3. | G | 3.95– 5.85 | 12.59– 6.08 | 7.59– 5.12 | 1.872 | 0.872 | ±0.005 | 3.152 | 9.510 | 1.40– 2.00 |
| 4. | C | 4.90– 7.08 | 9.37– 5.01 | 6.12– 4.25 | 1.590 | 0.795 | ±0.004 | 3.711 | 8.078 | —— |
| 5. | J | 5.85– 8.20 | 9.68– 4.29 | 5.66– 3.66 | 1.372 | 0.622 | ±0.004 | 4.301 | 6.970 | 0.56– 0.71 |
| 6. | H | 7.05–10.00 | 6.39– 3.52 | 4.25– 3.00 | 1.222 | 0.497 | ±0.004 | 5.259 | 5.700 | 0.35– 0.46 |
| 7. | X | 8.20–12.40 | 6.09– 2.85 | 3.66– 2.42 | 0.900 | 0.400 | ±0.003 | 6.557 | 4.572 | 0.20 – 0.29 |
| 8. | M | 10.00–15.00 | 4.86– 2.35 | 3.00– 2.00 | 0.750 | 0.375 | ±0.003 | 7.868 | 3.810 | —— |
| 9. | P | 12.40–18 00 | 3.75– 1.96 | 2.42– 1.67 | 0.622 | 0.311 | ±0.0025 | 9.486 | 3.160 | 0.12– 0.16 |
| 10. | N | 15.00–22.00 | 3.11– 1.60 | 2.00– 1.36 | 0.510 | 0.255 | ±0.0025 | 11.574 | 2.590 | —— |
| 11, | K | 18.00–26.50 | 2.66– 1.33 | 1.67– 1.13 | 0.420 | 0.170 | ±0.0025 | 14.047 | 2.134 | —— |
| 12. | R | 26.50–40.00 | 1.87– 0.88 | 1.13– 0.75 | 0.280 | 0.140 | ±0.0015 | 21.081 | 1.422 | 0.0822–0.031 |
| 13. | Q | 33.00–50.00 | 1.52– 0.702 | 0.91– 0.60 | 0.224 | 0.112 | ±0.0010 | 26.342 | 1.138 | 0.0140–0.020 |
| 14. | V | 50.00–75.00 | 1.002–0.468 | 0.60– 0.40 | 0.148 | 0.074 | ±0.0010 | 39.863 | 0.752 | 0.0063–0.009 |

*Although the operating frequencies for waveguide bands are standard, the latter designation varies with manufacturers. All except '*L*' and '*Q*' quoted here are Hewlett. Pacard designations.

which on using

$$E_\rho = E_\rho^o \, e^{j\omega t - \bar{\gamma}z} \text{ gives}$$

$$E_\rho = \frac{j\omega\mu}{\bar{\gamma}} H_\phi, \quad H_\phi = \frac{j\omega\epsilon}{\bar{\gamma}} E_\rho \tag{4.132}$$

and

$$E_\rho / H_\phi = \frac{j\omega\mu}{\bar{\gamma}} \times \frac{\bar{\gamma}}{j\omega\epsilon} = \sqrt{\mu/\epsilon} = \eta = Z_{\text{TEM}}.$$

Higher modes may also exist in coaxial waveguides if the separation of the conductors is greater than half the wavelength. The field distribution* of such modes is in terms of the Neumann and Bessel functions.

$$E_z^o = \{A_1 J_n (k_c\rho) + A_2 N_n (k_c\rho)\} \cos n\phi \qquad \text{(TM)}$$

$$H_z^o = \{B_1 J_n (k_c\rho) + B_2 N_n (k_c\rho)\} \cos n\phi. \qquad \text{(TE)}$$

For TM modes the boundary conditions require $E_z^o = 0$ at the metal surface. This yields

$$\frac{J_n (k_c a)}{J_n (k_c b)} = \frac{N_n (k_c a)}{N_n (K_c b)}. \tag{4.133}$$

Equation (4.133) gives the dispersion relation for higher TM modes in a coaxial wave guide.** The corresponding relation for TE modes will be

$$\frac{J_n' (k_c a)}{J_n' (k_c b)} = \frac{N_n' (k_c a)}{N_n' (K_c b)}. \tag{4.134}$$

For large values of $k_c a$ and $k_c b$, i.e. $k_c a \gg 1$ and $k_c b \gg 1$, the Bessel and Neumann functions may be replaced by their asymptotic values

$$J_n (k_c a) \simeq \sqrt{\frac{2}{\pi x}} \cos \left( k_c a - \frac{2n+1}{4} \pi \right)$$

$$N_n (k_c a) \simeq \sqrt{\frac{2}{\pi x}} \sin \left( k_c a - \frac{2n+1}{4} \pi \right).$$

For TM modes we have

$$E_z^o = A_1 \sin k_c (\rho - a) \cos \eta\phi \tag{4.135}$$

*For details, refer to N., Marcuvitz *Waveguide Hand Book*, New York: McGraw-Hill Book Co. (1951).

**We have written wave guide, (two separate words) to distinguish this from hollow waveguides.

Now $E_z^o$ must vanish at the other conductor, i.e. $\rho = b$.

Hence, $k_c(b-a) = m\pi$ or $k_c = \dfrac{m\pi}{(b-a)}$

$$m = 1,2,3. \ldots$$

The corresponding cut-off wavelength is

$$\lambda_c = 2\pi/k_c = \frac{2(b-a)}{m}. \tag{4.136}$$

Equation (4.135) has been depicted in Fig. 4.48(a) where clearly the propagation of a higher mode requires a large radius of the outer conductor than cylindrical waveguides.

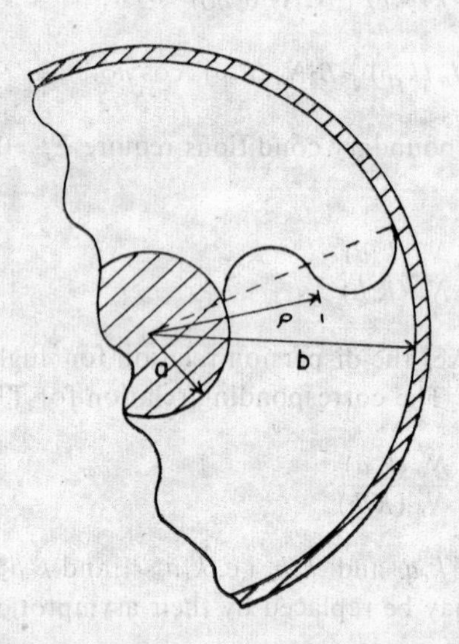


**Fig. 4.48** (a) Higher modes in coaxial lines.

An approximate estimation of the values of critical wavelengths for some simpler higher modes can be made by obtaining them from rectangular or circular waveguides.

Consider a rectangular waveguide operating in $TM_{11}$ mode as shown in Fig 4.49(a). Now if the broad wall of the waveguide is elongated infinitely the field pattern will be as in Fig. 4.49(b). If the walls are gradually curved and the joint is removed, what we obtain is the $TM_{01}$ mode in a coaxial wave guide. The cut-off wavelength will be $2b'$ or $2(b-a)$ where $b$ and $a$ are the outer and inner diameters of the inner and outer conductors respectively. The development of $TE_{01}$ mode in coaxial lines from $TE_{10}$ mode in

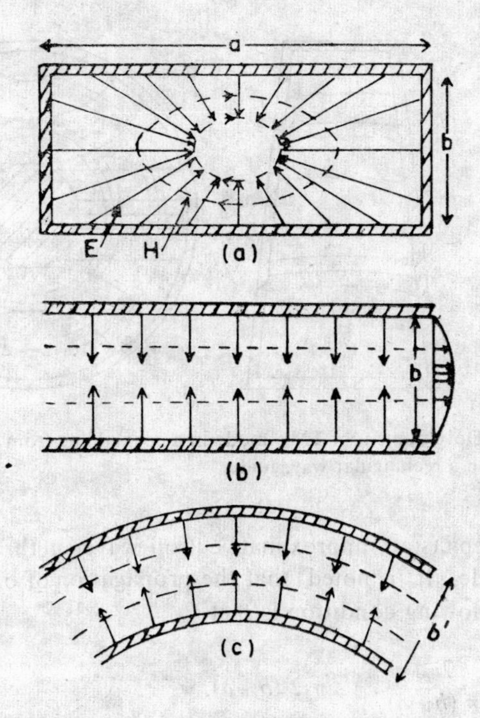


Fig. 4.49 Development of $TM_{01}$ mode in a coaxial wave guide from $TM_{11}$ mole in a rectangular waveguide.

rectangular waveguides is shown in Fig. 4.50 and the development of $TM_{11}$ mode in coaxial waveguides from $TE_{20}$ mode in rectangular waveguides is shown in Fig 4.51.

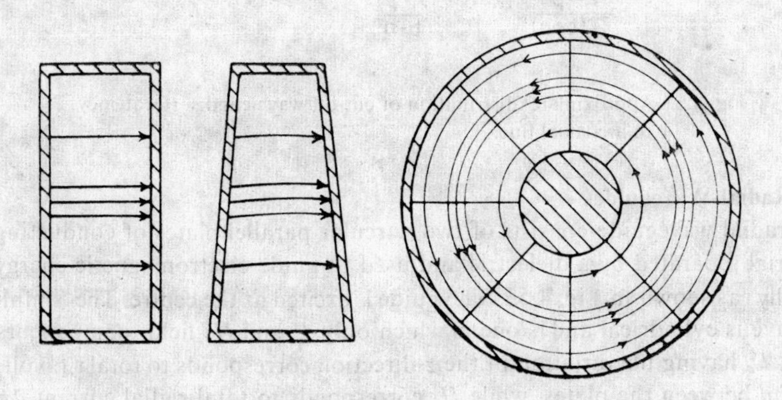

Fig. 4.50 Development of $TE_{10}$ mode in a coaxial waveguide from $TE_{10}$ mode in a rectangular waveguide.

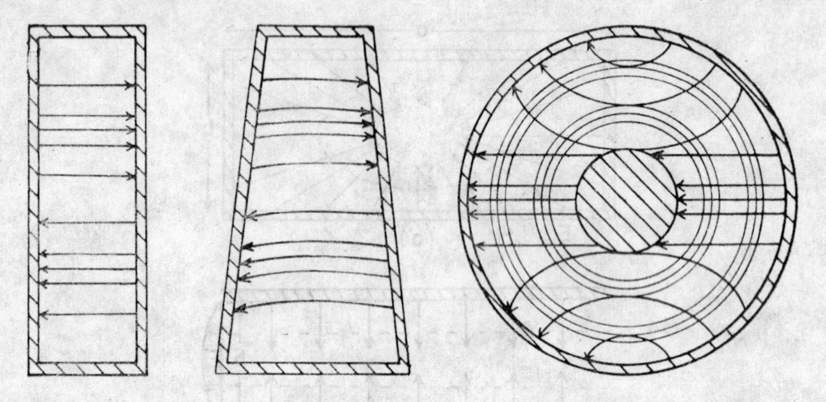

Fig. 4.51 Development of $TM_{11}$ mode in a coaxial line from $TE_{20}$ mode in a rectangular waveguide.

Figure 4.52 depicts an approximate cut-off wavelength spectrum for a coaxial waveguide. It is noted that the propagation of only one mode is ensured if the following condition is met

$$f < \frac{2v_p}{\pi (b+a)}, \lambda > \pi/2 (b+a). \tag{4.137}$$

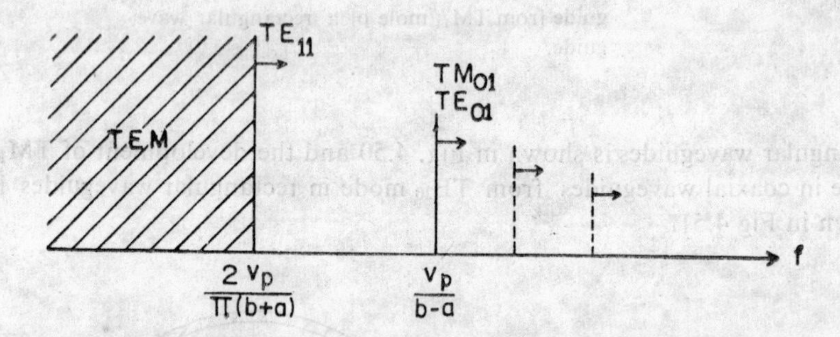


Fig. 4.52 Approximate distribution of cut-off wavelengths (frequency) in a coaxial line.

### (ii) Radial Waveguides

A radial waveguide consists of two circular parallel plates of conducting material separated by a dielectric and used to guide electromagnetic energy radially as shown in Fig. 4.53. The guide is excited at the centre. The simplest wave is cylindrical and is one in which only $E_z$ and $H_\phi$ field components exist. $E_z$ having no variation in the $z$-direction corresponds to total r.f. voltage $E_z d$ between the plates, while $H_\phi$ corresponds to total radial current $2\pi$ $H_\phi$, outward in one plate and inward in the other. Thus analysis of radial guides is similar to transmission line theory. However, since the cross-section of the radial guide is of a constant configuration but of variable dimensions, the guide is non-uniform.

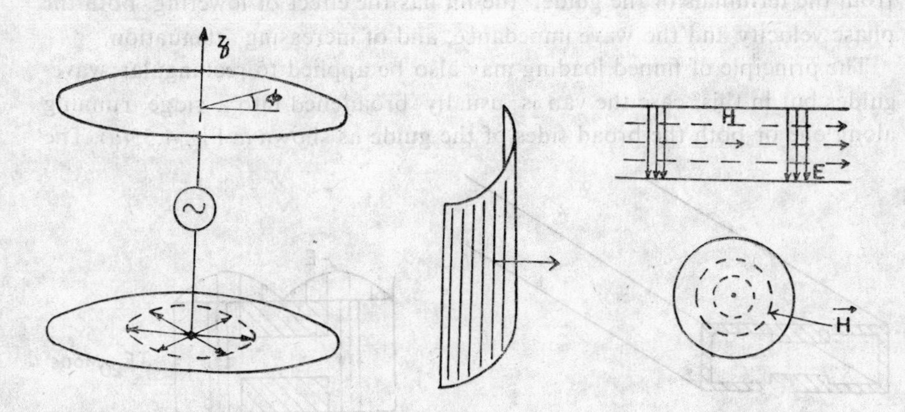


Fig. 4.53 Radial waveguide: Cylindrical wavefront.

Radial guides are seldom used as guiding structure, hence we omit analysis here and refer it to literature.*

### (iii) Fin and Ridge Waveguides**

If a rectangular waveguide having $a \gg b$ and $a$ supporting $TE_{10}$ mode, is gradually bent and the radius of the coaxial guide so formed is reduced to zero, leaving thereby a single longitudinal fin as shown in Fig. 4.54, what we obtain is known as a *fin waveguide*. The cut-off wavelength will clearly be,

$$\lambda_c = 2\pi b \tag{4.138}$$

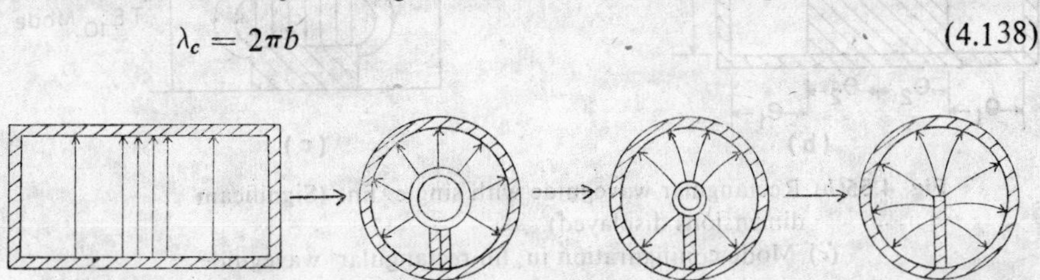

Fig. 4.54 Development of a fin-waveguide from a rectangular waveguide.

We see that cut-off wavelength of a fin waveguide is smaller than the cut-off wavelength of a cylindrical waveguide (without fin). Hence, the presence of a fin enhances the operating frequency range. In addition, it also tends to

*1. N. Marcuvitz, *Waveguide Hand Book*, New York: McGraw-Hill Book Co. (1951).
2. G. C. Montgomery and others, *Principles of Microwave Circuits*, New York: McGraw-Hill Book Co. (1947).

**For details consult: (i) S.B. Cohn, "Properties of ridged waveguides", *Proc. I.R.E.* Vol. 35 (Aug. 1947); (ii) W.L. Barrow, US Patent 2, 281, 252, filed Oct. 31, 1938; also *Elect. Engr.* Trans. Vol. 60 (March 1941).

maintain a fixed plane of polarization and avoid higher order waves. Thus, the effective band width of the guide is extended at both ends. As viewed from the terminals of the guide, the fin has the effect of lowering both the phase velocity and the wave impedance, and of increasing attenuation.

The principle of finned loading may also be applied to rectangular waveguides but in this case the van is usually broadened into a ridge running along one or both the broad sides of the guide as shown in Fig. 4.55(a). The

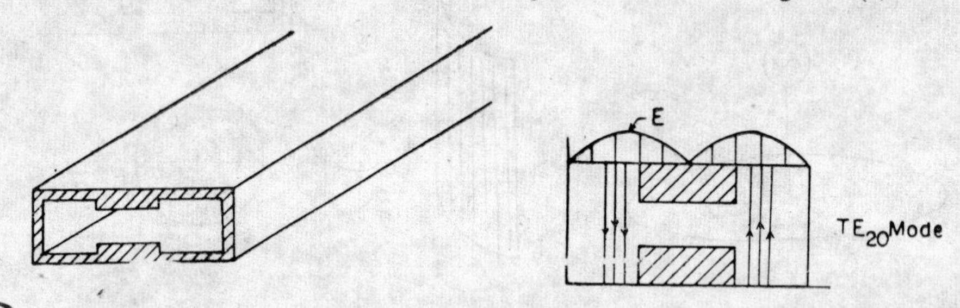


**Fig. 4.55(*a*)** Fin-rectangular waveguide,

presence of the ridge does not affect the higher modes as shown in Fig. 4.55 (a). but it lowers the cut-off wavelength of the dominant mode as desired. Referring to Fig. 4.55(c), the electrical distances $\theta_1$ and $\theta_2$, along the lateral

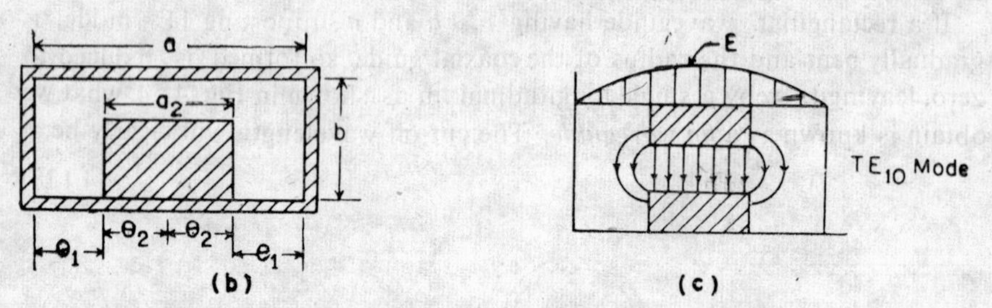


**Fig. 4.55(*b*)** Rectangular waveguide with single fin. (Significant dimensions displayed).
(*c*) Mode conifiguration in fin-rectangular waveguide.

dimension of the guide measured in cut-off wavelength $\lambda_c'$ units, can be written as

$$\theta_1 = \frac{a - a_1}{\lambda_c'} \pi, \ \theta_2 = \frac{a_1}{\lambda_c'} \pi.$$

The ratio of cut-off wavelengths of ridged and non-ridged waveguides is given by

$$\frac{1}{C} = \frac{\lambda_c'}{\lambda_c} = \frac{\pi}{2(\theta_1 + \theta_1)} \tag{4.139}$$

or

$$\frac{f_c'}{f_c} = C \text{ or } \left(\frac{f_c'}{f}\right) = \left(\frac{f_c}{f}\right) c.$$

The values of dimensional constant $C$ may vary from 1 to 5 and consequently, the cut-off frequency of a ridged waveguide is substantially more than the non-ridged rectangular waveguide. The guide wavelength and group velocity for a ridged waveguide respectively are given by

$$\lambda_g' = \frac{\lambda}{\sqrt{1-c^2(f_c/f)^2}} = \frac{\lambda}{\sqrt{1-c^2(\lambda/\lambda_c)^2}} \tag{4.140}$$

$$v_g' = \frac{v_p}{\sqrt{1-c^2(f_c/f)^2}} = \frac{\lambda}{\sqrt{1-c^2(\lambda/\lambda_c)^2}}.$$

The conductor attenuation for a ridged waveguide is given by

$$\alpha_c' = \frac{0.104}{\sqrt{\lambda}\sqrt{1-c^2(f_c/f)^2}}\left\{\frac{1}{b}+\frac{2c^2}{a}(f_c/f)^2\right\} \times \left\{\frac{cC'}{b\left(\sin\theta_2+\dfrac{b'}{b}\cos\theta_2\tan\dfrac{\theta_1}{2}\right)}\right\} \tag{4.141}$$

where $C'$ is another constant less than 1.5.

We see from Eq. (4.141) that $\alpha_c'$ is more than $\alpha_c$ as given by Eq. (4.116). This is because there exists high current density in the vicinity of the ridge.

Thus in ridged and finned waveguides the frequency band is increased at the cost of power.

The presence of a ridge may be interpreted in yet another way following the lumped constant analogy. The presence of a ridge presents a lumped capacitance at the middle of the waveguide wall connected in parallel to imaginary quarter wave stubs constituted by waveguide walls. The inward projection in the middle increases the capacity and hence the cut-off wavelength corresponding to the resonance tank circuit referred to above, increases.

### (*iv*) Surface Waveguides*

Consider a dielectric slab placed in a medium of low dielectric constant. If a wave travelling from the slab to the medium strikes the second medium at an angle greater than a certain critical value say, $\theta_c$, than it suffers total internal reflection and returns to the slab, reflected back as shown in Fig. 4.56. In this way energy propagates along a dielectric rod. Such guides are known as surface guides, i.e. energy flows along the surface of the guid-

*For details the reader is referred to:

1. C.H. Chandler, "An investigation of dielectric rod as waveguide", *J. Appl. Phys.* Vol. 20 (Dec. 1949).

2. M., "Cohn, TE modes of dielectric loaded through line" *IRE Trans.*, Vol MTT-8, (July 1960).

3. Young, L.. (edited) *Advances in Microwaves*, New York: Vol. 4, Academic Press, (1969).

ing structure. The fields excited on the rod decay exponentially in the natural direction away from the slab and the dielectric medium discontinuity acts as guiding boundary. However, it is not as sharp as the metal boundary.

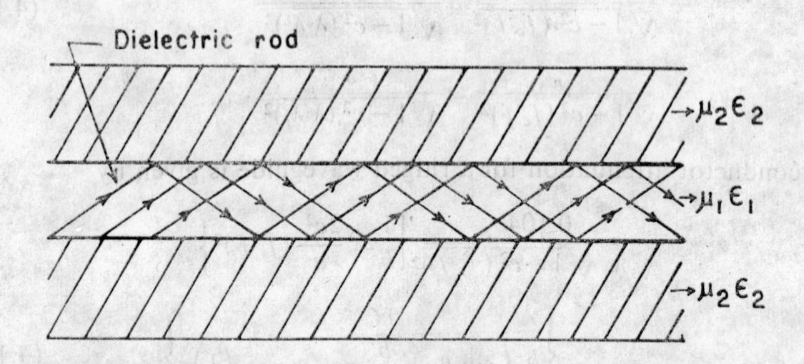


**Fig. 4.56** Dielective rod as a surfuce waveguide.

The critical angle of incidence is a function of frequency and hence there will be a cut-off wavelength. For the energy to be confined in the slab (or near the rod), its thickness must be exactly half the wavelength measured at a phase velocity transverse to the slab, i.e.

$$d=\frac{\lambda}{2\cos\theta}=\frac{1}{2f\sqrt{\mu_1\epsilon_1}\cos\theta}$$

The cut-off frequency is (put $\theta=\theta_c$)

$$f_c=\frac{1}{2d\sqrt{\mu_1\epsilon_1}\cos\theta_c}=\frac{1}{2d\sqrt{\mu_1\epsilon_1-\mu_2\epsilon_2}}. \tag{4.142}$$

If $\mu_1\epsilon_1 \gg \mu_2\epsilon_2$, the cut-off frequency of the slab is the same as if it were a conducting guide. This result is of immense importance because it suggests reduction in dimensions for higher frequency which is advantageous for design considerations.

To get an idea about the field pattern of various modes, we consider a dielectric rod ($\mu_1\epsilon_1$) placed in another medium ($\mu_2\epsilon_2$). The solutions of Maxwell's equations will be in terms of the Bessel and Hankel functions of the second kind. The continuity of the fields $E_z^o$ and $H_\phi^o$ at the dielectric interface requires

$$\frac{J_o(c_{c_1}a)}{J_1(c_{c_1}a)}=\frac{\epsilon_1 c_{c_2}}{\epsilon_2 c_{c_2}}\,\frac{H_o^{(2)}(c_{c_2}a)}{H_1^{(2)}(c_{c_2}a)} \tag{4.143}$$

a being the diameter of the rod.

If the waves are confined in the rod, i.e. in medium (1) and no energy is transmitted to medium (2), the fields must decay exponentially in medium (2), which is possible only when $k_{c_2}^2 < 0$. Hence, the critical condition is

$$k_{c_2} = 0. \tag{4.144}$$

The critical propagation constant is

$$J_c = j\beta_c = j\omega\sqrt{\mu_2.\epsilon_2} \tag{4.145}$$

This is propagation without attenuation and phase velocity equals characteristic velocity of medium 2.

Again for $k_{c_2} a = 0$, we find that $J_o (k_{c_1} a) = 0$. If $p_{ol}$ be the $l^{th}$ root of $J_o (k_{c_1} a)$, then $k_{c_1} a = p_{ol}$. Hence the cut-off frequency is

$$f_c = \frac{p_{ol}}{2\pi a\sqrt{\mu_1\epsilon_1 - \mu_2\epsilon_2}}. \tag{4.146}$$

The lowest root is 2.405 and so for $\mu_1\epsilon_1 \gg \mu_2\epsilon_2$, the cut-off frequency approaches the cut-off of a cylindrical waveguide supporting the $TM_{01}$ mode. It is instructive to note that large negative values of $k_{c_2}^2$ correspond to very high frequencies and fields are restricted within the rod, but negative small values of $k_{c_2}^2$ imply that the fields are radiated in medium (2). This property is utilized for designing dielectric antennas.

An analysis will show that not only do TE, TM and TEM modes propagate in a dielectric rod, but waves of the hybrid type TE $TM_{mn}$ or $HE_{mn}$ also propagate as shown in Fig 4.57. In this case all the six components of the fields exist. This type of mode features infinite cut-off wavelength with least radiation. This expression for the phase velocity, group velocity and attenuation constant is in terms of transcandental equations*.

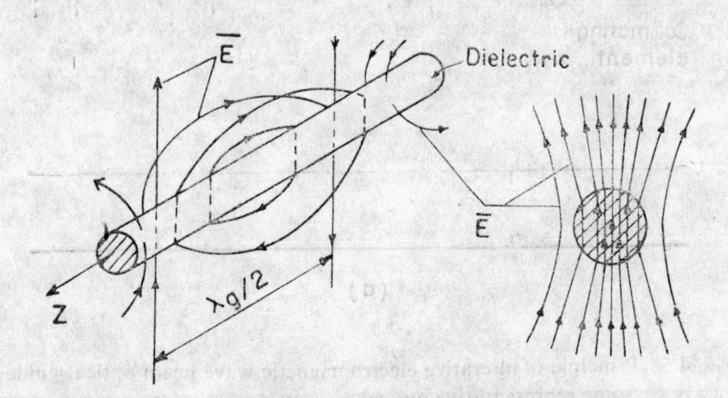


**Fig 4.57** Filed distribution dielectric rod with TE $TM_{11}$ hybrid mode.

Theoretically, surface waveguides may be used for the entire range of microwave frequency spectrum. However, in practice, low-loss materials

*S.A. Schelkunoff, *Electromagnetic Waves*, New York: D. Van Nostrand Co. (19.43)

with relative permittivity in excess of several thousands are not available; consequently, this type of guide is impracticable for frequencies below 100 $MH_z$. Similarly, dielectric or surface waveguides are not suitable for infra-red transmission.

**($v$) Optical Waveguides**

In communication engineering there has always been a tendency to exploit higher and higher frequencies of the electromagnetic spectrum for communication purposes and optical guides are step ahead of surface waveguides. These waveguides utilize a collimated laser beam for communication purposes. If available commercially, these waveguides may operate millions of channels at very low noise and attenuation (arising from diffraction, reflection and absorption) such as $\sim 10^{-3}$ $db/m$. Besides, these guides have a large bandwidth and the design of the antenna is greatly simplified. However, the non-availability of suitable moderators and demoderators is a negative point but with improving technology their development in the future is a distinct possibility.

(*a*) *Interative Electromagnetic Wave Beams.* Three representative waveguides of this class are shown in Fig. 4.58 (a, b and c). The collimated beam from the antenna (source) falls on an array of lenses which phase transform the bundle of rays thereby restoring the wavefront in a planer form so also the iris or mirror.

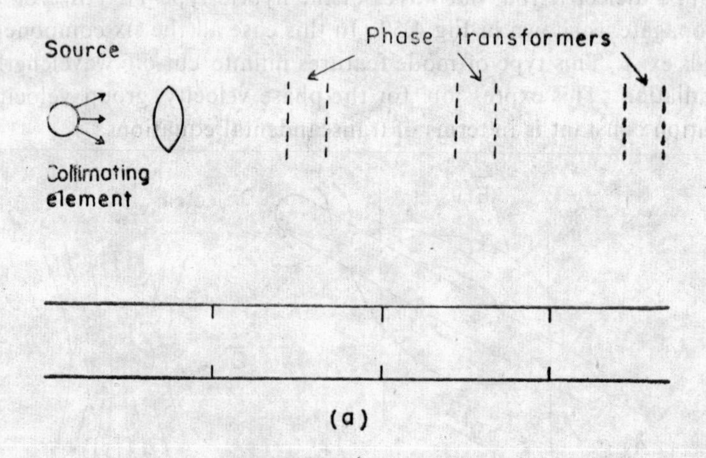


Fig. 4.58 Principle of interative electromagnetic wave beam optical guides. Some representative optical waveguides. (*a*) Ivis type optical waveguide.

Like other waveguides, optical waveguides support infinite modes each differing from the other in field distribution across the aperture (guide cross-section). The general form of *a* mode is

$$\psi_{mn}=f_{mn} \cos n\phi \, e^{\bar{\gamma}_n z} \tag{4.147}$$

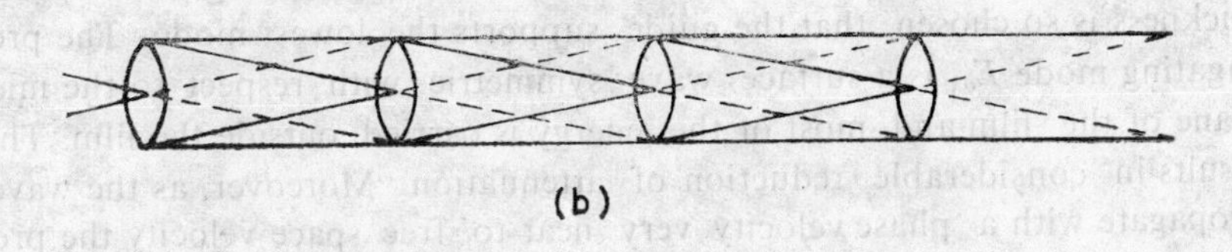

Fig. 4.58 (*b*) Lens type optical waveguide.

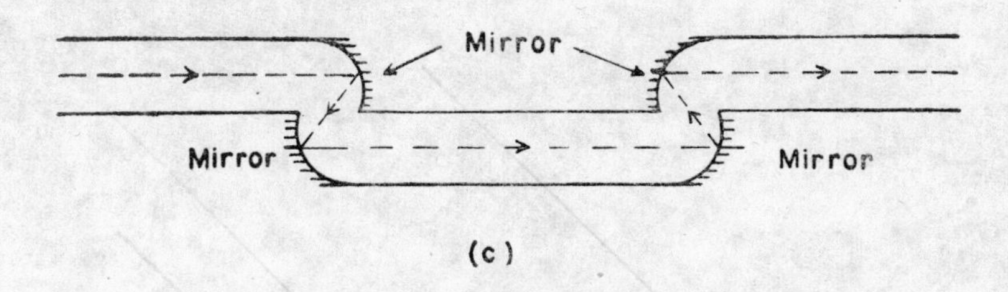


Fig. 4.58(*c*) Mirror type optical waveguide.

$f_{mn}$ is a function of the Laguerre polynomials and $\bar{\gamma}_n$ is the propagation factor for the $n^{th}$ mode. For lower order modes it differs slightly from the space propagation constant. The lowest mode is $f_{oo}$ and has a simple field distribution of the form $e^{-1/2 \frac{(\bar{\beta} r^2)}{l}}$ in the plane of every lense where $l$ is the spacing of lenses and $r$th radial distance. This mode has maximum energy concentration and is therefore of utmost practical importance.

(*b*) *Reflecting Pipes*. A silver metal pipe which has impedance and conductivity characteristics valid in the whole electromagnetic, spectrum, such as common hollow pipes, may be used to guide lasers if they have a perfect surface finish. The propagation characteristics of such optical guides are the same as for the guides discussed earlier. However, in optical guides, all the modes travel with almost the same velocity unlike metallic milimetre waveguides.

(*c*) *Fibre Guides* (*or Dielectric Rod Guides*). Two types of surface waveguides are available in the optical region. The first type support TE $TM_{11}$ mode while the second type have diameters ($10^3$ to $10^4$) $\lambda$ and therefore support numerous modes. Since phase velocities of these modes lie between $c$ and $c/n$ ($n$ being the refractive index) the dispersion effects are minimum. This results in minimum signal distortion. Gas fibres may also be synthesized by the application of suitable force of field and these carry lower attenuation as compared to dielectric fibres.

(*d*) *Optical Microguide.* A typical microguide in its simplest form, consists of a thin dielectric film mounted as illustrated in Fig. (4.59a). The thickness is so chosen that the guide supports the lowest mode. The propagating mode $E_{ox}$ is a surface wave symmetric with respect to the mid-plane of the film and most of the energy is carried outside the film. This results in considerable reduction of attenuation. Moreover, as the waves propagate with a phase velocity very near to free space velocity the problems of delay distortion, phase distortion, etc. are minimized. Figure 4.59(b) depicts an alternate way of supporting the microguide structure. The major disadvantage with these guides of Fig. 4.59(a and b) is that these can be bent in only one plane. Figure 4.59(c) illustrates another design in which the waveguide may be bent in any desired plane.

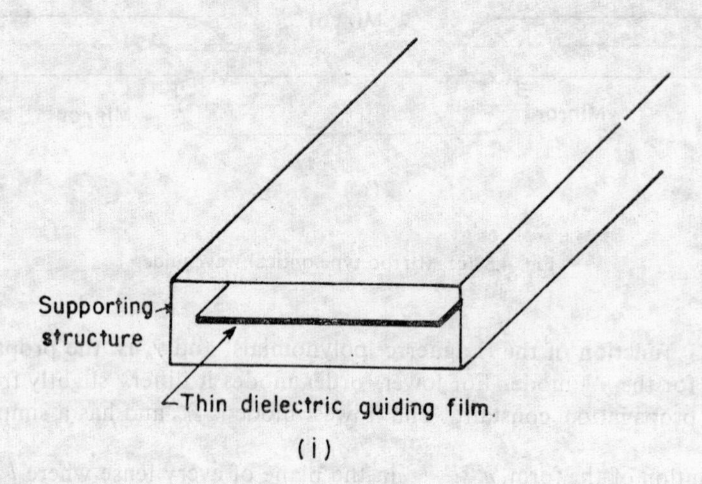


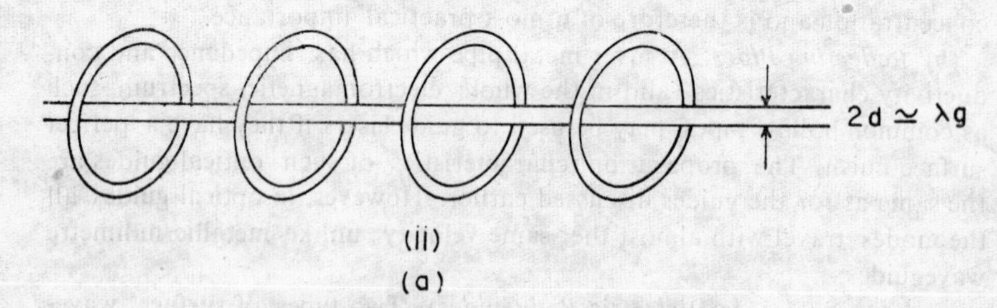


**Fig. 4.59**(*a*) Optical microguide supporting dominant mode.

The details of various optical guides and their characteristics may be found in literature*.

*L. Young, *Advances in Microwaves*, Vol. I, New York: Academic Press (1966).

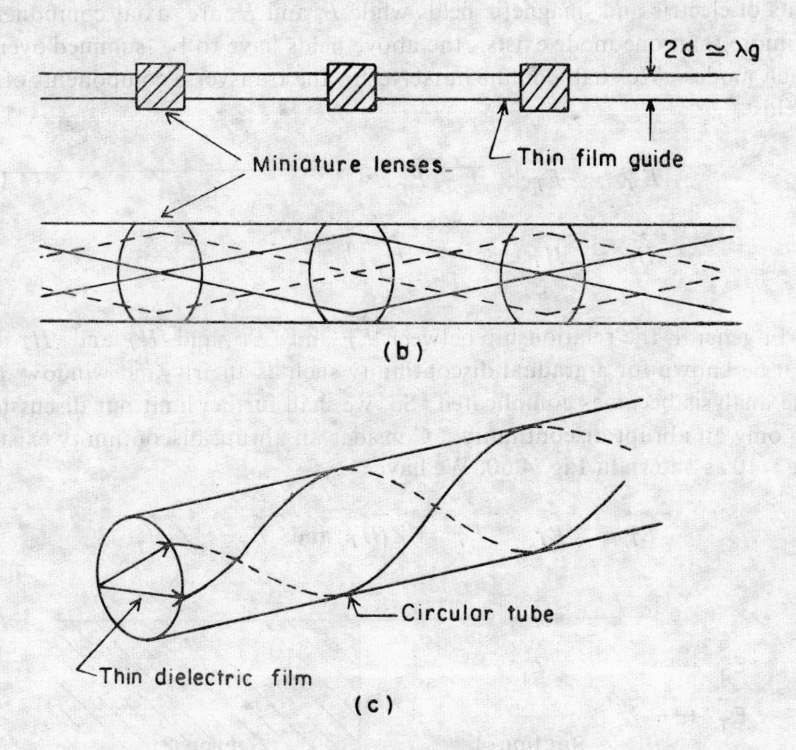


**Fig. 4.59**(*b*) Thin film optical guide supported by confocal system of cylindrical lenses. (*c*) The optical heliguide.

## 4.17 SIMPLE TRANSMISSION LINE TECHNIQUES APPLIED TO WAVEGUIDES

The total etectric and magnetic fields at any point of a waveguide are the sums of incident and reflected fields, and may be written as

$$\vec{E} = \{\vec{E}_T^+ \, e^{(j\omega t - \bar{\gamma} z)} + \vec{E}_T^- e^{(j\omega t + \bar{\gamma} z)}\} + \hat{z}\{\vec{E}_z^+ \, e^{(j\, t - \bar{\gamma} z)}$$

$$+ \, E_z^- \, e^{(j\omega t + \bar{\gamma} z)}\}$$

$$\vec{H} = \{\vec{H}_T^+ \, e^{(j\omega t - \bar{\gamma} z)} - \vec{H}_T^- \, e^{(j\omega t + \bar{\gamma} z)}\}$$

$$+ \hat{z} \{ \vec{H}_Z^+ \, e^{(j\omega t - \bar{\gamma} z)} + \vec{H}_Z^- \, e^{(j\omega t + \bar{\gamma} z)}\}$$

where subscripts + and − respectively indicate waves (fields) propagating in the positive and negative *z*-directions. $E_T$ and $H_T$ are the transverse components of electric and magnetic fields while $E_z$ and $H_z$ are axial components. If more than one mode exists, the above fields have to be summed over all such modes. We shall confine ourselves to the transverse components of the fields

$$\vec{E}_T^{+} = \vec{E}_T^{+} e^{j\omega t-\bar{\gamma}z} + \vec{E}_T^{-} e^{j\omega t+\bar{\gamma}z} \tag{4.148}$$

$$\vec{H}_T = \vec{H}_T^{+} e^{j\omega t-\bar{\gamma}z} - \vec{H}_T^{-} e^{j\omega t+\bar{\gamma}z}.$$

In general, the relationship between $\vec{E}_T^{+}$ and $\vec{E}_T^{-}$, and $\vec{H}_T^{+}$ and $\vec{H}_T^{-}$ may not be known for a gradual discontinuity such as an iris and window and the analysis becomes complicated. So, we shall further limit our discussions to only an abrupt discontinuity. Consider an abrupt discontinuity existing at $z=0$ as shown in Fig. 4.60. We have

$$\vec{E}_T^{+} = \vec{E}_T^{-}, \quad \vec{E}_T^{+} = Z_1\vec{H}_T^{+} \text{ and } \vec{E}_T^{-} = Z_1\vec{H}_T^{-}$$

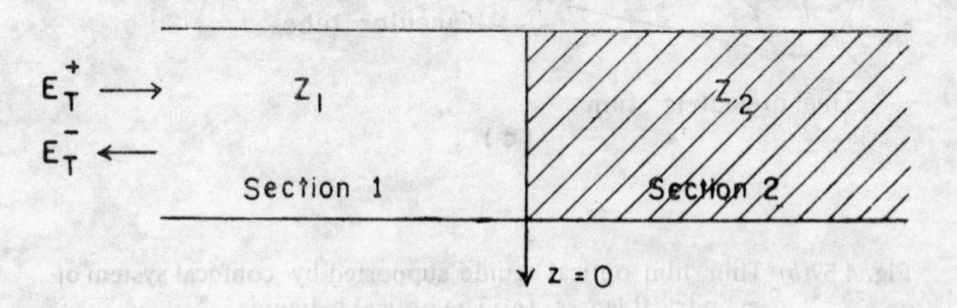


**Fig 4.60** Abrupt waveguide discontinuity at $z=0$.

where $Z_1$ is the wave impedance of section 1. If we omit time dependence, Eq. (4.148) may be written as

$$\vec{E}_T = \vec{E}_T^{+} e^{-\gamma z} + \vec{E}_T^{-} e^{\bar{\gamma} z} \tag{4.149}$$

$$\vec{H}_T = \frac{1}{Z_1} (\vec{H}_T^{+} e^{-\bar{\gamma} z} - \vec{E}_T^{-} e^{-\bar{\gamma} z}).$$

At $z=0$, we have

$$\left|\frac{E_T}{H_T}\right|_{z=0} = Z_1 \frac{1 + |\vec{E}_T^{-}/\vec{E}_T^{+}|}{1 - |\vec{E}_T^{-}/\vec{E}_T^{+}|}. \tag{4.150}$$

If waveguide 2 is well matched, i.e. terminated in its characteristic impedance or the guide is of infinite length, then there will be no reflected wave in the second guide. It follows that $|\vec{E}_T|/|\vec{H}_T|$ at $z=0$ is equal to the characteristic impedance of the second guide. Setting the R.H.S. of Eq. (4.150) equal to $Z_2$ and solving for $|\vec{E}_T^-|/|\vec{E}_T^+|$; we find that the ratio of the reflected electric field to the incident electric field called the reflection coefficient, is given by

$$\Gamma_E = |\vec{E}_T^+|/|\vec{E}_T^-| = \frac{(Z_2/Z_1) - 1}{(Z_2/Z_1) + 1}. \tag{4.151}$$

We can obtain the impedance at any plane of medium 1, say at $z=-l$, by setting $z=-l$ in Eq. (4.149), that is

$$Z_{(-l)} = Z_1 \times \frac{(Z_2/Z_1)\cos h\,\bar{\gamma}_1 l - \sin \bar{\gamma}_1 l}{\cos h\,\bar{\gamma}_1 l - (Z_2/Z_1)\sin h\,\bar{\gamma}_1 l}$$

$$= Z_1 \frac{Z_2 \cot h\,\bar{\gamma}_1 l - Z_1}{Z_1 \cot h\,\bar{\gamma}_1 l - Z_2}. \tag{4.152}$$

It is instructive to compare Eq. (4.152) with Eq. (2.49). The comparison reveals that the waveguide operating in a single mode may be treated like a transmission line with impedance relations as expressed in Eq. (4.152). However, if the waveguide operates at several modes, each mode may be taken as equivalent to a separate transmission line.

The voltage standing wave ratio is nothing but the ratio of maximum to minimum electric field, that is

$$\text{VSWR} = \rho = \frac{|E_{\max}|}{|E_{\min}|} = \frac{|E_T^+| + |E_T^-|}{|E_T^+| - |E_T^-|} \tag{4.153}$$

or

$$\rho = \frac{1 + |E_T^-/E_T^+|}{1 - |E_T^-/E_T^+|} = \frac{1+\Gamma_E}{1-\Gamma_E}. \tag{4.154}$$

Equation (4.154) is identical with Eq. (2.75).

To bridge the gap between the theory of transmission lines and waveguides we now proceed to infer an equivalent circuit for a waveguide.

Let us define equivalent voltage $V$ and current $I$ to exist on a waveguide that gives the desired field pattern of the mode. Then

$$V = V^+ e^{\bar{\gamma}z} + V^- e^{-\bar{\gamma}z} \tag{4.155}$$

$$I = I^+ e^{\bar{\gamma}z} + I^- e^{-\bar{\gamma}z}$$

where $V^+ = K_1E^+$, $V^- = K_1E^-$ and $I^+ = K_2H^+$, $I^- = K_2H^-$ and $K_1K_2$ are arbitrary amplitude constants that relate voltage and current to the electric and magnetic fields respectively that exist on a waveguide.

If $N$ modes exist simultaneously, then

$$\vec{E}_T = \sum_{n=1}^{N} (V_n^+ K_{ln}^{-1} e^{-\bar{\gamma}_n z} + V_n^- K_m^{-1} e^{+\bar{\gamma}_n z})$$

$$\vec{H}_T = \sum_{n=1}^{N} (I_n^+ K_{2n}^{-1} e^{-\bar{\gamma}_n z} - I_n^- K_{2n} e^{+\bar{\gamma}_n z}) \tag{4.156}$$

It is seen that the characteristic impedance of the waveguide may be written (following transmission line theory) as

$$Z_{TE} \text{ or } Z_{TM} = Z_0 = \frac{K_1}{K_2}$$

and it differs from wave impedance by a multiplicative constant.

To derive a distributed parameter equivalent circuit we consider TM waves on a rectangular waveguide. Then, since

$$H_z = 0, \quad \therefore (\nabla \times \vec{E})_z = -\left(\frac{\partial \vec{B}}{\partial t}\right)_z = 0$$

that is, in the $x$, $y$ plane the electric field has no curl. In this plane $\vec{E}$ may be written as the negative gradient of a scalar potential $V$.

$$E_x = -\frac{\partial V}{\partial x}, \; E_y = -\frac{\partial V}{\partial y}. \tag{4.157}$$

Using $\quad H_z = 0$ in $(\nabla \times \vec{H})_x = \left(\frac{\partial \vec{D}}{\partial t}\right)_x,$

we have $$\frac{\partial H_y}{\partial z} = j\omega\epsilon E_x. \tag{4.158}$$

Substituting the value of $H_y$ from Eq. (4.24b) and $E_x$ from Eq. (4.157), for TM waves ($H_z = 0$) we obtain

$$\frac{\partial}{\partial z}\left(\frac{j\omega\epsilon}{k_c^2}\frac{\partial E_z}{\partial x}\right) = -j\omega\epsilon\frac{\partial V}{\partial x}$$

where

$$\frac{\partial}{\partial z}\left(\frac{j\omega\epsilon}{k_c^2}E_z\right) = -j\omega\epsilon V. \tag{4.159}$$

Again from

$$(\nabla \times \vec{E})_y = -\left(\frac{\partial \vec{B}}{\partial t}\right)_y$$

$$\frac{\partial E_x}{\partial z} - \frac{\partial E_z}{\partial y} = -j\omega\mu H_y$$

which on using Eq. (4.24b) (with $H_z = 0$), gives

$$\frac{\partial E_x}{\partial z} - \frac{\partial E_z}{\partial x} = -\frac{\omega^2\mu\epsilon}{k_c^2}\frac{\partial E_z}{\partial x}$$

where
$$\frac{\partial V}{\partial z}=\left(\frac{\omega^2\mu\epsilon}{k_c^2}-1\right)E_z$$
$$=-\left(j\omega\mu+\frac{k_c^2}{j\omega\epsilon}\right)\left(\frac{j\omega\epsilon}{k_c^2}\right)E_z. \tag{4.160}$$

The quantity $j\omega\epsilon\, E_z$ is the longitudinal displacement current density and $1/k_c^2$ has dimensions of area, so $j\omega\epsilon\, E_z/k_c^2$ represents a current in the $z$-direction and is designated by $I_z$. Then Eqs. (4.159) and (4.160) become

$$\frac{\partial I_z}{\partial z}=-\,j\omega\epsilon\, V,\quad \frac{\partial V}{\partial z}=-\left\{j\omega\mu+\frac{k_c^2}{j\omega\epsilon}\right\}I_z. \tag{4.161}$$

Equations (4.161) may be identified as transmission line equations for a loss-less line having series impedance per unit length $Z=j\omega\mu+\dfrac{k_c^2}{j\omega\epsilon}$ and a shunt admittance per unit length $Y=j\omega\epsilon$. Thus, a waveguide operating in the TM mode has an equivalent circuit shown in Fig. 4.61(a). Similarly, it can be shown that the rectangular waveguide operating on the TE mode has an equivalent circuit shown in Fig 4.61(b).

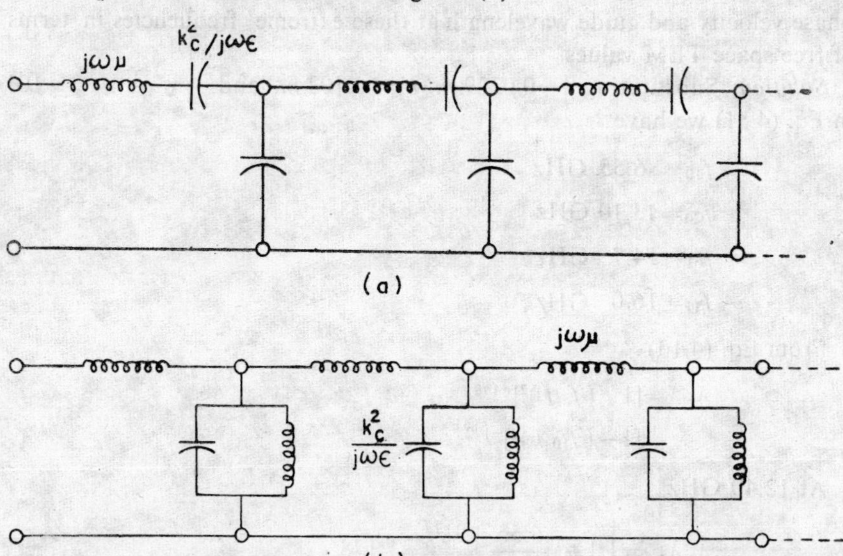


**Fig 4.61** Equivalent circuit for a rectangular waveguide operating in TE (*a*) TM and (*b*) TE mode.

The concept of waveguide as an equivalent transmission line with characteristic impedance

$$Z_{\text{TE}}=\sqrt{\frac{j\omega\epsilon}{j\omega\epsilon+\left(\dfrac{k_c^2}{j\omega\mu}\right)}}=\sqrt{\frac{\mu}{\epsilon}}\sqrt{\frac{1}{1-(\omega_c/\omega)^2}}$$
$$=\eta/\sqrt{1-(f_c/f)^2}$$

and $$Z_{TM} = \sqrt{\frac{j\omega\mu + (k_c^2/j\omega\epsilon)}{j\omega\epsilon}} = \eta\sqrt{1-(f_c/f)^2}$$

and propagation constant

$$\bar{\gamma}=j\bar{\beta}=j\sqrt{\omega^2\mu\epsilon-k_c^2}$$

is a powerful tool for solving many waveguide problems because it enables the engineer to obtain the solution by means of circuit and transmission line theory familiar to him. Concepts of stubs, terminated lines and transmission line charts are the indispensible aids available to him in transmission line theory.

## SOLVED EXAMPLES

**Example 4.1.** An air-filled waveguide of inside dimensions 22.9 mm$\times$10.2 mm is recommended for use in the dominant mode for frequencies 8.20 and 12.40 GHz. Find the four lowest cut-off frequencies. Also determine phase velocity and guide wavelength at these extreme frequencies in terms of free space TEM values.

*Solution*: Substituting $a=0.0229\ m$, $b=0.0102\ m$, and $1/\sqrt{\mu\epsilon} = 3\times10^8$ in Eq. (4.51) we have

$$f_{10}= 6.55 \text{ GHz}$$

$$f_{20}=13.10 \text{ GHz}$$

$$f_{01}=14.7 \text{ GHz}$$

$$f_{11}=16.0 \text{ GHz}.$$

From Eq. (4.13)

$$\lambda_g/\lambda=[1-(f_c/f)^2]^{-1/2}$$

$$=[1-(f_{10}/f)^2]^{-1/2}.$$

At 12.40 GHz,

$$\lambda_g/\lambda_g = \left[1-\frac{6.55}{12.40}\right]^{-1/2} =1.178.$$

At 8.20 GHz,

$$\lambda_g/\lambda = \left[1-\frac{6.55}{8.20}\right]^{-1/2} =1.662.$$

**Example 4.2.** Determine the attenuation constant for the waveguide in Ex. 4.1 operating at frequency 6 GHz.

*Solution*: From Eq. (4.28)

$$\bar{\gamma}=\alpha+j\beta=\sqrt{k_c^2-\omega^2\mu\epsilon}.$$

There is complete attenuation at 6 GHz because $f_{10}=6.55$ GHz

$$\bar{\gamma}=\alpha=2\pi\sqrt{\mu\epsilon}\,(f_{10}{}^2-f^2)^{\frac{1}{2}}$$

$$=2\times 3.14\times 3\times 10^8\,[6.55)^2-(6)^2]^{1/2}\cdot 10^9$$

$$=55 \text{ neper}/m$$

$$=478 \text{ db/m}.$$

The width of the waveguide is 22.9 mm

$$=10.9 \text{ db}.$$

**Example 4.3.** Calculate the propagation constants and phase velocities for the $TE_{10}$, $TE_{01}$, $TE_{11}$ and $TE_{02}$ modes for a rectangular waveguide with inside dimensions 7.214 cm $\times$ 3.404 cm operating at 5GHz.

*Solution*: From Eq. (4.30)

$$\beta_g=\sqrt{\beta^2-\beta_c^2}$$

$$\beta^2=\left(\frac{2\pi}{\lambda}\right)^2=4\pi^2f^2\epsilon\mu$$

$$\beta_c^2=\left(\frac{2\pi}{\lambda_c}\right)^2=\left(\frac{m\pi}{a}\right)^2+\left(\frac{n\pi}{b}\right)^2$$

$$\therefore\quad \beta_g=\frac{2\pi}{\lambda_c}\sqrt{\left(\frac{\lambda_c}{\lambda}\right)^2-1}.$$

In our case $f=5$ GHz,

$$\lambda=\frac{3\times 10^{10}}{5\times 10^9}=6 \text{ cm}.$$

*For* $TE_{10}$ *mode*

$$(\lambda_c)_{10}=2a=14.428 \text{ cm}$$

$$\beta_g=\frac{2\pi}{\lambda_{c10}}\sqrt{(2.404)^2-1}=0.954 \text{ rad/cm}$$

and

$$\lambda_g=\frac{2\pi}{\beta_g}=6.6 \text{ cm}$$

$$v_p=\frac{2\pi f}{\beta_g}=\frac{1}{\sqrt{\mu\epsilon}}\,\frac{\lambda_g}{\lambda}=\frac{1.10}{\sqrt{\mu\epsilon}}$$

$$=3.3\times 10^8 \text{ m/s}.$$

*For* $TE_{01}$ *mode*

$$(\lambda_c)_{01}=2b=6.808 \text{ cm}$$

$$\beta_g=\frac{2\pi}{\lambda_{c01}}\sqrt{(1.134)^2-1}=0.495 \text{ rad/cm}$$

so that $\lambda_g = 12.7$ cm

$$v_p = \frac{2.12}{\sqrt{\mu\epsilon}} = 6.36 \times 10^8 \text{ m/s.}$$

*For* TE$_{11}$ *mode*

$$(\lambda_c)_{11} = \frac{2ab}{\sqrt{a^2+b^2}} = 6.25 \text{ cm}$$

$$\beta_g = \frac{2\pi}{\lambda_{c11}} \sqrt{(1.025)^2 - 1} = 0.284 \text{ rad/cm}$$

$$\lambda_g = 22 \text{ cm}$$

$$v_p = \frac{3.68}{\sqrt{\mu\epsilon}}$$

$$= 11 \times 10^8 \text{ m/s.}$$

*For* TE$_{02}$ *mode*

$$\lambda_{c02} = b$$

$$\beta_g = \frac{2\pi}{b}\sqrt{\left(\frac{b}{\lambda}\right)^2 - 1.}$$

In our case $b < \lambda$

$$\therefore \qquad \beta_g = j\frac{2\pi}{b}\sqrt{1 - \left(\frac{b}{\lambda}\right)^2} = j\beta_g'.$$

There is thus only an evanescent mode with

$$\alpha = \beta_g' = \frac{2\pi}{a}\sqrt{1 - (0{\cdot}568)^2} = 1.51 \text{ neper/m.}$$

**Example 4.4.** An air-filled rectangular waveguide with inner dimensions 2 cm $\times$ 1 cm transports 0.5 h.p. energy in the dominant TE$_{10}$ mode at 30 GHz frequency. What is the peak value of the electric field occurring in the guide?

*Solution*: The field components of the dominant TE$_{10}$ mode are given by Eq. (4.62) where peak value of the electric field is

$$|E_{oy}| = \frac{\omega\mu\beta}{\pi/a} = \frac{\omega\mu\beta a}{\pi}.$$

The power flowing through the waveguide is given by Eq. (4.66)

$$P_T = \frac{ab\omega^2\mu^2\beta^2}{4Z_{TE}\,(\pi/a)^2}\text{ W.}$$

Thus

$$373=\frac{ab\omega^2\mu^2\beta^2}{4\left(\frac{\omega\mu}{\beta_z}\right)\left(\frac{\pi}{a}\right)^2}$$

giving us

$$\beta^2=\frac{373\times 4\times\left(\frac{\omega\mu}{\beta_z}\right)\left(\frac{\pi}{a}\right)^2}{ab\omega^2\mu^2}$$

$$=\frac{373\times 4\times\pi^2}{a^3b\omega\mu\beta_z}$$

Thus $|E_{oy}|=\frac{\omega\mu a\beta}{\pi}=\frac{\omega\mu a}{\pi}\sqrt{\frac{373\times 4\times\pi^2}{a^3b\omega\mu\beta_z}}.$

Now $\beta_z=\sqrt{\omega^2\mu\epsilon-\left(\frac{\pi}{a}\right)^2}$

$$=\pi\sqrt{\frac{4\times 9\times 10^{20}}{9\times 10^{16}}-\frac{1}{4\times 10^{-4}}}$$

$$=193.5\pi=608.81 \text{ rad/m}$$

giving us

$$|E_{oy}|=\sqrt{\frac{\omega\mu\times 373\times 4}{ab\beta_z}}$$

$$=\sqrt{\frac{2\times 3.14\times 3\times 10^{10}\times 4\pi\times 10^{-7}\times 373\times 4}{2\times 10^{-2}\times 10^{-2}\times 193.5\ \pi}}$$

$$=53.87 \text{ KV/m}.$$

**Example 4.5.** Classify the different cut-off frequencies upto a frequency four times the frequency of the dominant mode for $TE_{mn}$ and $TM_{mn}$ modes which can propagate in a rectangular guide with inner dimensions

$$a=0.282 \text{ cm and } b=0.564 \text{ cm}.$$

What is the range of frequencies for which dominant mode only propagates? What is the range for the $TE_{10}$ mode alone?

*Solution*: Here $b/a=2$. The cut-off wavelength is given by

$$\lambda_c=\frac{2}{\sqrt{\left(\frac{m}{a}\right)^2+\left(\frac{n}{b}\right)^2}} \quad \text{or} \quad f_c=\frac{c}{2}\sqrt{\left(\frac{m}{a}\right)^2+\left(\frac{n}{b}\right)^2}$$

Note that the lowest cut-off frequency is for $TE_{01}$ mode which is of course the dominant mode ($TM_{01}$ does not exist).

$$(f_c)_{01}=\frac{c}{2b}=26.4 \text{ GHz}.$$

Also

$$(f_c)_{0n}=\frac{nc}{2b}=n(f_c)_{01}.$$

And

$$(f_c)_{mn}=\frac{c}{2b}\sqrt{(2m)^2+(n)^2}$$

$$=(f_c)_{01}=\sqrt{4m^2+n^2}.$$

Modes for which frequencies are less than $4(f_c)_{01}$ are those for which

$$4\,m^2+n^2\leqslant 16.$$

That is,

$$m=1 \quad n=1 \;;\; (f_c)_{11}=(f_c)_{01}\sqrt{5}$$

$$m=1 \quad n=2 \;;\; (f_c)_{12}=(f_c)_{01}\sqrt{8}$$

$$m\ 01 \quad n=3 \;;\; (f_c)_{13}=(f_c)_{01}\sqrt{13}.$$

The lowest TM mode is $TM_{11}$. The results are schematically shown in the figure below.

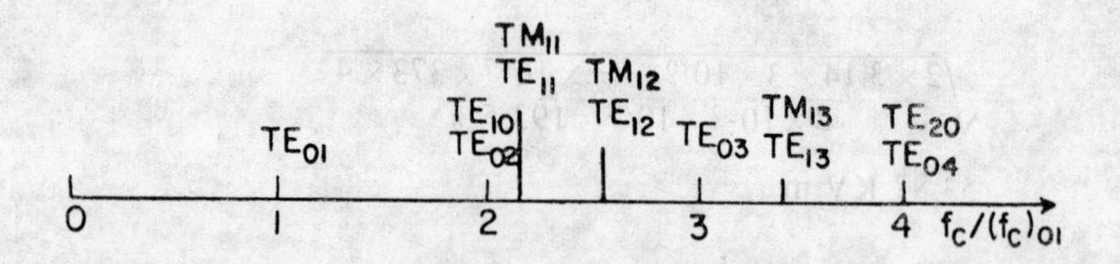


**Fig. Ex. 4.8**

For only the dominant mode $TE_{01}$ propagates

$$(f_c)_{01}\leqslant f\leqslant(f_c)_{02}$$

$$(f_c)_{01}\leqslant f\leqslant 2(f_c)_{01}$$

$$26.4 \text{ GHz}\leqslant f\leqslant 52.8 \text{ GHz}.$$

It is impossible to propagate the $TE_{10}$ mode alone since when

$$(f_c)_{10}<f<(f_c)_{11}$$

besides the $TE_{10}$, $TE_{02}$ mode will also propagate.

**Example 4.6.** Calculate the attenuation constant for the rectangular waveguide operating on $TE_{10}$ mode and having inner dimension 2.286 cm$\times$1.016 cm. The frequency of operation is 9.6 MHz and the conductivity of walls is $5.8\times10^7$ ℧/*m*.

*Solution*:

$$\alpha=\frac{R_s}{\eta b}\,\frac{1+\dfrac{2b}{a}\left(\dfrac{\lambda}{\lambda_c}\right)}{\eta b\sqrt{1+(\lambda/\lambda_c)^2}}$$

$$\lambda=\frac{3\times10^8}{9.6\times10^9}=0.0312 \text{ m}=3.12 \text{ cm}$$

$$\frac{\lambda}{\lambda_c}=\frac{\lambda}{2a}=\frac{3.12}{4.572}=0.684$$

which on substitution yields,

$$\alpha\simeq1.3\times10^{-2} \text{ neper/m}.$$

**Example 4.7.** It is required to propagate a single $TE_{10}$ mode in a rectangular waveguide of internal dimensions 2.686 cm$\times$1.016 cm. Calculate range of frequency at which the waveguide may be used. Also Calculate the variation of the guide wavelength over this frequency range.

*Solution*: The cut-off frequency for the $TE_{10}$ mode is

$$(f_c)_{10}=\frac{3\times10^8}{2a}=6.556 \text{ GHz}.$$

Writing the cut-off frequency in general form

$$(f_c)_{mn}=\frac{3\times10^8}{2a}\sqrt{(2.25\,m)^2+n^2}$$

(since $a/b=2.25$).

The next cut-off frequency is

$$(f_c)_{20}=2\,(f_c)_{10}<(f_c)_{01}=2.25\,(f_c)_{10}.$$

Fhus, the desired range of frequency is

$$6.556 \text{ GHz}\leqslant f\leqslant13.112 \text{ GHz}.$$

Again using

$$\lambda_g=\frac{\lambda_c}{\sqrt{\left(\dfrac{\lambda_c}{\lambda}\right)^2-1}}$$

$$=\frac{2a}{\sqrt{(f/f_c)^2-1}}.$$

Therefore,

$$f \to f_c, \ \lambda_g \to \infty$$

$$f = 1.1\, f_c, \qquad \lambda_g = 9.98 \text{ cm}$$

$$f = 2\, f_c, \qquad \lambda_g = \frac{2a}{\sqrt{3}} = 2.64 \text{ cm.}$$

Note that as expected, the guide wavelength moves drastically near cut-off frequency.

**Example 4.8.** An air-filled cylindrical waveguide of internal diameter 5 cm supports the $TE_{11}$ mode. Determine the cut-off frequency, guide wavelength and wave impedance at 3 GHz frequency.

*Solution*: Using Eq. (4.91), the cut-off frequency is

$$(f_c)_{TE_{11}} = \frac{p'_{11}}{2\,\pi a\,\sqrt{\epsilon\mu}}$$

$$= \frac{1.841 \times 3 \times 10^8}{2 \times 3.14 \times 5 \times 10^{-2}} = 1.758 \text{ GHz.}$$

Using Eq. (4.90), the phase constant is given by

$$\beta_{z,11} = \sqrt{\omega^2\mu\epsilon - p'_{11}/a}$$

$$= \sqrt{(2\pi \times 3 \times 10^9)^2\,(4\pi \times 10^{-7} \times 8.85 \times 10^{-12}) - (36.82)^2}$$

$$= 50.9 \text{ rad/m.}$$

Therefore

$$\lambda_g = \frac{2\pi}{\beta_{z,11}} = \frac{6.28}{50.9} = 12.3 \text{ cm.}$$

The wave impedance is given by Eq. (4.57)

$$Z_{TE} = \frac{\omega\mu}{\beta_z} = \frac{(2\pi \times 3 \times 10^9)\,(4\pi \times 10^{-7})}{50.9} = 465\ \Omega.$$

**Example 4.9.** An air-filled cylindrical waveguide of internal diameter 2 cm is to carry energy at 10 GHz. Find all possible modes of energy transmission in the guide.

*Solution*: Only those modes are possible for which the roots are less than that of

$$(\omega\sqrt{\mu\epsilon})a = k_c a = \frac{2 \times 3.14 \times 10^{10}}{3 \times 10^8}\ 2 \times 10^{-2}$$

$$= 4.18$$

that is,

$$k_c a \leqslant 4.18.$$

The possible modes are

| | |
|---|---|
| $TE_{11}$ (1.841) | $TM_{01}$ (2.405) |
| $TE_{21}$ (3.054) | $TM_{11}$ (3.832) |
| $TE_{01}$ (3.832). | |

Use Table 4.2 for $\lambda_c$ to calculate $(k_c = 2\pi/\lambda c)$

**Example 4.10.** A rectangular waveguide with internal dimensions 2.286 cm × 1.016 cm supporting $TE_{10}$ mode at 5 GHz is filled with a dielectric of relative permittivity $\epsilon_r$. What are the limits on $\epsilon_r$ if only the dominant mode $TE_{10}$ propagates? If $\epsilon_r = 2.25$ and loss tangent $\tan\delta = 10^{-3}$ calculate $\lambda_g$, $v_{pz}$ and $\alpha_d$.

*Solution*: In the absence of the dielectric, there will be propagation if $\lambda_c > \lambda$ or $\lambda < 2a = 4.572$ cm.
However,

$$\lambda = \frac{3 \times 10^{10}}{5 \times 10^{9}} = 6 \text{ cm}.$$

Thus propagation is not allowed.
When the guide is filled with dielectric $\lambda' = \lambda/\sqrt{\epsilon_r}$ and there will be a propagation for $\lambda' < \lambda_c$ or

$$\frac{\lambda}{\sqrt{\epsilon_r}} < 2a = 4.572 \text{ cm}.$$

Assuming that only the dominant mode propagates

$$(\lambda_c)_{20} < \lambda' < (\lambda_c)_{10}$$

or

$$a < \frac{\lambda}{\sqrt{\epsilon_r}} < 2a$$

or

$$\frac{\lambda}{2a} < \sqrt{\epsilon_r} < \frac{\lambda}{a}$$

$$\left(\frac{\lambda}{2a}\right)^2 < \epsilon_r < \left(\frac{\lambda}{a}\right)^2.$$

Substituting the values

$$\left(\frac{6}{2 \times 4.572}\right)^2 < \epsilon_r < \left(\frac{6}{4.572}\right)^2$$

or

$$1.72 < \epsilon_r < 6.88$$

For $\epsilon_r = 2.25$,

$$\lambda' = \frac{\lambda}{\sqrt{\epsilon_r}} = \frac{6}{\sqrt{2.25}} = 4 \text{ cm}.$$

Thus the guide wavelength in this case

$$\lambda_g = \frac{\lambda_c}{\sqrt{(1.14)^2 - 1}} = 8.35 \text{ cm}.$$

Again complex permittivity can be written as

$$\epsilon^* = \epsilon_o \epsilon_r (1 - j \tan \delta).$$

Now $\delta$ is small so

$$\tan \delta \simeq \delta$$

$$\therefore \epsilon^* = \epsilon_o \epsilon_r (1 - j\delta)$$

$$\therefore \quad \beta_{gz}^{*2} = \omega^2 \epsilon_o \mu_o \epsilon_r (1 - j\delta) - \beta_c^2$$

$$= \beta_{gz}^2 - j\delta\beta_z^2$$

or $$\beta_{gz}^* = \beta_{gz} \sqrt{1 - j\delta (\beta_z/\beta_{gz})^2}$$

$$\simeq \beta_{gz} \left[ 1 - \frac{j\delta}{2} \left( \frac{\beta_z}{\beta_{gz}} \right)^2 \right]$$

with $$\frac{\beta_z}{\beta_{gz}} = \frac{\lambda_g}{\lambda'} = \frac{8.35}{4}, \quad \delta = 10^{-3}$$

$$\beta_{gz}^* \simeq \beta_{gz} \left\{ 1 - j\frac{\delta}{\lambda} \left( \frac{\lambda_g}{\lambda'} \right)^2 \right\}.$$

The propagation constant is given by ($e^{-\beta_{gz}^*}$)

$$\beta_{gz}^* = -(j\beta_{gz} + \alpha_d)$$

$$\therefore \quad \alpha = \beta_{gz} \frac{\delta}{2} \left( \frac{\lambda_g}{\lambda'} \right)^2 = \pi\delta \frac{\lambda_{gz}}{\lambda'^2}$$

$$= 0.166 \times 10^{-2} \text{ neper/m}$$

and $$v_{pz} = \frac{\omega}{\beta_{gz}} = 4.17 \times 10^8 \text{ m/s}.$$

**Example 4.11.** A loss-less rectangular waveguide (2.286 cm × 1.016 cm), operating in the dominant $TE_{10}$ mode, is loaded with a dielectric plate of thickness $t$ and is followed by a matched load. Assuming $t$ to be small, show that the normalized input impedance just before the dielectric slab can be written as $z_1=1+jx_1$. Determine $x_1$ in terms of $\lambda$, $\lambda_g$, $t$ and $\epsilon_r$ and show that the dielectric plate behaves as a capacitance in parallel with the matched load. Calculate this capacitance for $f=9.8$ GHz, $\epsilon_r=2.25$ and $t=0.7$ mm.

*Solution*: Let the impedance of the matched load be $Z_o$. The desired input impedance is given by

$$Z_1=Z_o' \frac{Z_o+j\,Z_o' \tan \beta_{gz}' t}{Z_o'+j\,Z_o \tan \beta_{gz}' t}$$

where $Z_o'$ is the wave impedance of the guide loaded with the dielectric

$$Z_o=\frac{\omega\mu}{\beta_{gz}}=\eta\,\frac{\lambda_g}{\lambda_o}$$

$$Z_o'=\frac{\omega\mu}{\beta_{gz}'}=\eta\,\frac{\lambda_g}{\lambda_o}$$

for $f=9.8$ GHz, $Z_o=507\ \Omega$, and $Z_o=281\ \Omega$. Now,

$$z_1=\frac{Z_1}{Z_o}=\frac{Z_o'}{Z_o}\,\frac{1+j\left(\frac{Z_o'}{Z_o}\right)\tan \beta_{gz}' t}{\left(\frac{Z_o'}{Z_o}\right)+j\tan \beta_{gz}' t}$$

$$z\simeq K\,\frac{1+j\,K\,\beta_{gz}' t}{K+j\,\beta_{gz}' t}=\frac{1+j\,K\,\beta_{gz}' t}{1+j\,\beta_{gz}' t/K}$$

where $Z_o'/Z_o=\lambda_g'/\lambda_g$

$$z\simeq(1+jK\,\beta_{gz}' t)\left(1-j\frac{\beta_{gz}' t}{K}\right)$$

$$\simeq 1+\beta_{gz}'^2 t^2+j\,\beta_{gz}' t\left(K-\frac{1}{K}\right).$$

Since $\beta_{gz}' t \ll 1$, the second term may be neglected.

Thus, $z_1\simeq 1+jx$

where

$$x=\beta_{gz}' t\left(K-\frac{1}{K}\right)=\frac{2\pi}{\lambda_g'}\,t\left(\frac{\lambda_g'}{\lambda_g}-\frac{\lambda_g}{\lambda_g}\right)$$

$$=2\pi t\left(\frac{1}{\lambda_g}-\frac{\lambda_g}{\lambda_g'^2}\right). \tag{A}$$

$$\text{Now}\quad \left(\frac{\lambda_g}{\lambda_g'}\right)^2=\frac{\epsilon_r-(\lambda_o/\lambda_c)^2}{1-(\lambda_o/\lambda_c)^2}. \tag{B}$$

Eliminating $\lambda_g'$ between equations ($A$) and ($B$)

$$x_1=-2\pi t\left(\frac{\lambda_g}{\lambda_o^2}\right)(\epsilon_r-1) \qquad \text{(reactance is capacitive).}$$

The equivalent parallel capacitance is given by

$$Z_o x_1 = \frac{1}{\omega c} \quad \text{or} \quad c = \frac{1}{Z_o x_1\ \omega} \simeq 0.13 \text{ pF.}$$

**Example 4.12.** Two identical plates instead of one in Ex. 4.11, form an enclosure in the guide. Calculate the separation of the plates for the system to be matched at 9.8 GHz.

*Solution*: Equivalent transmission on line circuit of the system may be taken as shown below

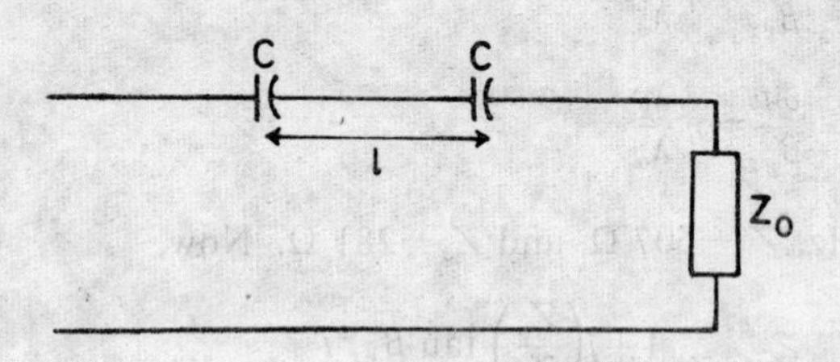


Fig. Ex. 4.9.

Impedance at the plane of the first plate

$$z_1 = 1 - j\,|x_1|.$$

The impedance at the plane of the second plate

$$z_2 = \frac{z_1 + j \tan \beta_{gz}\, l}{1 + j z_1 \tan \beta_{gz}\, l}.$$

Taking the second plate into account the input impedance of the system will be

$$z_i = -j\,|x_1| + z_2.$$

To make $z_i = 1$

$$1 = -j\,|x_1| + \frac{z_1 + j \tan \beta_{gz}\, l}{1 + j z_1 \tan \beta_{gz}\, l}.$$

The above equation is satisfied for

$$|x_1| = 0,\ \epsilon_r = 1 \quad \text{and} \quad 2 + |x_1| \tan \beta_{gz} l = 0$$

or $\tan \beta_{gz}\, l = \dfrac{-2}{|x_1|}.$

Since

$$|x_1| = 2\pi t \frac{\lambda_g}{\lambda_o^2} (\epsilon_r - 1) = 0.24$$

we have

$$\tan \beta_{gz} l = -8.33.$$

Then $\beta_{gz} l = \pi - 1.45 + n\pi$

which yields

$$l = 0.269\ \lambda_g + \frac{n\lambda_g}{2}$$

$$l = 1.108 + n \times 2.06 \text{ cm}.$$

**Example 4.13.** A matched load in a waveguide is made by placing a thin resistive sheet perpendicular to the axis of the guide at a suitable distance in front of a short circuit. What should the resistance be of the sheet per square metre and at what distance from the short circuit should the sheet be placed? Calculate these values for the waveguide of Ex. 4.10 at 10 GHz.

*Solution*: If a resistive sheet is placed in a waveguide operating in the dominant $TE_{10}$ mode a current sheet will flow and produce a discontinuity, in the magnetic fields on either side of the sheet because the fields are mutually perpendicular. If the sheet has a resistance equal to the wave impedance of the guide, the matching requires only to maintain $\vec{H} = 0$ on the far side of the sheet which can be achieved by placing the sheet at a distance of $\lambda_g/4$ from the short circuit.

Thus resistance per square metre

$$= E_T/H_T = \frac{\omega\mu_o}{\beta_g}$$

$$= 499\ \Omega$$

and distance $d = \frac{\lambda_g}{4} = 1$ cm.

## PROBLEMS

1. An air-filled rectangular waveguide with inside dimensions 3.5 cm × 7 cm is operating in the $TE_{10}$ mode. Determine cut-off frequency, phase velocity and guide wavelength at 3.5 GHz.

2. A rectangular guide with inner dimension $a = 2.286$ cm, and $b = 1.012$ cm is operating at 9 4 GHz in the $TE_{10}$ mode with incident power as 15 mW. Calculate the real components of the electromagnetic field.

3. It is customary to operate a waveguide of cut-off frequency $f_o$ at a normal frequency of $1.5 f_0$. Calculate the dimensions of (*i*) a rectangular waveguide whose sides are in the ratio 2:1 and which operates in the dominant mode; and (*ii*) a circular waveguide operating in the $TE_{11}$ mode, both suitable for working at 6 GHz.

4. Determine the modes which can propagate at 9 GHz in a circular waveguide of diameter 2 cm. Determine the values of the components of electric field strength at the centre of the waveguide (if they exist) when it carries a power of 2 mW.

5. Determine the cut-off frequency for the dominant mode in a rectangular waveguide having internal dimensions 5 cm × 3 cm. If the wavelength in unbounded medium (air) is 4 cm, find the phase velocity, the wavelength and the group velocity inside the guide. Determine the wave impedances corresponding to the $TE_{10}$ and $TE_{20}$ modes of the waveguide. Find also the angle that the constituent plane waves of the $TE_{10}$ mode make with the axis of the waveguide.

6. A rectangular waveguide with internal dimensions 2.50 cm × 1.25 cm is operated at 10 GHz. A pulse-modulated carrier of the above frequency is transmitted through the guide. How much pulse delay time is introduced by a guide of 100 m long?

If the modulated wave be $(1 + k \cos 40\pi t) \cos 2\pi \times 10^{10} t$ how long must the guide be so that the upper and lower side bands differ in phase by 180°.

7. A rectangular waveguide is to be designed to operate in the $TE_{10}$ mode at a frequency of 10 GHz. It is desired that the frequency of operation be at least 15% above the cut-off frequency of the propagating mode and 20% below the cut-off frequency of the next higher mode. Determine the dimensions of the rectangular waveguide.

8. In an air ($\mu=4\times10^{7}$ H/m, $\epsilon=10^{-9}/36\pi$ F/m) filled rectangular waveguide of internal dimensions 2.5 cm × 1 cm, if the electric voltage breakdown occurs at $10^{6}$ V/m, find the maximum time averaged power that can be transmitted in the $TE_{10}$ mode at a frequency 10 GHz without electric breakdown. Also calculate the surface charge and surface current densities on the walls of the waveguide under the condition of maximum power transport.

9. In a rectangular waveguide of Fig. 4.8, the surface current density $Js$ at the wall $y=b$ is given by

$$\vec{J_s}\Big|_{y=b} = -\hat{x}\, H_o \cos\left(\frac{\pi x}{a}\right) e^{-\bar{\beta}_z z} + \hat{z}\,\frac{j\bar{\beta} z}{\pi/a} \sin\left(\frac{\pi x}{a}\right) e^{-\bar{\beta}_z z}$$

Show that the lines of surface current density in the dominant $TE_{10}$ mode, at a particular instant of time $t=0$ are specified by the equation

$$\left[\cos\frac{\pi x}{a}\right]^n = \text{constant} \sin \bar{\beta}_z z$$

where $n=(f/f_c)^2-1$.

10. A dominant-mode wave of frequency 10 GHz propagating in a rectangular waveguide with broad dimension $a=2.5$ cm encounters a sudden transition to a long length of guide for which $a=1.2$ cm. What is the form of the dependence of the fields on $z$ in the narrower guide?

Calculate the component of the time-averaged Poynting vector in the direction of propagation in both waveguides. What is the magnitude of the reflection coefficient at the transition (complete attenuation)?

11. A rectangular waveguide extends from $x=0$ to $x=a$ and $y=0$ to $y=b$. In the region $0 \leqslant z \leqslant d$ the guide is filled with a dielectric of constitutive parameters $\mu$ and $\epsilon$. The remaining part ($-\infty < z < 0$ and $d < z < \infty$) of the waveguide is filled with air $\mu_o, \epsilon_o$). A wave is incident on the dielectric from the region $z < 0$. Determine the amplitude of the wave transmitted past the dielectric block in terms of the wave incident on the block if (*a*) the incident wave is in the $TE_{10}$ mode; (*b*) the incident wave is in the $TM_{11}$ mode.

12. A dielectric slab waveguide ($\mu=\lambda_o$, $\epsilon=2.69\,\epsilon_o$) is operated at 10 MHz in air. Find the minimum value of $d_{min}$ of the thickness of the dielectric slab for which an odd TE mode can be supported by the waveguide.

Hint: $d_{min} = \dfrac{\pi}{2\omega\sqrt{\mu\epsilon-\mu_0\epsilon_0}}$ for TE odd mode).

*REFERENCES*

ARGENCE E. AND KAHAN T., '*Theory of Waveguides and Cavity Resonators*, Blackie and Sons Ltd. London (1967).

ATWATER H.A., '*Introduction to Microwave Theory*'. McGraw-Hill Book Co. (1962).

BENDOW, B. AND MITTRA, S.S., '*Fibre Optics—Advances in Research and Development*'. Plenum (1979).

BROWN R.G. ET AL, *Lines, Waves and Antennas*'. Ronald Press New York (1973).

BRONWELL A.B., AND BEAM R.E., '*Theory and Applications of Microwaves*'. McGraw-Hill Book Co. (1947).

COLLIN R.E., '*Foundations for Microwave Engineering*'. McGraw-Hill Book Co. (1966).

DWORSKY, L.N., '*Modern Transmission Line Theory*'. Wiley (1979).

GELIN, P. ET AL, '*Regorous Analysis of Scaterng of Surface Waves in a Abruptly Ended Slab Dielectric Waveguide*'. *IEEE Trans.* MTT-29 (2) 1981.

GHOSE R. N., '*Microwave Circuit Theory and Analysis*'. McGraw-Hill Book Co. (1963).

HAMMOND, P., '*Electromagnetism for Engineers—An Introductory Course*'. Oxford Pregman (1978).

HARRIS, D.J. AND LEE, K.W., '*Theoretical and Experimental Characteristics of Double Groove Guide for 100* GHz *Operation*'. IEE Proc. 128 H (1) 1981.

HAYT W.H. Jr., '*Engineering Electromagnetics* (*2nd ed.*)'. McGraw-Hill Book Co. (1967).

HOWES, M.J. AND MORGAN, D.V., '*Optical Fibre Communication*'. Wiley (1980).

IEEE TRANS. SPECIAL ISSUE ON '*Microwave Communications*'. MTT-23 (4) 1975.

JANSEN, R.H., '*Hybrid Mode Analysis of End Effects of Planar Microwave and Millimeter—Wave Transmission Lines*' IEE Proc. 128 H (2) 1981.

JORDAN E.C., '*Electromagnetic Waves and Radiating Systems*' Prentice Hall of India (1971).

KATAYAMA, S. ET AL, '*Characteristics of Cocoon—Section Corrugated Waveguide*'. Trans, IECE J (Japan) J-62 (B)(4) 1979.

KEUSTER, E F. AND PATE, R.C., '*Fundamental Mode of Propagation on Dielectric Fibres of Arbitrary Cross-section*'. IEE Proc. 127 H (I) 1980.

KRAUS, J.D. AND CARVER K.R., '*Electromagnetics*,' McGraw-Hill Book Co. (1973).

LEVINE, P., '*Installation and RF Matching of Power Terminations for Microstrip and Stripline Applications*' Microwave J. 23 (9) 1980.

LEWIN, L. AND RUCHLE, T., '*Propagation in Twisted Square Waveguides*,. IEEE Trans. MTT—28(1) 1980.

LIN, W.G. '*Electromagnetic Wave Propagating in Uniform Waveguides Containing Inhomogeneous Dielectrics*,. IEEE Trans. MTT-28 (4) 1980.

LONGSFORD, P.A. AND PARKER, E.A., '*Cross Polarization of Electromagnetic Waves Transmitted through Waveguide Dichroic Plates.*' IEE Proc. 128 H (1) 1981

MAGNUSSON, P.C. '*Transmission Lines and Wave Propagation*'. Allyn and Bacon Boston Mass. (1970).

MATHERBE, J.A.G., '*A Transition from a Metal to Dielectric Waveguide.*' Microwave. J. 23 (11) 1980.

MAZUMDAR J., '*A Method for the Study of* TE *and* TM *Modes in Waveguides of Very General Cross-section*'. IEEE Trans. MTT-28 (11) 1980.

MIYAGI, M. AND NISHIDA, S., *Transmission Characteristics of Dielectric Tube Leaky Waveguide*'. IEEE Trans. MTT-8 (6) 1980.

NARASIMHAN, M.S., 'TE *and* TM *Modes Characteristics.*, IEEE Proc. MTT-22 (11) 1974.

PATRICIO, A.A., ET AL. '*Numerical Experiments on the Determination Cut-off Frequencies of Waveguides of Arbitrary Cross-section.* IEEE Trans. MTT-28 (6) 1980.

PREGLA, R. '*Determination of Conductor Losses in Planer Waveguide Structures*' IEEE Trans. MTT-28 (4) 1980.

Proceedings of the 7th European Microwave Conference, Denmark, Sept. 1977 (Sec. PR-1, PR-2) Tample House London (1977).

RAMO, S., WHINNERY J.R. AND VAN DUZER T., '*Fields and Waves in Communication Electronics.*' John Wiley and Sons, (1965).

RANGARAJAN, S R. AND LEWIS, J.E., '*Dielectric Loaded Elliptical Waveguides.*, IEEE Trans MTT-28 (10) 1980.

RAO, J.S. *et al*, '*Analysis of Small Aperture Coupling Between Rectangular Waveguide and Microstrip Line*'. IEEE Trans. MTT-29 (2) 1981.

RAO N.N. 'Basic *Electromagnetics with Applications*' Prentice Hall Inc. N.J. (1972).

RANGARAJAN, S.R. AND LEWIS, J.E., '*Propagation Characteristics of Elliptical Dielectric—Tubes Waveguide*'. IEE Proc. 127 H (3) 1980.

SAMUEL Y.L., '*Microwave Devices and Circuits.*' Prentice-Hall Inc. N.J. (1980).

SANDER K.F. AND REED G.A.L., '*Transmission and Propagation of Electromagnetic Waves*'. Cambridge University Press, London (1978).

SCHELKUNOFF S.A., '*Electromagnetic Fields*'. Blaisdell Pub. Co. New York (1963)

SESHADRI S.R., '*Fundamentals of Transmission Lines and Electromagnetic Fields*'. Addison-Wesley Pub. Co. (1971).

SHAFAI, L. AND HASSAN, E.E.M., '*Field Solution and Elliptical Characteristics of Slotted Waveguide*,. IEE Proc. 128 H (2) 1981.

SHEN, H.M., "*The Resistive Bifurcated Parallel Plate Waveguide*" IEEE Trans. MTT-28 (11) 1980.

SLATTER J.C. AND FRANK N.H., '*Electromagnetism*', McGraw-Hill Book Co. (1947).

SLIEDRIGHT, M.V., '*Cylindrical Waveguide Cut-off Wavelength—An Improved Algorithm for Handling Arbitrary Cross-section*'. Microwave J. 23 (5) 1980.

SOUTHWORTH G.C, '*Principles and Applications of Waveguide Transmission*'. D Van Nostrand New York (1950).

STEER, M.B. AND KAHN, P.J., '*Wide Band Equivalent Circuit for Radial Transmission Lines*'. IEE Proc. 128 H (2) 1981.

STONE, D.S., '*Mode Analysis in Multimode Waveguides Using Travelling Wave Ratios*'. IEEE Trans. MTT-29 (2) 1981.

TRANQUILLA, J.M. AND LEWIS, J.E., '*On the Propagation of Leaky Waves in a Longitudinally Slotted Rectangular Waveguides*'. IEEE Trans. MTT-28 (7) 1980.

YAMAMOTO, K., '*Fundamental Mode Transmission in Millimeter Wave Gas Confining Dielectric Waveguide*'. Trans. IECE J (Japan) J-62-B (6) 1979.

YAMASHITA, E. ET AL, '*Composite Dielectric Waveguides*'. IEEE Trans. MTT-28 (9) 1980.

# 5

# MICROWAVE RESONATORS

## 5.1 INTRODUCTION

A resonant tank circuit is of great importance in electronics. It is frequently used in designing oscillators, tuned amplifiers, frequency filters and many other networks. At microwave frequencies neither the lumped constant nor the distribution parameter tank circuit is of practical utility because at these frequencies the circuit dimensions become comparable to the wavelength of the exciting waves. Consequently, these circuits become potential sources of radiation losses. Moreover, the skin effect resistance also becomes very large and so conductor losses become appreciable. This results in a very low $Q$ value. The need was therefore felt for a high $Q$ low-loss tank circuit for use at microwave frequencies which could simultaneously also store the desirable amount of microwave power. The answer to this need was found in the form of a closed volume bounded by perfectly conducting walls*, forming thereby a cavity in which resonant electromagnetic field may exist.

In this chapter we shall present an analysis of different types of microwave resonators. However, before we proceed to make such an analysis, it is instructive to review the facts relating to low frequency lumped resonant circuits.

## 5.2 LUMPED ANALOGIES

In a series resonant circuit of inductance $L$, capacitance $C$, and resistance $R$, the current $i$ is related to the driving r.f. voltage $V$ by the following differential equation

$$L\,\frac{di}{dt}+Ri+\frac{1}{C}\int i\,dt=V. \tag{5.1}$$

When the deriving voltage is removed after the establishment of the current, current $i$ decays in amplitude as a function of time due to the presence of losses. Assuming $i \propto e^{j\omega t}$, we have from Eq. (5.1)

$$R+j\left(\omega L-\frac{1}{\omega C}\right)=0.$$

*The volume may also be bounded by dielectric walls.

Letting $\quad \dfrac{1}{LC}=\omega_r^2$ and $\dfrac{\omega_r L}{R}=Q,$

the above equation becomes

$$\frac{1}{Q}+j\left(\frac{\omega}{\omega_r}-\frac{\omega_r}{\omega}\right)=0. \tag{5.2}$$

Solving Eq. (5.2) for $\omega$, we obtain

$$\omega=\omega_r\sqrt{1-(1/2Q)^2}+j\,(\omega_r/2Q). \tag{5.3}$$

It is noted from Eq. (5.3) that frequency $\omega$ is complex. The physical interpretation of this complex frequency may be ascertained by letting $\omega=\omega_1+j\omega_2$. Using this in Eq. (5.3), we find

$$\omega_r=\sqrt{\omega_1^2+\omega_2^2} \tag{5.4a}$$

and

$$Q=\sqrt{\left(\frac{\omega_1}{2\omega_2}\right)^2+\frac{1}{4}}\,. \tag{5.4b}$$

Hence the decay of current $i$ can be written as

$$i \propto e^{-\omega_2 t}\; e^{j\omega_r\sqrt{1-(1/2Q)^2}}.$$

We can see from the above expression that in contrast to a case of purely real frequency, a frequency with a real as well as an imaginary part gives rise to an exponential decay with time of magnitude of current.

Equation (5.2) through (5.4b) can be obtained for a parallel resonant circuit if we define $Q=\omega_r CR$ rather than $Q=\omega_r L/C$ for the series circuit. Thus the definition of $Q$ for the series and parallel resonant circuit are the same when stated in terms of real and imaginary parts of the frequency. Therefore, this concept of complex frequency which is a natural consequence of our calculations, is a very useful concept and may be used to evaluate the $Q$ value of a microwave resonator.

## 5.3 RECTANGULAR RESONATORS

Consider a rectangular waveguide supporting a propagating mode of frequency $f$. If a short circuit is placed at any transverse plane it will result in a complete standing wave pattern of the fields. At the short there will be voltage minimum and this minimum is repeated at half guide wavelength intervals from the short circuit. If a short circuit is now placed at one of the voltage minima, there will be complete reflection back towards the first short and in phase with the original signal. The resultant configuration is a rectangular cavity or resonator that can support a signal which apparently bounces back and forth between two opposite walls. These considerations apply to any shape of hollow cavity.

Now we proceed to examine the field configuration within a rectangular cavity.

Consider a rectangular resonator oriented in a coordinate system shown in Fig. 5.1. The resonator is a rectangular waveguide shorted at $z = 0$ and $z = c$.

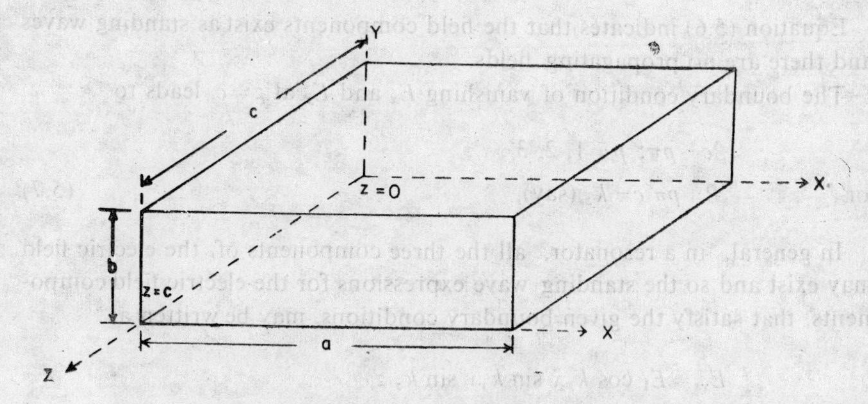


**Fig. 5.1** Rectangular resonator in rectangular coordinate system.

Assuming the walls of the resonator to be perfectly conducting and the dielectric inside to be loss-less, the electric field components of the $TE_{mn}$ mode propagating in the constituent waveguide, is given by* Eq. (4.48)

$$E_z = 0$$

$$E_x = \frac{j\omega\mu}{k_c^2} B\, k_y \cos k_x x \sin k_y y\; e^{-j\bar{\beta}z} \tag{5.5}$$

$$E_y = -\frac{j\omega\mu}{k_c^2} B\, k_x \sin k_x x \cos k_y y\; e^{-j\bar{\beta}z}$$

where $\quad k_x = \dfrac{m\pi}{a}, \quad k_y = \dfrac{n\pi}{b},$

$m$ and $n$ are integers, and $\bar{\beta}$ is the propagation constant.

As perfect conducting walls exist at $z=0$ and $z=c$ and the reflection coefficient at these walls is $-1$, the resultant field components arising from the superposition of incident and reflected waves, are given by*

$$E_z = 0$$

$$E_x = \frac{j\omega\mu}{k_c^2} B\, k_y \cos k_x x \sin k_y y\, (e^{-j\bar{\beta}z} - e^{j\bar{\beta}z})$$

$$= \frac{2\omega\mu B k_y}{k_c^2} \cos k_x x \sin k_y y \sin \bar{\beta}_z y$$

*Time factor $e^{j\omega t}$ has been dropped for convenience.

or $$E_x = E_1 \cos k_x x \sin k_y y \sin \beta z$$

$$E_y = E_2 \sin k_x x \cos k_y y \sin \beta z \tag{5.6}$$

where $E_1 = \frac{2\omega\mu B k_y}{k_c^2}$ and $E_2 = -\frac{2\omega\mu B k_x}{k_c^2}$ are amplitude constants.

Equation (5.6) indicates that the field components exist as standing waves and there are no propagating fields.

The boundary condition of vanishing $E_x$ and $E_y$ at $z=c$, leads to

$$\beta c = p\pi;\ p = 1, 2, 3 \ldots$$

or $$\beta = p\pi/c = k_z \text{ (say)}. \tag{5.7}$$

In general, in a resonator, all the three components of the electric field may exist and so the standing wave expressions for the electric field components, that satisfy the given boundary conditions, may be written as

$$E_{xs} = E_1 \cos k_x x \sin k_y y \sin k_z z$$

$$E_{ys} = E_2 \sin k_x x \cos k_y y \sin k_z z \tag{5.8}$$

$$E_{zs} = E_3 \sin k_x x \sin k_y y \cos k_z z$$

$E_3$ is another amplitude constant.

The corresponding magnetic field components may be obtained from Maxwell's curl equations

$$\nabla \times \vec{E} = j\omega\mu \vec{H}$$

$$\nabla \times \vec{H} = (\sigma + j\omega\epsilon)\vec{E}.$$

These are given by

$$H_{xs} = \frac{k_z E_2 - k_y E_3}{j\omega\mu} \sin k_x x \cos k_y y \cos k_z z$$

$$H_{ys} = \frac{k_x E_3 - k_z E_1}{j\omega\mu} \cos k_x x \sin k_y y \cos k_z z \tag{5.9}$$

$$H_{zs} = \frac{k_y E_1 - k_x E_2}{j\omega\mu} \cos k_x x \cos k_y y \sin k_z z$$

where

$$k_x = \frac{m\pi}{a},\quad k_y = \frac{n\pi}{b},\quad k_z = \frac{p\pi}{c},\ \text{each } m, n \text{ and } p \text{ are integers.} \tag{5.10}$$

However, $H_{zs} = 0$ for TM and $E_{zs} = 0$ for TE modes.

All the six field components satisfy similar wave equations. Let us consider any one component say, $E_{xs}$, then

$$\nabla^2 E_{xs} = \bar{\gamma}^2 E_{xs} = -\omega^2 \mu\epsilon E_{xs}$$

or
$$\omega^2 \mu\epsilon = \bar{\gamma}^2 = k_x^2 + k_y^2 + k_z^2.$$

At resonance $\omega = \omega_r = 2\pi f_r$ where $f_r$ is the resonance frequency.

Hence

$$f_r = \frac{1}{2\pi\sqrt{\mu\epsilon}} \sqrt{k_x^2 + k_y^2 + k_z^2}$$

or using Eq. (5.10)

$$f_r = \frac{1}{2\pi\sqrt{\mu\epsilon}} \sqrt{\left(\frac{m\pi}{a}\right)^2 + \left(\frac{n\pi}{b}\right)^2 + \left(\frac{p\pi}{c}\right)^2}$$

$$= \frac{1}{2\sqrt{\mu\epsilon}} \sqrt{(m/a)^2 + (n/b)^2 + (p/c)^2}. \tag{5.11}$$

The corresponding resonant wavelength is

$$\lambda_r = \frac{2}{\sqrt{(m/a)^2 + (n/b)^2 + (p/c)^2}}. \tag{5 12}$$

It may be noted that triply infinite resonant frequencies may exist corresponding to different field distributions or modes. Also there may be more than one field solution, pertaining to a given resonant frequency, that is, the modes are degenerate.

## 5.4 MODES IN A RECTANGULAR RESONATOR

A rectangular waveguide is capable of accommodating $TE_{mn}$ and $TM_{mn}$ propagating modes. Consequently, a rectangular resonator is capable of supporting $TE_{mnp}$ and $TM_{mnp}$ resonant modes. The third subscript indicates the number of half period field variations along the cavity length, i.e. it has the same physical significance as the other two subscripts.

Let us consider the case of $TE_{mno}$ and $TM_{mno}$ modes, i.e. $p=0$, then from Eq. (5.12),

$$\lambda_{r,\,mno} = \frac{2}{\sqrt{(m/a)^2 + (n/b)^2}} = \lambda_{c,mn}, \tag{5.13}$$

(see Eq. 4.52).

Equation (5.13) implies that resonance effects take place in this case precisely at the cut-off frequency of a waveguide with the same values of subscripts

$m$ and $n$. Since at cut-off there can be no variation of the field along the waveguide axis because guide wavelength approaches infinity, the presence of a perfectly conducting wall in any cross-section of a waveguide turns all the transverse electric field components into zero in all cross-sectional planes. Consequently, $TE_{mno}$ resonant modes cannot exist in a rectangular resonator. However, $TM_{mnp}$ modes feature at $E_{zs}=0$ and the field inside the cavity resonator will be present even when $p = 0$. Hence $TM_{mno}$ resonant modes exist in a rectangular resonator.

The impossibility of $TE_{mno}$ and possibility of $TM_{mno}$ modes holds for any resonator that can be treated as a section of uniform waveguide. The equallity of the resonant and cut-off wavelengths, Eq. (5.13), can be used to determine the cut-off wavelength of waveguides with complex configurations.

Further, if any two of the integers $m$, $n$ and $p$ vanish, all the field components in Eqs. (5.8) and (5.9) vanish. Hence there cannot be any resonant mode of this type.

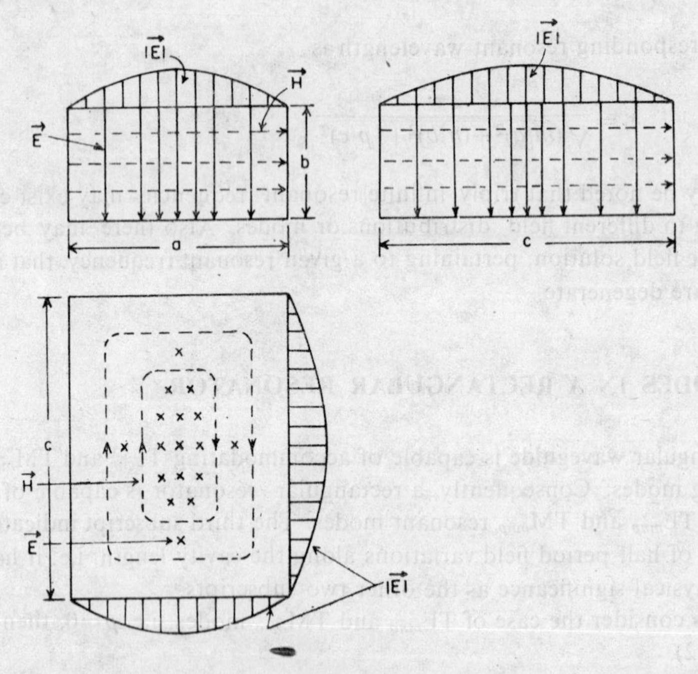


**Fig. 5.2** Field distribution in a rectangular cavity at $TE_{101}$ dominant mode.

Thus, the lowest or the dominant resonant mode corresponding to highest resonant wavelength and lowest resonant frequency in $TE_{mnp}$ modes is $TE_{101}$ because $c > a > b$. The field distribution of this mode is illustrated in Fig. 5.2. If we compare it with the field distribution of the $TE_{10}$ propagating mode in a rectangular waveguide (Fig. 4.18), we find that in both the cases, the electric field passes vertically entering top to bottom normally and becoming zero at the side walls as required by perfect conducting boundary walls. The magnetic field lines lie in the horizontal $(x-z)$ planes and surround the vertical displacement current resulting from the time rate of change of $E_{ys}$. However, the transverse component of the electric and magnetic fields in the resonator are displaced through a phase corresponding to a quarter guide wavelength. This phase shift arises from pure standing waves. There is no energy transfer in any direction, i.e. the Poynting vector $(\vec{E} \times \vec{H})$ is zero.

As the normal electric field ends at the top and bottom walls, equal and opposite charges are expected to exist on these walls. A current, therefore, flows between top and bottom, becoming vertical on the side walls. Thus, the top and bottom walls serve as the conventional capacitor and the side walls serve as a low value inductor and a rectangular resonator as the resonant tank circuit.

Higher resonant modes are of no great interest except modes of the type $TE_{102}$, $TE_{103}$, etc. These modes are characterized by two or more half-period field variations along cavity length. The field distribution may easily be drawn and resonant wavelengths obtained from Eq. (5.12).

The resonant frequency and resonant wavelength for a rectangular cavity excited in the dominant $TE_{101}$ mode are given by

$$f_r = \frac{1}{\sqrt{\mu\epsilon}} \frac{\sqrt{a^2+c^2}}{2ac} \tag{5.14}$$

$$\lambda_r = \frac{2ac}{\sqrt{a^2+c^2}} = \lambda_{c_{TM_{11}}} \tag{5.15}$$

It is seen that the resonant wavelength turns out to be equal to the cut-off wavelength of the $TM_{11}$ mode in a rectangular waveguide of transverse cross-section $(a \times b)$. For a cubical resonator,

$$\lambda_r = \sqrt{2}\, b \tag{5.16}$$

that is, resonant wavelength equals the diagonal of the rectangular resonator. Condon* represented resonant frequencies or wavelengths of a rectangular resonator by means of the lattice structure shown in Fig. 5.3. The basic equation used is (5.12) which can be rewritten as

$$\left(\frac{2f_r}{v_p}\right)^2 = \left(\frac{2}{\lambda_r}\right)^2 = \left(\frac{m}{a}\right)^2 + \left(\frac{n}{b}\right)^2 + \left(\frac{p}{c}\right)^2. \tag{5.17}$$

Plotting the values $m/a$, $n/b$ and $p/c$ along the $x$-, $y$-and $z$-axes yields Fig. 5.3.

*E.U. Condon, "Principles of microwave radio", *Rev. Mod. Phy.*, Vol. 15, (Oct. 1942).

In our discussions so far we have assumed the dielectric filling the cavity to be loss-less. However, it is permissible for a dielectric to have losses.

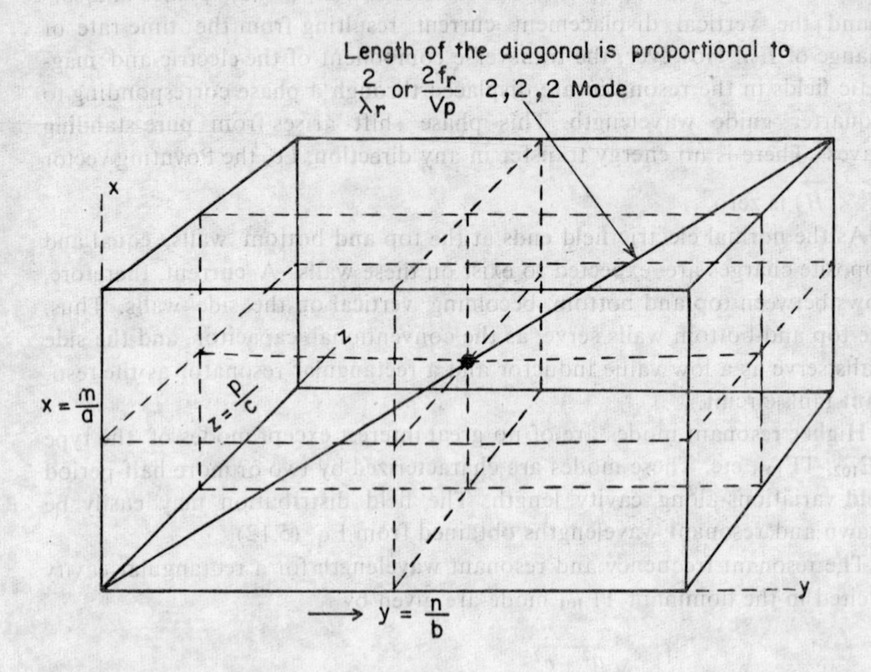


Fig. 5.3 Resonant frequencies of a rectangular resonator represented by a lattice structure.

Then $\bar{\gamma}$ cannot be equated to $j\bar{\beta}$. The problem is solved by the use of the concept of complex frequency developed in Sec. 5.2. Letting $\omega = \omega_1 + j\omega_2$ the propagation constant $\bar{\gamma}$ as given by Eq. (4.50), can be written as

$$\bar{\gamma} = \sqrt{\{[k_c^2 - (\omega_1^2 - \omega_2^2)\ \mu\epsilon - \sigma\omega_2 u)] + j\ [\omega_1\mu(\sigma - 2\ \omega_2\epsilon)]\}}.$$

For $\bar{\gamma}$ be purely imaginary ($\bar{\gamma} = j\bar{\beta}$), the expression in the first bracket must be negative while that in the second bracket must vanish, that is

$$[(\omega_1^2 - \omega_2^2)\ \mu\epsilon + \sigma\omega_2\mu] > k_c^2 \tag{5.18a}$$

and

$$\omega_2 = \sigma/2\epsilon. \tag{5.18b}$$

Since $\omega_r^2 = \omega_1^2 + \omega_2^2$, Eq. (5.18a) is a generalization of the statement that the operating frequency must be above the cut-off value for the propagation to occur in a waveguide containing a lossy dielectric. Equation (5.18b) on the other hand relates the imaginary part of the complex frequency to the conductivity and dielectric constant of the dielectric filling the cavity.

The fields in a lossy dielectric cavity will have the general form

$$E_y \propto e^{-\omega_2 t}\ e^{j\omega_1 t}\ e^{j\{[(\omega_1^2 - \omega_2^2)\mu\epsilon + \sigma\omega_2\mu] - k_c^2\}}$$

The above expression implies that as time passes, energy is absorbed in the dielectric and the amplitude of the waves and that of their linear combinations decrease with time.

Various losses present in a cavity are interpreted in terms of the quality factor $Q$ of the resonator.

## 5.5 Q OF A RECTANGULAR RESONATOR

Quality factor $Q$ of a resonator is defined in the same way a that of waveguide (See Eq. (4.124). However, in a cavity there may be many types of losses.

A cavity is a closed volume, but in actuality, of course, this is not the case, since energy must be coupled into or away from the cavity. This coupling is usually achieved via a small hole in one of the cavity walls. The hole radiates power and therefore lowers the $Q$ value. The new value is called the loaded $Q$ value of the cavity and is designated by $Q_L$. The unloaded $Q$ arises from the cavity wall conductor losses and dielectric losses and is designated by $Q_o$.

In general, $Q$ for a microwave resonator may be obtained by substituting Eqs. (5.18a) and (5.18b) in Eq. (5.4b)

$$Q_o = \omega_r \, \epsilon/\sigma. \tag{5.19}$$

It is noted from Eq. (5.19) that at microwave frequencies, very high values of $Q$ may be obtained. Moreover, since $Q \gg \frac{1}{2}$ from Eq. (5.3), $\omega_r \simeq \omega_1$ and hence

$$Q_o \simeq \frac{\omega_1}{2\omega_2}. \tag{5.20}$$

The definition of $Q_o$ expressed in Eq. (5.20) is equivalent to the general definition stated in Eq. (4.128) recalled for convenience here,

$$Q_o = \omega_r \frac{W_s}{P_L}. \tag{5.21}$$

$W_s$ is the energy stored in the resonator and $P_L$ is the time averaged power loss. In arriving from Eq. (5.20) to (5.21) the assumption $Q \gg \frac{1}{2}$ implies that dielectric losses are negligible as compared to finite wall conductivity losses in a cavity.

In general, if $P_{Ld}$, $P_{LW}$ and $P_{Ll}$ are the time average power losses arising from dielectric loss, wall losses and loading losses of a resonator, then total $Q$ of the resonator, $Q_T$ is given by

$$Q_T = \omega_r \frac{W_s}{(P_{Ld} + P_{Lw} + P_{Ll})}$$

or

$$\frac{1}{Q_T}=\frac{P_{Ld}}{\omega_r W_s}+\frac{P_{Lw}}{\omega_r W_s}+\frac{P_{Ll}}{\omega_r W_s}$$

$$=\frac{1}{Q_{\text{dielectric}}}+\frac{1}{Q_{\text{walls}}}+\frac{1}{Q_{\text{coupling}}}. \tag{5.22}$$

Equation (5.22) may be used to evaluate the loaded $Q_L$ of a resonator. The unloaded $Q_o$ is given by

$$\frac{1}{Q_o}=\frac{1}{Q_{\text{dielectric}}}+\frac{1}{Q_{\text{walls}}}.$$

The $Q_{\text{dielectric}}$ is generally interpreted in terms of complex permittivity, where

$$Q_{\text{dielectric}}=\omega_r\frac{W_s}{P_{Ld}}=\frac{\epsilon'}{\epsilon''} \tag{5.23}$$

and $\epsilon'$ and $\epsilon''$ respectively are the real and imaginary parts of the complex permittivity $\epsilon^*=\epsilon'-j\epsilon''$.

**$Q_w$ for a Rectangular Resonator**

To illustrate the method of evaluating the $Q_w$ of a rectangular waveguide, let us consider the case of the $TE_{mop}$ mode. This particular mode is chosen because of the lowest mode in this series, $TE_{101}$, besides being dominant, has a very high $Q$ value and is therefore of great practical interest.

The field components corresponding to resonant ($\omega=\omega_r$) $TE_{mop}$ mode may be obtained by substituting

$$k_y=\frac{n\pi}{b}=0 \text{ and } E_{zs}=0 \text{ in Eqs. (5.8) and (5.9).}$$

$$\begin{aligned}
E_{xs}&=E_{zs}=H_{ys}=0\\
E_{ys}&=E_2\sin k_x x\sin k_z z\\
H_{xs}&=\frac{k_z E_2}{j\omega_r\mu}\sin k_x x\cos k_z z\\
H_{zs}&=-\frac{k_x E_2}{j\omega_r\mu}\cos k_x x\sin k_z z
\end{aligned} \tag{5.24}$$

where $k_x=m\pi/a,\ k_z=p\pi/c$.

The peak energy stored is given by

$$W_s=\frac{\epsilon}{2}\int_v \vec{E}\cdot\vec{E}^*\,dv=\frac{\epsilon}{2}\int_v |E|^2\,dv.$$

Here $|E|^2=|E_{ys}|^2=E_2^2\sin^2 k_x x\sin^2 k_z z.$

Hence

$$W_s=\frac{1}{2}\;\epsilon\int_0^a\int_0^b\int_0^c E_2^2 \sin^2 k_x\, x \sin^2 k_z\, z\; dxdydz$$

or
$$W_s=\frac{\epsilon E_2^2 abc}{8}. \tag{5.25}$$

The time average power loss $P_L$ is the sum of the losses in all the six walls and can be obtained using the tangential component of the magnetic field at each wall.

Power loss $P_{L_1}$ in the front and back walls of a rectangular resonator depicted in Fig. 5.1 is given by

$$P_{L_1}=2\frac{R_s}{2}\int_0^a\int_0^c |H_t|^2_{\substack{z=0\\z=c}}\; dxdy+R_s\int_0^a\int_0^b |H_{xs}|^2\; dxdy.$$

Substituting the value of $H_{xs}$ from Eq. (5.24) we obtain,

$$P_{L_1}=R_s\frac{k_z^2E_2^2}{\omega_r^2\mu^2}\int_0^a\int_0^b \sin^2 k_x x\; dxdy=\frac{R_s}{2}\frac{p^2\pi^2E_2^2ab}{c^2\omega_r^2\mu^2}. \tag{5.26}$$

Similarly, the time average power loss in the top and bottom walls, $P_{L_2}$ is given by

$$P_{L_2}=2\frac{R_s}{2}\int_0^a\int_0^c |H_t|^2_{\substack{y=0\\y=b}}\; dxdz$$

$$=R_s\int_0^a\int_0^b |H_{xs}|^2\; dxdz+R_s\int_0^a\int_0^c |H_{zs}|^2\; dxdz$$

$$=R_s\frac{k_z^2E_2^2}{\omega_r^2\mu^2}\int_0^a\int_0^c \sin^2 k_x x\cos^2 k_z z\; dxdz$$

$$+\frac{k_x^2E_2^2}{\omega_r^2\mu^2}\int_0^a\int_0^c \cos^2 k_x\, x\sin^2 k_z z\; dxdz.$$

Using $k_x=m\pi/a$ and $k_z=p\pi/c$

$$P_{L_2}=R_s\left(\frac{E_2}{\omega_r\mu}\right)^2\left[\frac{\pi^2p^2}{c^2}\frac{ac}{4}+\frac{\pi m^2}{a^2}\frac{ac}{4}\right]$$

$$=R_s\left(\frac{\pi E_2}{\omega_r\mu}\right)^2\frac{ac}{4}\left[\frac{p^2}{c^2}+\frac{m^2}{a^2}\right]. \tag{5.27}$$

The time average power loss in the side walls is given by

$$P_{L_3}=\frac{R_s}{2}\int_0^b\int_0^c |H_t|^2_{\substack{x=0\\x=a}}\; dydz$$

$$=R_s\int_0^b\int_0^c |H_{zs}|^2\; dydz$$

$$=R_s \frac{k_x^2 E_2^2}{\omega_r^2 \mu^2} \int_0^b \int_0^c \sin^2 k_z z \, dy dz$$

or

$$P_{L3} = \frac{R_s k_x^2 E_2^2}{\omega_r^2 \mu^2} \frac{bc}{2} = R_s \left(\frac{\pi E_2}{\omega_r \mu}\right)^2 \frac{m^2}{a^2} \frac{bc}{2}. \tag{5.28}$$

The total time average power loss is

$$P_L = P_{L1} + P_{L2} + P_{L3}$$

$$= \frac{R_s}{2} \left(\frac{\pi E_2}{\omega_r \mu}\right)^2 \left[\frac{p^2 ab}{c^2} + \frac{ac}{12}\left(\frac{p^2}{c^2} + \frac{m^2}{a^2}\right) + \frac{m^2 bc}{a^2}\right]. \tag{5.29}$$

Hence $Q_w$ of the rectangular resonator is given by Eq. (5.21).

$$Q_w = \omega_r \frac{W_s}{P_L} = \omega_r \frac{\epsilon \, E_2^2 \, abc/8}{\frac{R_s}{2}\left(\frac{\pi E_2}{\omega_r \mu}\right)^2 \left[\frac{p^2 ab}{c^2} + \frac{m^2 bc}{a^2} + \frac{ac}{2}\left(\frac{p^2}{c^2} + \frac{m^2}{a^2}\right)\right]}.$$

Using $\quad \omega_r^2 \mu \epsilon = \left(\frac{2\pi}{\lambda_r}\right)^2$ and $\eta = \sqrt{\mu/\epsilon}$,

we have

$$Q_w = \frac{2\pi\eta \, abc}{R_s \lambda_c^3 \left[\frac{p^2 ab}{c^2} + \frac{m^2 bc}{a^2} + \frac{ac}{2}\left(\frac{p^2}{c^2} + \frac{m^2}{a^2}\right)\right]}. \tag{5.30}$$

For a cubical resonator $a=b=c$

$$Q_w = \frac{2\pi\eta \, a^3}{R_s \lambda_r^3 \, [p^2 + m^2 + p^2/2 + m^2/2]} = \frac{4\pi\eta a^3}{3 R_s \lambda_r^3 \, (p^2 + m^2)} \tag{5.31}$$

$Q_w$ for the dominant $TE_{101}$ mode can be obtained by substituting $m=1$, $n=0$ and $p=1$ in Eq. (5.31)

$$Q_{wTE_{101}} = \frac{4\pi\eta \, a^3}{3R \, R_s \lambda_r^3 \times 2} = \frac{\pi\eta}{3\sqrt{2} R_s} = \frac{0.741}{R_s} \eta \tag{5.32}$$

(The last step follows from $\lambda_r = \sqrt{2}\, a$).

For a typical resonator made of copper, $\sigma = 5.8 \times 10^7$ ℧/m with sides $a = b = c = 3$ cm; the resonant frequency is found to be 7.07 GHz, the skin effect surface resistance is 0.022 Ω, the $Q_w$ value is found to be

$$Q_w = \frac{0.471 \times 120\,\pi}{0.022} \simeq 12{,}700.$$

Clearly, this is a sufficiently high value.

Generally, the dielectric inside the resonant cavity is loss-less and hence $Q_w = Q_o$. Hence forth, we shall designate $Q_w = Q_o$ presuming that $Q_d = \infty$.

## 5.6 CYLINDRICAL RESONATORS

A cylindrical resonator is a section of cylindrical waveguide of length $d$ and radius a, with two plane *conducting* plates placed perpendicular to the axis of the cylinder at the ends as shown in Fig. 5.4. The analysis of such a resonator may be carried out along the lines used for a rectangular resonator.

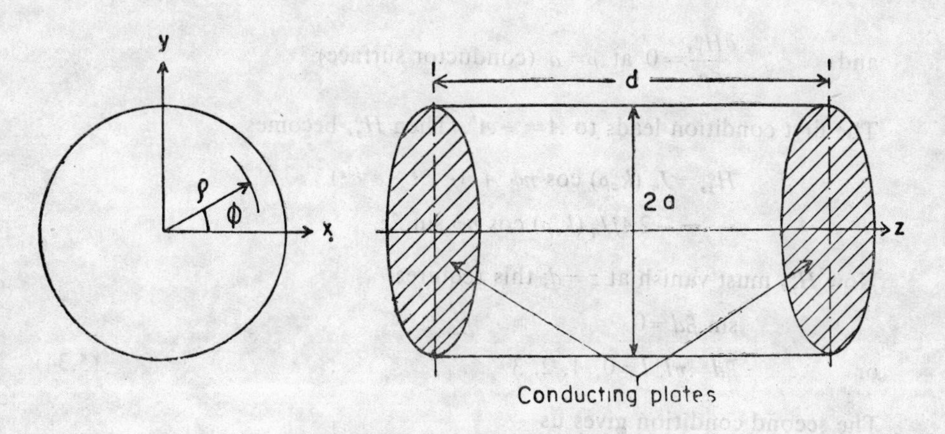


**Fig. 5.4** Cylindrical resonator.

To know the field configuration of a TE wave in a cylindrical resonator we recall Eq. (4.81) representing the propagating TE mode in a cylindrical waveguide.

$$H_z^o = AJ_n(k_c\rho)\cos n\phi\, e^{-\bar{\gamma}z}$$

$$H_\rho^o = -\frac{\bar{\gamma}}{k_c^2}\frac{\partial(H_z^o)}{\partial\rho} = -\frac{j\beta A}{k_c}J_n'(k_c\rho)\cos n\phi\, e^{-\bar{\gamma}z}$$

$$H_\phi^o = -\frac{\bar{\gamma}}{k_c^2}\frac{\partial H_z^o}{\sigma\phi} = -\frac{j\beta nA}{\rho k_c^2}J_n(k_c\rho)\sin n\phi\, e^{-\bar{\gamma}z} \tag{5.33}$$

$$E_\rho^o = -\frac{j\omega_r\mu}{k_c^2}\frac{\partial H_z^o}{\rho\partial\phi} = \frac{\omega_r\mu}{\beta}H_\phi^o = Z_{TE}H_\phi^o$$

$$H_\phi^o = \frac{j\omega_r\mu}{k_c^2}\frac{\partial H_z^o}{\partial\phi} = -\frac{\omega_r\mu}{\beta}H_\phi^o = -Z_{TE}H^o$$

where $\bar{\gamma}=j\beta$, the resonator is loss-less, $Z_{TE}=\eta/\sqrt{1-(f_c/f)^2}$ and $\omega=\omega_r$, i.e. the fields are at resonant frequency.

Propagating fields are reflected from the shorting plates at the ends, and forward and backward propagating fields combine to give the standing wave field configuration in a cylindrical resonator. Therefore, a standing wave of axial magnetic field $H^o_{zs}$, formed by combining forward and backward propagating TE waves is given by

$$H^o_{zs}=J_n(k_c\rho)\cos n\phi\,(A\,e^{-j\bar{\beta}z}+A'\,e^{j\bar{\beta}z}).$$

The boundary conditions that $H^o_{zs}$ must satisfy are

$$H^o_{zs}=0 \text{ at } z=0 \text{ and } z=d \text{ (metal plates)}$$

and $$\frac{\partial H^o_{zs}}{\partial \rho}=0 \text{ at } \rho=a \text{ (conductor surface).}$$

The first condition leads to $A=-A'$. Then $H^o_{zs}$ becomes

$$H^o_{zs}=J_n(k_c\rho)\cos n\phi\,A\,(e^{-j\bar{\beta}z}-e^{-j\bar{\beta}z})$$
$$=-2AjJ_n(k_c\rho)\cos n\phi\sin\bar{\beta}z.$$

Now $H^o_{zs}$ must vanish at $z=d$; this requires

$$\sin\bar{\beta}d=0$$

or $$\bar{\beta}d=\pi l,\ l=0,1,2,3\ldots \tag{5.34}$$

The second condition gives us

$$J'_n(k_c\rho)=0. \tag{5.35}$$

Hence the axial magnetic field is given by

$$H^o_{zs}=-2jA\,J_n(k_c\rho)\cos n\phi\sin\bar{\beta}z.$$

Similarly, we can obtain other standing wave field components:

$$H^o_{\rho s}=-\frac{2jA\bar{\beta}}{k_c}J'_n(k_c\rho)\cos n\phi\cos\bar{\beta}z$$

$$H^o_{\phi s}=-\frac{2jAn\bar{\beta}}{\rho k_c^2}J_n(k_c\rho)\sin n\phi\cos\bar{\beta}z$$

$$H^o_{\rho s}=\frac{j\,\omega_r\,\mu nA}{\rho k_c^2}J_n(k_c\rho)\sin n\phi\sin\bar{\beta}z \tag{5.36}$$

$$E^o_{\phi s}=\frac{j\omega_r\mu}{k_c}A\,J'_n(k_c\rho)\cos n\phi\sin\bar{\beta}z$$

$$E^o_{zs}=0$$

The field configuration of TM standing waves may be obtained from Eq. (4.83). In this case the boundary conditions are, $E^o_{\rho s}$ and $E^o_{\phi s}$, vanish at $z=0$

and $z=d$. This yields the following field configuration with the condition stated in Eq. (5.34)

$$J_n(k_c a)=0$$

$$H^o_{zs}=0$$

$$E^o_{zs}=B\,J_n(k_c\rho)\cos n\phi\cos\bar{\beta}z$$

$$E^o_{\rho s}=-\frac{j\beta B}{k_c}J'_n(k_c\rho)\cos n\phi\sin\bar{\beta}z \tag{5.37}$$

$$E^o_{\phi s}=\frac{j\beta nB}{k_c^2}J_n(k_c\rho)\sin n\phi\sin\bar{\beta}z$$

$$H^o_{\rho s}=-\frac{j\omega_r\epsilon}{\rho k_c^2}nB\,J_n(k_c\rho)\sin n\phi\cos\bar{\beta}z$$

$$H^o_{\phi s}=-\frac{j\omega_r\epsilon}{k_c}B\,J'_n(k_c\rho)\cos n\phi\cos\bar{\beta}z$$

with $\bar{\beta}=l\pi/d,\ l=0,1,2,3\ldots$

The resonant frequency is determined by the propagation constant $\bar{\gamma}$, that is

$$\bar{\gamma}=j\bar{\beta}=\sqrt{k_c^2-\omega_r^2\mu\epsilon}$$

or

$$-\bar{\beta}^2=k_c^2-\left(\frac{2\pi}{\lambda_r}\right)^2.$$

Using $\bar{\beta}=l\pi/d$ [Eq. (5.34)], the resonant wavelength is given by

$$\lambda_r=\frac{2\pi}{\sqrt{k_c^2+\left(\frac{l\pi}{d}\right)^2}}. \tag{5.38}$$

where cut-off wavenumber $k_c$ is determined from the root of the equation $J'_n(k_c a)=0$ and $J_n(k_c a)=0$ for TE and TM modes respectively.

The resonant frequency is given by

$$f_r=\frac{1}{2\pi\sqrt{\mu\epsilon}}\sqrt{\{k_c^2+(l\pi/d)^2\}}. \tag{5.39}$$

If $p'_{nm}$ be the $m^{\text{th}}$ root of Eq. $J'_n(k_c a)=0$ and $p_{nm}$ be the $m^{\text{th}}$ root of Eq. $J_n(k_c a)$ then the resonant frequencies for various modes are given by

$$f_{r,\text{TE}nml}=\frac{1}{2\pi\sqrt{\mu\epsilon}}\left\{\left(\frac{p'_{nm}}{a}\right)^2+\left(\frac{l\pi}{d}\right)^2\right\}^{1/2} \tag{5.40}$$

$$f_{r,\text{TM}nml}=\frac{1}{2\pi\sqrt{\mu\epsilon}}\left\{\left(\frac{p_{nm}}{a}\right)^2+\left(\frac{l\pi}{d}\right)^2\right\}^{1/2}. \tag{5.41}$$

## 5.7 $Q_o$ OF A CYLINDRICAL RESONATOR

Consider a cylindrical resonator filled with loss-less dielectric* and oriented in a coordinate system as shown in Fig. 5.4.

### $TM_{nml}$ Mode

Let us first consider the case of transverse magnetic waves whose standing wave field distribution is given by Eq. (5.37).

The peak energy stored in the cavity resonator is given by

$$W_s = \frac{\mu}{2}\int_0^d\int_0^{2\pi}\int_0^a \{|H^o_{\rho s}|^2 + |H^o_{\phi s}|^2\}\, dz\, \rho d\phi \cdot d\rho.$$

Using the values $H^o_{\rho s}$ and $H^o_{\phi s}$ from Eq. (5.37)

$$W_s = \frac{\mu}{2}\int_0^d\int_0^{2\pi}\int_0^a \left\{\frac{\omega_r^2\epsilon^2 n^2 B^2}{\rho^2 k_c^4} J_n^2(k_c\rho)\sin^2 n\phi \cos^2 \beta z + \frac{\omega_r^2\epsilon^2 B^2}{k_c^2} J_n'^2(k_c\rho)\cos^2 n\phi\cos^2 \beta z\right\} dz\, \rho\, d\phi\, d\rho$$

$$= \frac{\mu}{2}\left\{\frac{\omega_r^2\epsilon^2 B^2}{k_c^4}\cdot\frac{d}{2}\cdot\pi\int_0^a \frac{n^2}{\rho^2} J_n^2(k_c\rho) + k_c^2 J_n'(k_c\rho)\right\}\rho d\rho. \qquad (5.42)$$

To evaluate the integral in the above equation, we recall some properties of the Bessel function**. We know that the solutions of Bessel's equation

$$\frac{d^2y}{d\rho^2} + \frac{1}{\rho}\frac{dy}{d\rho} + \left(k_c^2 - \frac{n^2}{\rho^2}\right)y = 0$$

are Bessel functions of the first kind and order $n$ and written as $J_n(k_c\rho)$ Therefore, $J_n(k_c\rho)$ must satisfy the above equation. This yields

$$-\{(k_c\rho)^2 - n^2\} J_n = \rho^2 \frac{d^2}{d\rho^2} J_n(k_c\rho) + \rho\frac{d}{d\rho} J_n(k_c\rho). \qquad (5.43)$$

Multiplying the above equation on both the sides by $dJ_n/d\rho$ and carrying out some mathematical manipulation, we reduce the LHS and RHS of above equation to the following form

$$-\{(k_c\rho)^2 - n^2\} J_n \frac{dJ_n}{d\rho} = \left\{\rho^2\frac{d^2}{d\rho^2} J_n(k_c\rho) + \rho\frac{d}{d\rho} J_n(k_c\rho)\right\}\frac{dJ_n}{d\rho}$$

or

$$k_c^2\rho J_n^2 - \frac{1}{2}\frac{d}{d\rho}\{(k_c^2\rho^2 - n^2) J_n^2\} = \frac{k_c^2}{2}\frac{d}{d\rho}(\rho J_n'^2)^2\} \qquad (5.44)$$

*When dielectric is loss-less $Q_0 = Q$ walls.

**Pipes and Harvill *Applied Mathamatics for Engineers; and Physists*, McGraw-Hill (1970).

where on the RHS we have used

$$J_n' = \frac{dJ_n(\rho k_c)}{d(\rho k_c)} = \frac{1}{k_c}\frac{dJ_n}{d\rho}.$$

The integration of Eq. (5.44) is known as Lommel's Integral.

$$2\,k_c^2 \int_0^a \rho\, J_n^2\, d\rho = \Big[ (k_c^2\,\rho^2 - n^2)\, J_n^2 + (k_c\rho\, J_n')^2 \Big]_0^a \tag{5.45}$$

Let us now consider the identity

$$\frac{d}{d\rho}\left(\rho J_n \frac{dJ_n}{d\rho}\right) = \rho\left(\frac{dJ_n}{d\rho}\right)^2 + J_n\frac{dJ_n}{d\rho} + \rho\, J_n\,\frac{d^2J_n}{d\rho^2}.$$

The RHS of the above equation may be reduced using Eq. (5.43) to the following form

$$\frac{d}{d\rho}\left(\rho\, J_n\,\frac{dJ_n}{d\rho}\right) = \rho\left[(kJ_n')^2 + \left(\frac{n^2}{\rho^2} - k_c^2\right) J_n^2\right].$$

The integration of above equation leads to

$$\Big[k_c J_n J_n'\Big]_0^a = \int_0^a \rho\left[(k_c J_n')^2 + \left(\frac{nJ_n}{\rho}\right)^2\right] d\rho$$

$$-k_c^2 \int_0^a \rho\, J_n^2\, d\rho. \tag{5.46}$$

Using Lommel's Integral Eq. (5.45) to evaluate the last term in the above equation and noting that the common boundary conditions we come across are $J_n'(k_c\rho)=0$ or $J_n(k_c\rho)=0$ we obtain a very important integral.

$$\int_0^a \rho\left[(k_c\, J_n')^2 + \left(\frac{nJ_n}{\rho}\right)^2\right] d\rho = \begin{cases} \frac{1}{2}\,(k_c^2a^2 - n^2)\, J_n^2(k_ca);\ J_n'(k_c\, a) \\ \qquad = 0 \text{ (TE waves)} \\ \frac{1}{2}\,[k_caJ_n'\,(k_ca)]^2;\ J_n\,(k_ca) \\ \qquad = 0 \text{ (TM waves)} \end{cases} \tag{5.47}$$

Use of Eq. (5.47) in Eq. (5.42) gives

$$W_s = \frac{\mu}{2}\,\frac{\omega_r^2\epsilon^2B^2}{k_c^4}\cdot\frac{d}{2}\cdot\pi\cdot\frac{1}{2}\,k_c^2a^2\,J_n'^2\,(k_ca). \tag{5.48}$$

The time average power dissipated in the cavity walls is given by

$$P_L = \frac{R_s}{2}\Big\{2\int_0^{2\pi}\Big[|H_{\rho s}^o|^2 + |H_{\phi s}^o|^2\Big]_{z=0 \text{ and } z=d} \rho d\phi\cdot d\rho$$

$$+\int_0^{2\pi}\int_0^d\Big[|H_{\phi s}^o|^2 + |H_{zs}^o|^2\Big]_{\rho=a} a d\phi\cdot dz\Big\}$$

$$=\frac{R_s}{2}\Big\{2\int_0^{2\pi}\int_0^a\Big[\frac{\omega_r^2\epsilon^2n^2B^2}{\rho^2k_c^4}\,J_n^2(k_c\rho)\sin^2 n\phi$$

$$+\frac{\omega_r^2\epsilon^2B^2}{k_c^2}\,J_n'^2(k_c\rho)\cos^2 n\phi\Big]\rho d\phi d\rho$$

$$+\int_0^{2\pi}\int_0^d\frac{\omega_r^2\epsilon^2B^2}{k_c^2}\,J_n'^2(k_c\,a)\cos^2 n\phi\cos^2\bar{\beta}z\;ad\phi\,dz\Big\}.$$

On integration $\cos^2\bar{\beta}z$ gives $d/2$ and $\cos^2 n\phi$ gives $\pi$

$$P_L=\frac{R_s}{2}\frac{\omega_r^2\epsilon^2B^2}{k_c^4}\pi\Big\{2\cdot\frac{1}{2}\;k_c^2\,a^2\,J_n'^2(k_ca)+k_c^2\;J_n'^2(k_ca)\cdot\frac{ad}{2}\Big\}$$

$$=\frac{R_s\,\pi\omega_r^2\epsilon^2B^2}{k_c^4}\,k_c^2\,J_n'^2\,(k_ca)\cdot\Big(a^2+\frac{ad}{2}\Big). \tag{5.48a}$$

Substituting values of $W_s$ and $P_L$ from Eqs. (5.48) and (5.48a) in Eq. (5.21) gives us $Q_o$,

$$Q_{o,\,\mathrm{TM}_{nml}}=\frac{\omega_r\mu d}{4R_s}\frac{1}{1+d/2a}. \tag{5.49}$$

Using $\quad \omega_r=\dfrac{2\pi}{\sqrt{\mu\epsilon}}\Big\{\Big(\dfrac{p_{nm}}{a}\Big)^2+\Big(\dfrac{l\pi}{d}\Big)^2\Big\}^{1/2}$ and $\delta=\sqrt{\dfrac{2R_2}{\omega_r\mu}}$,

Eq. (5.49) may be written as

$$Q_o=\frac{\lambda_r}{\delta}\,\frac{\Big\{p_{nm}^2+\Big(\frac{l\pi a}{d}\Big)^2\Big\}^{1/2}}{2\pi\,(1+2a/d)}\quad(\text{for } l\neq 0) \tag{5.50}$$

$\delta$ is the skin depth of the metal used in designing guide walls. However, for $l=1$,

$$Q_o=\frac{\lambda r}{\delta}\,\frac{p_{nm}}{2\pi\,(1+a/d)}. \tag{5.51}$$

For an air-filled cavity of radius $a=2$ cm and length $d=10$ cm made of brass, we have for the $TM_{011}$ mode,

$$\sigma=1.45\times10^7\,\mho/\text{m},\;k_{01}a=2.405$$

$$f_r=f_{011}=6.47\text{ GHz},\;\mu=\mu_0\text{ for brass.}$$

The value of $Q_o$ comes out to be

$$Q_o\simeq 8681$$

which is indeed a high value.

**$TE_{nml}$ Mode**

Now we return to the $TE_{nml}$ mode whose field configuration is given by Eq. (5.36).

The peak energy stored in the cavity is given by

$$W_s = \frac{\epsilon}{2}\int_0^a\int_0^{2\pi}\int_0^d \left\{|E^o_{\rho s}|^2 + |E^o_{\phi s}|^2\right\} d\rho \cdot \rho d\phi\ dz.$$

Substituting the values of $E^o_{\rho s}$ and $E^o_{\phi s}$ from Eq. (5.36), we obtain

$$W_s = \frac{\epsilon}{2}\int_0^0\int_0^{2\pi}\int_0^d \left\{\frac{\omega_r^2\mu^2}{\rho^2 k_c^4}\, n^2A^2\, J_n^2\,(k_c\rho)\,\sin^2 n\phi\,\sin^2\bar{\beta}z\right.$$

$$\left.+\frac{\omega_r^2\mu^2A^2}{k_c^2}\, J_n'^2\,(k_c\rho)\cos^2 n\phi\,\sin^2\bar{\beta}z\right\}\rho d\rho\ d\phi\ dz.$$

The integrals over $\phi$ and $z$ respectively contribute $\pi$ and $d/2$.

$$W_s = \frac{\epsilon}{2}\,\frac{d}{2}\,\pi\,\frac{\omega_r^2\mu^2A^2}{k_c^4}\int_0^a \left\{\left(\frac{n^2}{\rho^2}\right) J_n^2\,(k_c\rho) + k_c^2\, J_n'^2\,(k_c\rho)\right\}\rho d\rho.$$

On using Eq. (5.47), the above equation reduces to

$$W_s = \frac{\epsilon}{2}\,\frac{d}{2}\,\pi\,\frac{\omega_r^2\mu^2A^2}{k_c^4}\,\frac{1}{2}\,(k_c^2\,a^2 - n^2)\,J_n^2\,(k_ca). \tag{5.52}$$

The time average power loss in the guide walls is given by

$$P_L = \frac{2R_s}{2}\int_0^a\int_0^{2\pi}\Big[|H^0_{\phi s}|^2 + |H^0_{\rho s}|^2\Big]_{z=0 \text{ and } z=d} d\rho\cdot\rho d\phi$$

$$+\frac{R_s}{2}\int_0^{2\pi}\int_0^d \left\{|H^0_{zs}|^2 + |H^0_{\phi s}|^2\right\}_{\rho=a} a d\phi dz$$

$$P_L = R_s\int_0^a\int_0^{2\pi}\left\{\frac{4A^2n^2\bar{\beta}^2}{\rho^2\,k_c^2}\,J_n^2\,(k_c\rho)\,\sin^2 n\phi\right.$$

$$\left.+\frac{4\,A^2\bar{\beta}^2}{k_c^2}\,J_n'^2\,(k_c\rho)\cos^2 n\phi\right\}\rho d\rho\ d\phi$$

$$+\frac{R_s}{2}\int_0^{2\pi}\int_0^d\left\{4\,A^2\,J_n^2\,(k_ca)\cos^2 n\phi\,\sin^2\bar{\beta}z\right.$$

$$\left.+\frac{4\,A^2n^2\bar{\beta}^2}{a^2\,k_c^4}\,J_n^2\,(k_ca)\,\sin^2 n\phi\,\cos^2\bar{\beta}z\right\} a d\phi dz$$

$$= R_s\,\frac{4\,A^2\bar{\beta}^2}{k_c^4}\,\pi\int_0^a\left\{\frac{n^2}{\rho^2}\,J_n^2\,(k_c\rho) + k_c^2\,J_n'^2\,(k_c\rho)\right\}\rho d\rho$$

$$+\frac{R_s}{2}\,4\cdot A^2J_n^2\,(k_c\rho)\left\{\frac{\pi d}{2} + \frac{n^2\bar{\beta}^2}{a^2k_c^4}\,\pi\,\frac{d}{2}\right\}.$$

The integral in the first term may be evaluated using Eq. (5.47)

$$P_L = \frac{R_s\,4\,A^2\bar{\beta}^2\pi}{k_c^4}\,\frac{1}{2}\,(k_c^2\,a^2 - n^2)\,J_n^2\,(k_c\,a)$$

$$+\frac{R_s}{2}\,4\,A^2\,J_n^2\,(k_c\,a)\,\frac{\pi da}{2}\left(1 + \frac{n^2\bar{\beta}^2}{a^2k_c^4}\right)$$

Rearranging the terms

$$P_L = R_s A^2 J_n^2 (k_c\, a)\, \pi \left\{ ad\left(1 + \frac{n^2\beta^2}{a^2k_c^2}\right) + 2\,\frac{\beta^2 k_c^2 a^2}{k_c^4}\left(1 - \frac{n^2}{k_c^2a^2}\right)\right\} \tag{5.53}$$

Hence $Q_o$ of a cylindrical resonator excited to $TE_{nml}$ mode, may be obtained by substituting Eqs. (5.52) and (5.53) in Eq. (5.21).

$$Q_{o,\ TE\ nml} = \omega_r \frac{\frac{\epsilon}{2}\,\pi\,\frac{d}{2}\,\frac{\omega_r^2\mu^2A^2}{k_c^4}\,\frac{1}{2}\,(k_c^2\,a^2 - n^2)\,J_n^2\,(k_c\,a)}{R_sA^2J_n^2(k_ca)\pi\left[ad\left(1 + \frac{n^2\beta^2}{a^2k_c^2}\right) + 2\,\frac{\beta^2k_c^2a^2}{k_c^4}\left(1 - \frac{n^2}{k_c^2a^2}\right)\right]} \tag{5.54}$$

Using $\quad p'_{nm} = k_c a,\ \delta = \sqrt{\frac{2\,R_s}{\omega_r\mu}},\ \omega_r = 2\pi\, f_r,\ \lambda_r = \frac{1}{\sqrt{\mu\epsilon}\, f_r}$

$$\beta = l\pi/d,\ \text{and}\ f_r = \frac{1}{2\pi\,\sqrt{\mu\epsilon}}\left\{\left(\frac{p'_{nm}}{a}\right)^2 + \left(\frac{l\pi}{d}\right)^2\right\}^{1/2},$$

Eq. (5.54) may be reduced to the following form

$$Q_{o,\ TE\ nml} = \frac{\lambda_r}{\delta}\,\frac{\left\{1 - \left(\frac{n}{p'_{nm}}\right)^2\right\}\left\{(p'_{nm})^2 - \left(\frac{l\pi}{d}\,a\right)^2\right\}^{3/2}}{2\pi\left\{(p'_{nm})^2 + \frac{2a}{d}\left(\frac{l\pi a}{d}\right)^2 + \left(1 - \frac{2a}{d}\right)\left(\frac{nl\pi a}{p'_{um}\,d}\right)^2\right\}} \tag{5.55}$$

It is noted from Eq. (5.55) that all the terms on the RHS except $\lambda_r/\delta$ are independent of frequency hence $Q_o$ of a cylindrical cavity varies as $\lambda_r/\delta$ and thus decreases as $1/\sqrt{f}$.

## 5.8 MODES IN A CYLINDRICAL RESONATOR

It is seen from Eqs. (5.40) and (5.41) that the resonant frequency is determined by three integers $n$, $m$ and $l$. It is therefore expected that the modes in a cylindrical cavity are triply infinite and are designated as $TE_{nml}$ (or $H_{nml}$) and $TM_{nml}$ (or $E_{nml}$). Of course, indices $n$, $m$ and $l$ carry the same physical interpretation, i.e. these indicate the number of half period variations of the microwave field along the azimuthal, number of zeros of electric field intensity in the radial direction (exclusive of zero on the axis), and a number of half period variations in the longitudinal direction respectively. The properties of any of these modes can be analyzed by means of the corresponding modes in the cylindrical waveguide. However, in general $p'_{nm} \neq p_{nm}$, and consequently cylindrical resonators do not possess the type of degeneracy found in rectangular waveguides.

For $n=0$, the electric intensity has a radial distribution corresponding to zero order Bessel function and there is no variation of the field in the azimuthal ($\phi$) direction. For TM modes in addition to $n=0$, we may have $p=0$ corresponding to uniform intensity in the axial direction. Thus the lowest modes in TM series are $TM_{omo}$. The dominant mode in a cylindrical waveguide in TM series is $TM_{01}$. Hence the lowest, dominant resonant mode in a cylindrical resonator will be $TM_{010}$. The resonant wavelength corresponding to this mode is obtained by substituting $n=l=0$ in Eq. (5.41)

$$\lambda_r\ r(TM_{010})=2.62a. \tag{5.56}$$

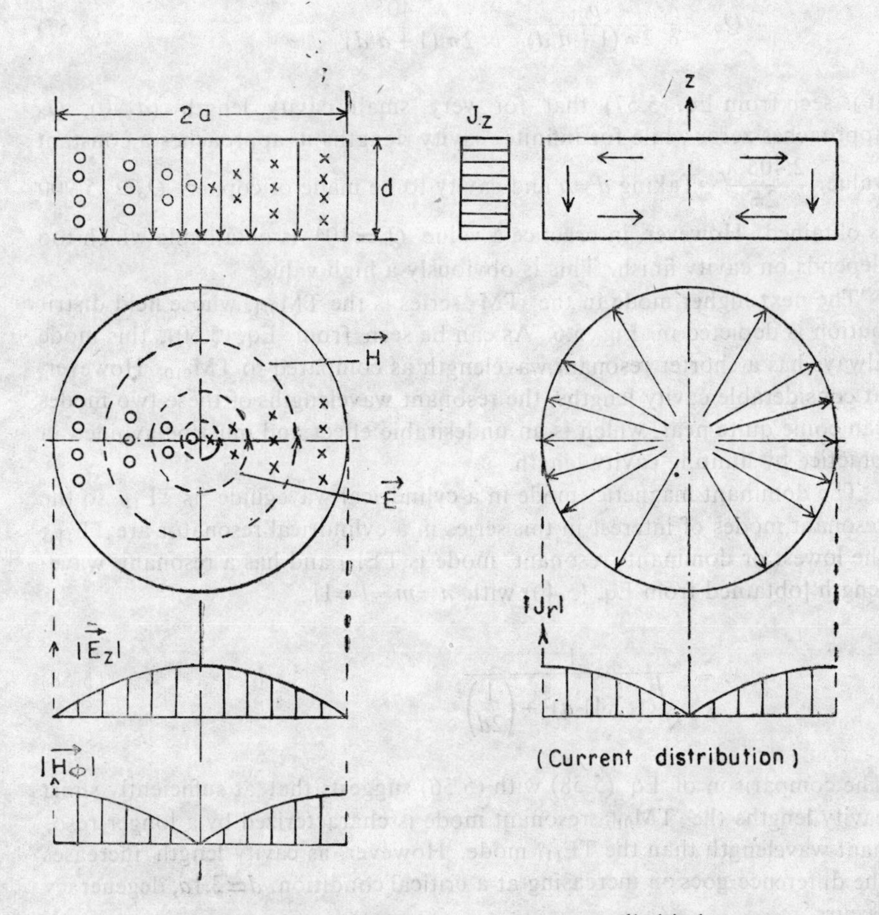


**Fig. 5.5** Field pattern and current distribution in a cylindrical resonator at $TM_{010}$ resonant mode.

It is seen that this resonant wavelength is the same as the cut-off wavelength of the propagating $TM_{01}$ mode in a cylindrical waveguide and is independent of the length of the cavity. The field pattern and current distribution of this mode are depicted in Fig. (5.5). They resemble those of the propagating $TM_{01}$ mode of a cylindrical waveguide operating at cut-off. Physically, the existence of the $TM_{010}$ resonant mode can be explained by the fact that the electric field is everywhere parallel to the axis of the circular waveguide when the wavelength becomes equal to the cut-off value. Consequently, the plates short circuiting the waveguide butt ends turn out to be normal to the lines of electric field and therefore do not preclude the existence of a microwave field irrespective of the resonant cavity length.

The unloaded $Q_o$ for this mode may be obtained by substituting $n=l=0$ and $m=1$ in Eq. (5.51). It is given by

$$Q_o=\frac{\lambda_r}{\delta}\frac{p_{01}}{2\pi\,(1+a/d)}=\frac{\lambda_r}{\delta}\frac{2.405}{2\pi\,(1+a/d)}. \tag{5.57}$$

It is seen from Eq. (5.57) that for very small cavity length $(d\to 0)$, $Q_o$ approaches zero, while for infinite cavity lengths it approaches a constant value. $\frac{2.405\,\lambda_r}{2\pi\,\delta}$. Taking $d=a$ and cavity to be made of copper, $Q_o\simeq 15{,}900$ is obtained. However, in practice a value $Q_o\simeq 10^4$ is obtainable which too depends on cavity finish. This is obviously a high value.

The next higher mode in the TM series is the $TM_{011}$, whose field distribution is depicted in Fig. 5.6. As can be seen from Eq. (5.50), this mode always has a shorter resonant wavelength as compared to $TM_{010}$. However, at considerable cavity lengths, the resonant wavelengths of these two modes can come quite near, which is an undesirable effect and must be avoided in practice by limiting cavity length.

The dominant magnetic mode in a cylindrical waveguide is $TE_{11}$ so the resonant modes of interest in this series in a cylindrical resonator are $TE_{111}$; the lowest or dominant resonant mode is $TE_{111}$ and has a resonant wavelength [obtained from Eq. (5.40) with $n=m=l=1$]

$$\lambda_r=\frac{1}{\sqrt{(1/3.41\,a)^2+\left(\frac{1}{2d}\right)^2}} \tag{5.58}$$

The comparison of Eq. (5.58) with (5.56) suggests that at sufficiently short cavity lengths the $TM_{010}$ resonant mode is characterized by a longer resonant wavelength than the $TE_{111}$ mode. However, as cavity length increases the difference goes on increasing at a critical condition, $d\simeq 2.1a$, degeneracy occurs.

Thereafter $(d>2.1a)$ $TE_{111}$ resonant mode has longer resonant wavelength than $TM_{010}$. Thus, the dominant resonant mode may correspond to the

$TM_{01}$ propagation mode despite the fact that in a circular waveguide the dominant propagation mode is $TE_{11}$.

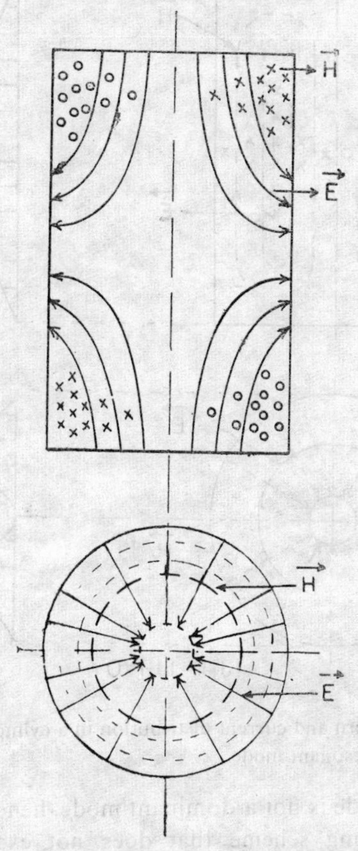


**Fig. 5.6** Field distribution in a cylindrical cavity at $TM_{011}$ resonant mode.

The higher magnetic (TE) resonant modes correspond to $TE_{01}$, $TE_{02}$, $TE_{21}$, $TE_{12}$ . . . propagation modes in cylindrical waveguides. The nearest higher mode, i.e. $TE_{011}$ resonant mode in a cylindrical resonator is of special practical interest. The field pattern and current distribution corresponding to this mode are depicted in Fig. 5.7. Since in this mode $H^o_{\phi s}=0$, there are no axial currents. This means that the end plates of the cavity can be free to move to adjust the cavity length $d$ for tuning purposes without introducing any significant loss since no currents flow across the gap, i.e. the gap between the circular end plate and the cylinder wall is parallel to current flow lines. Further, Fig. 5.8 depicts the variation in $Q$ values of some important modes in a cylindrical resonator. It is noted that the $Q_o$ value for the $TE_{011}$ mode is practically upto $10^5$ (because of low dissipation in cavity walls) which is two to three times the $Q_0$ of the $TE_{111}$ resonant mode. These advantages of the $TE_{011}$ mode over $TE_{111}$ and $TM_{010}$ modes makes it specially suitable in designing high resolution wavemeters, echo-boxes, spectrum analyzers, etc.

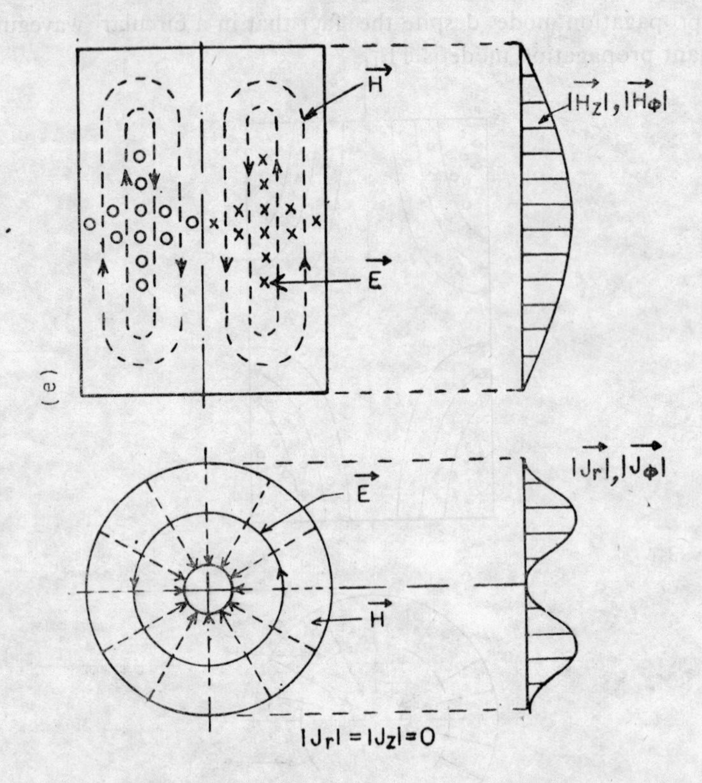


**Fig. 5.7** Field pattern and current distribution in a cylindrical resonator at $TE_{011}$ resonant mode.

However, the $TE_{011}$ mode is not a dominant mode, hence care must be exercised to choose a coupling scheme that does not excite any mode other than $TE_{011}$ or mode supressors are used to supress spurious resonances.

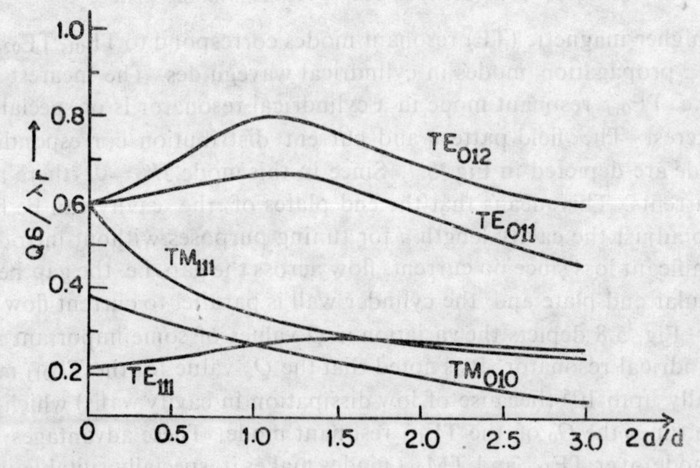


**Fig. 5.8** Frequency variation of Q values for circular-cylindrical resonator modes.

A mode chart shown in Fig. 5.9 is therefore constructed for a particular cavity resonator. This chart shows possible modes that can be excited. It shows that for $2a/d$ ratios between 2 and 3, only $TE_{011}$ and $TM_{111}$ modes can

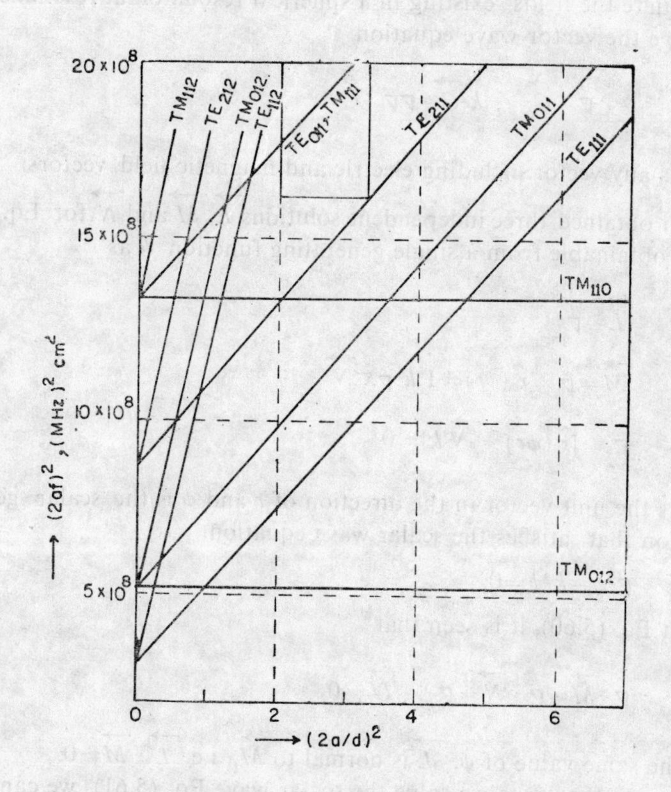


**Fig. 5.9** Mode chart for a cylindrical cavity of circular cross-section.

be excited (see the rectangle) in the resonant frequency range corresponding to $(2af_r)^2$ between $16.3 \times 10^8$ and $20.4 \times 10^8$. Thus for $2a/d \simeq 2$ to 3, the only mode needed to be suppressed is $TM_{111}$.

## 5.9 SPHERICAL RESONATORS

We know that the peak energy stored in a resonator is proportional to the volume enclosed with in the cavity and the time average power loss is proportional to the conducting wall area. Hence using the general definition of $Q$ (Eq. 5.21), it can be inferred that a spherical resonator that has maximum volume to surface ratio, offers attractive possibilities as a high $Q_o$ resonator.

To analyze a spherical resonator, Maxwell's equations have to be solved under the boundary conditions that $E_{tan}$ and $H_{normal}$ vanish at the conductor walls. A number of scientists* have taken up the problem, the details of which are beyond the scope of present book. However, we shall follow the treatment given by Stratton.

To evaluate the fields existing in a spherical resonator at resonance, one has to solve the vector wave equation

$$\nabla \times \nabla \times \vec{C} = -k^2 \vec{C} - \nabla \nabla \cdot \vec{C} \tag{5.59}$$

where $\vec{C}$ is any vector including electric and magnetic field vectors.

Stratton obtained three independent solutions $\vec{L}$, $\vec{M}$ and $\vec{N}$ for Eq. (5.59) which are obtainable from a single generating function $\psi$ as

$$\vec{L} = \nabla \psi$$

$$\vec{M} = \nabla \times \hat{r}\,(\psi r) = 1/k\ \nabla x\ \vec{N} \tag{5.60}$$

or $$\nabla \times\ [\hat{r}\,(\psi r) - \vec{N}/k] = 0$$

where $\hat{r}$ is the unit vector in the direction of $r$ and $\psi$ is the scalar generating function that satisfies the scalar wave equation

$$\nabla^2 \psi - k^2 \psi = 0. \tag{5.61}$$

Also from Eq. (5.60), it is seen that

$$\nabla \cdot \vec{M} = \nabla \cdot \vec{N} = \nabla \times \vec{L} = 0 \tag{5.62}$$

and for the same value of $\psi$, $\vec{L}$ is normal to $\vec{M}$, i.e. $\vec{L} \cdot \vec{M} = 0$.

Thus we see that once we solve the scalar wave Eq. (5.61) we can determine $\vec{L}$, $\vec{M}$ and $\vec{N}$ and hence solve the vector wave Eq. (5.59).

As there exists spherical symmetry assuming the solution of Eq. (5.57) to be

$$\psi = R(r)\ \Theta(\theta)\ \Phi(\phi). \tag{5.63}$$

Substitution of Eq. (5.63) in Eq. (5.61) when the laplacian is expressed in polar form and division of the equation so obtained by Eq. (5.63) results in the separation of variables. Taking integers $m$ and $n$ as separating constants we obtain three independent equations which are given by

$$\frac{d^2R}{dr^2} + \frac{2}{r}\frac{dR}{dr} + \left[k^2 - \frac{n(n+1)}{r^2}\right] R = 0 \tag{5.64}$$

*1. J.A. Stratton, *Electromagnetic Theory*, New York: McGraw-Hill Book Co. Inc., (1941).

2. E.U. Condon, "Principles of microwave radio", *Rev. Modern Phys.*, Vol. 14 (Oct. 1942).

$$\frac{d^2\oplus}{d\theta^2}+\cot\theta\,\frac{d\oplus}{d\theta}+\left[n(n+1)-\frac{m^2}{\sin^2\theta}\right]\oplus=0 \tag{5.65}$$

$$\frac{d^2\Phi}{d\phi^2}+m^2\Phi=0. \tag{5.66}$$

The solutions to the above equations are respectively given by

$$R(r)=j_n\,(kr),\ n_n(kr) \tag{5.67}$$

$$\oplus\,(\theta)=P_n^m\,(\cos\theta),\ \ Q_n^m\,(\cos\theta) \tag{5.68}$$

$$\Phi(\phi)=\cos m\phi,\ \sin m\phi \tag{5.69}$$

where $j_n(kr)$ and $n_n(kr)$ are spherical Bessel functions of the first and second kind, and $P_n^m$ (cos θ) and $Q_n^m$ (cos θ) are associated Legendre functions. Since the centre of the cavity has finite fields and, $n_n\,(kr)\;Q_n^m$ (cos θ) become infinite at the origin, these cannot be included in the acceptable solution for the resonant fields in a spherical cavity. Thus, the general solution for the generating function is given by

$$\psi_{mn}=A_{mn}\ P_n^m(\cos\theta)\ J_n\,(kr)\ (\cos m\phi+\Gamma_{mn}\sin m\phi) \tag{5.70}$$

where $A_{mn}$ and $\Gamma_{mn}$ are constants depending on the values of integers $m$ and $n$.

Assuming that there is no field variation in $\phi$-direction,

$$\frac{\partial\psi}{\partial\phi}=0 \tag{5.71}$$

and hence $m=0$. Equation (5.70) when subjected to Eq. (5.71), yields

$$\psi_n=A_nP_n^0(\cos\theta)\ j_n\,(kr). \tag{5.72}$$

Now we can assign a physical interpretation to the solutions $\vec{L}$, $\vec{M}$ and $\vec{N}$ which can be evaluated from $\psi_n$. If $\vec{E_a}$ and $\vec{H_a}$ respectively represent generalized electric and magnetic fields corresponding to the $a^{th}$ resonant mode in a spherical resonator then, since $\nabla\cdot\vec{E_a}=\nabla\cdot\vec{H_a}=0$, we can assume two functions $\vec{F_a}$ and $\vec{G_a}$ such that

$$\nabla\times\vec{F_a}=\nabla\times\vec{G_a}=0. \tag{5.73}$$

It then follows that

$$\vec{F_a}=\nabla\phi_a \text{ and } \vec{G_a}=\nabla\psi_a$$

The comparison of Eq. (5.73) with Eq. (5.60) leads us to conclude that $\vec{M}$ and $\vec{N}$ correspond to the generylized electric and magnetic fields respectively and $\vec{L}$ corresponds the vector $\vec{F_a}$ and $\vec{G_a}$ defined in Eq. (5.73) in terms of the scalar functions $\phi_a$ and $\psi_a$.

## 5.10 MODES IN A SPHERICAL RESONATOR

For the $a^{th}$ TE mode, $\vec{E_{ar}}=0$ while for the $a^{th}$ TM mode, $\vec{H_{ar}}=0$ or in terms of generalized fields $\vec{M_{ar}}=0$. Hence the field components corresponding to the $a^{th}$ mode are given by

$$\vec{M_{ar}}=0$$

$$\vec{M_{a\theta}}=\frac{1}{r\sin\theta}\frac{\partial}{\delta\phi}(\psi_a r)$$

$$\vec{M_{a\phi}}=-\frac{1}{r}\frac{\partial}{\partial\theta}(\psi_a r) \tag{5.74}$$

where $\vec{M_{ar}}$ may be $\vec{E_{ar}}$ or $\vec{H_{ar}}$.

Now for the $a^{th}$ TE mode $\vec{M_a}=\vec{E_a}$. Assuming that there exists no variation of fields in the $\phi$-direction, the field configuration for the $a^{th}$ TE mode is given by*

$$\vec{E_r}=0 \tag{5.75a}$$

$$\vec{E_\theta}=\frac{A}{\sin\theta}\frac{\partial}{\partial\phi}[A_nP_n^o(\cos\theta)\,j_n(kr)]=0 \tag{5.75b}$$

$$\vec{E_\phi}=-A\frac{d\psi}{d\theta}=-A\,j_n(kr)\,P_n'(\cos\theta) \tag{5.75c}$$

The magnetic field components may be obtained from Maxwell's equations (M−1) and (M−2), expressed in spherical polar coordinates. These are given by

$$\vec{H_r}=-\frac{A}{j\omega\mu}\left(\frac{n^2+n}{r}\right)\psi=-\frac{A}{j\omega\mu}n(n+1)\,k\left\{\frac{j_n(kr)}{kr}\right\}P_n(\cos\theta) \tag{5.75d}$$

$$\vec{H_\theta}=-\frac{A}{j\omega\mu r}\frac{\partial^2(\psi_\gamma)}{\partial r\,\partial\theta}=-\frac{A}{j\omega\mu}\frac{\partial}{\partial(kr)}\cdot[j_n(kr)\cdot(kr)]\,P_n'(\cos\theta) \tag{5.75e}$$

*Suffix a has been omitted for convenience.

$$\vec{H}_\phi = -\frac{A}{j\,\omega\mu}\,\frac{1}{r\sin\theta}\,\frac{\partial^2(\psi r)}{\partial r\,\partial\phi}=0. \tag{5.75 f}$$

Similarly, for TM modes $\vec{M}_a = \vec{H}_a$, there is no variation of fields in $\phi$-direction; we have

$$\begin{aligned}
&\vec{E}_\phi = \vec{H}_r = \vec{H}_\theta = 0\\
&\vec{E}_r = \frac{B}{j\,\omega\epsilon}\,n(n+1)\,k\left\{\frac{j_n(kr)}{kr}\right\}P_n(\cos\theta)\\
&\vec{E}_\theta = \frac{B}{j\,\omega\epsilon}\left\{\frac{\partial}{\partial(kr)}\,j_n(kr)\right\}(kr)\,P_n'(\cos\theta)\\
&\vec{H}_\phi = -B\,j_n(kr)\,P_n'(\cos\theta)
\end{aligned} \tag{5.76}$$

where $A$ and $B$ are constants, and $k=\omega^2\mu\epsilon$ is the wave vector.

To satisfy the boundary condition it is necessary that $E_\theta = E_\phi = 0$ at the metal surface where $r=a$. This requires that

$$j_n(k_a)=0; \qquad \text{TE modes} \tag{5.77}$$

$$\frac{\partial}{\partial(kr)}\,[(kr)\,j_n(kr)]'_{r=a}=0; \qquad \text{TM modes} \tag{5.78}$$

The roots of Eqs. (5.77) and (5.78) determine the values of resonant frequency. If $(ka)$ be the root of Eq. (5.77) or (5.78) as the case may be, then resonant frequency is given by

$$\omega_r = 2\pi f_r = \frac{1}{\sqrt{\mu\epsilon}}\,\frac{(ka)}{a} \tag{5.79}$$

and the corresponding resonant wavelength is

$$\lambda_r = \frac{2\pi a}{(ka)}. \tag{5.80}$$

The modes in the spherical resonator are designated by $TE_{nmp}$ and $TM_{npm}$ where $n$ is the order of spherical Bessel function, $p$ is an integer denoting the rank of the roots of Eq. (5.77) or (5.78) for a given value of $n$, and $m$ is periodicity in $\phi$-direction.

It is interesting to note from Eq. (5.79) that $\omega_r$ resonant frequency is independent of integer $m$, but since $P_n^m(\cos\theta)$ vanishes when $m>n$, hence, for a given value of $n$ the integer $m$ must have values from 0 to $n$ (inclusive). Each of these values corresponds to a separate mode but all modes have the same resonant frequency. Furthermore, the intensity distribution in the $\phi$-direction is of the form $\cos m\phi$ or $\sin m\phi$ or any linear combination of the two. Hence a spherical resonator has degenerate modes.

For $n=0$, the field components vanish, hence, there cannot be any allowed mode of the type $TE_{omp}$ or $TM_{omp}$. The lowest or dominant mode in both TE and TM series occurs when $n=p=1$ and has resonant wavelength

$$\lambda_r(TE_{11m})=1.40a \text{ as } (k_a)_{11}=4.49 \text{ (for TE)}$$

$$\lambda_r(TM_{11m})=2.29a. \text{ as } (k_a)_{11}=2.75 \text{ (for TE)} \tag{5.81}$$

We have already assumed no variation in $\phi$-direction, i.e. $m=0$. Hence Eqs. (5.75) and (5.76) represent field configurations of $TE_{110}$ and $TM_{110}$ modes respectively, which are the dominant modes and only these are of interest. The field patterns of these modes in the meridian plane have been shown in Fig. 5.10 (a & b). It is noted from Eqs. (5.75) and (5.76) that the electric and magnetic fields for either mode are in time quadrature. The current flowing through the resonator walls is proportional to $H_\phi$, and perpendicular to it. In case of the $TM_{110}$ mode magnetic lines of force are parallel to the equator and hence the current flows along the meridian planes from pole to pole.

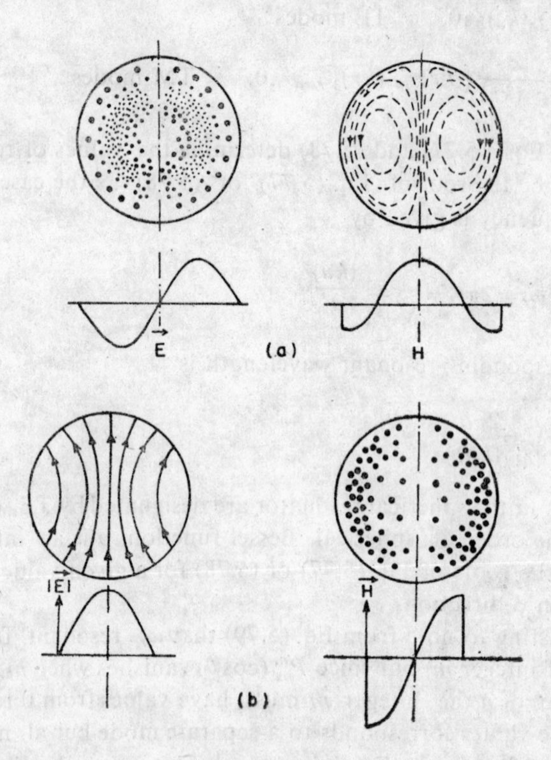


Fig. 5.10 (*a*) Field pattern in the meridian plane of a spherical resonator at $TE_{110}$ mode.
(*b*) $TM_{110}$ mode in a spherical resonator.

## 5.11 $Q$ OF A SPHERICAL RESONATOR

For this resonator too, assuming that the dielectric inside the resonator is loss-less, the $Q_o$ value may be obtained from the general definition (Eq. 5.21). Since the $TM_{110}$ mode has the longest resonant wavelength, this mode is of practical importance and so it is considered for evaluation of $Q_o$.

The peak energy stored in the cavity is given by

$$W_s = \frac{\mu}{2}\int_v |H_\phi|^2\, dv$$

$$= \frac{\mu}{2} B^2 \int_0^\pi \int_0^a \int_0^{2\pi} j_1^2(k_{11}r)\sin^2\theta \,.\, r\sin\theta d\theta\, rd\theta$$

$$= \frac{\mu B^2}{2} 2\pi \cdot \frac{4}{3}\int_0^a r^2 j_1^2(k_{11}r)\, dr.$$

Integrating above integral by parts and using the boundary condition of Eq. (5.78)

$$W_s = \frac{4\pi\mu B^2}{3k_{11}^2}\frac{(k_{11}a)^3}{2}\left[j_1^2(k_{11}a) - j_0(k_{11}a)\, j_2(k_{11}a)\right] \ldots \tag{5.82}$$

The time average power loss in the cavity wall is given by

$$P_L = \frac{R_s}{2}\int_0^\pi \int_0^{2\pi} B^2 j_1^2(k_{11}a)\sin^2\theta\, a^2 \sin\theta\, d\theta\, d\phi$$

$$P_L = \frac{R_s}{2} B^2 j_1^2(k_{11}a)\, a^2 \,.\, 2\pi \,.\, \frac{4}{3} = \frac{4}{3}\pi\, a^2 B^2 j_1^2(k_{11}a) \ldots \tag{5.83}$$

Hence substituting Eqs. (5.82) and (5.83) in Eq. (5.21), we have

$$Q_o = \frac{\omega_{r11}\,\mu a}{2R_s}\left[1 - \frac{j_0(k_{11}a)\, j_2(k_{11}a)}{j_1^2(k_{11}a)}\right]$$

using Eq. (5.79) for the value of $\omega_{r11}$

$$Q_o = \frac{1.37\,\eta}{R_s}\left[1 - \frac{j_0(k_{11}a)\, j_2(k_{11}a)}{j_1^2(k_{11}a)}\right] \tag{5.84}$$

where $\eta = \sqrt{\mu/\epsilon}$ is the free space impedance.

Values of the spherical Bessel function involved in Eq. (5.84) may be seen from the standard tables*. For the $TM_{110}$ mode we have $(ka) = 2.75$ and $j_0(ka) = 0.139$, $j_1(ka) = 0.386$, and $j_2(ka) = 0.282$. These values when substituted in Eq. (5.84) give

$$Q_o = \frac{1.01\eta}{R_s}. \tag{5.85}$$

*H.B. Dwight, *Mathematical Tables*, Dover Pub. Co. (1958).

For a silver plated spherical resonator $R_s \simeq 0.0139$ and for the $TM_{110}$ mode $f_r = 3 \times 10^9$ $H_z$. These results yield a theoretical value of $Q_o$ as high as 27,400.

## 5.12 COAXIAL RESONATORS

Coaxial resonators are the sections of coaxial lines with both or one end shorted by a conductor of very high conductivity, as shown in Fig. 5.11 (a) and (b) respectively. Any field excited in the resonator is reflected from the

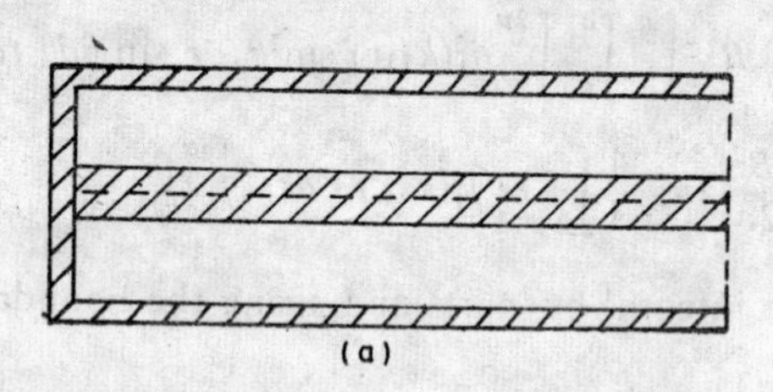

(a)

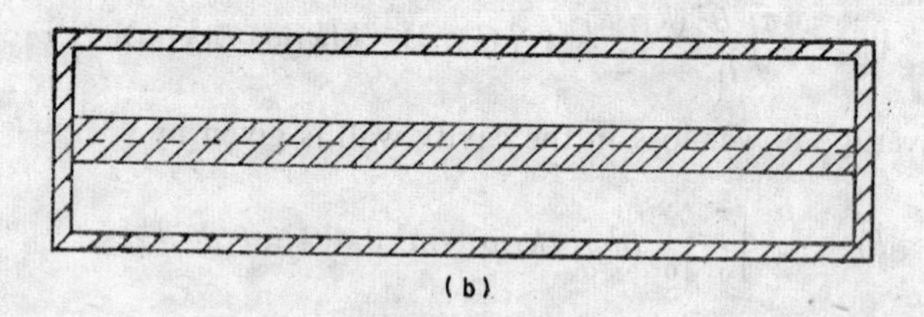

(b)

**Fig. 5.11.** (*a*) Coaxial quarterwave resonator
(*b*) Coaxial halfwave resonator.

ends. The waves add in phase or the resonator resonates when at every point of the line the waves add in phase, i.e. the total phase shift in the propagating waves must be $2\pi$ or its multiple. If $\phi_1$ and $\phi_2$ are the phases introduced by the reflections at the two ends and $l$ is the length of the resonator, then the condition for the resonance will be

$$\phi_1 + \phi_2 + 2\beta d = 2n\pi \tag{5.86}$$

$$n = 1, 2, 3 \ldots$$

**Open End Resonator**

For an open end resonator (Fig. 5.11a) $\phi_1 = 0$, $\phi_2 = \pi$; so

$$2\beta d = (2n-1)\,\pi$$

or
$$\lambda_r = \frac{4d}{2n-1}. \tag{5.87}$$

Thus, the length of the coaxial line section with one end open, acts as a resonator when its length is an odd multiple of quarter wavelength. Conversely, for a resonator of definite length $l$, various resonant modes are possible having resonant wavelengths $\lambda_r = 4d, 4d/3, 4d/5 \ldots$ or cavity lengths $\lambda/4$, $3\lambda/4$, $5\lambda/4 \ldots$ and so on. The lowest mode corresponds to a resonator length$=\lambda/4$ and hence this type of resonator is also designated as the *quarter wave coaxial resonator* or *open end resonator*.

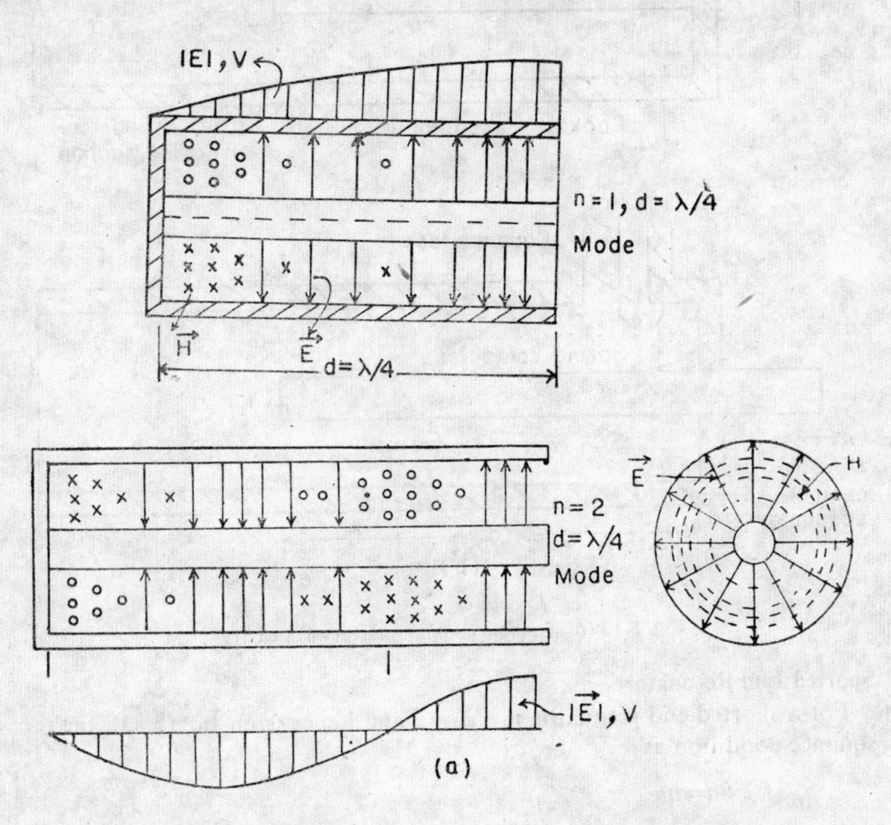


Fig. 5.12 (*a*) Field distribution in a quarlir-wave coaxial resonator.

The field configuration for these modes may easily be inferred from the propagating TEM mode in a coaxial line. Figure 5.12(a) depicts the field patterns of the first two modes $n=1$ and $n=2$, of a quarter wave coaxial resonator. However, care must be exercised so that the higher modes, TE and TM, are not excited in such a resonator. The basic condition is given by Eq. (4.137), i.e. $\lambda_r > \pi/2\ (b+a)$. Further, in a quarter wave resonator one end is left open which is a potential source of radiation loss and results in a low value of $Q_o$. To overcome this the outer conductor is elongated beyond the end of the inner conductor such that it forms a cylindrical waveguide operating below cut-off for a given resonant frequency band.

The length of such a resonator may be varied for tuning purposes by replacing the shorting conductor by a movable shorting plunger as shown in Fig. 5.12(b).

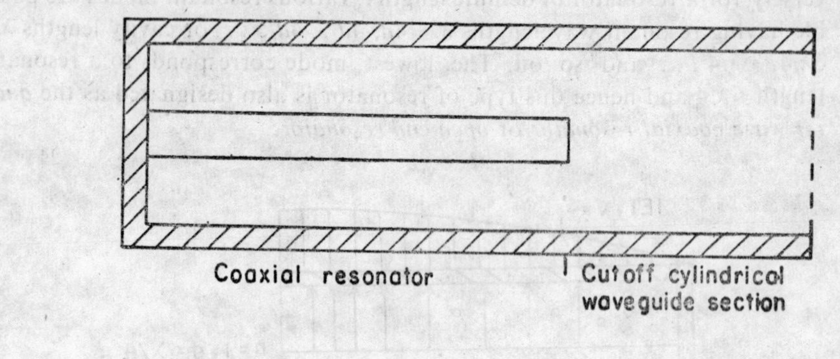


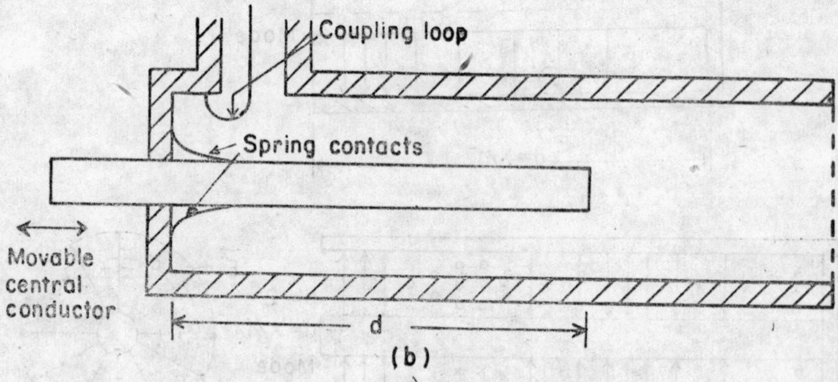


Fig. 5.12 (*b*) Tunable coaxial quarterwave cavity.

**Shorted End Resonator**

For a shorted end resonator $\phi_1=\phi_2=\pi$ and hence from Eq. (5.83), the resonance condition is

$$\beta d=n\pi$$

or

$$\lambda_r=\frac{2d}{n}; \qquad n=1, 2, 3 \ldots \tag{5.88}$$

Thus both end shorted coaxial line sections act as a resonator when their length corresponds to an even multiple of quarter wavelength. Conversely, a coaxial resonator will resonate various modes corresponding to $n=1, 2, 3 \ldots$ having resonant wavelengths as $\lambda_r=2d, d, d/2 \ldots$ and so on. The lowest mode is $n=1$, having a resonant wavelength $\lambda_r=2d$, or length of the cavity as half the wavelength. That is why this type of resonator is generally called the *half wave coaxial resonator* or *shorted end coaxial resonator*. The field pattern corresponding to the dominant mode is depicted in Fig. 5.13(a). The resonant wavelength of a half wave coaxial resonator may be varied, i.e. the resonator may be tuned by replacing one shorted end by a shorting plunger as shown in Fig. 5.13(a).

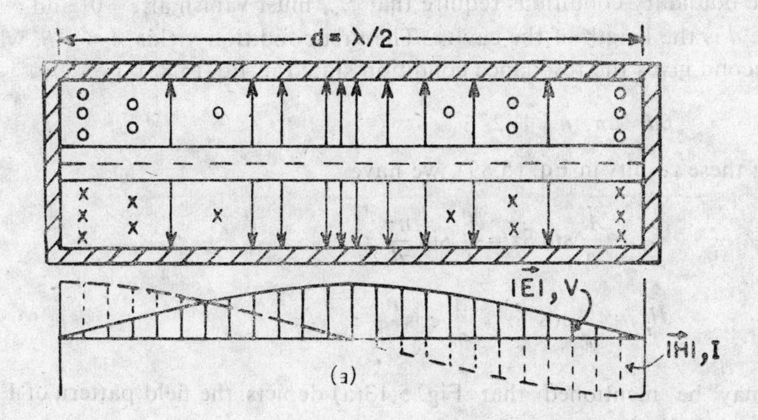


Fig. 5.13 (*a*) Field pattern and current distribution in a half-wave coaxial resonator.

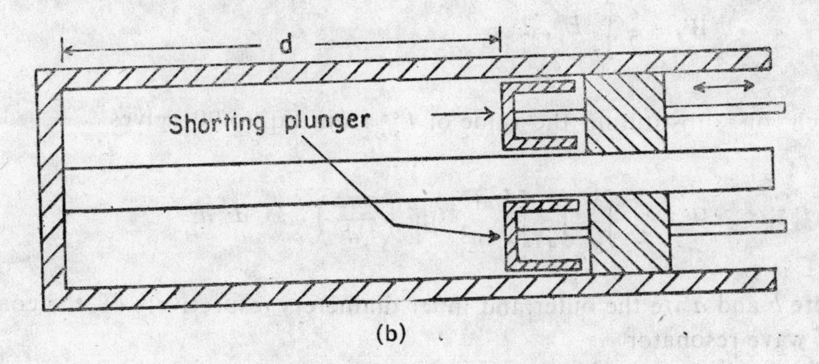


Fig. 5.13 (*b*) Tunable coaxial halfwave resonator.

A half wave coaxial resonator is easy to design and is commonly used in designing microwave wavemeters.

## 5.13 $Q_o$ OF A COAXIAL RESONATOR

As a half wave coaxial resonator is only of practical use, we now consider it for the evaluation of unloaded $Q_o$ value.

The standing wave field components of the resonant TEM mode are obtained by combining the forward and backward propagating waves

$$E^o_{\rho s} = \frac{A}{\rho}\, e^{j\bar{\beta}z} + \frac{B}{\rho}\, e^{-j\bar{\beta}z}$$

$$H^o_{\phi s} = -\frac{A}{\eta\rho}\, e^{j\bar{\beta}z} + \frac{B}{\eta\rho}\, e^{-j\bar{\beta}z} \tag{5.89}$$

where $\eta$ is the free space impedance.

The boundary conditions require that $E^o_{\rho s}$ must vanish at $z=0$ and $z=d$. where $d$ is the length of the cavity. The first condition yields $A=-B$, while the second gives the resonance condition stated in Eq. (5.88), i.e.

$$\beta d=n\pi;\ n=1, 2, 3\ldots$$

Using these results in Eq. (5.89), we have

$$E^o_{\rho s}=\frac{A}{\rho}\sin\beta z=\frac{A}{\rho}\sin\frac{n\pi}{d}z$$

$$H^o_{\phi s}=\frac{A}{\eta\rho}\cos\beta z=\frac{A}{\eta\rho}\cos\frac{n\pi}{d}z. \tag{5.90}$$

It may be mentioned that Fig. 5.13(a) depicts the field pattern of Eqs. (5.90) when $n=1$.

Now the peak energy stored in the half wave resonator is

$$W_s=\frac{\epsilon}{2}\int_v E^o_{\rho s}\,dv$$

which on substituting the value of $E^o_{\rho s}$ from Eq. (5.90), gives

$$W_s=\frac{\epsilon}{2}\int_o^{2\pi}\int_a^b\int_o^d\frac{A^2}{\rho^2}\sin^2\left(\frac{n\pi z}{d}\right)\rho d\phi,\ d\rho\ dz$$

where $b$ and $a$ are the outer and inner diameters respectively of the coaxial half wave resonator

or
$$W_s=\frac{\epsilon}{2}\cdot 2\pi\cdot\frac{d}{2}\cdot A^2\int_a^b\frac{1}{\rho}\,d\rho=\frac{\pi\epsilon A^2 d}{2}\log_e(b/a). \tag{5.91}$$

The time average power loss is given by

$$P_L=\frac{R_s}{2}\int_{\text{curved surface}}|H_{\phi s}|^2\,ds+2\times\frac{R_s}{2}\int_{\text{end plates}}|H_{\phi s}|^2\,ds$$

$$=\frac{R_s}{2}\left\{\underbrace{\int_o^d\frac{A^2}{\eta^2a^2}\cos^2\frac{n\pi z}{d}\,2\pi a dz}_{\text{(outer conductor)}}+\underbrace{\int_o^d\frac{A^2}{\eta^2b^2}\cos^2\frac{n\pi z}{d}\,2\pi b\,dz}_{\text{(inner conductor)}}\right\}$$

$$+2\times\frac{R_s}{2}\int_a^b\int_o^{2\pi}\frac{A^2}{\eta^2\rho^2}\rho d\rho\,d\phi$$

$$=\frac{R_s}{2}\frac{\pi}{\eta^2}A^2d\left(\frac{1}{a}+\frac{1}{b}\right)+2\pi R_s\frac{A^2}{\eta^2}\log_e(b/a)$$

$$=\frac{R_s}{\eta^2}\pi A^2\left\{\frac{d}{2}\left(\frac{1}{a}+\frac{1}{b}\right)+2\log_e b/a\right\}. \tag{5.92}$$

Substituting Eqs. (5.91) and (5.92) in Eq. (5.21), we obtain

$$Q_o = \omega_r \frac{\frac{\pi\epsilon A^2 d}{2}\log_e(b/a)}{\frac{R_s}{\eta^2}\pi A^2\left\{\frac{d}{2}\left(\frac{1}{a}+\frac{1}{b}\right)+2\log_e(b/a)\right\}}. \tag{5.93}$$

At resonance

$$\omega_r = 2\pi f_n = \frac{2\pi}{\lambda_r\sqrt{\mu\epsilon}} = \frac{2\pi}{\frac{d}{2}\sqrt{\mu\epsilon}} = \frac{4\pi}{d\sqrt{\mu\epsilon}}$$

and for a coaxial line the characteristic impedance is given by

$$Z_o = \frac{\eta}{2\pi}\log_e(b/a).$$

Using these values Eq. (5.93) is reduced to

$$Q_o = \frac{4\pi^2 Z_0}{R_s\left\{\lambda_r\left(\frac{1}{a}+\frac{1}{b}\right)+8\log_e(b/a)\right\}}.$$

## 5.14 OTHER TYPES OF RESONATORS

So far we have discussed only resonators that have regular shapes. Theoretically, any cavity (bounded volume) can behave as a resonator and such resonators may be analyzed by generalized theory*. In this section as shall discuss some of the practical resonators of uncommon shape. The shapes of such resonators are determined by the basic resonator characteristics such as field configuration, $Q$ values, resonant frequency, conductance, tunning arrangement, etc.

### (*i*) Resonator with Shorting Capacitance and Inductive Iris

To obtain high field density in a desired region of the cavity, the regular resonators are provided with a shorting capacitance positioned at the cavity axis, as shown in Fig. 5.14(a, b and c). Such resonators are specially suitable in microwave tubes and other appliances. For example, in a klystron the shorting capacitance in a rectangular resonator may be used as an inter-electrode capacitance and the tuning of the cavity may be accomplished by

*1. J.C. Statter, *Microwave Electronics*, D. Van Nostrand Co. Inc. (1950).
2. R.N. Ghosh, *Microwave Circuit Theory and Analysis*, McGraw-Hill Book Co. (1963).
3. J.A. Statton, *Electromagnetic Theory*, McGraw-Hill Book Co. (1941).

means of two ganged plungers as shown in Fig. 5.14(c). The rectangular shorting capacitance cavity is coupled to waveguide system by means of in-

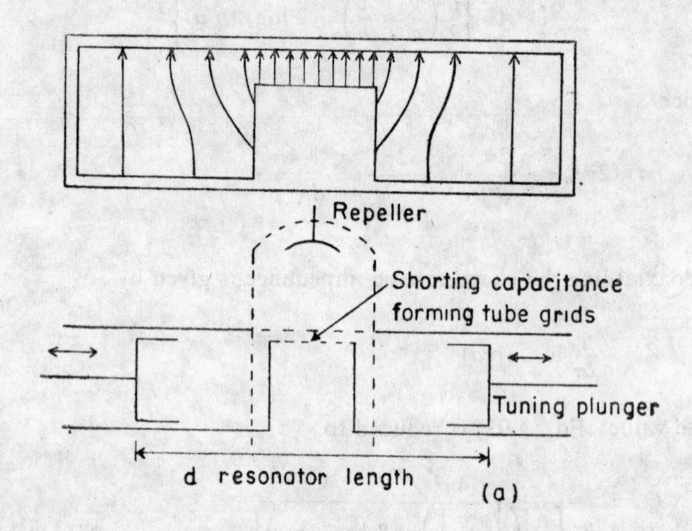


**Fig. 5.14** (*a*) Rectangular resonator with shorting capacitance and its application in microwave reflux klystron.

ductive iris coupling. This is shown in Fig. 5.14(d), along with the equivalent circuit. The resonance condition requires that the total susceptance of the system (iris+cavity) must vanish, that is, if $b$ is the normalized susceptance of the iris, then

$$b-\tan^{-1}\left(\frac{2\pi d}{\lambda_g}\right)=0 \tag{5.94}$$

where $d$ is the cavity length. Hence the resonant wavelength is given by

$$\lambda_r=\frac{1}{\left\{\frac{1}{\lambda_c^2}+\left(\frac{\text{arc}\tan^{-1}(b)}{2\pi d}\right)^2\right\}^{1/2}}. \tag{5.94a}$$

Assuming that the energy dissipated in the cavity is infinitesimally small as compared to the energy dissipated in the load, the loaded $Q$ of the cavity is given by

$$Q_L=\frac{1}{2}\,(1+b^2)\,\frac{\text{arc}\tan^{-1} b}{1-(\lambda_r/\lambda_c)^2}. \tag{5.94b}$$

Microwave cavities with two inductive iris ports are used in high power klystrons amplifiers

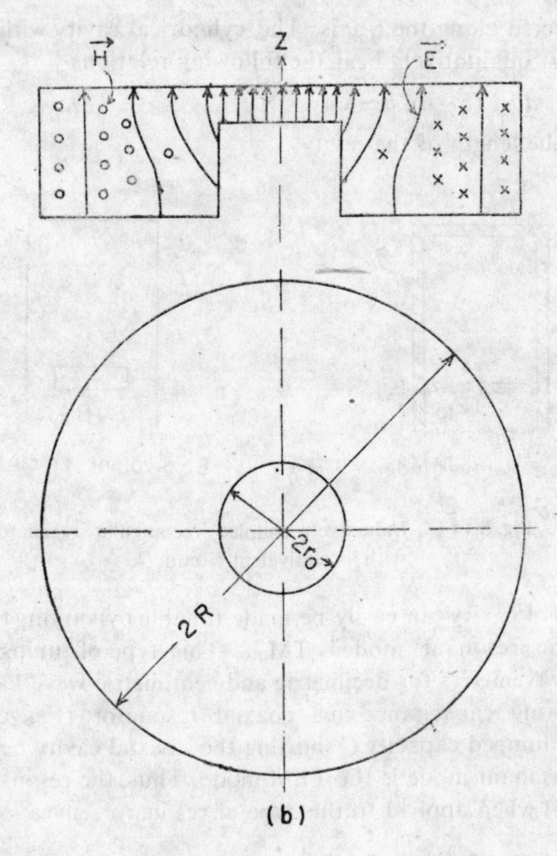


Fig. 5.14 (*b*) Cylindrical resonator with shorting capacitance at TMoto resonant mode.

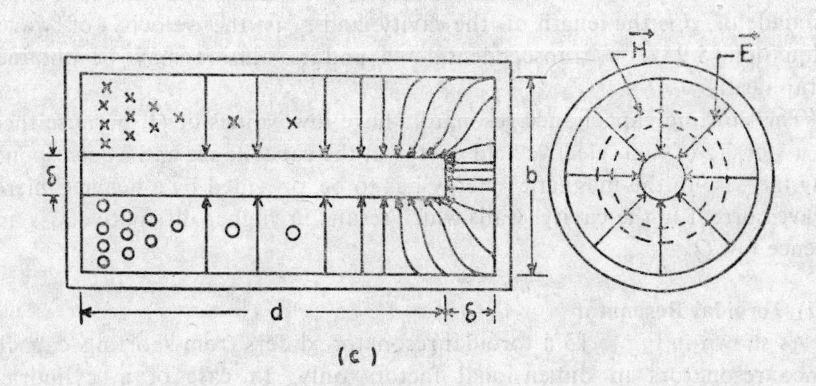


Fig. 5.14 (*c*) Coaxial halfwave resonator with shorting capacitance at dominant mode.

The shorting capacitance (ridge at the centre) in a cylindrical resonator (Fig. 5.14b), reduces the cavity radius. The electric field in this cavity is mainly directed along the $z$-axis. The cylindrical cavity with shorting capacitance must qualitatively bear the following relations

$$d < (b-a);\ b \simeq \lambda/4$$

where $d$ is the length of the cavity.

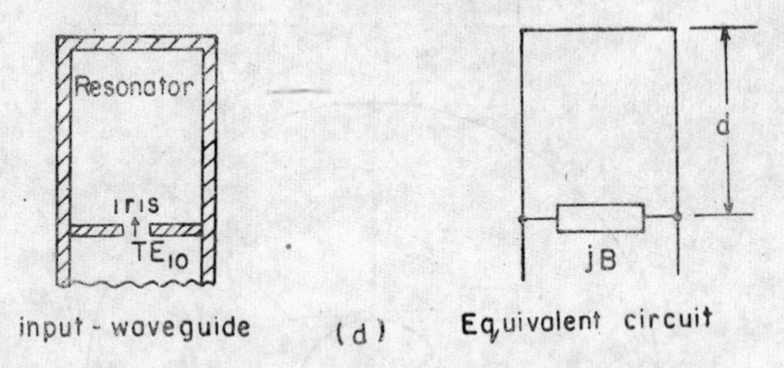


**Fig. 5.14** *(d)* Inductively coupled rectangular resonator with its equivalent circuit.

This type of cavity can easily be made tunable by varying the ridge length provided the resonant mode is $TM_{010}$. This type of tuning is commonly utilized in wavemeters for decimetric and centimetric wave bands.

The shorting capacitance in a coaxial resonator (Fig. 5.14c) may be treated as a lumped capacity $C$ shorting the coaxial cavity because the fundamental resonant mode is the TEM mode. Thus, the resonance condition of Eq. (5.94) when applied to this type of resonator, gives

$$\omega_r C - \frac{1}{Z_o} \tan^{-1}\left(\frac{\omega_r d}{v_p}\right) \tag{5.95}$$

where $Z_o$ is the characteristic impedance of the coaxial line which the cavity is made of, $d$ is the length of the cavity, and $v_p$ is the velocity of waves. Equation (5.95) is a transcendental one and its solution may be obtained graphically.

The shorting capacitance resonators have low values of $Q_o$ because there is a storage of high electric field energy at the capacitance and a corresponding increase in the magnetic energy has to be provided by a heavier microwave current in the cavity walls which results in higher dissipative loss and hence low $Q_o$.

### (*ii*) Toroidal Resonator

As shown in Fig. 5.15 a toroidal resonator differs from shorting capacitance resonators in dimensional factors only. In case of a cylindrical toroidal resonator the dimensional restrictions are

$$b \ll \lambda/4,\ d \ll \lambda/4 \text{ and } \delta \ll d \tag{5.96}$$

where $\delta$ is the shorting capacitive gap. Under these conditions, there can be no significant variation of the electric or magnetic fields along the radial, azimuthal and axial directions. This leads to such cavities to be treated as

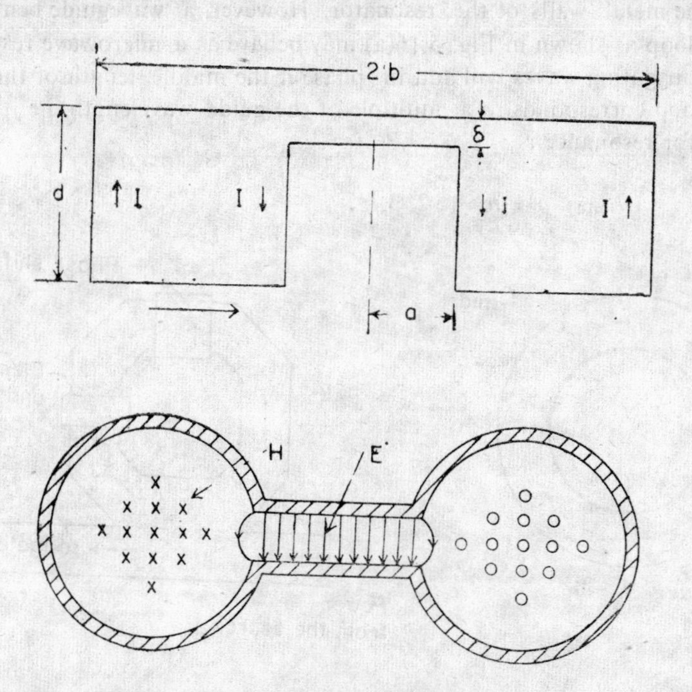


**Fig. 5.15** Toroidal resonator with field pattern.

circuits with lumped constants. The plane gap at the cavity centre functions as the lumped capacitance, while the inductance is concentrated mainly on the toroidal surface, of the resonator, forming a single loop. The resonant frequency of these cavities may be determined from the relation

$$\omega_r = \frac{1}{\sqrt{LC}} \tag{5.97}$$

with $\quad L = \frac{\mu d}{2\pi} \log (b/a)$ and $C = \epsilon \frac{\pi a^2}{8}$.

The corresponding resonant wavelength is

$$\lambda_r = \pi a \sqrt{\frac{2d}{\delta} \log \left(\frac{b}{a}\right)}. \tag{5.98}$$

Such resonators find applications in microwave tubes, high QTR and ATR gas discharge tubes. The tuning may be provided by changing the gap between the ridge. Magnetrons utilize cavities that may be considered as a series of such cavities and may, therefore, be similarly analyzed.

***(iii)* Travelling Wave Resonator**

So far we have confined ourselves to resonators which resonate under standing wave conditions resulting from the multiple reflections of the fields from the metal walls of the resonator. However, a waveguide bent into a closed loop as shown in Fig. 5.16(a) may behave as a microwave resonator. The propagating waves will add in phase if the middle length of the waveguide loop corresponds to a multiple of the guide wavelength, i.e. the condition for resonance is

$$L_{mid} = n\lambda_g, \; n = 1, 2, 3, \ldots \tag{5.99}$$

L mid
(a)
Phase shifter
Element under test
$P_2$
$P_1$
To matched load
Power from the source
(b)

**Fig. 5.16(*a*)** Travelling wave resonator.
(*b*) Practical form of Travelling wave resonator with its application in testing of waveguide elements under high power travelling condition.

The resonant wavelength is therefore given by

$$\lambda_r = \frac{1}{n\sqrt{\left\{(1/\lambda_c)^2 + \left(\frac{n}{L_{mid}}\right)^2\right\}}}. \tag{5.100}$$

The unloaded $Q_o$ for such a resonator is given by

$$Q_o = 2\pi n \; \frac{(\lambda_g/\lambda_r)^2}{(1 - e^{-2\alpha L\, mid})} \tag{5.101}$$

where $\alpha$ is the attenuation constant of the waveguide.

Practically, such resonators are designed with rectangular waveguides of standard cross-sectional dimensions and directional couplers as shown in Fig. 5.16(b). The phase shifter used here serves as a tuner.

These resonators are used as power simulators for testing various elements of waveguides at very high power ratings.

**(iv) Open Resonator Structures: Optical Resonators**

A typical open resonator system consists of two reflecting surfaces (mirrors) $M_1$ and $M_2$ between which a TEM wave undergoes multiple reflections as shown in Fig. 5.17(a). The resonance occurs when the separation of the mirrors $d$ corresponds to an integral multiple of half wavelength, i.e.

$$d = p\,\frac{\lambda_r}{2} \text{ or } \lambda_r = \frac{2d}{p};\ p = 1, 2, 3 \ldots \tag{5.102}$$

Various modes in these resonator systems are represented by $TEM_{mnp}$, where $m$ and $n$ represent half period field variation in the transverse plane while $p$ represents half period field variation in the longitudinal plane. The lowest dominant mode corresponding to the longest resonant wavelength which is of maximum interest, is the $TE_{oop}$. The field pattern of this mode features one maximum (Fig. 5.17a) and minimum diffraction losses.

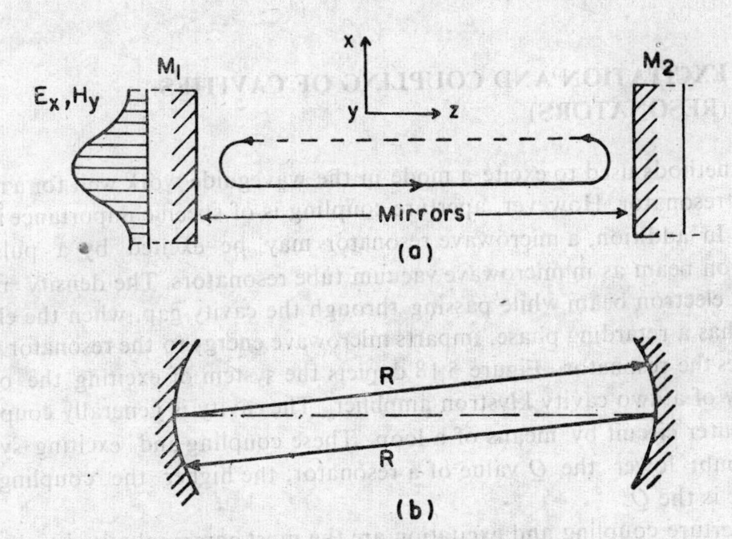


Fig. 5.17 *(a, b)* Open resonator structure.

In these resonators there are no side walls, which besides eliminating a large number of higher modes through damping due to the radiation from the open side, also result in low wall losses. However, radiation loss corresponding to the dominant mode cannot be ruled out. Such loss can be minimized by selecting transverse dimensions of the mirrors several times the operating wavelength. The $Q$ of such a resonator is given by*

$$Q_o = \frac{P\pi Z_o}{4R_s} \tag{5.103}$$

*R.E., Collin, *Foundations for Microwave Engineering* McGraw-Hill Book Co. (1966).

where $Z_o$ is the characteristic impedance corresponding to the $TEM_{oop}$ mode. The typical $Q_o$ values fall in the range $10^4$ to $10^5$.

In practice, plane mirrors are ordinarily replaced by concave spherical surfaces and the resonator is made confocal as shown in Fig. 5.17(b). The focal distance is half the radius of curvature $R$ and therefore the centre of curvature of each mirror is located on the surface of the opposite mirror. The main advantage of the confocal resonator is that it has smaller losses due to diffraction effects as compared to any other mirror configuration. At microwave frequencies metallic arrays or perforated mirrors are employed instead of solid reflecting surfaces, to provide a convenient means for coupling these structures to other elements.

The resonators which resemble the Fabry-Perot interferometers, generally called the Febry-Perot resonators, are specially suitable in optics (for lasers) and are sometimes called optical resonators. The details may be found in the literature.*

## 5.15 EXCITATION AND COUPLING OF CAVITIES (RESONATORS)

The methods used to excite a mode in the waveguide work well for a microwave resonator. However, aperture coupling is of specific importance in this case. In addition, a microwave resonator may be excited by a pulsating electron beam as in microwave vacuum tube resonators. The density modulated electron beam while passing through the cavity gap, when the electric field has a retarding phase, imparts microwave energy to the resonator which excites the resonator. Figure 5.18 depicts the system of exciting the output cavity of a two cavity klystron amplifier. The cavity is generally coupled to the outer circuit by means of a loop. These coupling and exciting systems no doubt lower the $Q$ value of a resonator; the higher the coupling, the lower is the $Q$.

Aperture coupling and excitation are the most commonly used techniques for microwave resonators and, therefore, need some analysis. However, more detailed, comprehensive and general treatment may be found in the literature**.

*1. G.D. Boyd and J.P. Gordon "Confocal multimode resonator for millimeter through optical masers", *Bell Syst. Tech. Jour.*, 40 (March 1961).

2. G. Goupau and F., Schwering "On the guided propagation of electromagnetic wave beams", *IRE Trans* PGAP, AP-9 (May 1961).

3. S,, Ramo, J.R., Whinnery, and T.V., Duzer, '*Fields and Waves in Communication Electronics*', New Delhi Wiley Eastern Private Limited (1970).

**1 R.N., Ghosh *Microwave Circuit Theory and Analysis*, McGraw-Hill Book Co. (1963).

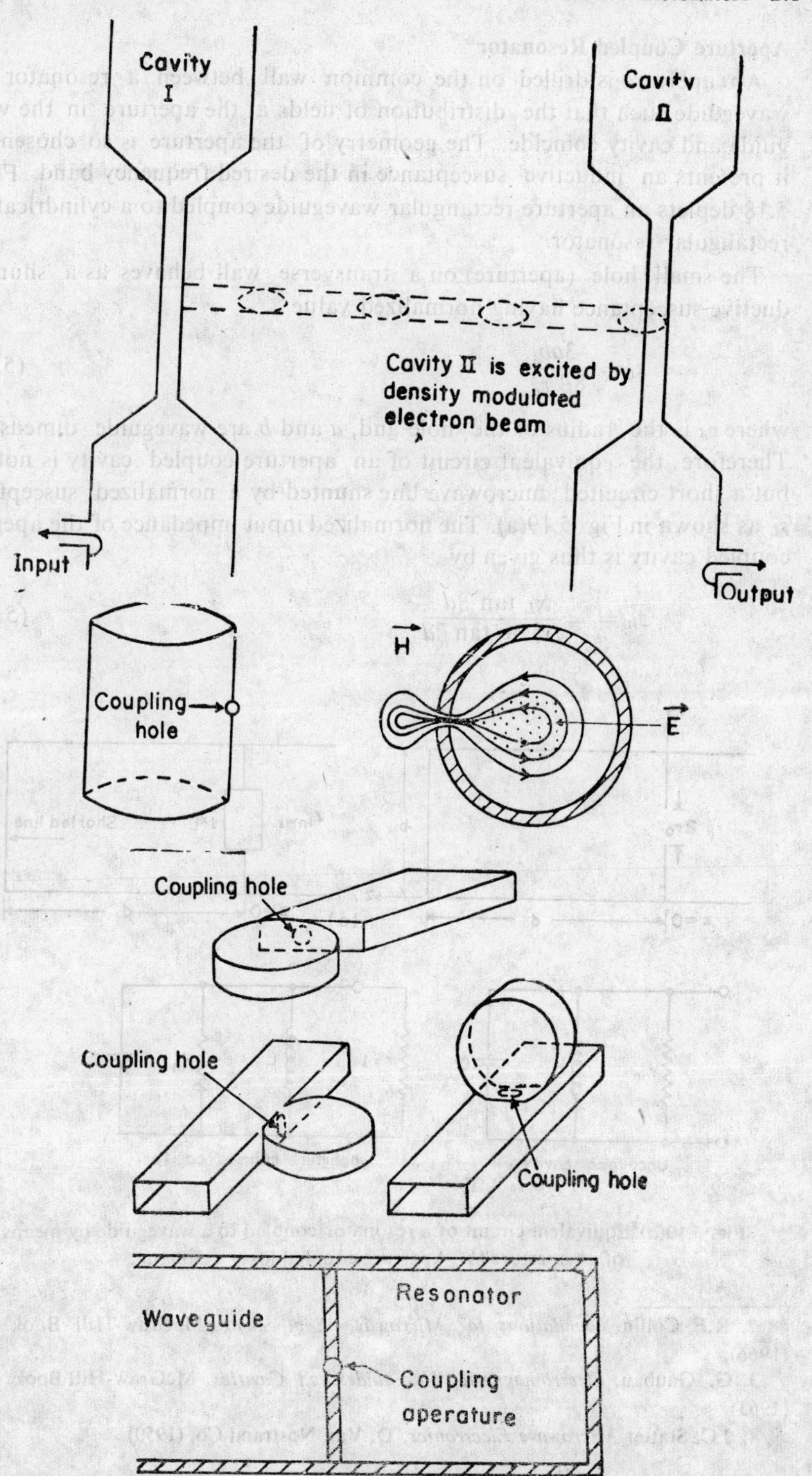


Fig. 5.18 Aperture coupled resonators.

**Aperture Coupled Resonator***

An aperture is drilled on the common wall between a resonator and waveguide such that the distribution of fields at the aperture in the waveguide and cavity coincide. The geometry of the aperture is so chosen that it presents an inductive susceptance in the desired frequency band. Figure 5.18 depicts an aperture rectangular waveguide coupled to a cylindrical and rectangular resonator.

The small hole (aperture) on a transverse wall behaves as a shunt inductive susceptance having normalized value

$$s_L = \frac{3ab}{8\beta\, r_o^3} \tag{5.104}$$

where $r_o$ is the radius of the hole and, $a$ and $b$ are waveguide dimensions. Therefore the equivalent circuit of an aperture coupled cavity is nothing but a short-circuited microwave line shunted by a normalized susceptance $s_L$ as shown in Fig. 5.19(a). The normalized input impedance of the aperture coupled cavity is thus given by

$$z_{in} = -\frac{x_L \tan \beta d}{jx_L + j\tan \beta d} \tag{5.105}$$

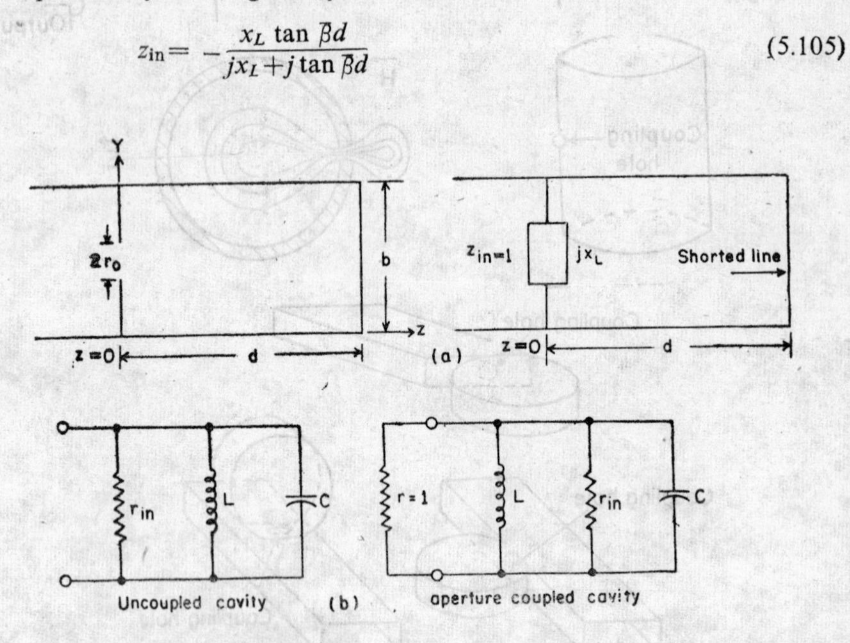


Fig. 5.19(*a*) Equivalent circuit of a resonator coupled to a waveguide by means of an aperture (*b*) Aperture coupled lossy resonater.

---

2. R.E. Collin, *Foundations for Microwave Engineering*, McGraw-Hill Book Co. (1966),

3. G., Gaubau, *Electromagnetic Waveguides and Cavities*, McGraw-Hill Book Co. (1963).

4. J.C. Statter *Microwave Electronics*, D. Van Nostrand Co. (1950).

where $x_L$ is the normalized reactance of the aperture and $d$ is the cavity length. For the cavity to resonate, $z_{in} = \infty$. This yields

$$x_L = -\tan \bar{\beta}d \, \frac{8r_o^3 \, \bar{\beta}d}{3abd} \tag{5.106}$$

Equation (5.106) can be solved graphically by plotting values of and $\bar{\beta}d$ and $\bar{\beta}d$ versus $x_L$. The point of intersection (such as point $P$ in Fig. 5.20) gives the possible values of $\bar{\beta}$. The corresponding resonant frequency is given by

$$f_r = \frac{\omega_r}{2\pi} = \frac{1}{2\pi\sqrt{\mu\epsilon}} \{\bar{\beta}^2 + (\pi/a)^2\}^{1/2}. \tag{5.107}$$

Though an infinite number of solutions is possible, for the fundamental mode $\bar{\beta}d - \pi$, and the solution near $\pi$ say $\bar{\beta}_1$ is the solution for the dominant mode However, propagation constants for the higher modes in an aperture coupled rectangular cavity are given by

$$\bar{\beta}_n d \doteq (n - \tfrac{1}{2})\, \pi; \; n = 1, 2, 3 \tag{5.108}$$

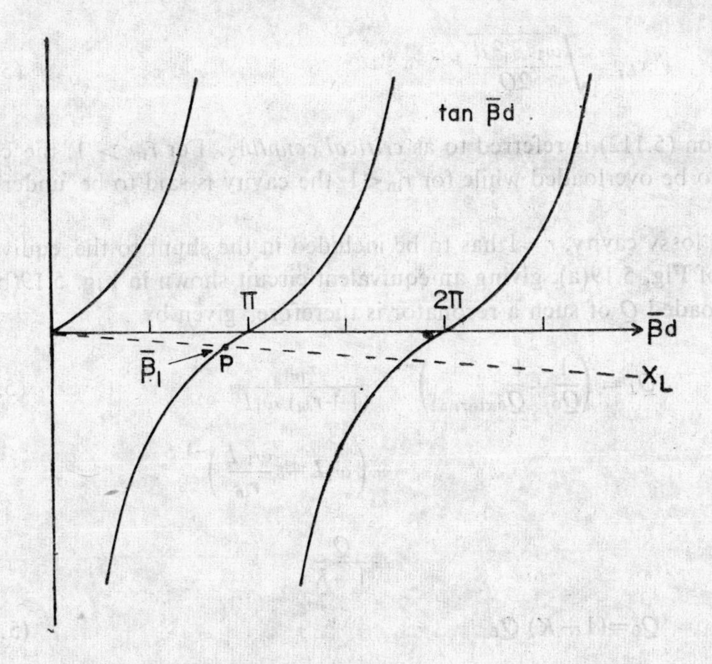


**Fig. 5.20** Graphical solution for resonant frequeney of aperture coupled resonator.

In practice, there are a number of resonances in the vicinity of $\bar{\beta}_1$ and hence an infinite number of equivalent lumped parameter network circuits has to be used in the vicinity of resonant frequency $\omega_1$. However, total series inductance $L$ and total shunt capacitance may be deduced using perturbation methods, and these are given by* $C$

$$L = \frac{2x_{L_1}^2}{\omega_1^2 \beta_1^2 d}, \quad C = \frac{\beta_1^2 d}{2x_{L_1}^2} \tag{5.109}$$

To extend the discussion to lossy resonators, the frequency $\omega_1$ has to be replaced by the complex frequency $\omega_1 (1+j/2Q)$. Then, the normalized input impedance can be written as

$$Z_{in} = -j \frac{x_{L_1}^2}{\beta_1' d (\omega - \omega_1 - 2\omega_1/2Q)} \tag{5.110}$$

where $\beta_1'$ is the derivative of $\beta$ with respect to $\omega$.

At resonance $\omega \simeq \omega_1$, the impedance is purely resistive and is given by

$$r_{in} = z_{in} = \frac{2x_{L_1}^2 Q}{\omega \beta_1'^2 d}. \tag{5.111}$$

For well-matched coupling $Z_{in} = r_{in} = 1$, that is

$$x_{L_1} = \sqrt{\frac{\omega_1 \beta_1'^2 d}{2Q}}. \tag{5.112}$$

Condition (5.112) is referred to as *critical coupling*. For $r_{in} > 1$, the cavity is said to be overloaded while for $r_{in} < 1$ the cavity is said to be underloaded.

For a lossy cavity, $r = 1$ has to be included in the shunt to the equivalent circuit of Fig. 5.19(a), giving an equivalent circuit shown in Fig. 5.19(b).

The loaded $Q$ of such a resonator is therefore, given by

$$Q_L = \left(\frac{1}{Q_0} + \frac{1}{Q_{\text{external}}}\right)^{-1} = \frac{r_{in}}{(1 + r_{in})\, \omega_1 L} \tag{5.113}$$

$$= \left(\omega_1 L + \frac{\omega_1 L}{r_{in}}\right)^{-1}$$

$$= \frac{Q}{1+K}$$

or

$$Q_0 = (1+K)\, Q_L \tag{5.114}$$

where $K$ is the coupling parameter. It may be defined as

$$K = \frac{Q_0}{Q_{ext}}$$

*R.E. Collin, *op. cit.*

and is given by

$$K = \frac{r_{in}\ \omega_1\ L}{\omega_1 L \cdot 1} \tag{5.115}$$

for the circuit of Fig. 5.19(b).

This agrees with our earlier definition, that $r_{in} = 1$ corresponds to critical coupling.

## 5.16 DESIGN AND APPLICATIONS OF RESONATORS

The design of resonators greatly depends upon the use to which they are to be put. The utility in turn, is determined by the cavity characteristics and parameters such as field configuration, $Q$ values, resonant frequency band, input impedance, resonator conductance and susceptance, resonator efficiency and induced loss, etc. Before we turn to the cavities in use, it is necessary to define some of the parameters which have not been referred earlier.

### Resonator Conductance and Susceptance

The equivalent conductance of a resonator is very useful in microwave tube designs and may be defined and calculated with the aid of the obvious relation between the power dissipated in the resonator and the voltage amplitude $|V_m|$ applied to the input terminals of the cavity equivalent circuit of Fig. 5.21

$$G = \frac{2\ P_L}{|V_m|^2} \tag{5.116}$$

where $P_L$ is the time average power loss given by Eq. (4.108).

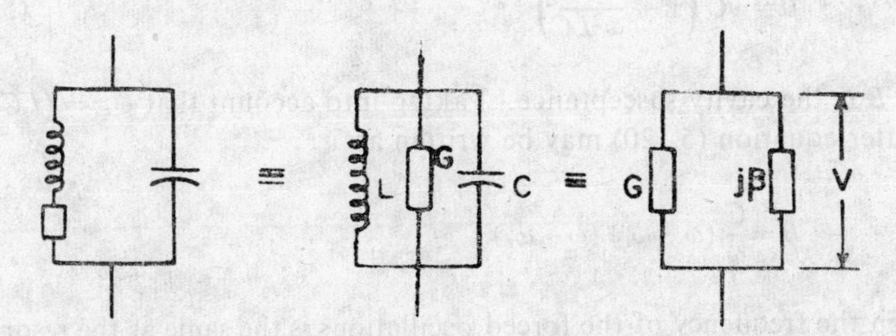


**Fig. 5.21** Equivalent circuit of a resonator in terms of equivalent conductance.

The peak voltage $|V_m|$ cannot uniquely be defined in a resonator and so resonator conductance, strictly speaking, remains undefined. However, specifying any two points $a$ and $b$ on the inner surface of a resonant cavity or in the feeder line and integrating along a specified path between these points, we obtain a quantity that for any fixed moment of time can be used as the equivalent of voltage $|V_m|$, to calculate the cavity conductance

$$|V_m| = -\int_a^b \vec{E_m} \cdot \vec{dl} \tag{5.117}$$

where $\vec{E_m}$ is the amplitude of the electric field vector. Hence, using Eqs. (5.116), (4.108) and (5.117) we find,

$$G = R_s \int_s |H_t|^2 \, ds \Bigg/ \left\{ \int_a^b \vec{E_m} \cdot \vec{dl} \right\}^2 \tag{5.118}$$

It may be noted from Eq. (5.118) that $G$ of a resonator depends on the way the 'reference points' are specified, in contrast to unloaded $Q_o$ which represents losses and is *invariable* for a given resonator.

To connect $G$ to $Q_o$, we recall the general definition of Eq. (5.21), which for the circuit of Fig. 5.19(a) can be written as

$$Q_o = \omega_r \frac{C |V_m|^2/2}{G |V_m|^2/2} = \omega_r \frac{C}{G} \tag{5.119}$$

where $C$ is the lumped capacitance of the resonant cavity. Since $C$ cannot precisely be defined, hence Eq. (5.119) is not a precise relation and so $C$ has to be excluded.

The admittance of a parallel resonant circuit in the vicinity of the resonant frequency $\omega_r$ varies as follows:

$$Y = G + jB = G + j\left(\omega C - \frac{1}{\omega L}\right)$$

or

$$B = \omega C \left(1 - \frac{1}{\omega^2 LC}\right) \tag{5.120}$$

where $B$ is the cavity susceptance. Taking into account that $\omega_r = (LC)^{-1/2}$, this latter equation (5.120) may be written as

$$B = \frac{C}{\omega} (\omega - \omega_r)(\omega + \omega_r).$$

When the frequency of the forced oscillations is the same as the resonance frequency, we can write,

$$\omega + \omega_r \simeq 2\omega \text{ and}$$

$$(B)_{\omega \simeq \omega_r} = 2C(\omega - \omega_r)$$

or

$$\left(\frac{dB}{d\omega}\right)_{\omega \simeq \omega_r} = 2C$$

or $$C=0.5\left(\frac{dB}{d\omega}\right)_{\omega\simeq\omega_r} \tag{5.121}$$

Replacing lumped capacitance in Eq. (5.119) by Eq. (5.121), we obtain

$$Q_o=\frac{\omega_r}{2G}\left(\frac{dB}{d\omega}\right)_{\omega\simeq\omega_r}=-\frac{\lambda_r}{2G}\left(\frac{dB}{d\lambda}\right)_{\lambda=\lambda_r} \tag{5.122}$$

where $\left(\frac{dB}{d\omega}\right)_{\omega\simeq\omega_r}$ is the rate of circuit susceptance variation at frequencies in the vicinity of circuit resonance which can easily be measured at microwave frequencies. $Q_o$ is also precisely known hence $G$ is precisely measurable.

**Resonator Efficiency and Induced Loss**

When a resonator is connected to a load via a loss-less line (waveguide) the ratio of the power dissipated in a given load to the total power dissipated in the system is defined as the resonator efficiency.

$$\eta_{cav}=\frac{P_{L,load}}{P_{L,total}}\times 100 \quad \text{percent} \tag{5.123}$$

If $W_s$ is the peak energy stored in a resonator, then loaded value of '$Q$' is

$$Q_L^{-1}=\frac{1}{\omega_r}\frac{P_{L,load}}{W_s} \quad \text{and} \quad Q_{ext}^{-1}=\frac{1}{\omega_r}\frac{P_{L,total}}{W_s}.$$

Since $\quad Q_L^{-1}=Q_o^{-1}+Q_{ext}^{-1}$,

$$\eta_{cav}=\frac{Q_L}{Q_{ext}}=\frac{Q_o}{Q_o+Q_{ext}}=1-\frac{Q_L}{Q_o} \tag{5.124}$$

where $Q_{ext}$ is the value of $Q$ arising from the losses in external coupling. Equation (5.124) suggests that the higher the value of $Q_o$ the higher is the efficiency.

When the cavtiy excited to a one desired mode, is inserted in a microwave circuit it causes some loss of power. The ratio of the maximum power delivered by a matched power source to a matched load in the absence of a resonant cavity in the connecting guide, to the power delivered to the same load by the same source with resonant cavity of $f=f_r$ in the connecting guide, is termed as *insertion* or *induced loss* in the cavity. Mathematically, induced loss is given by

$$L=10\log_{10}(P_{in}/P_{out})=10\log_{10}(1/\eta_{cav})\ db. \tag{5.125}$$

For general input-output connection $L$, can be writte asn

$$L=10\log_{10}\frac{Q_{in}\,Q_{out}}{2\,Q_L^2} \tag{5.126}$$

A low insertion loss cavity is of specific use in a TR-switch whose function is to pass the received signal during the "receive" period with as low-loss as possible.

**Resonators in Use**

Resonators used in filter networks are designed by placing two symmetrical irises in a section of a rectangular waveguide. The size of the iris is calculated from the desired loaded $Q_L$ value. Inductive as well as capacitive irises may be used. However, for the same value of $Q_L$, the inductive iris has larger dimensions than the capacitive iris, hence, the former is commonly employed. For example, for a normalized susceptance of 0.1 Ω, a 0.5 inch waveguide requires a gap size in a capacitive iris to be 0.004 inch whereas the opening in the symmetrical inductive iris with the same susceptance is approximately 0.25 inch. Further, the thickness of the iris alters the input impedance so due care has to be taken. Moreover, the presence of the iris lowers resonant frequency, hence, the distance between the irises has to be taken slightly less than the resonant length for a particular mode. The tuning is provided by a capacitive tuning screw as shown in Fig. 5.22. The screw is placed at the centre of the cavity where the current is minimum so as to cause minimum additional loss. The susceptances produced by various apertures are listed in table 5.1.

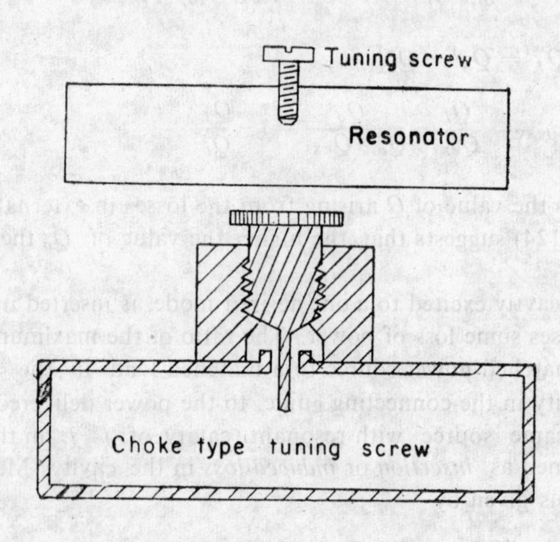


**Fig. 5.22** Tuning screw type resonant cavity.

**TABLE 5.1. Impedances offered by various coupling apertures at its plane.**

| *Transverse outline of the aperture* | *Circuit character* | *Normalised impedance at the plane of the aperture* $z_{in}$ |
|---|---|---|
| Resonant aperture | | $z_{in} = \infty$ when $\frac{a'}{b}\sqrt{1-(\lambda/2a')^2} = \frac{a}{b}\sqrt{1-\left(\frac{\lambda}{2a}\right)^2}$ |
| Small circular aperture | | $j\frac{2\pi d^3}{ab\lambda_g}$ ; $d \ll b$ |
| Symmetrical capacitive aperture | | $-j\frac{\lambda_g}{4b}\left(\log_e \frac{2b}{\pi d}+\frac{b^2}{2\lambda_g^2}\right)^{-1}$ ; $\frac{d}{b} \ll 1$<br>$-j\frac{2b\lambda_g}{\pi^2 (b-d)^2}$; $\frac{b-d}{b} \ll 1$ |
| Symmetrical Inductive aperture | | $j\frac{a}{\lambda_g}\tan^2\frac{\pi d}{2a}\left(1+\frac{1}{6}\frac{\pi^2 d^2}{\lambda^2}\right)$ ; $\frac{d}{a} \ll 1$<br>$j\frac{a}{\lambda_g}\tan^2 2\pi\frac{a-d}{a}\left[1+\frac{8\pi^2}{3}\frac{(a-d)^2}{\lambda^2}\right]$;<br>$\frac{a-d}{a} \ll 1$ |
| Asymmetric capacitive aperture | | $-j\frac{\lambda_g}{8b}\left(\log_e \frac{2b}{\pi d}+\frac{2b^2}{\lambda_g^2}\right)^{-1}$ ; $\frac{d}{b} \ll 1$<br>$-\frac{jb\lambda_g}{\pi^2 (b-d)^2}$ ; $\frac{b-d}{b} \ll 1$ |
| Asymmetric Inductive aperture | | $\simeq j\frac{2a^3}{\pi^2\lambda_g (a-d)^2}\left\{1+\frac{\pi^2}{\lambda^2}(a-d)^2 \log_e \frac{\pi}{2}\cdot\frac{a-d}{a}\right\}^{-1}$ ; $\frac{a-d}{a} \ll 1$. |

The resonators used in microwave tubes are re-entrant or capacitive shorted type. At the shorted gap a very high field is obtained without reducing the volume of the cavity appreciably. The tuning is provided by a flexible wall.

The conductance of such a cavity is related to the cavity plane carrying the electron beam; typically, the magnitude of the cavity conductance determined from Eq. (5.122), amounts to $10^{-4} - 10^{-5}$ ℧ and the equivalent dynamic resistance is very high—10 to 10 K Ω.

Cavities used in measuring equipments such as wavemeters require a wide range of tuning frequency. This is achieved by varying the length of the cavity by means of a movable shorting plunger forming one of the shorted ends. The plunger is so designed that the losses are minimum. A coaxial resonator in a coaxial wavemeter is shown in Fig. 5.23(a). The penetration of the inner conductor into the chamber changes the resonant frequency. The central conductor is usually an extension of the plunger of a micrometer so that its motion can be measured and calibrated in terms of frequency. Figure 5.23(b) depicts a tunable coaxial cavity with a coaxial shorting plunger. In order to increase the contact pressure between the plunger insert providing a short circuit between the inner and outer conductor, the plunger is slotted axially so as to form fingers. The length of the plunger is made approximately a quarter wave in the centre of the frequency tuning range so that the ends of the finger are put near a voltage antinode. The contacting plungers have several disadvantages. The small insulating film developed on finger contacts lowers the $Q_o$ value. The small metal particles along with poor finger contact causes noise and erratic tuning. Mechanical hysterisis caused by friction between the sliding surfaces together with the backlash error in the driving mechanism cause the readings to be different when the plunger is moved in two different directions.

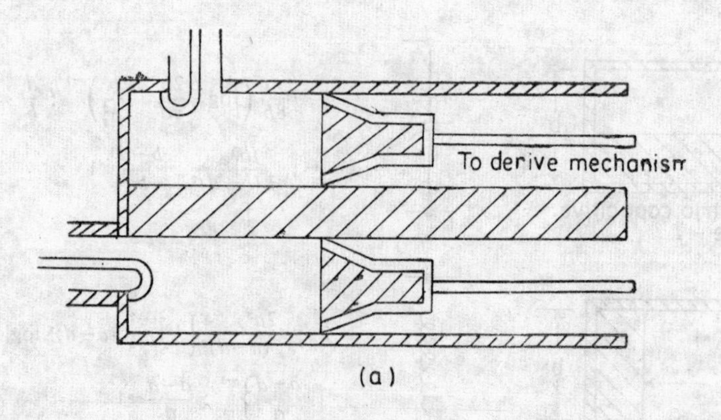


**Fig. 5.23** (*a*) Coaxial resonator (wavemeter) and coaxial shorting plunger.

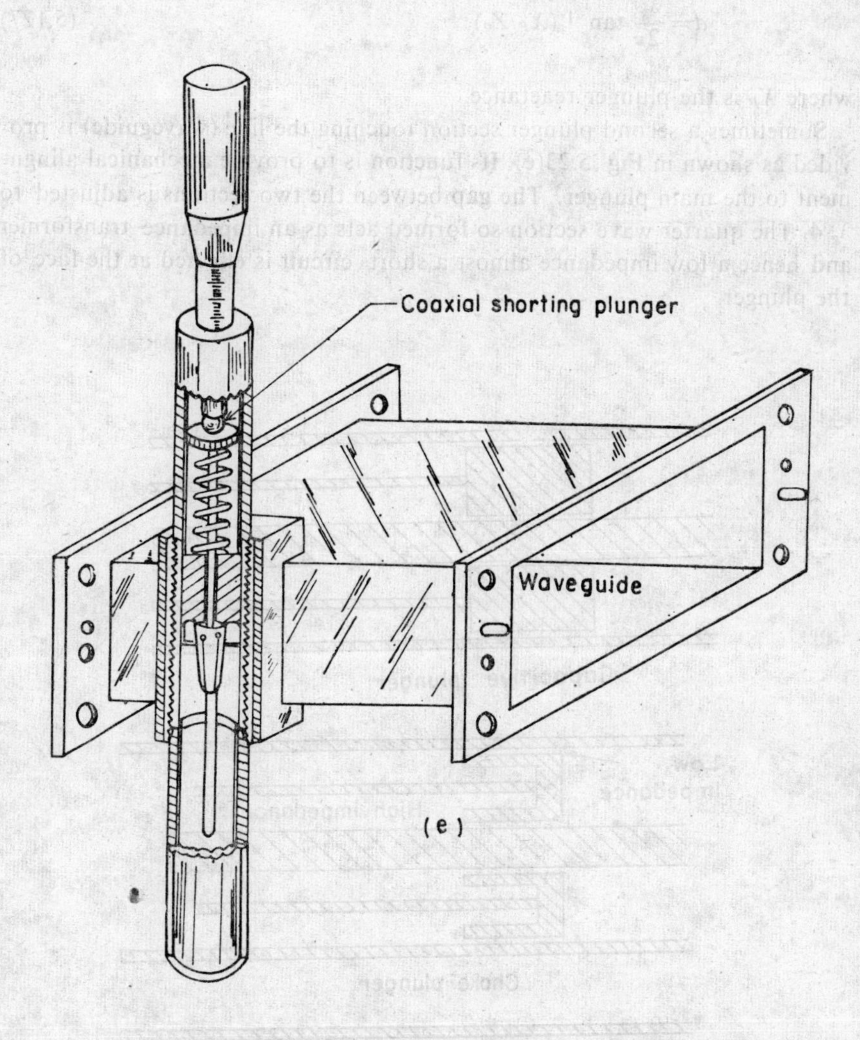


**Fig. 5.23** (*b*) Coaxial wavemeter with waveguide output; Tunable coaxial cavity.

These disadvantages are removed in non-contacting plungers shown in Fig. 5.23(c) and 5.23(d). Such plungers produce a low impedance plane at the plane shunting the line. Though there will be some loss of power at the junction, the satisfactory performance of the plunger does not require it to produce a perfect short circuit but only such that the power loss and associated input resistance be negligible. Since a reactance terminating the line is equivalent to an additional short-circuited section on a loss-less line, the plunger input reactance merely shifts the position of the short circuit from the plunger face by a distance *d*, given by

$$d = \frac{\lambda_g}{2\pi} \tan^{-1} (X_P/Z_o) \tag{5.127}$$

where $X_P$ is the plunger reactance.

Sometimes a second plunger section touching the line (waveguide) is provided as shown in Fig. 5.23(c). Its function is to provide mechanical alingnment to the main plunger. The gap between the two sections is adjusted to $\lambda_g/4$. The quarter wave section so formed acts as an impedance transformer and hence a low impedance almost a short circuit is effected at the face of the plunger.

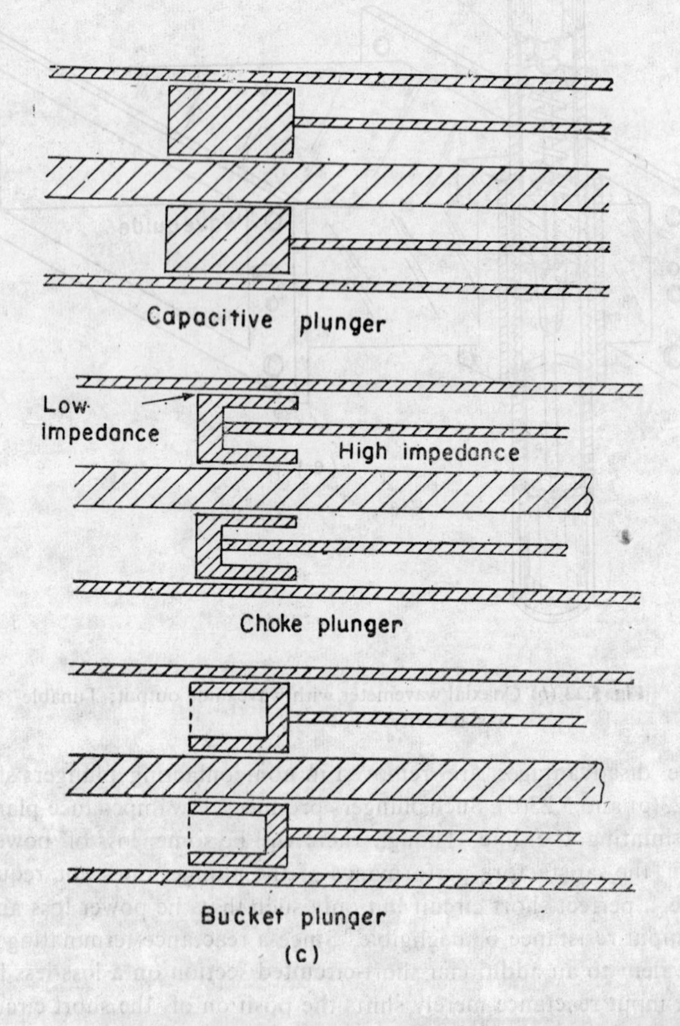


Fig. 5.23 (c) Non-contacting plungers.

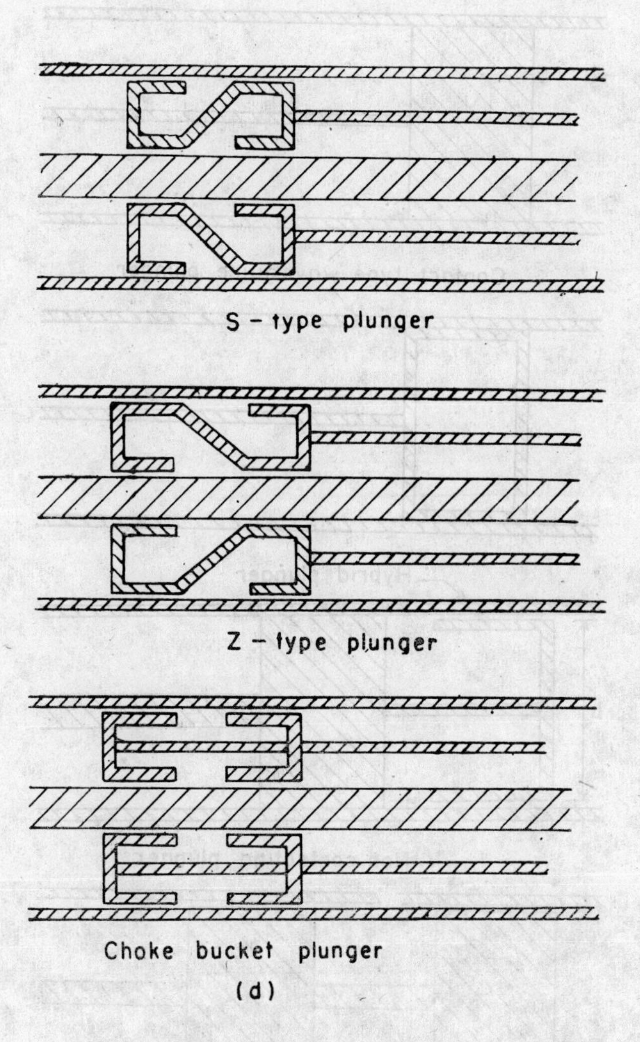


Fig. 5.23(*d*) Non-contacting plungers.

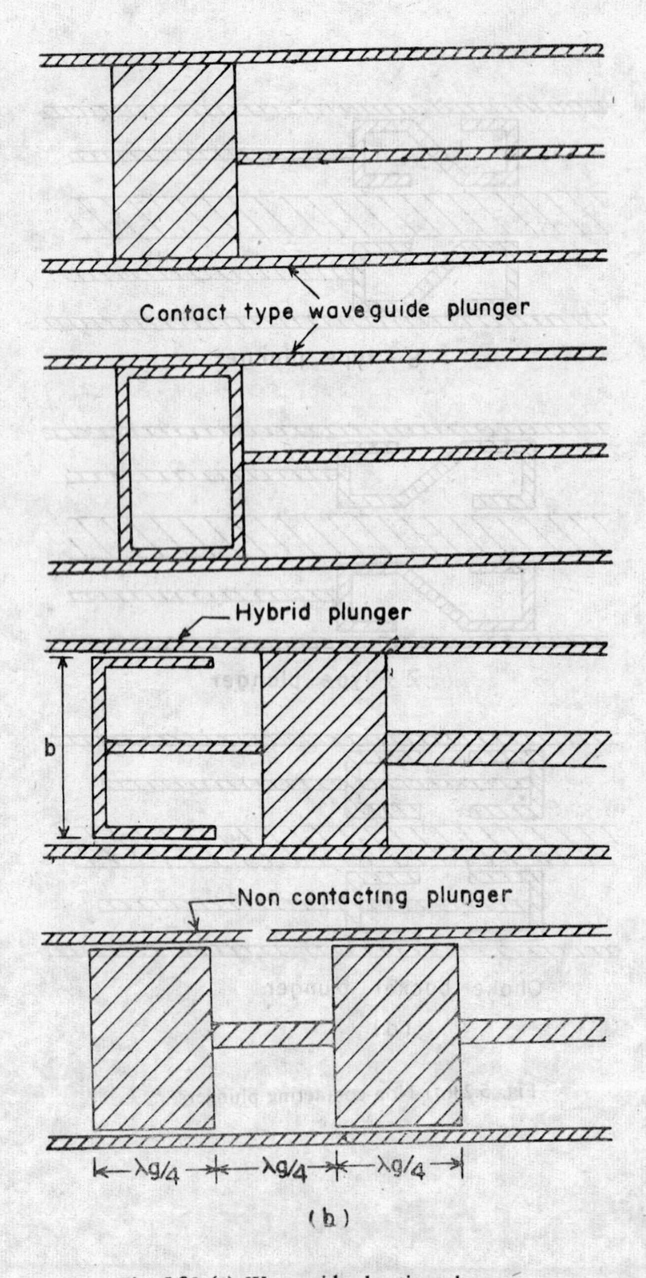


Fig. 5.23 (*e*) Waveguide shorting plungers.

## SOLVED EXAMPLES

**Exapmle 5.1**. Determine the length of a cavity which will resonate at 10 GHz. The cavity is made from WG 16 of the shortest length. Calculate the $Q$ factor.

$$[\sigma=6\times10^7 \text{ sm}^{-1},\ \mu_0=4\pi\times10^{-7},\ \epsilon_0=(36\pi\times10^9)^{-1}.$$

*Solution*: The dominant mode in WG 16 is $TE_{10}$ and the resonant frequency is $f_{TE_{10}}=6.55$ GHz.
At 10 GHz,

$$\lambda_g=\frac{C}{(f^2-f_{10}^2)}$$

$$=\frac{3\times10^8}{(10^2-6.55^2)\ 10^{18}}$$

$$=0.0397 \text{ m.}$$

The shortest cavity length is

$$\tfrac{1}{2}\lambda_g=\frac{0.0397}{2} \text{ m}$$

$$=1.99 \text{ cm.}$$

The $Q$ factor is given by Eq. (5.32)

$$Q=\frac{0.741\ \eta}{R_s}.$$

Now $$R_s=\sqrt{\frac{\pi f\mu_0}{\sigma}}=25.6\times10^{-3}.$$

Hence $$Q=\frac{0.741}{25.6\times10^{-3}}\sqrt{\frac{\epsilon}{\mu}}$$

$$=7.3\times10^3.$$

**Example 5.2**. A rectangular cavity has cross-section $0.38\times0.76$ cm. If it oscillates in the $TE_{102}$ mode at 50 GHz, calculate the length ($l$) of the cavity with air as a dielectric. Are there other modes with the same resonance frequency?

*Solution*: From Eq. (5.17) for the $TE_{102}$ mode [$m_2=1$, $n=0$, $p=2$]

$$\left[\frac{2f_r}{v_p}\right]^2=\left(\frac{1}{a}\right)^2+\left(\frac{2}{c}\right)^2.$$

Cavity length parallel to $z$-axis-unit

$$(c)=\frac{1}{\sqrt{\left(\frac{f_r}{v_1}\right)^2-\frac{1}{4a^2}}}=0975.$$

Since $a=2b$, the $TE_{022}$ mode will have the same resonant frequency as the $TE_{102}$, that is, the two modes are degenerate.

**Example 5.3.** A rectangular cavity of cross-sectional dimensions $a=2.286$ cm and $b=1.016$ cm is closed by a perfect conductor with a small hole at $z=0$ and the other end is closed by a perfect short circuiting plunger. What must the range of movement of the piston be, if the cavity is to resonate at frequencies from 9.3 GHz to 10.2 GHz?

*Solution*: The resonate frequency is given by Eq. (5.17)

$$f_r=\frac{1}{\sqrt{\mu\epsilon}}\left\{\left(\frac{m}{2a}\right)^2+\left(\frac{n}{2b}\right)^2+\left(\frac{p}{2c}\right)^2\right\}^{1/2}.$$

For the $TE_{102}$ mode

$$m=1,\quad n=0,\quad p=2$$

$$f_r=\frac{1}{\sqrt{\mu\epsilon}}\left\{\left(\frac{1}{2a}\right)^2+\left(\frac{1}{c}\right)^2\right\}^{1/2}$$

or

$$c=\frac{2a}{\sqrt{(2a/\lambda_r)^2-1}}$$

for

$$f_r=9.3\text{ GHz},\quad \lambda_r=3.25\text{ cm}\quad\text{and}\quad c_1=4.57\text{ cm}$$

$$f_r=10.2\text{ GHz},\quad \lambda_r=2.94\text{ cm}\quad\text{and}\quad c_2=3.84\text{ cm}.$$

The range of movement is thus 0.73 cm.

**Example 5.4.** A klystron oscillator cavity has a $Q$ factor 5000 and coupling coefficient $\alpha=1$. What is the cavity length if it is tuned to 9.6 GHz? Through what distance must the piston be moved to vary the frequency of resonance over the cavity pass band? ($a=1.016$, $b=2.286$ cm).

*Solution*: Using Eq. (5.15), $\lambda_r=3.125$ cm

$$\text{cavity length } c_r=\frac{2b}{\left\{\left(\frac{2b}{\lambda_r}\right)^2-1\right\}^{1/2}}=4.281\text{ cm.}\qquad(1)$$

Now half-width and cavity $Q$ are related as

$$\frac{2\delta f}{f_r}Q_L=1.$$

Using

$$Q_L=\frac{Q_0}{1+\alpha}=2500$$

$$f_1-f_r=\frac{f_r}{2Q_L}=1.92\text{ MHz.}\qquad(2)$$

Differentiating Eq. 1 and using $c_0=f_r\ \lambda_r$, $c_0$ is velocity of light in air

$$dc=-\frac{2b}{2}\left[\left(\frac{2b}{\lambda_r}\right)-1\right]^{-3/2}\cdot\frac{4b^2}{c_o^2}f_r\ 2df$$

or
$$dc=-\ c_r^3\frac{fdf}{c_o^2}$$

($c_r$ being resonant cavity length at $f_r$).

If $df$=1.92 MHz, $dc=-1.6\times10^{-3}$ cm.

Assuming that whole resonance curve to be absorbed in the frequency width 4 $df$, a movement 4 $dc$ will cover the whole resonance band. Thus the required movement is

$$4\ dc\simeq\frac{1}{10}\text{ mm.}$$

**Example 5.5.** A air-filled cavity resonates at 10.6 GHz. What will be the resonant frequency of the cavity when it is filled with a dielectric of relative permittivity $\epsilon_r$=1.63? The $Q$ factor of the cavity is $Q_0$=8200 and the loss tangent of the dielectric is $10^{-3}$.

What will be the new $Q$ factor?

*Solution*: The new resonant frequency will be

$$f_r'=\frac{f_r}{\sqrt{\epsilon_r}}=8.30\text{ GHz.}$$

The new $Q$ factor is given by

$$\frac{1}{Q_0'}=\frac{1}{Q_0}+\frac{1}{Q}.$$

Since $Q_\epsilon$ = cot $\phi$,

$$\therefore\quad\frac{1}{Q_0'}=\frac{1}{8200}+10^{-3}$$

$$\therefore\quad Q_0'=890.$$

*Note.* It is worth noting that $Q$ is reduced to ~ 10% of its value due to losses in the dielectric; hence, dielectric losses are more important than conductor losses.

**Example 5.6.** Consider a resonant cavity connected to two rectangular waveguides propagating the $TE_{10}$ mode. The first (1) guide is matched to a microwave source while the second (2) is terminated in its characteristic impedance $Z_0$. The source supplies constant power. Taking the equivalent circuit shown in Fig. Ex. 5.6, find the admittance at plane $E$ and deduce the equivalent circuit for this plane. Calculate the loaded $Q$-factor $Q_L$, of the cavity. Also deduce the external $Q$ factors, $Q_{E1}$ and $Q_{E2}$, and express these in terms of the equivalent circuit.

*Solution:*

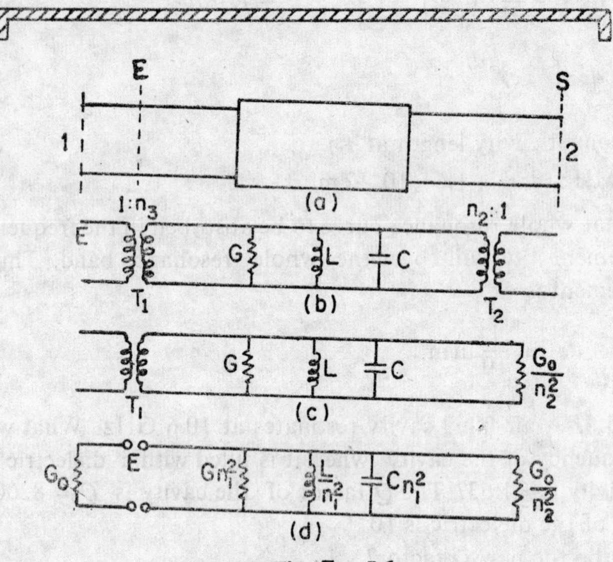


**Fig. Ex. 5.6.**

The effective admittance across the primary of the transformer $T_2$ is $G_0/n_2^2$. The equivalent circuit will thus be as in Fig. 5.6(c). The total admittance for the secondary of $T_1$ is

$$Y=G+\frac{G_0}{n_2^2}+j\left(\omega C-\frac{1}{\omega L}\right).$$

The admittance across the primary of $T$ is then $Y'=Yn_1^2$ and

$$Y'=n_1^2\left[G+\frac{G_0}{n_2^2}+j\left(\omega C-\frac{1}{\omega L}\right)\right].$$

The equivalent circuit for the input can thus be taken as Ex. 5.6 (d).

The $Q$ factor for the loaded resonant circuit including the conductance $G_0$ due to the generator, is given by

$$Q_L=\frac{n_1^2\omega_0 C}{G_0+n_1^2\,G+G\frac{n_1^2}{n_2^2}}$$

$$=\frac{\omega_0 C}{G}\frac{1}{1+\frac{G_0}{n_1^2 G}+\frac{G_0}{n_2^2 G}}$$

$$=\frac{\omega_0 C}{G}\frac{1}{1+\alpha_1+\alpha_2}$$

$$=\frac{Q_0}{1+\alpha_1+\alpha_2}\ \text{(say)}$$

where $Q_0=\frac{\omega_0 C}{G}$ is the $Q$ factor for the empty cavity.

$$\therefore \quad \frac{1}{Q_L}=\frac{1}{Q_0}+\frac{1}{Q_{E_1}}+\frac{1}{Q_{E_2}}=\frac{1}{Q_0}+\frac{\alpha_1}{Q_0}+\frac{\alpha_2}{Q_0} \cdot \text{ etc.}$$

**Example 5.7.** Calculate the power dissipated in the system of Ex. 5.6 as a function of $P_0$, $\alpha_1$, $\alpha_2$, $Q_L$ and $x=\omega/\omega_0$. What is the form of this expression in the neighbourhood of resonance when $\alpha_1=\alpha_2=\alpha$?

Can there be match for $\alpha_1=\alpha_2$?

*Solution.* The power dissipated in the cavity and load is

$$P=P_0(1-|\Gamma|^2), \quad \Gamma=\frac{1-y'}{1+y'}$$

with
$$y'=\frac{n_1^2}{G_0}\left[G+\frac{G_0}{n_2^2}+j\left(\omega C-\frac{1}{\omega L}\right)\right].$$

Take $G$ as factor

$$y'=\frac{1}{\alpha_1}\left[1+\alpha_2+j\,Q_0\left(x-\frac{1}{x}\right)\right]$$

which gives

$$\Gamma\Gamma^*=\frac{[\alpha_1-(1-\alpha_2)]^2+Q_o^2\left(x-\frac{1}{x}\right)}{(1+\alpha_1+\alpha_2)^2+Q_o^2\left(x-\frac{1}{x}\right)^2}$$

so that

$$P_1=\frac{4P_0\,\alpha_1\,(1+\alpha_2)}{(1+\alpha_1+\alpha_2)^2}\,\frac{1}{1+Q_L^2\left(x-\frac{1}{x}\right)^2}.$$

In the neighbourhood of resonance

$$Q_L\left(x-\frac{1}{x}\right)\cong 2\,Q_L\frac{\delta\omega}{\omega_0}=u \quad \text{(say)}$$

and for $\alpha_1=\alpha_2=\alpha$

$$P=\frac{4P_0\,\alpha\,(1+\alpha)}{(1+2\alpha)^2}\,\frac{1}{1+u^2}.$$

The system will be matched at resonance if

$$P=P_o, \Gamma=0 \text{ or } Y'=1, \text{ that is}$$

$$Y'(\omega_0)=\frac{1+\alpha_2}{\alpha_1}=1.$$

Consequently there cannot be any matching at resonance for $\alpha_1=\alpha_2$.

## PROBLEMS

1. An ideal air-filled rectangular cavity has the following dimensions: $a=4$ cm, $b=2$ cm, and $c=5$ cm. Designate the first three (*i*) TE and (*ii*) TM modes of oscillations. Find their resonant frequencies.

2. Show that in a rectangular resonator of Fig. 5.1 operating on the $TM_{mnp}$ mode, neither $m$ nor $n$ can be zero but $p$ can be zero.

3. A rectangular cavity resonator is filled with a lossy dielectric. Show that for the $TE_{mnp}$ mode of oscillator, the $Q$ of the cavity due to dielectric loss is given by $Q_d = \omega\epsilon/\sigma$, where $\epsilon$ and $\sigma$ are respectively the permittivity and the conductivity of the dielectric. Determine the resonant frequency of oscillation.

4. How will the resonant frequency and $Q$ of a cubical cavity of length 10 cm change if its air is replaced by a lossy dielectric of relative permittivity 2.25?

5. Show that the number of oscillations in a frequency interval $\Delta\omega$ is $\frac{abc\omega^2\Delta\omega}{\pi^2 c^3}$ for a rectangular resonator of dimensions $a$, $b$ and $c$ and with ideally conducting walls.

6. Show that in a cylindrical resonator the total stored energy is proportional to the cavity volume and is just twice the mean stored electric or magnetic energy.

7. The end of a rectangular waveguide is closed by a perfect short circuit. At what distance $l$ from this short circuit must another short circuit be placed if the resulting cavity is to resonate at a frequency $f$ (*i*) in case the guide is loss-less, (*ii*) in case there are losses in the guide but $\alpha l \ll l$? What is the cavity conductance in each case?

8. If in Prob. 5.7, the second short-circuit is an iris which acts as a negative susceptance $B=-1/\omega_L$, what is the new resonance conditions in the loss-less case? If depending on the diameter of this iris, the value $B/Y_0$ varies between 2 and 400, what will the range of values of $l$ be?

9. A cylindrical cavity of radius 1 cm and length 3 cm is filled with dielectric of relative permittivity $\epsilon_r=(25-j\,0.0001)$. Find the resonant frequency if the resonator is made of copper. How will the resonant frequency change if the walls of the resonator are made of dielectric?

10. Show that the peak energy stored in the electric and magnetic fields of a rectangular resonator at resonance are equal.

11. Calculate the dimensions at which a (*i*) rectangular, (*ii*) cylindrical, and (*iii*) spherical hollow resonator will oscillate at 10 GHz. Assume walls of the resonator to be perfectly conducting.

12. If the diameter of a cylindrical resonator is equal to its height, calculate resonator dimensions to resonate at 10 GHz. Calculate $Q$ value if the resonator is made of silver.

## *REFERENCES*

ARGENCE, E. and Kahan T., *Theory of Waveguides and Cavity Resonators*, Blackie and Sons Ltd. London (1967).

ATWATER, H.A., *Introduction to Microwave Theory*, McGraw-Hill Book Co. (1962).
Bell Systems Technical Journal 33 Nov. 1954.

BRONWELL, A.B. and BEAM R.E., *Theory and Applications of Microwaves*, Mc-Graw Hill Book Co. (1947).

CAP F. and DEUTSCH R., *Toroidal Resonator for Electromagnetic Waves—I and II*, IEEE Trans. MTT-28 (7) 1980.

COLLIN, R.E., *Foundations for Microwave Engineering*, McGraw-Hill Book Co. (1966).

GHOSE R.N., *Microwave Circuit Theory and Analysis*, McGraw-Hill Book Co. (1963).

GOPINATH A., *Maximum 'Q' Factor of Microstrip Resonators*, IEEE Trans. MTT-29 (2) 1981.

GOUBAU, G., *Electromagnetic Waveguides and Cavities*, Pregman Press, New York (1961).

HAMMOND, P., *Electromagnetism for Engineers—An Introductory Course*, Oxford Pregman (1978).

HARVEY, A.F., *Microwave Engineering*, Academic Press (1963).

HAYT, W.H. JR., *Engineering Electromagnetics*, (2nd Ed.) McGraw-Hill Book Co. (1967).

KRAUS, J.D. and CARVER K.R., *Electromagnetics*, McGraw-Hill Book Co. (1973).

LEGIER J.F. *et. al.*, *Resonant Frequencies of rectangular Dielectric Resonators*, IEEE Trans. MTT-28 (9) 1980.

MAGNUSSON, P.C., *Transmission lines and Wave Propagation*, (2nd Ed.) Allyn and Bacon, Boston Mass. (1970).

MC. DONALD N.A. AND MAGEWSKI M.L., *Dielectric Resonators and their Applications*, J. Elect. Electro. Engg., Australia 1 (1) 1981.

MONTGOMERY G.C. *et. al.*, *Principles of Microwave Circuits*, McGraw-Hill Book Co. (1951).

POSPIESZALSKI, M., *Cylindrical Dielectric Resonators and their Applications in TEM—Line Microwave Circuits* Proc. 7th European Microwave Conference, Denmark, Sept. 1977, Temple House, London.

RAMO, S., WHINNERY, J.R. and VAN DUZER, T., *Fields and Waves in Communication Electronics*, John Wiley (1965).

RAO, N.N., *Basic Electromagnetics with Applications*, Prentice Hall Inc. (1972).

REICH H.J. *et. al. Microwave Principles*, D. Van Nostran Princeton, N.J. (1966).

SAMUEL, Y.L., *Microwave Devices and Circuits*, Prentice-Hall Inc. (1980).

SANDER, K.F. and REED G.A.L., *Transmission and Propagation of Electromagnetic Waves*, Cambridge University Press, London (1978).

SESHADRI, S.R., *Fundamentals of Transmission Lines and Electromagnetic Fields*, Addition-Wesley (1971).

SHARMA A.K., '*Spectral Domain Analysis of Elliptical Microstrip Disk Resonators*' IEEE Trans. MTT-28 (6) 1980.

SLATTER, J.C. and FRANK N.H., *Electromagnetism*, McGraw-Hill Book Co. (1947).

VAN BLADEL, J., *Electromagnetic Fields*, McGraw-Hill Book Co., New York (1964).

WALDRON, R.A., *Theory of Guided Electromagnetic Waves*, Van Nostrand Co., London (1970).

# 6

# MICROWAVE COMPONENTS

## 6.1 INTRODUCTION

In the preceding chapters we discussed wave propagating systems such as transmission lines, waveguides and cavity resonators. In microwave engineering, one generally encounters components that make use of small lengths of propagating systems. These include waveguide tees, directional couplers, impedance transformers, phase shifters, attenuators and terminations. Our aim in this chapter, is to present a description of these components. Since these components play the same role in microwave engineering as their low frequency counterparts, they are also known as *passive microwave components*.

Section *A* of this chapter is devoted to the description methods of such components while section *B* deals with the characteristics of such components.

*SECTION A*

## MATRIX DESCRIPTION OF A MICROWAVE CIRCUIT

## 6.2 SCATTERING MATRIX

Microwave components and networks are usually designated by the number of ports entering and leaving the network. For example, a section of a waveguide terminated by some means (Fig. 6.1a) or a cavity resonator coupled to a single transmission line (Fig. 6.1b) can be regarded as one port networks. A section of a waveguide or a microwave component that has one input and one output is a two port network. Couplers and tees form three and/or four port networks. It may be noted that in a microwave circuit there may be more "electrical ports" than actual physical ports. Different field polarizations may be regarded as separate ports.

The relationship between input and output parameters such as currents, voltages, impedances admittances, electric and magnetic fields describe the system or network completely. Consider an $n$-port microwave junction as

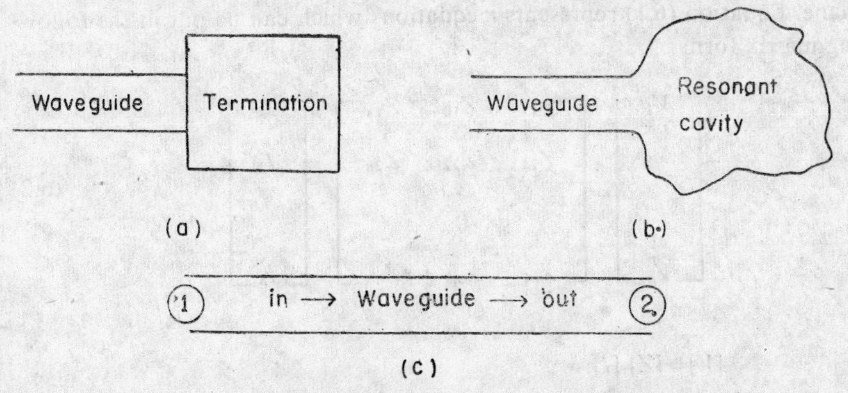


**Fig. 6.1** Microwave one-port networks (*a*) Terminated waveguide (*b*) Resonant cavity with one output (*c*) waveguide section two port network.

shown in Fig. 6.2. The voltage at port $p$ may be regarded as the resultant contribution from the respective currents at ports 1, 2, 3. . . $p$. . . $n$. Then*

$$V_p = \sum_{q=1}^{n} Z_{pq} I_q, \tag{6.1}$$

where $Z_{pq}$ is an impedance coefficient such that $(Z_{pq}\, I_q)$ gives the contribution of the current in port $q$ to the voltage at port $p$ at a given reference

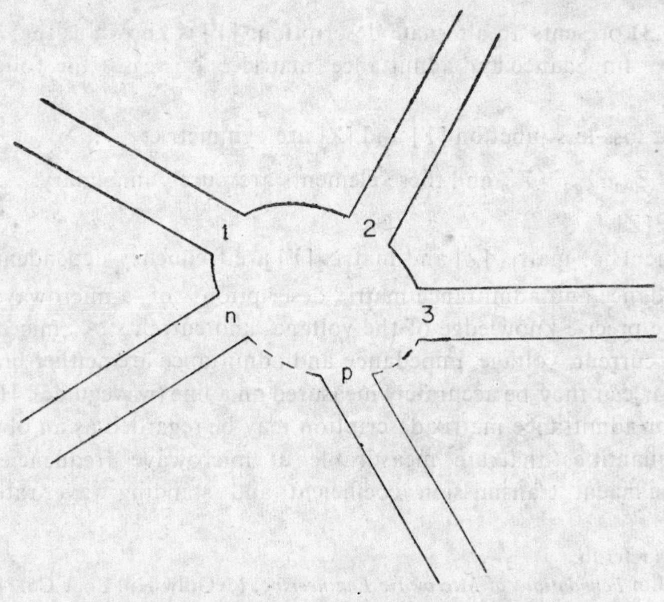


**Fig. 6.2** A general $n$-port microwave junction.

*Two assumptions are implied in Eq. (6.1): (*i*) number of physical ports is equal to the number of electrical ports; (*ii*) the junction is linear, which is generally true.

plane. Equation (6.1) represents $n$ equations which can be put in the following matrix form

$$\begin{bmatrix} V_1 \\ V_2 \\ \cdot \\ \cdot \\ \cdot \\ V_n \end{bmatrix} = \begin{bmatrix} Z_{11} & Z_{12} & Z_{1n} \\ Z_{21} & Z_{22} & Z_{2n} \\ \cdot & \cdot & \cdot \\ \cdot & \cdot & \cdot \\ \cdot & \cdot & \cdot \\ Z_{n1} & Z_{n2} & Z_{nn} \end{bmatrix} \begin{bmatrix} I_1 \\ I_2 \\ \cdot \\ \cdot \\ \cdot \\ I_n \end{bmatrix} \qquad (6.2)$$

or

$$[V]=[Z]\,[I].$$

Matrix $[Z]$ is called the *impedance matrix* and is a characteristic of the network because Eq. (6.2) represents a relation between what is "fed in", $[Z]$ a "processing device", and $[V]$ what "comes out".

If we consider $V_p$, the quantity "fed in" and $I_p$ the outcome, then we may write

$$I_p=\sum_{q=1}^{n} Y_{pq}\,V_q$$

or

$$[I]=[Y]\,[V]. \qquad (6.3)$$

Equation (6.3) presents an alternate description. $[Y]$ is known as the *admittance matrix*. Impedance and admittance matrices possesses the following properties*:

(*i*) For a loss-less junction $[Y]$ and $[Z]$ are symmetric,

$Z_{pq}=Z_{qp}$, $Y_{pq}=Y_{qp}$ and these elements are purely imaginary.

(*ii*) $[Y]=[Z]^{-1}$.

(*iii*) Elements of matrix $[Z]$ and matrix $[Y]$ are frequency dependent.

The impedance and admittance matrix descriptions of a microwave network require precise knowledge of the voltages and currents. At microwave frequencies, current, voltage, impedance and admittance are neither precisely defined nor can they be accurately measured on a line (waveguide). Hence, impedance or admittance matrix description may be regarded as an obstraction. The quantities that are measurable at microwave frequencies are reflection coefficient, transmission coefficient and standing wave ratio. A

*For details refer to:

1. R.E. Collin *Foundations af Microwave Engineering*, McGraw-Hill Book Co. (1966).
2. G.C. Montgomery, *Principles of Microwave Circuits*, McGraw-Hill Book Co. (1948) (Rad. Lab. Series, Vol. 8).
3. J., Helszajn, *Passive and Active Microwave Circuits*, New York, John Wiley & Sons (1978).

linear relationship between "input" and "output" may be established that involves precisely measurable parameters. Such a relationship forms the basis for an alternative description, called the *Scattering matrix* description.

Consider the $n$-port network of Fig. 6.2. Let us define a *normalized incident wave*, $a_p$, at a given reference plane of port $p$, as the complex scalar quantity proportional to the complex magnitude of the incident transverse electric field (wave) say $E_T^+$. Similarly, we define a *normalized scattered wave*, $b_p$, at the same reference plane of port $p$, as a complex scalar quantity proportional to the complex magnitude of the scattered transverse electric field, $E_T^-$. The incident and scattered waves are so normalized that $\frac{1}{2}a_p a_p^* = \frac{1}{2}a_p^2$ and $\frac{1}{2}b_p b_p^* = \frac{1}{2}|b_p|^2$ respectively giving total incident and total scattered power at the given reference plane of port $p$.

In general, the normalized scattered wave at port $p$ will be a function not only of the incident wave at port $p$ but also of the waves incident at other ports. In particular, if all the ports of the junction except the $p^{th}$ are terminated in matched loads, then the normalized scattered wave is related to the normalized incident wave as

$$b_p = \Gamma_p a_p \tag{6.4}$$

where $\Gamma_p$ is the reflection coefficient at the given reference plane of port $p$. In general, if $a_1, a_2, \ldots a_n$ are the normalized incident waves at ports 1, 2,. . . . $n$ respectively, then the normalized scattered wave coming out of port $p$, can be written as

$$b_p = S_{p1}a_1 + S_{p2}a_2 + S_{p3}a_3 \ldots S_{pn}a_n$$

or

$$b_p = \sum_{q=1}^{n} S_{pq}\, a_q \tag{6.5}$$

where $S_{pq}a_q$ represents the contribution to the normalized scattered wave coming out of part $p$ due to a normalized wave $a_q$ incident at port $q$. Clearly, $S_{pp} = \frac{b_p}{a_p} = \Gamma_p =$ reflection coefficient at port $p$.

Equation (6.5) may be expanded for each port to give $n$ simultaneous equations, which can be put in the following matrix form:

$$\begin{bmatrix} S_{11} & S_{12} & \ldots & S_{1n} \\ S_{21} & S_{22} & \ldots & S_{2n} \\ \vdots & \vdots & & \vdots \\ S_{n1} & S_{n2} & & S_{nn} \end{bmatrix} \begin{bmatrix} a_1 \\ a_2 \\ \vdots \\ a_n \end{bmatrix} = \begin{bmatrix} b_1 \\ b_2 \\ \vdots \\ b_n \end{bmatrix}$$

or

$$[S]\,[a] = [b] \tag{6.6}$$

where $[a]$ is a vector (column matrix) the input, $[b]$ a vector (column matrix) output and $[S]$ is by definition the *scattering matrix*. The scattering coefficient $S_{pq}$ represents the output contribution at port $p$ when the input at port $q$ is unity. That is, $S_{pq}$ is the transmission coefficient at port $q$ when the wave is incident at port $p$. Consequently, $S_{pq}$ is a measure of *coupling* between ports $q$ and $p$. The diagonal terms as $S_{ii}$ represent output contribution at port $i$ due to unit input at port $i$. If all $a_p$s except $a_i$ are zero, then $b_i = S_{ii} a_i$. Therefore, the diagonal terms $S_{ii}$ assume the role of the reflection coefficient at the given reference plane of port $i$.

It sometimes becomes necessary to know the nature of voltages and currents existing on a given system, for example, when drawing an equivalent circuit. Therefore, one should establish a relation between the scattering and impedance matrices.

Letting* the incident voltage $V^+ = a$ and scattered voltage $V^- = b$, then the total voltage $V(z) = a + b$ and total current $I(z) = a - b$. Hence

$$\frac{b}{a} = \left\{\frac{V(z)}{I(z)} - 1\right\} \Big/ \left\{\frac{V(z)}{I(z)} + 1\right\} = \frac{Z(z) - 1}{Z(z) + 1}$$

or in matrix form

$$[b] = \{[Z] - [I]\} \{[Z] + [I]\}^{-1} [a]. \tag{6.7}$$

A comparison of Eq. (6.7) with (6.6) establishes a relation between scattering and impedance matrices

$$[S] = \{[Z] - [I] \{[Z] + [I]\}^{-1}. \tag{6.8}$$

Similarly, the admittance matrix is related to the scattering matrix as

$$[S] = \{[I] - [Y]\} \{[I] + [Y]\}^{-1}. \tag{6.9}$$

Since $b/a$ is the reflection coefficient in a given port, hence, the standing wave ratio in that port is given by

$$S = \frac{1 + [\Gamma]}{1 - [\Gamma]} = \frac{1 + b/a}{1 - b/a}. \tag{6.10}$$

Equation (6.8) connects the scattering coefficients to the measurable parameter $S$, the voltage standing wave ratio (VSWR).

## 6.3 PROPERTIES OF A SCATTERING MATRIX

The following are the important properties of a scattering matrix:

(*i*) **[S] is Symmetric and Unitary for any Loss-less Microwave Junction**

To prove the above property let us define

$$[Z] - [I] = [G]$$
$$[Z] + [I] = [H] \tag{6.11}$$

*The assumption is true because voltages are proportional to respective electric field and, $a$ and $b$ are also proportional to the respective electric fields. The definition of normalized waves reduces proportionality constant to equality.

Then from Eq. (6.8), we have

$$[S]=[G]\,[H]^{-1}. \tag{6.12}$$

Since $[Z]$ is a symmetric matrix, the junction being reciprocal and loss-less, and $[I]$ is the unit matrix, the addition of [1] to $[Z]$ merely affects the diagonal elements so that each $[G]$ and $[H]$ will be symmetric and commutative, i.e.

$$[G]\,[H]=[H]\,[G]. \tag{6.13}$$

Multiplying Eq. (6.13) on the left and right by $[H]^{-1}$ we have,

$$[H]^{-1}\,[G]\,[H]\,[H]^{-1}=[H]^{-1}\,[H]\,[G]\,[H]^{-1}$$

or

$$[H]^{-1}\,[G]=[G]\,[H]^{-1}. \tag{6.14}$$

It therefore follows from Eqs. (6.12) and (6.14) that

$$[S]=[H]^{-1}\,[G]=[G]\,[H]^{-1}. \tag{6.15}$$

Designating the transpose of $[S]$ by $[S]^T$, we can write

$$[S]^T=\{[H]^{-1}\,[G]\}^T=[G]^T\,\{[H]^{-1}\}^T$$
$$=[G]\,[H]^{-1}=[H]^{-1}\,[G].$$

Since $[G]$ and $[H]$ are symmetric, so

$$[S]^T=[S]. \tag{6.16}$$

Thus, $S_{ij}=S_{ji}$ and therefore $[S]$ is symmetric.

To prove the unitary property of $[S]$ we recall the Poynting energy theorem* for an $n$-port microwave junction

$$\sum_{p=1}^{n} v_p i_p^* = 2P + 4j\omega(W_H - W_E) \tag{6.17}$$

where $v_p$ and $i_p$ are the equivalent voltage and current respectively at the given reference plane of the $i^{th}$ port. $P$ is the power dissipated in the junction and $W_H$ and $W_E$ are the average energies stored in the magnetic and electric fields respectively.

Using $V=a+b$ and $I=a-b$, Eq. (6.17) may be written as

$$\sum_{p=1}^{n} (a_p+b_p)\,(a_p^*-b_p^*) = 2P+4j\omega\,(W_H-W_E).$$

Equating the real and imaginary parts in the above equation

$$\sum_{p=1}^{n} (a_p a_p^* - b_p b_p^*) = 2P$$

*J.L. Altman, *Microwave Circuits*, D. Van Nostrand Co. Inc. (1964).

and

$$\sum_{p=1}^{n} (a_p b_p^* - a_p b_p^* ) = 4j\ \omega(W_H - W_E)$$

or, in matrix notation

$$[a]^{T*}\ \{[I]-[S]^*\ [S]\}\ [a]=2P \tag{6.18}$$

and

$$[a]^{T*}\ \{[S]-[S]^*\}\ [a]=4j\omega\ (W_H-W_E). \tag{6.19}$$

For a loss-less junction, $p=0$, so Eq. (6.18) yields

$$[I]-[S]^*\ [S]=0$$

or

$$[S]^{-1}=[S]^*.$$

Moreover, $[S]$ is symmetric, i.e. $[S]^*=[S]^{T*}$. Hence

$$[S]^{-1}=[S]^{T*}=[S]^T. \tag{6.20}$$

Equation (6.20) establishes that $[S]$ is unitary.

Unitary matrix $[S]$ has the additional property

$$\sum_{p=1}^{n} S_{pq}S_{pq}^* = \delta_{pq} = \begin{cases} 1 \text{ for } p=q \\ 0 \text{ for } p\neq q \end{cases}$$

or

$$\sum_{p=1}^{n} S_{pq}S_{pq}^* = \sum_{p=1}^{n} |S_{pq}|^2 = 1. \tag{6.21}$$

It should, however, be noted that for a lossy junction $P>0$, $[S]$ satisfies the condition

$$|\ [I]-[S]^*\ [S]| \geqslant 0.$$

(*ii*) **[S] is Transformed under a Shift in the Position of the Terminal Reference Plane**

Suppose the reference plane in the $p^{th}$ port is shifted by a distance $l_p$. This corresponds to an electrical phase shift $\theta_p=\beta_p l$; $\beta_p$ is the propagation constant corresponding to a given mode in the $p^{th}$ port. Now, if we assume that the normalized incident and scattered waves are still given by $a$ and $b$ then each of the scattering coefficients $S_{pq}$ $(p\neq q)$, for transmission into line $p$ from line $q$, should be multiplied by a phase term $e^{j\theta_p}$ to account for the additional phase shift caused in travelling a distance $l_p$. The reflected wave travels a distance $2l_p$ more as compared to the incident wave at the new terminal plane. Hence, a new scattering coefficient, $S'_{pq}$, arising due to the shift in the reference plane can be written as, $S'_{pq}=e^{j\theta_p} S_{pq}$. Similar arguments follow for all other scattering coefficients, hence the transformed scattering matrix is given by

$$[S'] = \begin{bmatrix} e^{j\theta_1} & \cdot & \cdot & \cdot \\ & e^{j\theta_2} & & 0 \\ & & e^{j\theta_3} & \\ & & & e^{j\theta_4} \\ 0 & & \cdot & \cdot \; e^{j\theta_n} \end{bmatrix} \begin{bmatrix} S_{11} & S_{12} & \ldots & S_{1n} \\ S_{21} & S_{22} & \ldots & S_{2n} \\ \vdots & & & \\ S_{n1} & S_{n2} & \ldots & S_{nn} \end{bmatrix}$$

$$\begin{bmatrix} e^{-j\theta_1} & & & \\ & e^{-j\theta_2} & & 0 \\ & & \cdot & \\ 0 & & \cdot & \\ & & & e^{-j\theta_n} \end{bmatrix} \qquad (6.22)$$

## 6.4 SCATTERING MATRIX CONSIDERATION

The following are the important considerations to be borne in mind while using a scattering matrix description.

(*i*) The scattering matrix approach consists of two steps, viz. (*a*) obtaining the scattering matrix for a given device, and (*b*) extracting its useful properties from the matrix. The first step may be accomplished either by measurement or from the symmetry properties of the device or both. In case of many practical devices it is possible to derive the complete scattering matrix from symmetry considerations. Though all methods of obtaining the scattering matrix are fair and give the desired results, the mixed procedure is most often used.

Once the scattering matrix is established and each of its elements is well defined, the second step is trivial in case of a single source junction. In case of a multisource junction, the normalized input waves from each source, both in phase as well as in magnitude, have to be known. Simple addition is required when the sources are synchronized. However, if the sources are not synchronized (have different frequencies) separate ([*b*]=[*S*] [*a*]) representations have to be used for individual frequencies. The results are then added up in the time domain to obtain the complete information.

(*ii*) The scattering matrix is the intrinsic property of the junction or device and uniquely supplies all its circuit properties. Therefore, the change in initial conditions does not affect, in any way, the scattering matrix. However, the change in reference plane does vary the scattering coefficient but only in phase and the phase change can be inferred and appropriate transformation carried out. The change in the operating frequency of the source also varies the scattering coefficients and so appropriate transformations* of the scattering matrix should be done.

If the scattering matrix is known then the desired device may be designed and as such the scattering matrix may prove to be inventive.

(*iii*) The scattering matrix for a loss-less junction is symmetric and unitary. These properties of the scattering matrix stem from the given definitions and power considerations. As such these are valid for all such junctions.

For a lossy junction $[S]$ is neither symmetric nor unitary. Due care must, therefore, be taken in transformation etc.

## 6.5 MEASUREMENT OF SCATTERING COEFFICIENTS

To get an idea of how scattering coefficients are determined experimentally, consider a simple example of two port junctions

$$\begin{bmatrix} b_1 \\ b_2 \end{bmatrix} = \begin{bmatrix} S_{11} & S_{12} \\ S_{21} & S_{22} \end{bmatrix} \begin{bmatrix} a_1 \\ a_2 \end{bmatrix}$$

or

$$b_1 = S_{11}a_1 + S_{12}a_2$$

$$b_2 = S_{21}a_1 + S_{22}a_2. \tag{6.23}$$

Suppose port 2 is terminated in a load which gives a reflection coefficient

$$\Gamma_2 = \frac{a_2}{b_2},$$

then
$$b_1 = S_{11}a_1 + S_{12}(\Gamma_2 b_2)$$

$$b_2 = S_{21}a_1 + S_{22}(\Gamma_2 b_2). \tag{6.24}$$

From the second of the above equation (6.24),

$$\frac{b_2}{a_1} = \frac{S_{21}}{1 - S_{22}\Gamma_2}. \tag{6.25}$$

Substituting the value of $b_2$ from Eq. (6.23) in the first of Eq. (6.24) and solving for $\frac{b_1}{a_1}$,

$$\frac{b_1}{a_1} = S_{11} + \frac{S_{12}S_{21}\Gamma_2}{1 - S_{22}\Gamma_2} = \Gamma_1. \tag{6.26}$$

*G.C. Montgomery, *Principles of Microwave Circuits*, McGraw-Hill Book Co. (1948) (Rad. Lab. Series Vol. VIII).

$\Gamma_1$ is the reflection coefficient at port 1.

Equation (6.26) forms the basis for the experimental measurement of $S_{11}$, $S_{12}$, $S_{21}$ and $S_{22}$. Known values of $\Gamma_2$ are substituted and the ratio $\frac{b_1}{a_1}$ i.e., $\Gamma_1$ is measured experimentally using a standard VSWR and position minimum methods (or by reflectometer techniques). This gives us a set of simultaneous equations from which scattering coefficients are readily determined. This may be further clarified by the following practical example:

Let $\Gamma_2=0$, i.e. matched load is connected at part 2, then

$$\left.\frac{b_1}{a_1}\right|_1 = S_{11}. \tag{6.27a}$$

If the matched load is now replaced by a short circuit then $\Gamma_2=-1$,

$$\left.\frac{b_1}{a_1}\right|_2 = S_{11} - \frac{S_{12}S_{21}}{1+S_{22}}. \tag{6.27b}$$

If there is an open circuit at port 2 (or the short circuit reference plane is shifted by a distance $\lambda_g/4$), then $\Gamma_2=1$;

$$\left.\frac{b_1}{a_1}\right|_3 = S_{11} + \frac{S_{12}S_{21}}{1-S_{22}}. \tag{6.27c}$$

$S_{11}$ has been determined in Eq. (6.27a), $S_{22}$ and the product $S_{12}\,S_{21}$ can be determined by solving Eqs. (6.27b) and (6.27c) as simultaneous equations. Hence we have

$$S_{22} = \frac{\left(\left.\frac{b_1}{a_1}\right|_3 - S_{11}\right) + \left(\left.\frac{b_1}{a_1}\right|_2 - S_{11}\right)}{\left(\left.\frac{b_1}{a_1}\right|_3 - S_{11}\right) - \left(\left.\frac{b_1}{a_1}\right|_2 - S_{11}\right)} \tag{6.28}$$

$$S_{12}S_{21} = (1-S_{22})\left\{S_{11} - \left.\frac{b_1}{a}\right|_2\right\}. \tag{6.29}$$

In this type of measurement it is not possible to separate $S_{12}$ from $S_{21}$. But for a reciprocal device $S_{12}=S_{12}$, and so $S_{12}\,S_{21}=S_{12}^2$ can be measured which gives us $S_{12}=S_{21}=\sqrt{S_{12}^2}$.

There may be infinite ways of measuring the scattering parameters. For example, by changing generator and load and terminating 1 in a matched load $S_{22}$ can be obtained directly. Similarly, it is possible to measure coupling from port 1 to port 2 by means of a matched detector and directional coupler to yield,

$$\frac{b_2}{a_1} = S_{21}.$$

In case of anisotropy (non-reciprocal device) the interchange of the ports in the above set up gives

$$\frac{b_1}{a_2}=|S_{21}|.$$

To measure phases of $S_{21}$ and $S_{12}$ obtained above, it is necessary to compare the phases of $a_1$ and $b_2$ in a hybrid junction by means of direction couplers. In many practical cases, the amplitude measurements suffice.

Experimentally, it is quite impossible to place a matched load at port 2 and it is always necessary to check on the measurements to get self-consistent and statistically true results. To fulfil the above requirements, Deschamps (for lossy devices) and Weissfloch (for loss-less junctions) have given us methods which can be found in the literature*.

## 6.6 SCATTERING MATRIX FOR SOME COMMON SYSTEMS

**(*a*) Transmission Line**

Consider a section of a uniform transmission line of length $L$ as shown in Fig. (6.3). Let $a_1$, $a_2$ and $b_1$, $b_2$ be the normalized incident and normalized scattered (reflected) waves. Then, for this two port device, we may write

$$\begin{bmatrix} b_1 \\ b_2 \end{bmatrix}=\begin{bmatrix} S_{11} & S_{12} \\ S_{21} & S_{22} \end{bmatrix}\begin{bmatrix} a_1 \\ a_2 \end{bmatrix}. \tag{6.30}$$

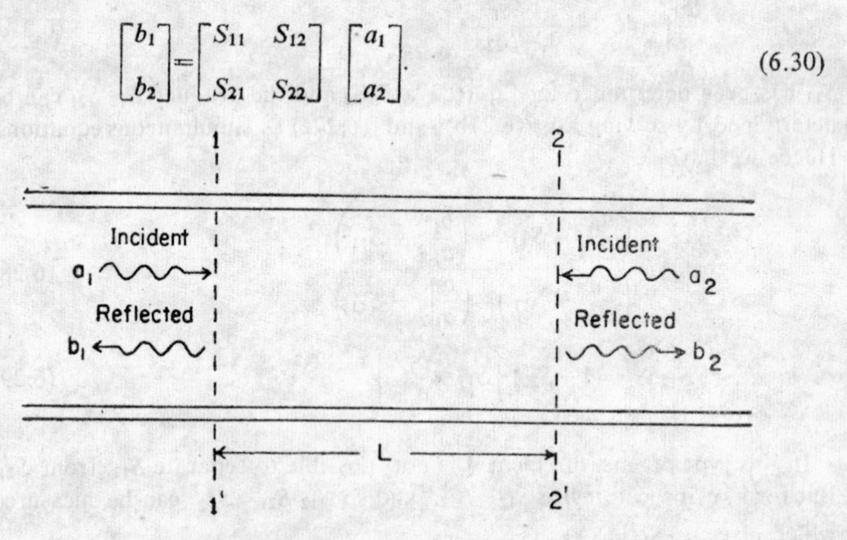


**Fig. 6.3** Normalized incident and reflected waves in a section of transmission line.

From Fig. (6.3) it is clear that for a loss-less line of propagation canstant $\beta$, we have

$$b_1=a_2\, e^{-j\beta L} \text{ and } b_2=a_1\, e^{-j\beta L}.$$

*J.L. Altman, *Microwave Circuits*, D. Van Nostrand Co. Inc. (1964).

Therefore, the scattering coefficients of the system are given by

$$\begin{bmatrix} b_1 \\ b_2 \end{bmatrix} = \begin{bmatrix} 0 & e^{-j\beta L} \\ e^{-j\beta L} & 0 \end{bmatrix} \begin{bmatrix} a_1 \\ a_2 \end{bmatrix}.$$

The scattering matrix is

$$[S] = \begin{bmatrix} 0 & e^{-j\beta L} \\ e^{-j\beta L} & 0 \end{bmatrix}. \tag{6.31}$$

If the reference plane is shifted by a distance $l_1$ then the transformed scattering matrix can be obtained using Eq. (6.29), that is

$$[S'] = \begin{bmatrix} e^{-j\beta l_1} & 0 \\ 0 & e^{-j\beta l_1} \end{bmatrix} \begin{bmatrix} 0 & e^{-j\beta L} \\ e^{-j\beta L} & 0 \end{bmatrix} \begin{bmatrix} e^{-j\beta l_1} & 0 \\ 0 & e^{-j\beta l_1} \end{bmatrix}. \tag{6.32}$$

### (*b*) Transition Between a Coaxial Line and Waveguide

Consider a typical coaxial to waveguide transition shown in Fig. (6.4). The two reference planes 1 and 2 are considered on the left of the transition.

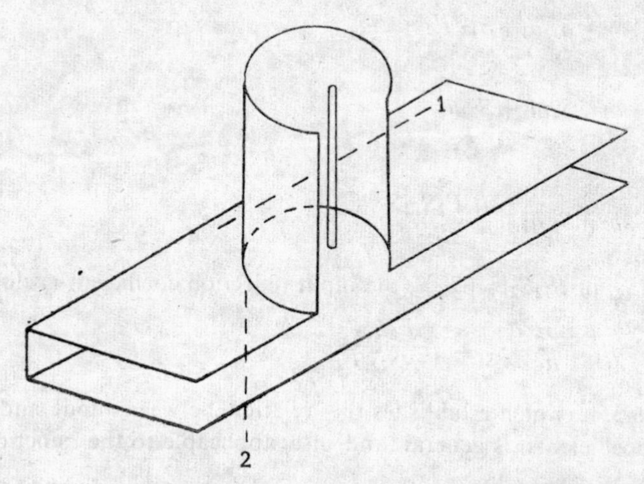


**Fig. 6.4** Half-sectional view of a coaxial line to waveguide transition.

Ignoring the losses in the extended probe and in the coaxial and waveguide lines, we may write from the unitary property of the scattering matrix that,

$$[S]\,[S]^{\dagger} = 1$$

or

$$\sum |S_{ij}|^2 = 1$$

or

$$|S_{11}|^2 + |S_{12}|^2 = 1$$

$\dagger[S] = [S]^{*T}$.

$$|S_{11}|^2+|S_{22}|^2=1. \tag{6.33}$$

Further, since $[S]$ is symmetric

$$|S_{12}|^2=|S_{21}|^2.$$

Hence,
$$|S_{11}|=|S_{22}|. \tag{6.34}$$

Thus the voltage standing wave ratio (VSWR) in either port, when a matched load is connected to the other, is

$$S_1=\frac{1+|S_{11}|}{1-|S_{11}|}=\frac{1+|S_{22}|}{1-|S_{22}|}=S_2.$$

The fractional power reflected from the coaxial line is $|S_{11}|^2$, the incident power being unity. However, if reference plane 2 is terminated in a reflection coefficient $\Gamma_2$, then

$$\Gamma_2=\frac{a_2}{b_2} \text{ or } a_2=\Gamma_2 b_2$$

Again

$$b_1=S_{11}a_1+S_{12}a_2$$
$$=S_{11}a_1+S_{12}\,\Gamma_2 b_2$$

and

$$b_2=S_{21}a_1+S_{22}a_2.$$

Therefore,

$$b_2=S_{11}a_1+\frac{a_1\,\Gamma_2 S_{12}S_{21}}{1-S_{22}\,\Gamma_2}.$$

Since $S_{12}=S_{21}$ or $S_{12}S_{21}=(S_{12})^2$, the input reflection coefficient is close

$$\Gamma_1=\frac{b_1}{a_1}=S_{11}+\frac{\Gamma_2 S_{12}^2}{1-S_{22}\,\Gamma_2}. \tag{6.35}$$

Equation (6.35), which establishes the relation between input and output reflection coefficients, is general and also applicable to the junction with losses.

*SECTION B*

# WAVEGUIDE COMPONENTS AND TEE JUNCTIONS

## 6.7 WAVEGUIDE TEES

Waveguide tee junctions are used to split the line power into two or combine the power from two lines *with proper consideration of the phase.* The junctions that are widely encountered in the microwave techniques are *E*- and *H*-plane tees. *An E-plane tee* is designed by fastening a piece of a similar waveguide to the broader wall of a waveguide section. The fastened waveguide, called the auxiliary arm, is parallel to the plane of the electric field of the dominant $TE_{10}$ mode in the main waveguide as shown in Fig. 6.5

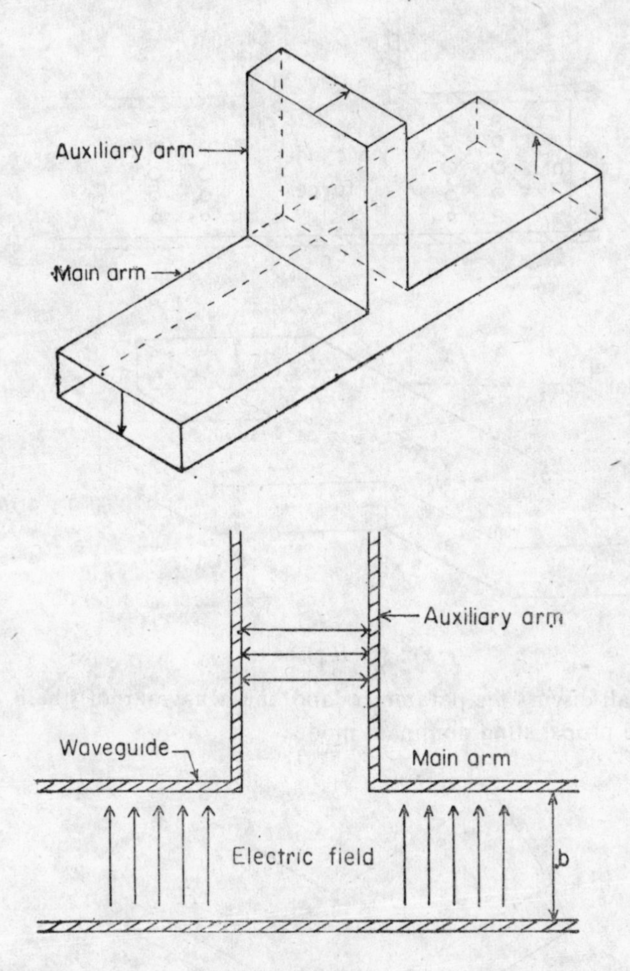


**Fig. 6.5 *E*-plane tee.**

and hence this type of junction is known as an *E*-plane tee. *An H-plane* tee is obtained by fastening the auxiliary waveguide perpendicular to the narrow arm of the main waveguide section, i.e. the auxiliary arm should lie in the *H-plane* of the dominant $TE_{10}$ mode in the main waveguide as shown in Fig. 6.6. Other important microwave junctions include symmetric *Y* junction, formed by joining three waveguide sections at 120° in a plane (Fig. 6.7a) and a hybrid junction or a magic tee which is a combination of *E*- and *H*-plane tees (Fig. 6.7b). Before we go into the details of each tee junc-

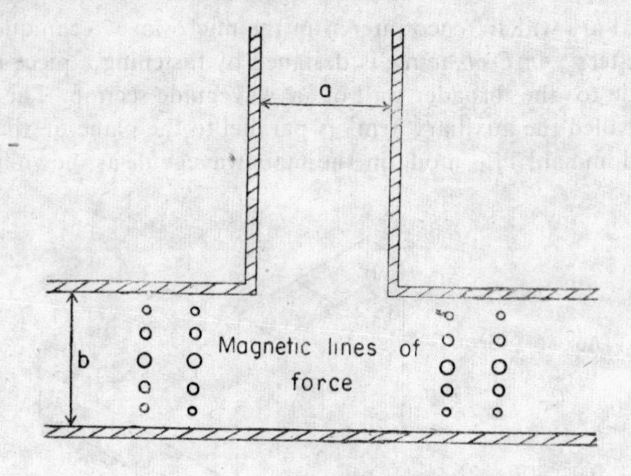


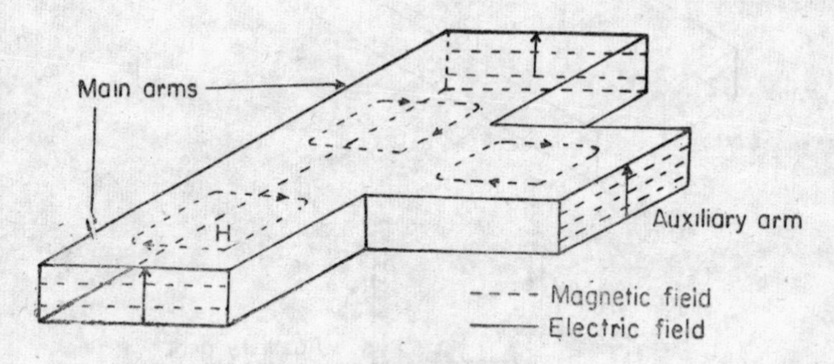


Fig. 6.6 *H*-plane tee

tion, we shall discuss tee parameters and the behaviour of these junctions towards the propagating dominant mode.

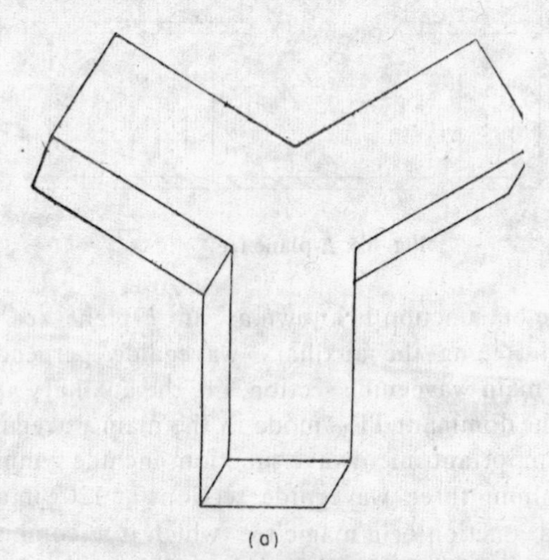


Fig. 6.7 (*a*) Symmetric *H*-plane *Y*-junction

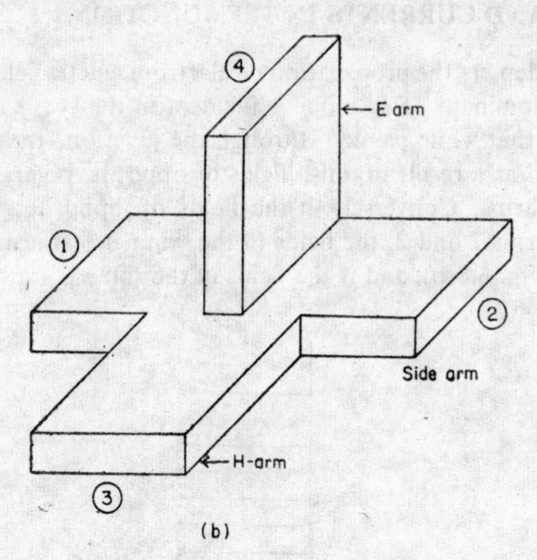


**Fig. 6.7** (*b*) *E-H* tee or Magic tee.

## 6.8 TEE JUNCTION PARAMETERS

The following are the parameters that define coupling characteristics of tee junctions and can be measured experimentally.

(*i*) **Iodation**

The isolation between $p$ and $q$ arms of a tee junction is defined as the ratio of the power measured in decibels supplied by a matched generator to arm $p$ and the power dectected by a matched detector in arm $q$ when all other ports are matched terminated,

$$I_{pq} = \log \frac{P_p}{P_q} \, db. \tag{6.36}$$

($p$ and $q$ each have values, 1, 2, 3 . . . )

(*ii*) **Coupling Coefficient**

When $\alpha_{pq}$ is the attenuation of power flowing from port $p$ to port $q$, then the coupling coefficient between these ports is defined as

$$C_{pq} = 10^{-\alpha_{pq}/20} \tag{6.37}$$

where $\alpha_{pq}$ is measured in decible.

(*iii*) **Input VSWR**

The input VSWR in any port of a microwave junction is the ratio of maximum voltage to minimum voltage existing on that port when all other ports are matched terminated.

## 6.9 FIELDS AND CURRENTS IN TEE JUNCTIONS

Figure 6.8(a) depicts the propagation of electromagnetic fields in an *E*-plane tee when the dominant $TE_{10}$ mode is incident in the symmetrical port 3. It may be noted that while passing through the junction, the electric lines of force bend and as a result of this, fields of opposite polarity emerge from the two main arms. Conversely, if the fields of opposite polarity are fed into the two arms 1 and 2, the fields in the symmetrical arm 3 are added in as shown in Fig. 6.8(b); and if the fields of the same polarity are fed into

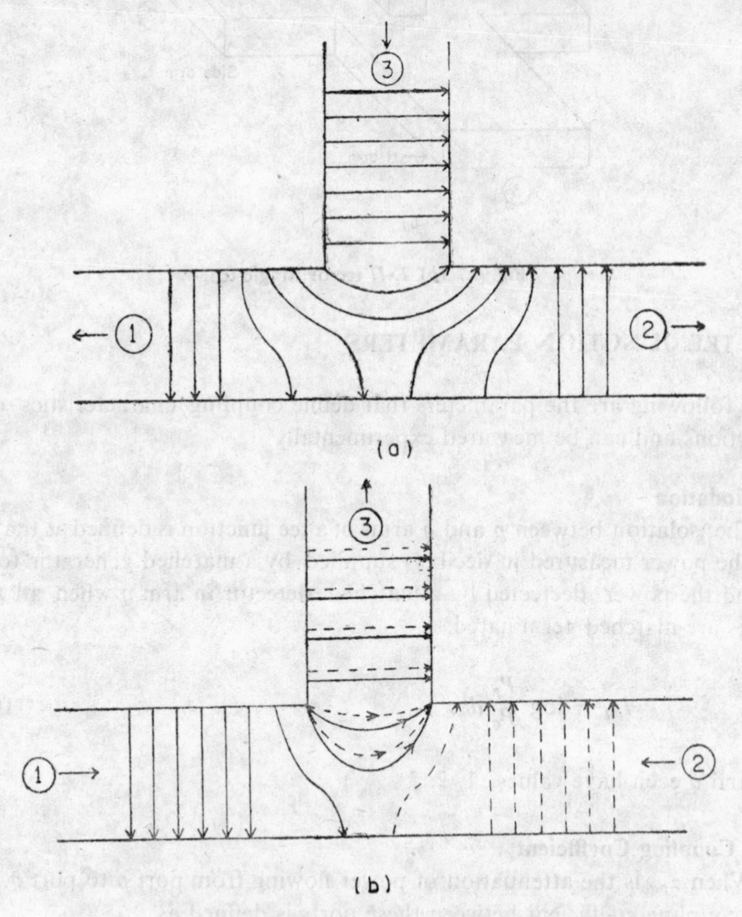


**Fig. 6.8** (*a*) Field entering in auxiliary arms 3 divides equally into two equal but opposite phase waves emitting out of arms 1 and 2. (*b*) Field of opposite polarity fed in symmetric arms 1 and 2 are added in arm 3.

the two side arms these emerge out the symmetric arm in opposite phase and therefore cancel out as shown in Fig. 6.8(c). It should be clear that the field patterns shown in the figures above represent propagating wave fronts (fields) and not the field pattern at any instant of time.

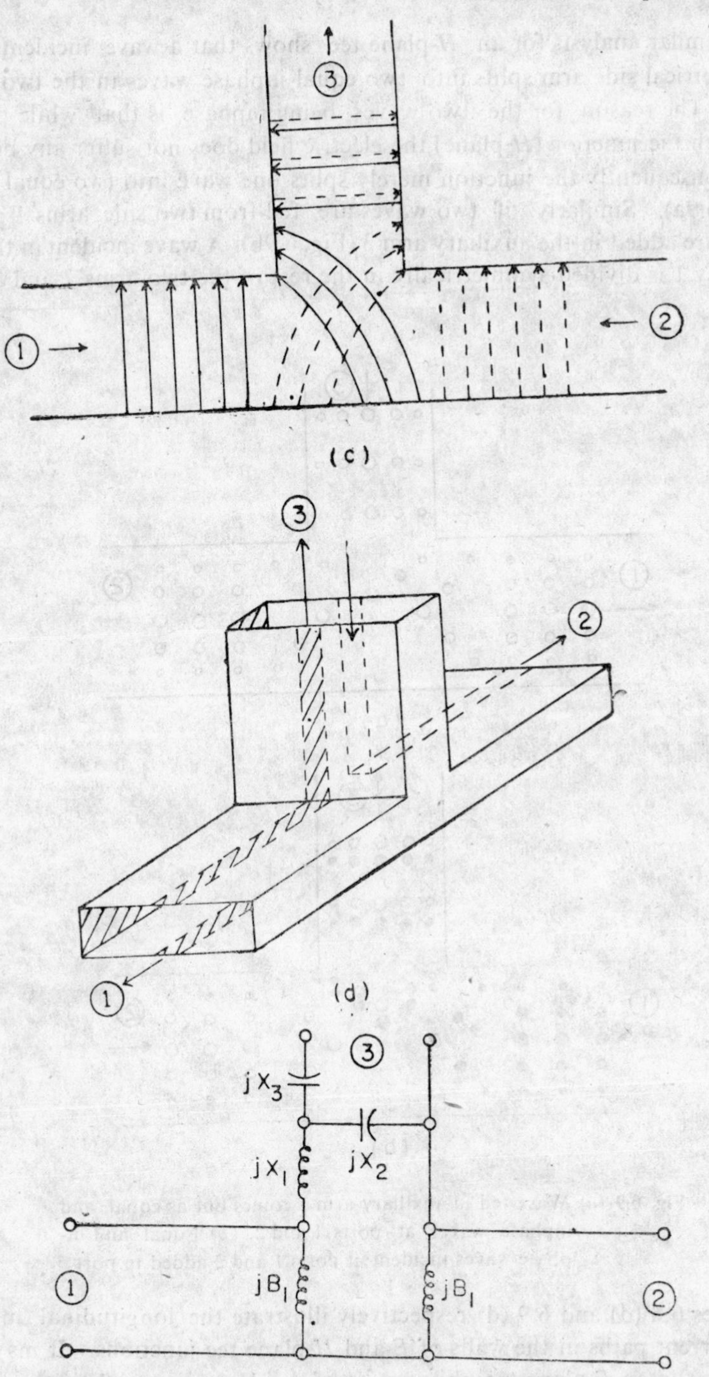


**Fig. 6.8** (*c*) Fields of the same polarity are fed in symmetric arms 1 and 2 they add out of phase in arm 3 and nothing comes out. (*d*) Microwave current path in *E*-plane tee junction. (*e*) equivalent circuit of an *E*-plane tee.

A similar analysis for an *H*-plane tee shows that a wave incident in the symmetrical side arm splits into two equal inphase waves in the two main arms. The reason for the two waves being inphase is that while passing through the junction (*H*-plane) the electric field does not suffer any bending and consequently the junction merely splits one wave into two equal waves (Fig. 6.9a). Similarly, if two waves are fed from two side arms 1 and 2, these are added in the auxiliary arm 3 (Fig. 6.9b). A wave incident in the side arm say 1 is divided symmetrically in the rest of the two arms 2 and 3 (Fig. 6.9c).

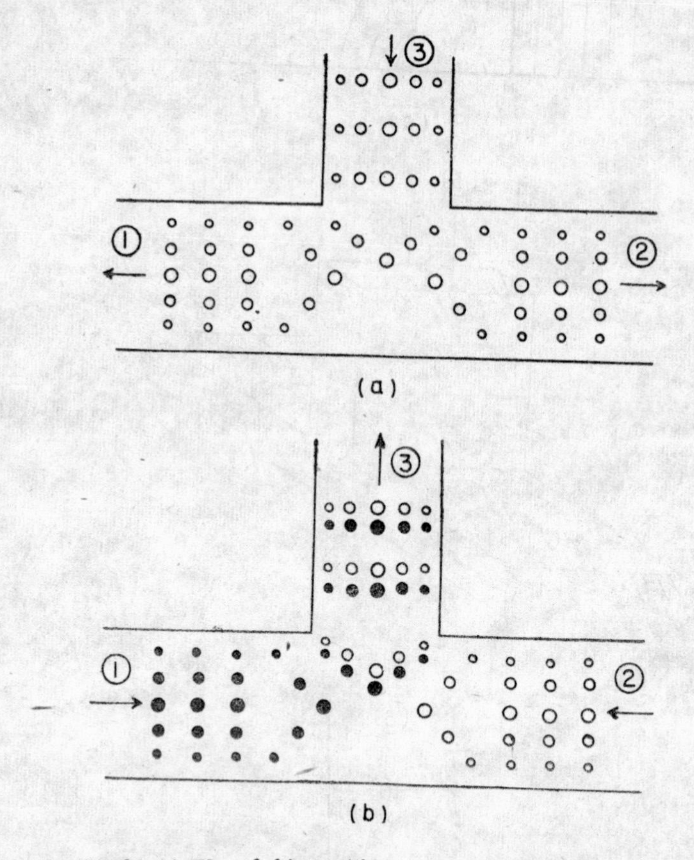


Fig. 6.9 (*a*) Wave fed in auxiliary arm 3 comes out as equal and inphase waves at ports 1 and 2. (*b*) Equal and inphase waves incident at ports 1 and 2 added in port 3.

Figures 6.8 (d) and 6.9 (d) respectively illustrate the longitudinal microwave current paths in the walls of *E*-and *H*-plane tee junctions. It may be noted that in an *E*-plane tee, the current of the dominant mode is the same for all the three arms while in an *H*-plane tee the current is distributed among the junction arms. The reason for this is that the magnetic field determines the current in the walls and it deteriorates at the junction in case

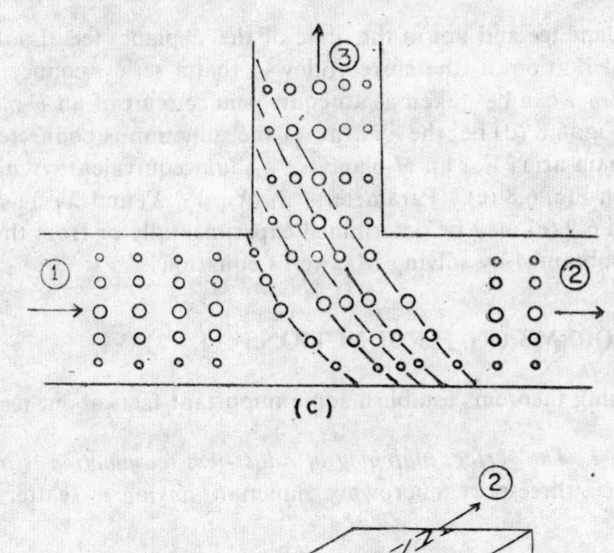


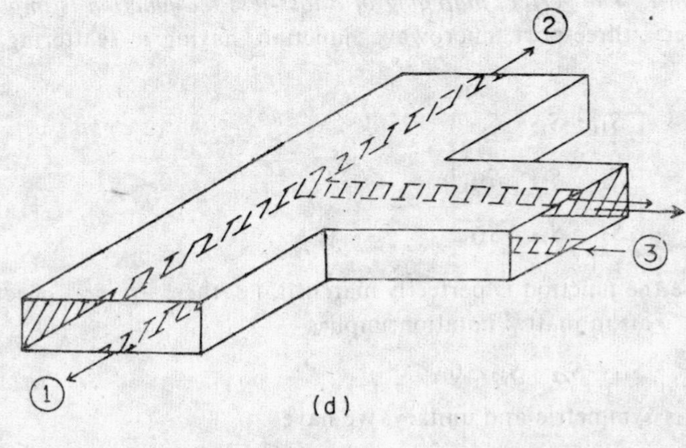


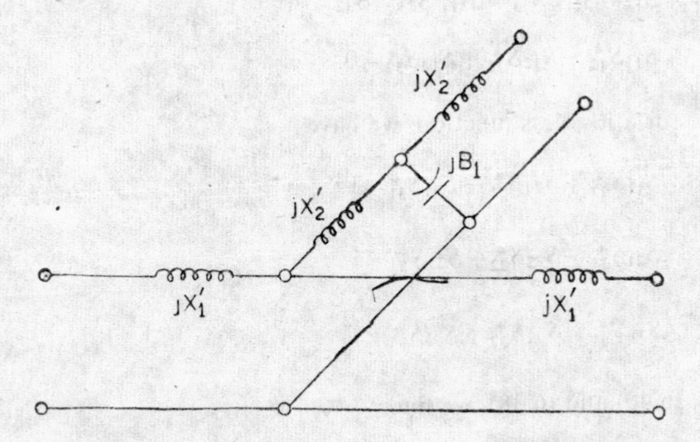


**Fig. 6.9** (*c*) Wave incident in side arm divides smmetrically in rest two arms.
(*d*) Microwave current path in an *H*-plane tee junction.
(*e*) Equivalent circuit of an *H*-plane tee.

of the $H$-plane tee and not in the case of the $E$-plane tee. Looking at the current distribution, it therefore follows, that a series connection of two two-wire lines can be taken as an equivalent circuit of an $E$-plane tee as shown in Fig. 6.8 (d) i.e. the $E$-arm of the junction is connected in series with the main arm. For an $H$-plane tee a shunt equivalent circuit is needed as shown in Fig. 6.8 (e). Parameters $B_1$, $X_1$, $X_2$, $X_3$ and $X_1'$ $X_2'$ $X_3'$ in Figs. 6.8 (d) and 6.9 (e) may be determined experimentally or from the field configuration obtained by solving Maxwell's equations*.

## 6.10 THEOREMS ON TEE JUNCTIONS

The following theorems establish some important facts about tee junctions.

*Theorem I: The perfect matching of a loss-less tee junction is impossible.*

Consider a three-port microwave junction having a scattering matrix given by

$$[S] = \begin{bmatrix} S_{11} & S_{12} & S_{13} \\ S_{21} & S_{22} & S_{23} \\ S_{31} & S_{32} & S_{33} \end{bmatrix}.$$

Suppose the junction is perfectly matched, i.e. there are no reflections at the ports. This in matrix notation implies

$$S_{11}=S_{22}=S_{33}=0. \tag{6.38}$$

Since $[S]$ is symmetric and unitary, we have

$$S_{12}=S_{21},\ S_{13}=S_{31},\ S_{23}=S_{32} \tag{6.39}$$

$$S_{11}S_{12}^*+S_{12}S_{22}^*+S_{13}S_{32}^*=0. \tag{6.40}$$

Moreover, for a loss-less junction, we have

$$S_{11}S_{11}^*+S_{21}S_{21}^*+S_{31}S_{31}^*=1 \tag{6.41a}$$

$$S_{21}S_{21}^*+S_{22}S_{22}^*+S_{23}S_{23}^*=1 \tag{6.41b}$$

$$S_{31}S_{31}^*+S_{32}S_{32}^*+S_{33}S_{33}^*=1. \tag{6.41c}$$

From Eqs. (6.40) and (6.38), we find

$$S_{13}S_{32}^*=0. \tag{6.42}$$

*N., Marcuvitz *Waveguide Hand Book*, McGraw-Hill Book Co. (1951) (Rad. Lab. Series, Vol. 10).

Also, using Eqs. (6.38) in Eqs. (6.41a and b) we find,

$$|S_{12}|^2 = 1 - |S_{13}|^2 = 1 - |S_{23}|^2. \tag{6.43}$$

The last step follows from Eq. (6.39). Equation (6.42) implies that either $S_{13}$ or $S_{32}^*$ is zero. Suppose $|S_{13}|=0$ then it follows from Eq. (6.43) that $|S_{23}|=0$. The disappearance of $|S_{23}^*|$ questions the validity of Eq. (6.41b) because of Eq. (6.38). Similarly, if $|S_{32}|=0$ it follows from Eq.(6.43) that $|S_{13}|=0$, which questions the validity of Eq. (6.41a) because of Eq. (6.38). We therefore conclude that Eqs. (6.41) and (6.43) are inconsistent if diagonal elements in $[S]$ vanish. In other words, $[S]$ has non-zero diagonal elements, i.e., the junction cannot be a perfectly matched one.

Following similar arguments it can be easily proved that any two of the diagnoal elements of $[S]$ can vanish only when the third arm of the junction is completely decoupled from the other two.

*Theorem II: It is always possible to isolate two ports of a tee junction by placing a short circuit in the third arm at an appropriate reference plane.*

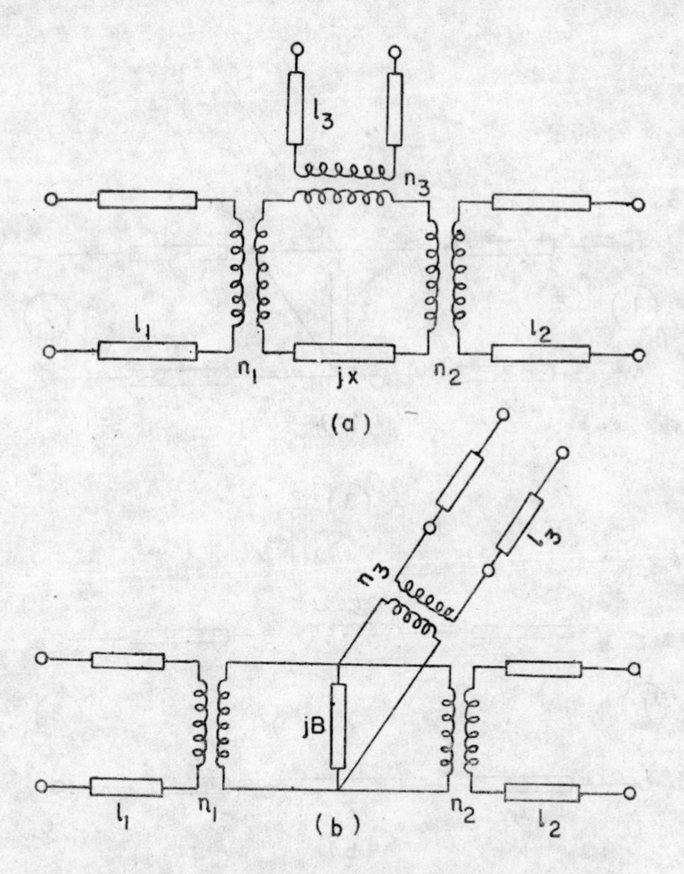


Fig. 6.10(*a*) Transformer equivalent circuit of *E*-plane tee.
(*b*) Transformer equivalent circuit of *H*-plane tee.

The theorem can easily be proved using transformer equivalent circuits* of *E*-and *H*-plane tees as shown in Fig. 6.10 (a) and (b) respectively. It may be mentioned that these circuits are equivalent to those in Figs. 6.8 (e) and 6.9 (e) $l_1$, $l_2$ and $l_3$ are the distances respectively of the reference plane on the ports from the junction.

Consider the series circuit of Fig. (6.9a). If the distance of the short circuit from the terminal plane is such that the total line length from the short circuit to the terminals of the transformer is an odd number of quarter wavelengths, the circuit is open at the point and no transmission occurs. In the case of a shunt tee, however, the transformer must be short-circuited to prevent transmission. This proves the theorem.

*Theorem III: It is always possible to place a short-circuit in the symmetrical port (auxiliary arm) of a tee junction to have perfect transmission between the rest of the two ports.*

This theorem may be proved by reducing the equivalent circuits of Fig. 6.10 (a) and (b) to the equivalent circuit of Fig. 6.11(a) and (b) respectively.

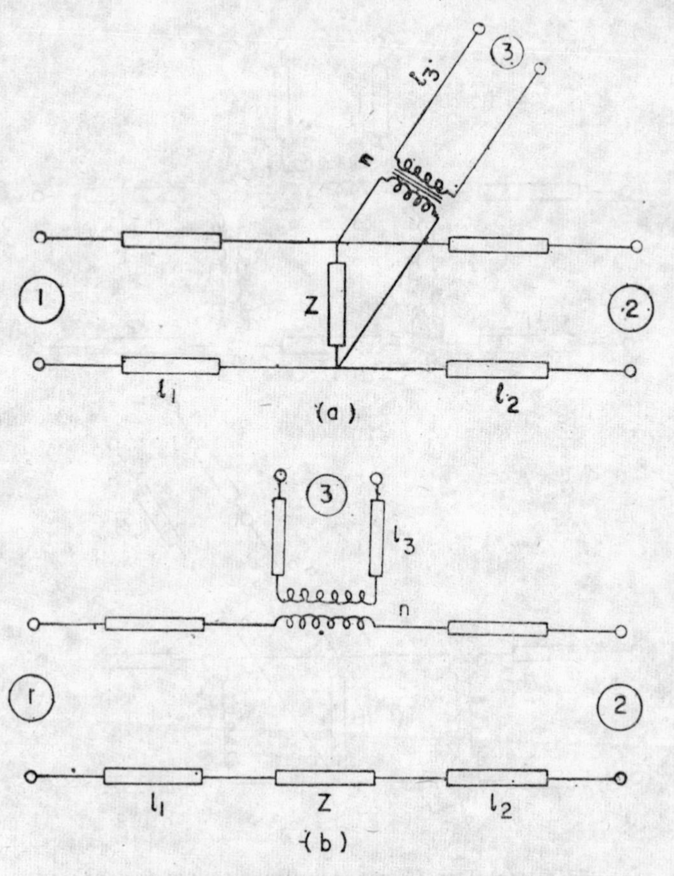


**Fig. 6.11**(*a*) **Transformer equivalent circuit of *E*-plane tee.**
(*b*) **Transformer equivalent circuit of *H*-plane tee.**

Perfect transmission from terminals (1) to (2) results if proper impedance is placed on arm 3 to resonate impedance $Z$. The circuit then reduces to a transmission line with a normalized characteristic impedance unity and length $l_1+l_2$. This proves the theorem.

From theorems II and III, we find that by placing a shorting plunger in the symmetric arm 3, we can have maximum and minimum transmission of power between ports 1 and 2. These positions of the short circuit in port 3 are called *characteristic planes* and are experimentally measurable just as input VSWR, reflection coefficient, coupling, isolation, etc. If we plot admittances offered by the short circuit on a Smith chart by measuring input VSWR and the position of voltage minimum, the resultant diagram is known as the *admittance locus*.

Furthermore, theorem III suggests nothing more than a tuner, i.e. if a shorting plunger is placed in the symmetric arm, by varying the position of this short circuit, ports 1 and 2 may be matched or tuned.

The theorems discussed above are useful for understanding the general behaviour of microwave tee junction. However, individual varieties of tee junctions often have other special properties which are best described by their scattering matrices.

## 6.11 SHUNT OR *H*-PLANE TEE

It is a three-port device, so the scattering matrix may, in general, be written as

$$[S]=\begin{bmatrix} S_{11} & S_{12} & S_{13} \\ S_{21} & S_{22} & S_{23} \\ S_{31} & S_{32} & S_{33} \end{bmatrix}.$$

Since $[S]$ is symmetric, we have

$$S_{12}=S_{21},\ S_{13}=S_{31},\ S_{23}=S_{32}. \tag{6.44}$$

Moreover, an *H*-plane tee is symmetric about ports 1 and 2 (symmetric arms),

$$S_{13}=S_{23},\ S_{11}=S_{22}. \tag{6.45}$$

Combining Eqs. (6.44) and (4.45) and defining,

$$\begin{aligned} S_{11}&=S_{22}=\alpha \\ S_{33}&=\beta \\ S_{13}&=S_{31}=S_{23}=S_{32}=\gamma \\ S_{21}&=S_{12}=\delta. \end{aligned} \tag{6.46}$$

We have,

$$[S]=\begin{bmatrix} \alpha & \delta & \gamma \\ \delta & \alpha & \gamma \\ \gamma & \gamma & \beta \end{bmatrix}. \tag{6.47}$$

We note that from purely isotropy and symmetry properties, the number of unknown scattering coefficients is reduced to four.

Furthermore, the device is loss-less hence $[S]$ must be unitary or

$$[S][S]^{\dagger}+[S][S]^{T*}=[S][S]^{*}=[1].$$

Thus we have,

$$(1, 1) \text{ term of } [S][S]^{*}=|\alpha|^2+|\delta|^2+|\gamma|^2=1 \tag{6.48a}$$

$$(3, 3) \text{ term of } [S][S]^{*}=2|\gamma|^2+|\beta|^2=1 \tag{6.48b}$$

$$(2, 1) \text{ term of } [S][S]^{*}=\delta\alpha^{*}+\alpha\delta^{*}+|\gamma|^2=0 \tag{6.48c}$$

$$(3, 1) \text{ term of } [S][S]^{*}=\gamma\alpha^{*}+\gamma\delta^{*}+\beta\gamma^{*}=0. \tag{6.48d}$$

Much information may be extracted from Eqs. (6.48a to d) by solving* these as simultaneous equations for $\alpha$, $\beta$, $\gamma$ and $\delta$ and substituting these values in $[S]$. Once $[S]$ is known, the junction is completely defined. However, we consider the following cases of practical interest.

**(*i*) Matching of Ports 1 and 2 Simultaneously**

For matching ports 1 and 2 simultaneously, in the language of a scattering matrix we imply that $S_{11}=S_{22}=\alpha=0$. It then follows from Eq. (6,48c) $\gamma=0$, and from Eq. (6.48b) $\beta=1$ and from Eq. (6.48b) $\delta=0$. The scattering matrix for such a circuit is thus given by

$$[S]=\begin{bmatrix} 0 & 1 & 0 \\ 1 & 0 & 0 \\ 0 & 0 & 1 \end{bmatrix}. \tag{6.49}$$

It is seen from Eq. (6.49) that arm 3 is completely decoupled from arms 1 and 2. Hence arm 3 must contain a metallic wall (shorting plunger) at the proper location, or a waveguide below cut-off with a proper matching structure. Conversely, the power tapped off at port 3, regardless of how small it is, will result in a simultaneous mismatch at ports 1 and 2 if the structure is symmetrical, or at least at port 1 or 2 if either port is matched by itself.

*J.L. Altman, *Microwave Circuits*, D. Van Nostrand and Co. Inc. (1964).

(*ii*) **Matching of the Symmetrical Port** (3)

When port 3 is matched there are no reflections in this port, hence $S_{33}=\beta=0$. It then follows from Eq. (6.48b) that $\gamma=1/\sqrt{2}$. Using this value of $\gamma$ in Eqs. (6.48a) and (6.48d) and solving for $\alpha$ and $\delta$ we obtain,

$$\alpha=\frac{1}{2} \text{ and } \delta=+\frac{1}{2}.$$

Thus, the scattering matrix of the *H*-plane tee, with its symmetrical port matched, is given by

$$[S]=\tfrac{1}{2}\begin{bmatrix} -1 & 1 & \sqrt{2} \\ 1 & -1 & \sqrt{2} \\ \sqrt{2} & \sqrt{2} & 0 \end{bmatrix}. \tag{6.50}$$

The normalized scattered and incident waves bear the following relationship:

$$\begin{bmatrix} b_1 \\ b_2 \\ b_3 \end{bmatrix}=\tfrac{1}{2}\begin{bmatrix} -1 & 1 & \sqrt{2} \\ 1 & -1 & \sqrt{2} \\ \sqrt{2} & \sqrt{2} & 0 \end{bmatrix}\begin{bmatrix} a_1 \\ a_2 \\ a_3 \end{bmatrix}. \tag{6.51}$$

From Eq. (6.51), we find that $b_1/a_1=b_2/a_2=1/\sqrt{2}$, i.e. the device provides a 3*db* power split from arm 3, with the waves in phase at the symmetrical ports 1 and 2. It should, however, be noted that (*i*) if power is applied to port 1, 25 percent of the power is reflected back to port 1, 25 percent of the power is transmitted to arm 2, and the remaining 50 percent of the power is transmitted to the matched arm 3; (*ii*) if equal and in phase powers are supplied to ports 1 and 2 then

$$\begin{bmatrix} b_1 \\ b_2 \\ b_3 \end{bmatrix}=\tfrac{1}{2}\begin{bmatrix} -1 & 0 & \sqrt{2} \\ 1 & -1 & \sqrt{2} \\ \sqrt{2} & \sqrt{2} & 0 \end{bmatrix}\begin{bmatrix} a_1 \\ a_1 \\ 0 \end{bmatrix}=\begin{bmatrix} 0 \\ 0 \\ \sqrt{2}\,a_1 \end{bmatrix}.$$

That is, all the power comes out of the matched port 3. Many of the microwave devices utilize this property of the *H*-plane tee.

(*iii*) **Shunting an Impedance Across the Line**

Consider an *H*-plane tee with port 2 matched terminated and port 3 shunted by a normalized impedance $z$. We may write

$$\begin{bmatrix} b_1 \\ b_2 \\ b_3 \end{bmatrix}=\begin{bmatrix} \alpha & \delta & \gamma \\ \delta & \alpha & \gamma \\ \gamma & \gamma & \beta \end{bmatrix}\begin{bmatrix} a_1 \\ 0 \\ \Gamma_3 b_3 \end{bmatrix}. \tag{6.52}$$

where $\Gamma_3$ is the reflection coefficient at port 3. $\Gamma_3 = \frac{z-1}{z+1}$. Solving Eq, (6.52) by Cramer's rule, for $b_1$, $b_2$ and $b_3$ we have.

$$b_1 = \left(\alpha + \frac{\gamma^2 \Gamma_3}{1-\Gamma_3 \beta}\right) a_1 \tag{6.53a}$$

$$b_2 = \left(\delta + \frac{\gamma^2 \Gamma_3}{1-\Gamma_3 \beta}\right) a_1 \tag{5.53b}$$

$$b_3 = \left(\frac{\gamma}{1-\Gamma_3 \beta}\right) a_1. \tag{6.53c}$$

Now, suppose we want to use the tee junction as a symmetrical two port microwave circuit, then the scattering matrix or rather the "reduced scattering matrix" operating on $\begin{bmatrix} a_1 \\ o \end{bmatrix}$ must yield $\begin{bmatrix} b_1 \\ b_2 \end{bmatrix}$, i.e.

$$[S]_R = \begin{bmatrix} S_{11} & S_{12} \\ S_{21} & S_{22} \end{bmatrix} \begin{bmatrix} a_1 \\ o \end{bmatrix} = \begin{bmatrix} \left(\alpha + \frac{\gamma^2 \Gamma_3}{1-\Gamma_3 \beta}\right) a_1 \\ \left(\delta + \frac{\gamma^2 \Gamma_3}{1-\Gamma_3 \beta}\right) a_1 \end{bmatrix}$$

or by inspection

$$[S]_R = \begin{bmatrix} \alpha + \frac{\gamma^2 \Gamma_3}{1-\Gamma_3 \beta} & \delta + \frac{\gamma^2 \Gamma_3}{1-\Gamma_3 \beta} \\ \delta + \frac{\gamma^2 \Gamma_3}{1-\Gamma_3 \beta} & \alpha + \frac{\gamma^2 \Gamma_3}{1-\Gamma_3 \beta} \end{bmatrix}. \tag{6.54}$$

We therefore conclude from Eq. (6.54) that it is always possible to find a position for the short circuit in arm 3 that gives rise to any predetermined reactance at the junction. When port 3 is matched the equivalent reactance of the junction is equal to $z/2$.

**(*iv*) Complete Electrical Symmetry**

The choice of complete electrical symmetry requires

$$S_{11} = S_{22} = S_{33} \text{ or } \alpha = \beta \text{ and } \gamma = \delta.$$

It then follows from Eqs. (6.48a to d) that

$$\alpha = \beta = -1/3 \text{ and } \gamma = \delta = 2/3.$$

Hence, the scattering matrix becomes

$$[S] = \tfrac{1}{3} \begin{bmatrix} -1 & 2 & 2 \\ 2 & -1 & 2 \\ 2 & 2 & -1 \end{bmatrix}. \tag{6.55}$$

## 6.12 SYMMETRICAL *H*-PLANE *Y*-JUNCTION

From the symmetry of the junction (Fig. 6.7) it is noted that

$$S_{11}=S_{22}=S_{33} \text{ and } S_{12}=S_{23}=S_{31}.$$

It therefore follows that $\alpha=\beta$ and $\gamma=\delta$. Hence the scattering matrix is given by

$$[S]=\begin{bmatrix} \alpha & \gamma & \gamma \\ \gamma & \alpha & \gamma \\ \gamma & \gamma & \alpha \end{bmatrix}. \tag{6.56}$$

Since $[S]$ is unitary, $[S]\,[S]^{\dagger}=[1]$ or from Eq. (6.56)

$$\alpha^2+2\,|\gamma|^2=1 \tag{6.57a}$$

$$\gamma\alpha^*+\alpha\gamma^*+|\gamma|^2=0. \tag{6.57b}$$

The following are cases of practical interest.

(*i*) $S_{11}=S_{22}=S_{33}=\alpha=0$, i.e. all ports are matched. Then from Eq. (6.57a). $|\gamma|=\frac{1}{\sqrt{2}}$, while from Eq. (6.57b) $|\gamma|=0$, which is in contradiction to above result. Hence, all the three ports of a symmetrical *H*-plane *Y*-junction cannot be matched simultaneously.

(*ii*) If $S_{11}=0$, then $\alpha=0$ and hence $|\gamma|=1/\sqrt{2}$ from Eq. (6.57 a) but $|\gamma|=0$ from Eq. (6.57b). Thus, none of the ports of a *Y*-plane tee can be matched terminated.

(*iii*) If any of the arms is terminated in a given impedance, the reduced scattering matrix is still given by Eq. (6.54) and hence the behaviour of the *Y*-junction is similar to the *H*-plane tee.

## 6.13 SERIES OR *E*-PLANE TEE

From the symmetry of the junction, it may be noted that

$$S_{11}=S_{22}=\alpha,\ S_{31}=S_{13}=-S_{23}=S_{32}=\gamma,$$

$$S_{33}=\beta \text{ and } S_{12}=S_{21}=\delta.$$

Hence the scattering matrix is given by

$$[S]=\begin{bmatrix} \alpha & \delta & \gamma \\ \delta & \alpha & -\gamma \\ \gamma & -\gamma & \beta \end{bmatrix}. \tag{6.58}$$

It may also be noted from Eq. (6.58) that the transmission coefficient from fort 3 to port 2 is no longer equal to that from port 3 to port 1 but suffers a change of sign.

The device is loss-less; $[S]$ should be unitary, i.e. $[S]\,[S]^{\dagger}=[1]$. Hence,

$$(1,1) \text{ term of } [S]\,[S]^{*}=|\alpha|^2+|\gamma|^2+|\delta|^2=1, \tag{6.59a}$$

$$(3,3) \text{ term of } [S]\,[S]^{*}=|\beta|^2+2|\gamma|^2=1, \tag{6.59b}$$

$$(2,1) \text{ term of } [S]\,[S]^{*}=\alpha^{*}\delta+\alpha\delta^{*}-|\gamma|^2=0 \tag{6.59c}$$

and

$$(3,1) \text{ term of } [S]\,[S]^{*}=\alpha^{*}\gamma-\gamma\delta^{*}+\beta\gamma^{*}=0. \tag{6.59d}$$

The values of $\alpha$, $\beta$, $\gamma$ and $\delta$ may be determined from Eqs. (6.59a to d). However, the following are the case of practical interest.

**(*i*) Matching Port 1 and 2 Simultaneously**

Matching of ports 1 and 2 requires $S_{11}=S_{22}=\alpha=0$. This when substituted in Eq. (6.59) yields $\gamma=0$ which in turn gives $\beta=\pm 1$ from Eq. (6.59b) and $\delta=\pm 1$. Choosing proper reference planes in ports we may have $\beta=1$ and $\delta=-1$. The scattering matrix is then given by

$$[S]=\begin{bmatrix} 0 & -1 & 0 \\ -1 & 0 & 0 \\ 0 & 0 & 1 \end{bmatrix}. \tag{6.60}$$

From Eq. (6.60), it can be inferred that simultaneous matching of ports 1 and 2 is possible only if arm 3 is completely decoupled from arms 1 and 2. Hence arm 3 must contain a metallic wall at the proper location, or a waveguide below cut-off with the proper-matching structure. It therefore follows that if power is tapped off at port 3, a simultaneous mismatch will appear at ports 1 and 2 (if the structure is symmetrical, or at least a mismatch will appear at port 1 or port 2 if either port is matched by itself).

**(*ii*) Matching Port 3**

It is always possible to match port 3, i.e. $S_{33}=\beta=0$; then Eqs. (6.59a to d) yield $\alpha=\frac{1}{2}$, $\delta=\frac{1}{2}$ and $\gamma=1/\sqrt{2}$. Hence,

$$[S]=\frac{1}{2}\begin{bmatrix} 1 & 1 & \sqrt{2} \\ 1 & 1 & -\sqrt{2} \\ \sqrt{2} & -\sqrt{2} & 0 \end{bmatrix}. \tag{6.61}$$

Equation (6.61) suggests that the device provides a 3*db* power split from arm 3, with the waves 180° out of phase at the symmetrical ports 1 and 2. On the other hand, if power is fed to port 1, 25 percent of the power is reflected back to arm 1, 25 percent of the power is transmitted to arm 2 and the remaining 50 percent of the power is transmitted to arm 3.

However, if equal out of phase powers are fed to ports 1 and 2, that is

$$[a]=\begin{bmatrix} a_1 \\ -a_1 \\ 0 \end{bmatrix}$$

then

$$[b]=\begin{bmatrix} 1 & 1 & \sqrt{2} \\ 1 & 1 & \sqrt{2} \\ \sqrt{2} & -\sqrt{2} & 0 \end{bmatrix}\begin{bmatrix} a_1 \\ -a_1 \\ 0 \end{bmatrix}=\begin{bmatrix} 0 \\ 0 \\ \sqrt{2}a_1 \end{bmatrix}.$$

That is, all the power comes out of port 3. It therefore follows that the net power coming out of port 3 is the difference of the powers fed in arms 1 and 2.

### (*iii*) Placing an Impedance in Series with the Line

Suppose a normalized impedance $z$ is placed in arm 3 at the given reference plane and port 2 is matched. Then, following the procedure of section 6.11 (*iii*), we can obtain the following reduced matrix

$$[S]_R=\begin{bmatrix} \left(\alpha+\dfrac{\gamma^2\Gamma_3}{1-\beta\Gamma_3}\right) & \left(\delta-\dfrac{\gamma^2\Gamma_3}{1-\beta\Gamma_3}\right) \\ \left(\delta-\dfrac{\gamma^2\Gamma_3}{1-\beta\Gamma_3}\right) & \left(\alpha+\dfrac{\gamma^2\Gamma_3}{1-\beta\Gamma_3}\right) \end{bmatrix} \quad (6.62)$$

where

$$\Gamma_3=\frac{z-1}{z+1}.$$

It can be shown* from Eq. (6.62) that it is possible to match port 3 but the necessary condition is that the equivalent junction reactance equals double the normalized series impedance, that is

$$z_e=2z. \quad (6.63)$$

### (*iv*) Complete Electrical Symmetry

If there is complete electrical symmetry in the device then $\alpha=\beta$ and $\gamma=\delta$. Hence

$$[S]=\frac{1}{3}\begin{bmatrix} 1 & 2 & 2 \\ 2 & 1 & -2 \\ 2 & -2 & 1 \end{bmatrix}. \quad (6.64)$$

*J.L. Altman, *Microwave Circuits*, D. Van Nostrand Co. Inc. (1964).

It should be noted that all the above considerations apply not only to the colinear series tee but also to any three port device having the same type of antisymmetry of the electric fields with respect to reference plane $P_1$ of Fig. 6.12. However, they do not apply to coaxial and strip-line 3-port devices whose fields are symmetrical with respect to plane $P_1$ when operating in the TEM mode.

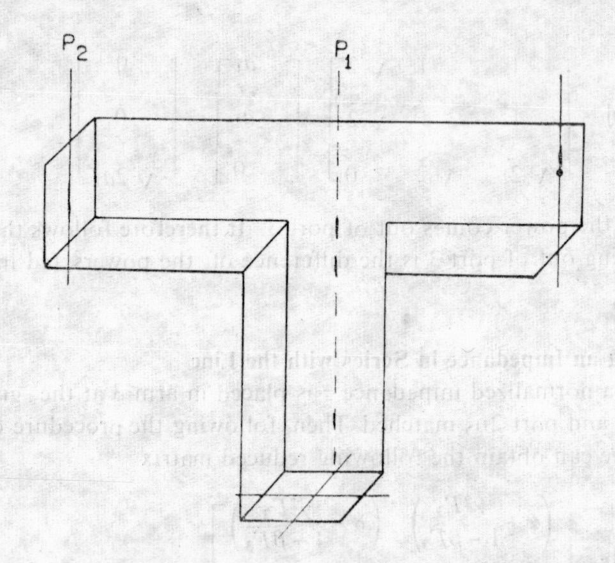


**Fig. 6.12** Symmetry in a Tee junction.

Waveguide tees besides being used as tuners (by placing a short circuit in the symmetrical arm), are also used as power dividers and adders. An interesting case of tee junction application is in the duplexing assemblies of radar installations where they are used in conjunction with *TR* and *ATR* switches. A typical duplexing system is shown in Fig. 6.13. When the sou-

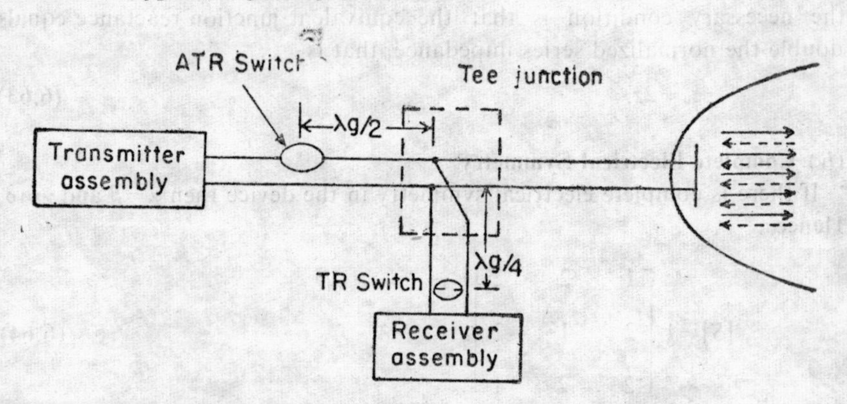


**Fig. 6.13** Duplexer circuit using a tee junction.

rce emits power, it is supplied to the antenna but when the antenna receives some power, the tee supplies less power to the switch connected with source and more to the receiver switch and hence antenna is connected to the receiver. The same antenna can therefore be used for the transmitter and the receiver.

## 6.14 MAGIC TEE

A magic tee is widly used in impedance bridges, antenna duplexer circuits, balanced mixers, phase detectors, frequency discriminators and microwave modulators. Before we go into details of all these let us discuss the behaviour of a magic tee in some detail.

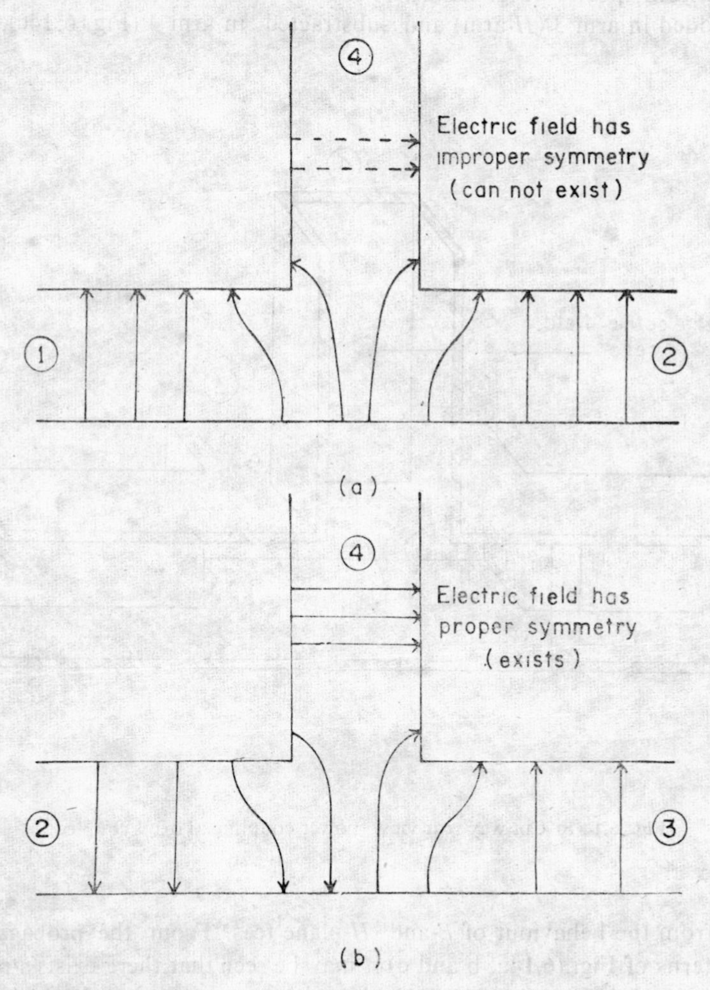


**Fig. 6.14**(*a*) Electric field pattern (Electric) when power (wave) is incident in port 3.
(*b*) Electric field pattern when power (wave) incident in port 4.

A magic tee is an interesting variation to tee junctions. It is the combination of *E*- and *H*- plane tees. Arm 3, (Fig. 6.7b) which is parallel to the lines of magnetic field, in combination, with arms 1 and 2 forms an *H*-plane tee while arm 4 which is parallel to the lines of electric field, forms an *E*-plane tee with arms 1 and 2. Common arms 1 and 2 are called colinear or side arms. The name "magic tee" is derived from the manner in which the power divides among its various arms. If power is fed in arm 3 (*H*-arm), it divides equally between two arms 1 and 2, and the electric field in these arms happens to be in phase but arm 4 remains uncoupled. It can be easily inferred from the symmetry of the magic tee that no field parallel to the narrow dimension is excited in arm 4 when the dominent $TE_{10}$ mode is incident in arm 3 (Fig. 6.14 a). Similarly, port 3 remains uncoupled when power is incident in port 4 (Fig. 6.14b). However, powers fed in arms 1 and 2 are added in arm 3 (*H*-arm) and substracted in arm 4 (Fig. 6.14c). This

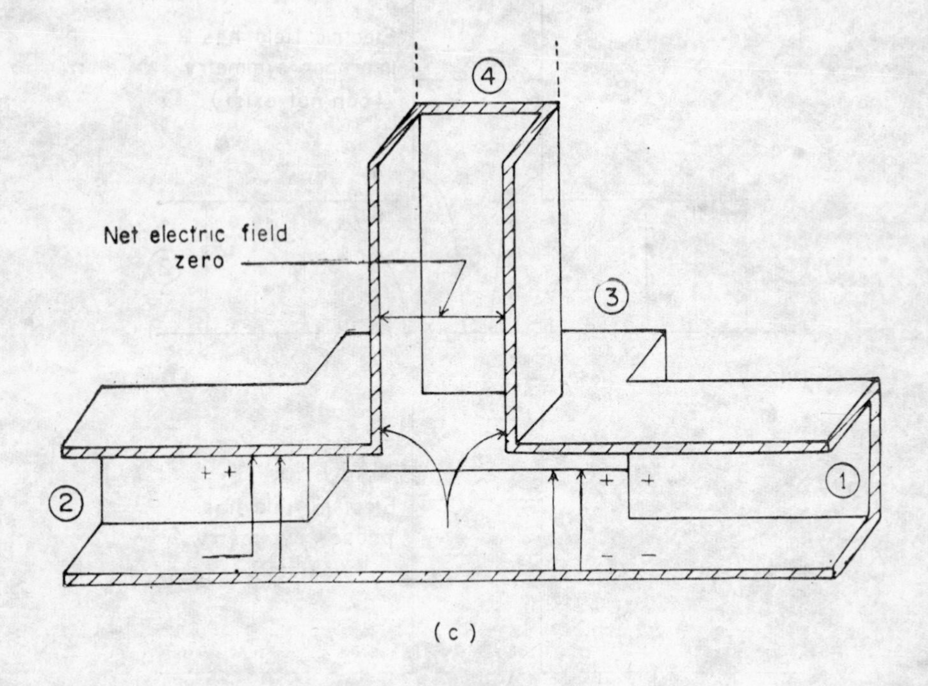


**Fig. 6.14**(*c*) Cut way rear view: power coupling. Port 3 coupled.

follows from the behaviour of *E*-and *H*-plane tees. From the propagating field patterns of Fig. (6.14a, b and c) it may be seen that there exists a plane of symmetry (*P in* Fig. 6.14d), the field pattern on two sides of which are mirror images of one another.

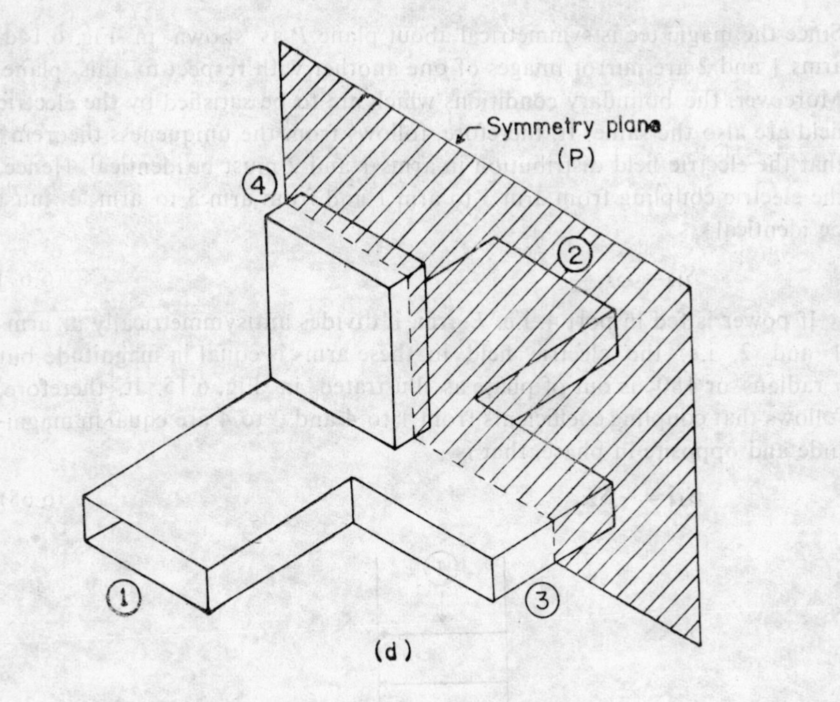


Fig. 6.14($d$) Plane of symmetry of magic tee.

## 6.15 SCATTERING MATRIX FOR MAGIC TEE

A magic tee is a four-port device, so in general,

$$[S]=\begin{bmatrix} S_{11} & S_{12} & S_{13} & S_{14} \\ S_{21} & S_{22} & S_{23} & S_{24} \\ S_{31} & S_{32} & S_{33} & S_{34} \\ S_{41} & S_{42} & S_{43} & S_{44} \end{bmatrix}. \tag{6.65}$$

Assuming the tee to be loss-less and medium inside to be isotropic, the symmetry of the tee requires

$S_{12}=S_{21}$, $S_{13}=S_{31}$, $S_{14}=S_{41}$, $S_{23}=S_{32}$, $S_{24}=S_{42}$,

$S_{34}=S_{43}$, thus

$$[S]=\begin{bmatrix} S_{11} & S_{12} & S_{13} & S_{14} \\ S_{12} & S_{22} & S_{23} & S_{24} \\ S_{13} & S_{23} & S_{33} & S_{34} \\ S_{14} & S_{24} & S_{34} & S_{44} \end{bmatrix}. \tag{6.66}$$

Since the magic tee is symmetrical about plane $P$ as shown in Fig. 6.14d. arms 1 and 2 are mirror images of one another with respect to this plane. Moreover, the boundary conditions which are to be satisfied by the electric field are also the same. It, therefore, follows from the uniqueness theorem* that the electric field distribution in arms 1 and 2 must be identical. Hence, the electric coupling from arm 3 to arm 1 and from arm 3 to arm 2 must be identical.

$$S_{13}=S_{23}. \tag{6.67}$$

If power is fed in port 4, i.e. $E$-arm, it divides antisymmetrically in arms 1 and 2, i.e. the electric field in these arms is equal in magnitude but $\pi$ radians or 180° is out of phase as illustrated in Fig. 6.15. It, therefore, follows that coupling coefficients from 1 to 4 and 2 to 4 are equal in magnitude and opposite in phase, that is

$$S_{14}=-S_{24}. \tag{6.68}$$

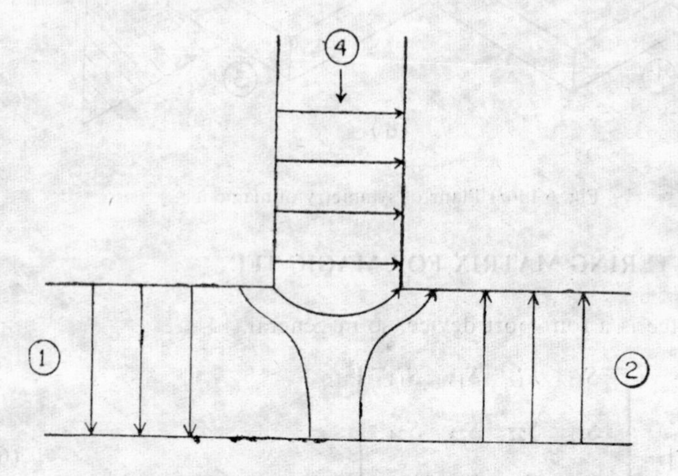


**Fig. 6.15** Power incident in port 4 of magic tee is divided into equal and out of $\pi$ phase powers in ports 1 and 4.

The symmetry of the magic tee, looking into port 1, with all other ports matched terminated, is equivalent to looking into port 2, with all other ports matched terminated. Hence

$$S_{11}=S_{22}. \tag{6.69}$$

Moreover, ports 3 and 4 are uncoupled because of cross-polarizations of the electric field.

$$S_{34}=0. \tag{6.70}$$

*The solution of Maxwell's equations under given boundary conditions is unique G.C. Montgomery R.H. Dicke, E.M. Purcell, *Principles of Microwave Circuits*, Mc-Graw-Hill Book Co. (1948).

Using Eqs. (6.67) through (6.70), Eq. (6.66) can be reduced to

$$[S]=\begin{bmatrix} S_{11} & S_{12} & S_{13} & S_{14} \\ S_{12} & S_{11} & S_{13} & -S_{14} \\ S_{13} & S_{13} & S_{33} & 0 \\ S_{14} & -S_{14} & 0 & S_{44} \end{bmatrix}. \tag{6.71}$$

Equation (6.71) is the scattering matrix of any symmetrical, 4 port tee of the form shown in Fig. 6.14(d) It does not matter what is inside the junction provided it is symmetrical and propagates the $TE_{10}$ mode. However, in practice the device is made more useful by making ports 3 and 4 well-matched, i.e.

$$S_{33}=S_{44}=0. \tag{6.72}$$

This can be done by having a proper choice of the reference planes on ports 3 and 4. Using Eq. (6.72) in Eq. (6.71), we obtain

$$[S]=\begin{bmatrix} S_{11} & S_{12} & S_{13} & S_{14} \\ S_{12} & S_{11} & S_{13} & -S_{14} \\ S_{13} & S_{13} & 0 & 0 \\ S_{14} & -S_{14} & 0 & 0 \end{bmatrix}. \tag{6.73}$$

Since the device is assumed to be loss-less,

$$[S]\,[S]^{\dagger}=[1].$$

whose

$$(1, 1)\ \text{term}=|S_{11}|^2+|S_{12}|^2+|S_{13}|^2+|S_{14}|^2=1 \tag{6.74a}$$

$$(3, 3)\ \text{term}=2\,|S_{13}|^2=1 \tag{6.74b}$$

$$(4, 4)\ \text{term}=2\,|S_{14}|^2=1. \tag{6.74c}$$

From Eqs. (6.74b) and (6.74c), we have respectively,

$$S_{13}=\frac{1}{\sqrt{2}}\,e^{j\phi}\ \text{and}\ S_{14}=\frac{1}{\sqrt{2}}\,e^{j\psi}. \tag{6.75}$$

Phases $\phi$ and $\psi$ in Eq. (6.75) are the choice of the reference planes and can be made each equal to zero, i.e. $\phi=\psi=0$. It then follows from Eq. (6.74a) and (6.75) that

$$S_{11}=S_{12}=0. \tag{6.76}$$

because each of $|S_{11}|^2$ and $(S_{12})^2$ is positive or zero. Therefore, using Eqs. (6.75) and (6.76), Eq. (6.73) can be reduced to

$$[S] = \frac{1}{\sqrt{2}} \begin{bmatrix} 0 & 0 & 1 & 1 \\ 0 & 0 & 1 & -1 \\ 1 & 1 & 0 & 0 \\ 1 & -1 & 0 & 0 \end{bmatrix}. \tag{6.77}$$

## 6.16 MAGIC TEE WITH VARIOUS TERMINAL CONDITION

**CASE I: The ports of the tee are well matched**

Consider a magic tee supplied with normalized waves $a_1$, $a_2$, $a_3$ and $a_4$ by matched synchronized generators at its ports 1, 2, 3 and 4 respectively. The scattered wave is given by

$$\begin{bmatrix} b_1 \\ b_2 \\ b_3 \\ b_4 \end{bmatrix} = \frac{1}{\sqrt{2}} \begin{bmatrix} 0 & 0 & 1 & 1 \\ 0 & 0 & 1 & -1 \\ 1 & 1 & 0 & 0 \\ 1 & -1 & 0 & 0 \end{bmatrix} \begin{bmatrix} a_1 \\ a_2 \\ a_3 \\ a_4 \end{bmatrix}$$

$$= \frac{1}{\sqrt{2}} \begin{bmatrix} a_3 + a_4 \\ a_3 - a_4 \\ a_1 + a_2 \\ a_1 - a_2 \end{bmatrix}. \tag{6.78}$$

The particular cases of interest are:

(*a*) $a_1 = a_2 = a_4 = 0$, i.e. power is incident on port 3 only. Then

$$b_1 = \frac{1}{\sqrt{2}} a_3,\ b_2 = \frac{1}{\sqrt{2}} a_3 \text{ and } b_3 = b_4 = 0.$$

The corresponding power output is given by

$$P_1 = \tfrac{1}{4}|a_3|^2,\ P_2 = \tfrac{1}{4}|a_3|^2 \text{ and } P_3 = P_4 = 0. \tag{6.79}$$

That is, the input at port 3 splits up equally and in phase at port 1 and port 2, with no power back to port 3 or out of port 4.

(*b*) $a_1 = a_2 = a_3 = 0$, i.e. power is incident on port 4 only. Then

$$b_1 = \frac{a_4}{\sqrt{2}},\ b_2 = -\frac{a_4}{\sqrt{2}} \text{ and } b_3 = b_4 = 0.$$

The corresponding power output is

$$P_1 = \tfrac{1}{4}\,|a_4|^2,\; P_2 = \tfrac{1}{4}\,|a_4|^2 \text{ and } P_3 = P_4 = 0. \tag{6.80}$$

That is, the input at port 4 splits up equally, but is 180° out of phase at ports 1 and 2, with no power back to port 4 or out of port 3.

(*c*) $a_3 = a_4$ and $a_1 = a_2 = 0$, i.e. equal and in phase waves are fed in ports 3 and 4 and ports 1 and 2 are matched terminated. Then $b_1 = \sqrt{2}\,a_3$, $b_2 = b_3 = b_4 = 0$ and corresponding power

$$P_1 = |a_3|^2,\; P_2 = P_3 = P_4 = 0. \tag{6.81}$$

That is, the two equal signals combine at port 1 and cancel out at port 2. No power is reflected back to ports 3 and 4.

(*d*) $a_3 = -a_4$ and $a_1 = a_2 = 0$, i.e. equal and 180° out of phase waves are incident at ports 3 and 4. Then

$$b_2 = \sqrt{2}\,a_3 \text{ and } b_1 = b_3 = b_4 = 0 \tag{6.82}$$

or

$$P_2 = |a_3|^2 \text{ and } P_1 = P_3 = P_4 = 0. \tag{6.82}$$

That is, the two signals combine at port 2 and cancel out at port 1. No power is reflected back to ports 3 and 4.

(*e*) $a_2 = a_3 = a_4 = 0$, i.e. power is fed to port 1 only. Then

$$b_1 = b_2 = 0,\; b_3 = \frac{a_1}{\sqrt{2}} \text{ and } b_4 = \frac{a_1}{\sqrt{2}}$$

or

$$P_1 = P_2 = 0,\; P_3 = \frac{1}{4}\,|a_1|^2 \text{ and } P_4 = \frac{1}{4}\,|a_1|^2. \tag{6.83}$$

That is, the input at port 1 is split equally and in phase at ports 3 and 4, with no power reflected back to port 1 or out of port 2.

(*f*) $a_1 = a_3 = a_4 = 0$, i.e. power is fed in port 2. Then,

$$b_1 = b_2 = 0,\; b_3 = \frac{a_2}{\sqrt{2}} \text{ and } b_4 = -\frac{a_2}{\sqrt{2}}$$

or

$$P_1 = P_2 = 0,\; P_3 = \frac{1}{4}\,|a_2|^2 \text{ and } P_4 = \frac{1}{4}\,|a_2|^2. \tag{6.84}$$

That is, the input at port 2 is split up equally but is 180° out of phase at ports 3 and 4, with no power reflected back to port 2 or out of port 1.

(*g*) $a_1=a_2$ and $a_3=a_4=0$, i.e. equal and in phase waves are incident at ports 1 and 2. Then

$$b_1=b_2=0,\ b_3=\sqrt{2}\ a_1 \text{ and } b_4=0$$

or

$$P_1=P_2=0,\ P_3=|a_1|^2 \text{ and } P_4=0. \tag{6.85}$$

That is, the equal and in phase inputs to ports 1 and 2 combine at port 3 but cancel out at port 4, with no power reflected back to port 1 or to port 2.

(*h*) $a_2=-a_1$ and $a_3=a_4=0$, i.e. equal and out of 180° phase waves are incident at ports 1 and 2. Then

$$b_1=b_2=0,\ b_3=0 \text{ and } b_4=\sqrt{2}\ a_1$$

or

$$P_1=P_2=0,\ P_3=0 \text{ and } P_4=|a_1|^2. \tag{6.86}$$

That is, the equal but out of 180° phase inputs at ports 1 and 2 combine at port 4, cancel out at port 3, with no power reflected back to port 1 or to port 2.

**Case II: The ports of the tee are not necessarily matched**

The practical cases are:

(*a*) $a_3$ is fed to port 3 and ports 1, 2 and 4 are terminated in the reflection coefficients $\Gamma_1$, $\Gamma_2$ and $\Gamma_4$ respectively. Then

$$a_1=\Gamma_1 b_1,\ a_2=\Gamma_2 b_2 \text{ and } a_4=\Gamma_4 b_4. \tag{6.87}$$

Substituting Eq. (6.87) in Eq. (6.78) and expanding the resulting equation we have four simultaneous equations in four unknowns $b_1$, $b_2$, $b_3$ and $b_4$. Solving these four simultaneous equations by Cramer's rule we, have

$$b_1=\sqrt{2}\ \frac{1-\Gamma_2\Gamma_4}{2-\Gamma_4\,(\Gamma_1+\Gamma_2)}\ a_3$$

$$b_2=\sqrt{2}\ \frac{1-\Gamma_1\Gamma_4}{2-\Gamma_4\,(\Gamma_1+\Gamma_2)}\ a_3$$

$$b_3=\frac{\Gamma_1+\Gamma_2-2\,\Gamma_1\Gamma_2\Gamma_4}{2-\Gamma_4\,(\Gamma_1+\Gamma_2)}\ a_3 \tag{6.88}$$

$$b_4=\frac{\Gamma_1-\Gamma_2}{2-\Gamma_4\,(\Gamma_1+\Gamma_2)}\ a_3.$$

The power absorbed at each termination is the difference between the power incident on the termination and the power reflected from the termination,

$$P_1 = \frac{1}{2}\,(|b_1|^2 - |a_1|^2)$$

$$= \frac{1}{2}\,(1-|\Gamma_1|^2)\left|\sqrt{2}\,\frac{1-\Gamma_2\Gamma_3}{2-\Gamma_4\,(\Gamma_1-\Gamma_2)}\,a_3\right|^2 \tag{6.89a}$$

$$P_2 = \frac{1}{2}\,(|b_2|^2 - |a_2|^2)$$

$$= \frac{1}{2}\,(1-|\Gamma_2|^2)\left|\sqrt{2}\,\frac{1-\Gamma_1\Gamma_4}{2-\Gamma_4\,(\Gamma_1+\Gamma_2)}\,a_3\right|^2 \tag{6.89b}$$

$$P_4 = \frac{1}{2}\,(|b_4|^2 - |a_4|^2)$$

$$= \frac{1}{2}\,(1-|\Gamma_4|^2)\left|\frac{\Gamma_1-\Gamma_2}{2-\Gamma_4\,(\Gamma_1+\Gamma_2)}\,a_3\right|^2 \tag{6.89c}$$

For port 3, the total power absorbed by the system is, however,

$$P_3 = \frac{1}{2}\,(|a_3|^2 - |b_3|^2)$$

$$= \frac{1}{2}\left\{|a_3|^2 - \left|\frac{\Gamma_1+\Gamma_2-2\,\Gamma_1\Gamma_2\Gamma_4}{2-\Gamma_4(\Gamma_1+\Gamma_2)}\,a_3\right|^2\right\}. \tag{6.89d}$$

The following are the important conclusions that can be drawn from Eqs. (6.89a to d).

(*a*) If $\Gamma_1 = \Gamma_2$, i.e. $P_4 = 0$ from Eq. (6.89d), it follows that no power comes out of port 4 regardless of $\Gamma_4$ while the reflection coefficient at port 3 (input) as can be seen from Eq. (6.88), is $\Gamma_3 = \Gamma_1 = \Gamma_2$.

(*b*) If $\Gamma_1 = \Gamma_2 = 0$, i.e. ports 1 and 2 are matched, it follows from Eq. (6.88) that $\Gamma_3 = b_3/a_3 = 0$, i.e. port 3 is matched regardless of $\Gamma_4$.

(*c*) If $\Gamma_4 = 0$ and $\Gamma_1 = -\Gamma_2$, i.e. port 4 is terminated in a matched load and ports 1 and 2 are terminated in arbitrary loads such that $\Gamma_1 = -\Gamma_2$, then it follows from Eq. (6.88) that

$$\Gamma_3 = b_3/a_3 = 0.$$

(*d*) $\Gamma_4 = 0$, i.e. port 4 is matched terminated, then from Eq. (6.88) $b_1 = b_2$ regardless of whether $\Gamma_2 = \Gamma_1$ or not.

Now we consider the case when $a_4$ is fed to port 4 and ports 1, 2 and : are terminated with reflection coefficients $\Gamma_1$, $\Gamma_2$ and $\Gamma_3$ respectively. Then

$$a_1 = \Gamma_1 b_1,\; a_2 = \Gamma_2 b_2 \text{ and } a_3 = \Gamma_3 b_3. \tag{6.90}$$

Substituting Eq. (6.90) in Eq. (6.78), expanding the equation so obtained in four simultaneous equations and then solving these equations for $b_1$, $b_2$, $b_3$ and $b_4$, we have

$$b_1=\sqrt{2}\ \frac{1-\Gamma_2\Gamma_3\cdot}{2-\Gamma_3\ (\Gamma_1+\Gamma_2)}\ a_4$$

$$b_2=-\sqrt{2}\ \frac{1-\Gamma_1\Gamma_3}{2-\Gamma_3\ (\Gamma_1+\Gamma_2)}\ a_4$$

$$b_3=\frac{\Gamma_1-\Gamma_2}{2-\Gamma_3\ (\Gamma_1+\Gamma_2)}\ a_4 \tag{6.91}$$

$$b_4=\frac{\Gamma_1+\Gamma_2-2\ \Gamma_1\Gamma_2\Gamma_3}{2-\Gamma_3\ (\Gamma_1+\Gamma_2)}\ a_4.$$

Hence the power absorbed at each termination is given by

$$P_1=\frac{1}{2}\ (1-|\Gamma_1|^2)\ |b_1|^2$$

$$P_2=\frac{1}{2}\ (1-|\Gamma_2|^2)\ |b_2|^2 \tag{6.91a}$$

$$P_3=\frac{1}{2}\ (1-|\Gamma_3|^2)\ |b_3|^2.$$

The total power absorbed in the system is

$$P_4=\frac{1}{2}\ (|a_4|^2-|b_4|^2)=P_1+P_2+P_3. \tag{6.91b}$$

The following conclusions may be drawn from Eq. (6.91) (6.91 a), and (6.91b)

(*a*) If $\Gamma_1=\Gamma_2$ then $b_3=0$, i.e. no power comes out of port 3 regardless of $\Gamma_3$, while the reflection coefficient at the input (port 4) is effectively

$$\Gamma_4=\frac{b_4}{a_4}=\Gamma_1=\Gamma_2 \text{ if } \Gamma_3=0.$$

(*b*) If $\Gamma_1=\Gamma_2=0$ then $b_4=0$. Hence, if ports 1 and 2 are matched, port 4 is matched regardless of $\Gamma_3$ and $\Gamma_4$.

(*c*) If $\Gamma_3=0$, then $b_1=-b_2$ regardless of whether or not $\Gamma_1=\Gamma_2$.

(*d*) If $\Gamma_3=0$ then port 4 will be matched, i.e. $b_4=0$ if $\Gamma_1=-\Gamma_2$.

The concepts developed in this section are directly applicable to mixers, modulators and other hybrids; however, some modifications may be needed.

## 6.17 EQIUVALENT CIRCUIT OF A MAGIC TEE

Using Eqs (6.8) and (6.77), an impedance matrix for a magic tee can be obtained.

$$[Z]=\frac{j}{\sqrt{2}}\begin{bmatrix} 0 & 0 & 1 & 1 \\ 0 & 0 & -1 & 1 \\ 1 & -1 & 0 & 0 \\ 1 & 1 & 0 & 0 \end{bmatrix}. \tag{6.92}$$

In Eq. (6.92) $j$ arrives from the proper choice of reference planes. Zeros on the principal diagonal of [Z] indicate that the waveguide voltage at any point is independent of the current flowing at that port. The voltage in port 1, for example, depends only on the current flowing in ports 3 and 4, the $H$- and $E$-plane ports respectively. If ports 3 and 4 are open-circuited so that no current flows in them, the voltage at port 1 is zero, irrespective of the current flowing at port 1. The equivalent circuit, shown in Fig. 6.16(a) satisfies the above properties and hence is the equivalent circuit for a matched magic tee.

The equivalent circuit for a non-matched magic tee is no doubt complicated but may be derived from experimental measurements*. A typical circuit is shown in Fig. 6.16(b).

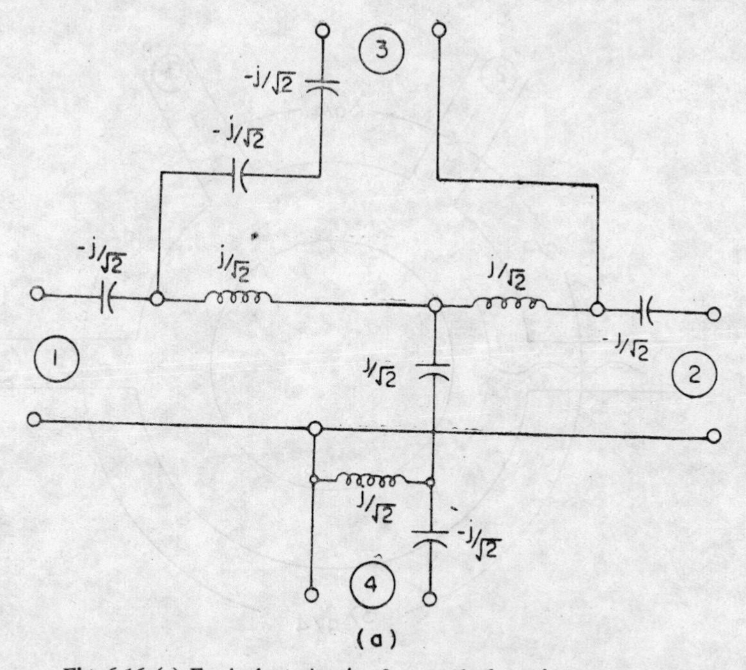


Fig. 6.16 (*a*) Equivalent circuit of a matched magic tee.

*G.C. Montgomery, *op. cit*.

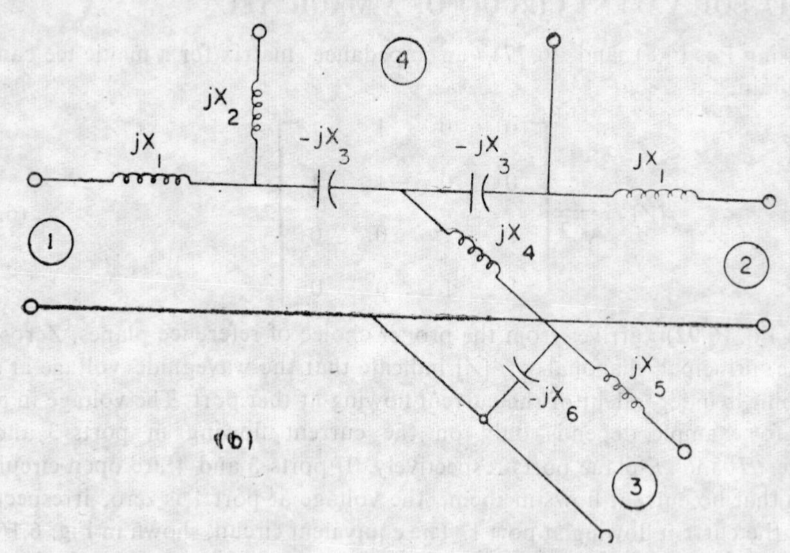


Fig. 6.16 (*b*) Equivalent circuit of an unmatched magic tee.

## 6.18 RAT-RACE

As shown in Fig. 6.17, rat-race is another form of hybrid tee junction. The operation of the tee is very simple. The wave incident at port 1 splits into

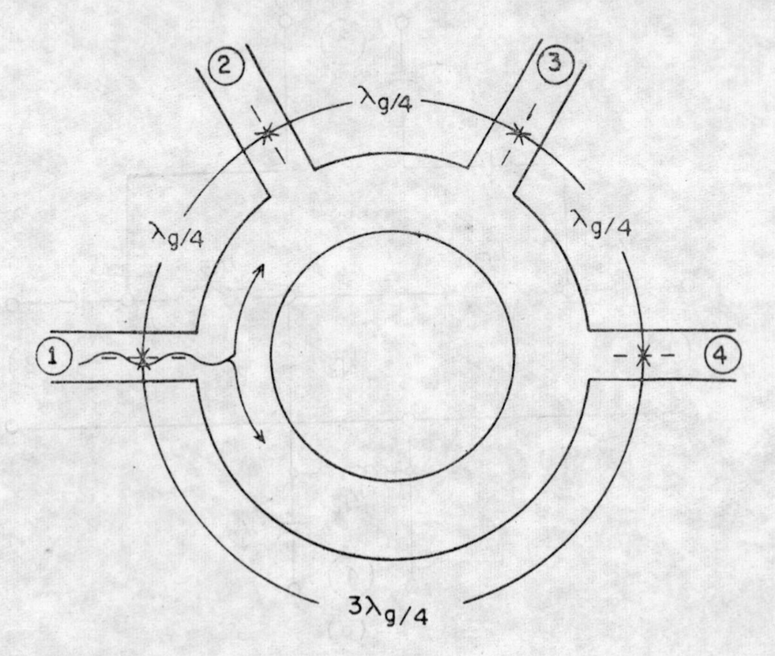


Fig. 6.17 Rat-race magic tee.

two waves travelling around the ring circuit in opposite directions. The waves will be in phase at ports 2 and 4 and out of phase (180°) at port 3. Hence, ports 2 and 4 are coupled to port 1 but port 3 is left uncoupled. Similarly, it can be inferred from Fig. 6.17 that if power is fed in port 2, ports 2 and 4 are uncoupled because the paths travelled by the two waves arriving at port 2, differ by $\lambda_g/2$. Thus, rat-race works as a magic tee.

## 6.19 APPLICATIONS* OF THE MAGIC TEE

### Impedance Bridges

A magic tee is frequently employed in microwave impedance measuring bridges. These bridges are similar to low frequency Wheatstone bridges. Figure 6.18 depicts a typical microwave impedance bridge designed by uti-

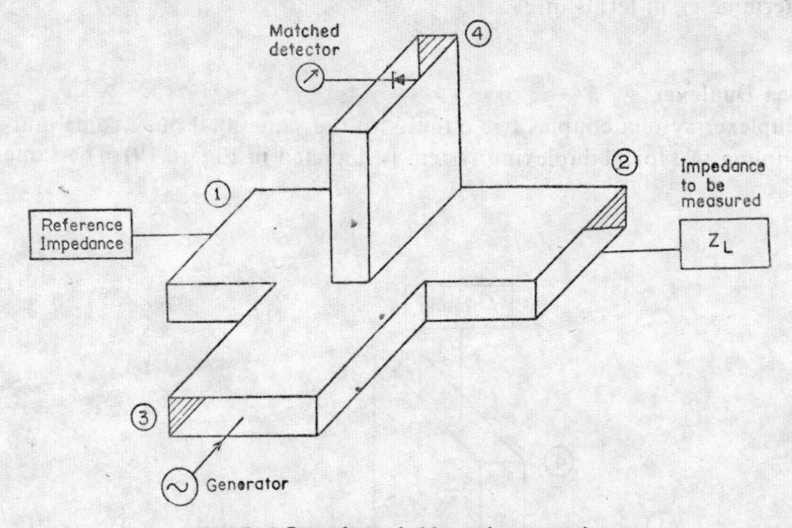


Fig. 6.18 Impedance bridge using a magic tee.

lizing a magic tee. Power from a matched source is fed in the *H*-arm (3) of the magic tee. A standard variable impedance is connected to port 1 as reference impedance and arm 2 is terminated in the impedance to be measured. A matched detector is connected to port 4 (*E*-arm) to receive power reflected from arms 1 and 2. These powers will be out of 180° phase.

The reference impedance is adjusted so as to have no signal in the detector. Under this condition the power reflected from the reference impedance and reaching the detector (half of the reflected power) equals the power reflected from the unknown impedance reaching the detector. Because the

*Detailed discussions may be found in the following literature:

1. J.L. Altman, *op. cit.*
2. G.C. Montgomery, *op. cit.*
3. H.J. Reich and others, *Microwave Principles,* New Delhi: Affiliated East-West Press, (1972).
4. J, Helszain, *Passive and Active Microwave Circuits,* New York: John Wiley & Sons (1978).

two powers are out of 180° phase and the lengths of the two ports are equal (or in other words, when the bridge is balanced) then,

$$\Gamma_R = \Gamma_L$$

or

$$\frac{z_R - 1}{z_R + 1} = \frac{z_L - 1}{z_L + 1} \tag{6.93}$$

$\Gamma_R$ and $z_R$ are respectively the reflection coefficient and normalized impedance for the reference impedance whereas $\Gamma_L$ and $z_L$ are the respective quantities for the given unknown impedance. Equation (6.93) may be used to determine $z_L$ in terms of $z_R$.

**Antenna Duplexer**

A duplexer system couples two circuits to the same load but avoids mutual coupling. A typical duplexing system is depicted in Fig. (6.19). The same

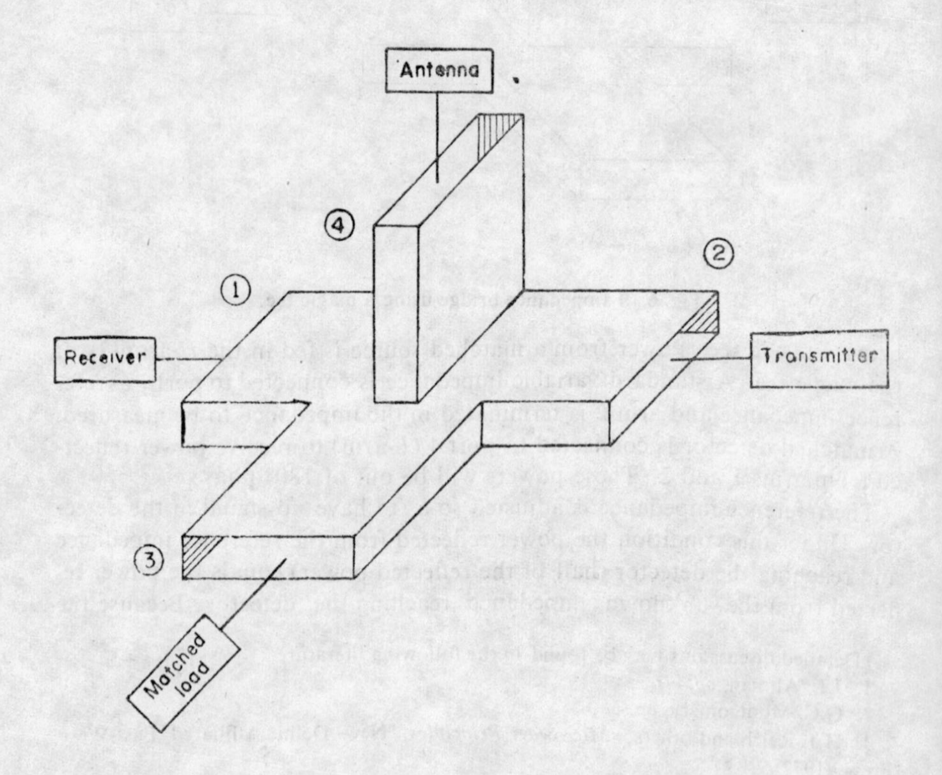


**Fig. 6.19** Magic tee – antenna duplexer system.

antenna is used for the transmitter as well as receiver but the transmitter power is kept out of the receiver and vice-versa. In this system the matched generator and matched detector are connected to arms 1 and 2 rsspectively. *H*-arm (3) is terminated in a matched load and *E*-arm (4) is connected to the matched antenna. In this case, power received by the antenna is coupled to the detector due to the coupling properties of the magic tee.

In this tee, since half of the power is transmitted, it is useful at low power levels.

**Balanced Mixer**

A balanced microwave mixer is used in superhetrodyne receivers to balance out the local oscillator noise at the input of the intermediate frequency (i.f.) amplifier. Figure (6.20) depicts a typical arrangement for i.f. frequencies 1

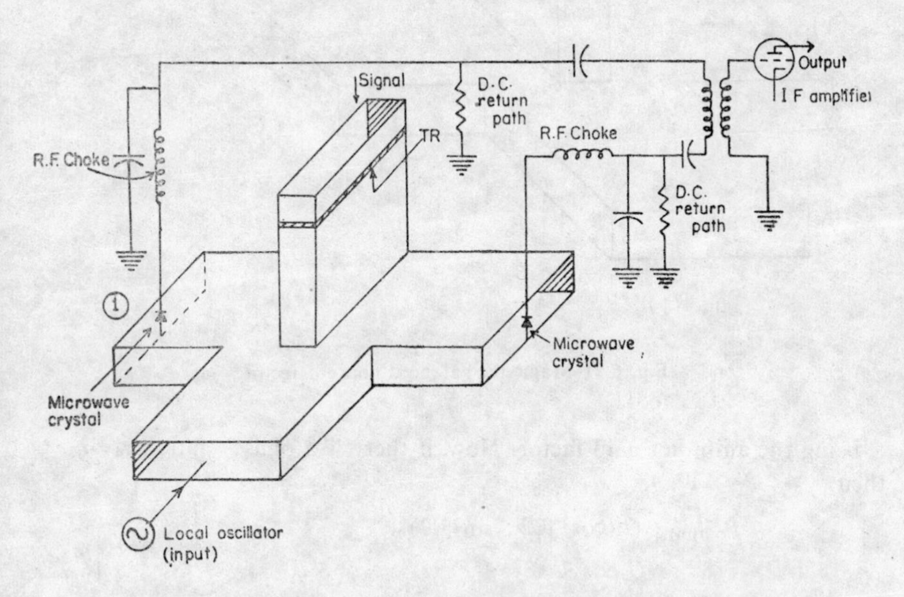


**Fig. 6.20** Magic tee balanced microwave mixer (I.F. 1—100 MHz).

to 100 MHz. The superhetrodyne mixer receives signals from both the antenna and the local oscillator of the receiver but the local oscillator power is prevented from reaching the antenna by the isolation properties of *E*- and *H*-arms.

**Balanced Phase Detector**

Figure 6.21 shows a balanced phase detection system utilizing a magic tee. The two signals whose relative phase has to be detected, are fed from the *E*- and *H*-arms. The powers detected by matched microwave crystals at ports 1 and 2 are fed to the differential amplifier. If the power fed at ports

1 and 2 is equal, i.e. $a_3=a_4$, then the output power recorded by the differential amplifier will be

$$P_{\text{output}}=G\left[\left\{\frac{1}{\sqrt{2}}(a_3+a_4)\right\}^2-\left\{\frac{1}{\sqrt{2}}(a_3-a_4)\right\}^2\right] \tag{6.94}$$

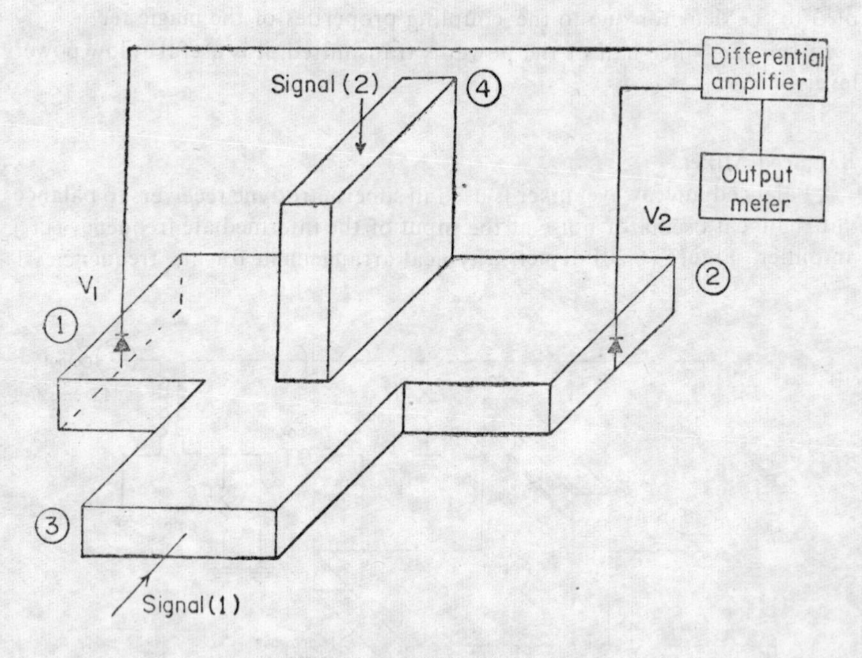


**Fig. 6.21** Magic tee balanced phase-detector.

$G$ being the amplifier gain factor. Now, if there is a phase relation $a_4=a_3e^{j\theta}$, then

$$P_{\text{output}}=G\,(\cos^2\theta/2-\sin^2\theta/2)$$
$$=G\cos\theta. \tag{6.95}$$

Thus, phase can be detected and a meter can be calibrated in $\cos\theta$ to indicate the phase difference directly on the scale.

**Frequency Discriminator**

The device that produces a signal proportional to the departure of the frequency of its microwave input from the reference value is called a frequency discriminator. A typical frequency discriminator arrangement developed by Pound* is depicted in Fig. 6.22. The microwave cavity (may be of a source like klystron) is connected to port 1 while port 2 is shorted by a mo9ing plunger at a distance $\lambda_g/8$ that provides a reflection coefficient $\Gamma_2$. A pad is

*R.V. Pound, Rad. Lab. Series Vol. 11, pp. 58-78, McGraw-Hill Book Co. (1947).

placed in the $E$-arm (4). Powers detected by microwave crystal detectors placed at each $E$- and $H$-arms are fed to the differential amplifier and to the

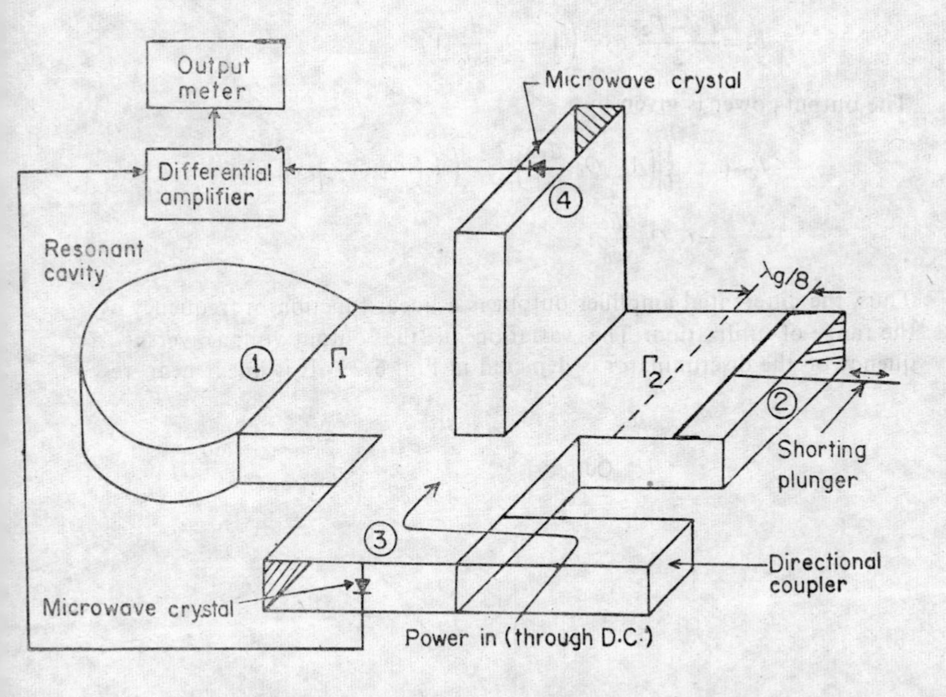


**Fig. 6.22** Magic tee frequency discriminator (due to Pound).

output meter in turn. The directional coupler connected to the $H$-arm supplies input power. At port 1 the reflection coefficient produced by the cavity at its plane of (detuned frequency different from its resonant frequency) short is given by

$$\Gamma_1 = Q_L \left(\frac{1}{Q_E} - \frac{1}{Q_u}\right) - jQ_L \frac{2d\omega}{\omega_0}$$

$$\simeq -jQ_u \frac{d\omega}{\omega_0} \tag{6.96}$$

because $Q_u = Q_E =$ external $Q$;

where $Q_L =$ loaded $Q$;

$Q_u =$ unloaded $Q$;

$d\omega =$ change in frequency; and

$\omega_0 =$ resonant frequency.

At port 2

$$\Gamma_2 = -e^{-j\pi/2} = j. \tag{6.97}$$

Therefore from Eqs. (6.89) ($\Gamma_3=\Gamma_4=0$)

$$b_3=\frac{\Gamma_1+\Gamma_2}{2}\ a_3=\left(1-Q_u\,\frac{d\omega}{\omega_0}\right)j$$

$$b_4=\frac{\Gamma_1-\Gamma_2}{2}\ a_3=\left(1+Q_u\,\frac{d\omega}{\omega_0}\right)j.$$

The output power is given by*

$$P_{\text{out}} \propto \left\{\left|\left(1-Q_u\,\frac{d\omega}{\omega_0}\right)\right|^2-\left|\left(1+Q_u\,\frac{d\omega}{\omega_0}\right)\right|^2\right\}$$

$$=G\,Q_u\,\frac{d\omega}{\omega_0}. \tag{6.98}$$

Thus, the differential amplifier output is a linear function of frequency over the range of utilization. The variation of the output voltage versus frequency of the discriminator is depicted in Fig. 6.23. It is linear near reso-

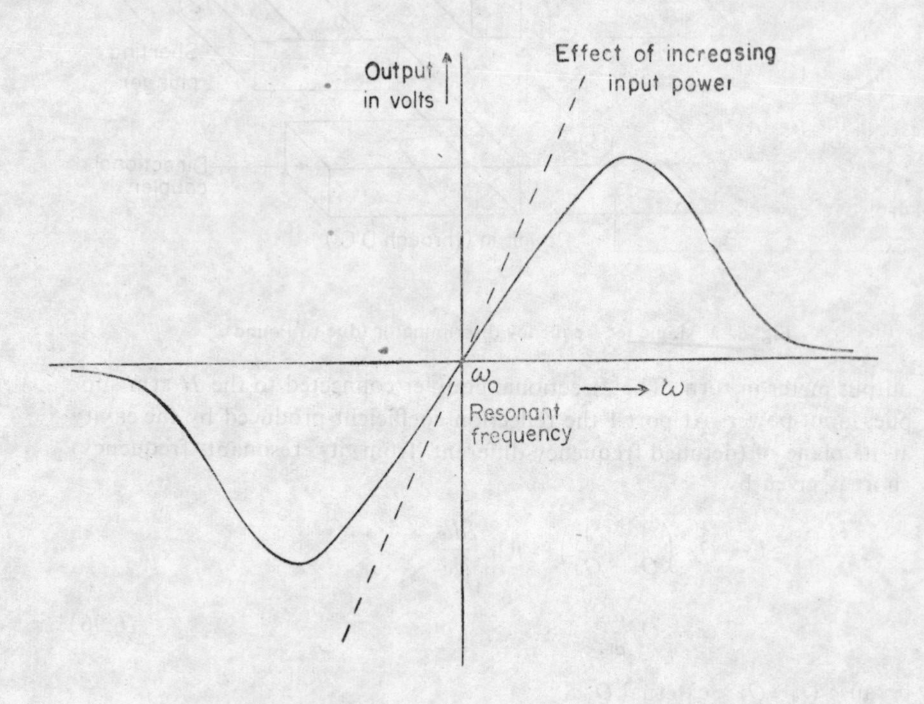


**Fig. 6.23** Pound frequency discriminator output vs frequency curve.

nance being negative for frequencies below resonance and positive for frequencies above resonance. If the signal is frequency modulated the output of the differential amplifier will be proportional to the instantaneous frequency deviation, and the frequency of this output will be equal to the rate

*Square law operation of the crystal detector is assumed here.

of frequency modulation. However, if the signal is amplitude modulated and if the rate of modulation is small, the output may be obtained from Fig. 6.23 of course the slope of the curve is a function of amplitude.

**Modulators**

A magic tee may be utilized to design both single side baed and double side band suppressed carrier modulators. Figure 6.24 depicts an arrangement for double side band suppressed carrier modulation. The microwave signal enters at port 3; ports 1 and 2 are terminated in time varying reflection coefficients $\Gamma_1$ and $\Gamma_2$. The output from port 4 emerges as a modulated one.

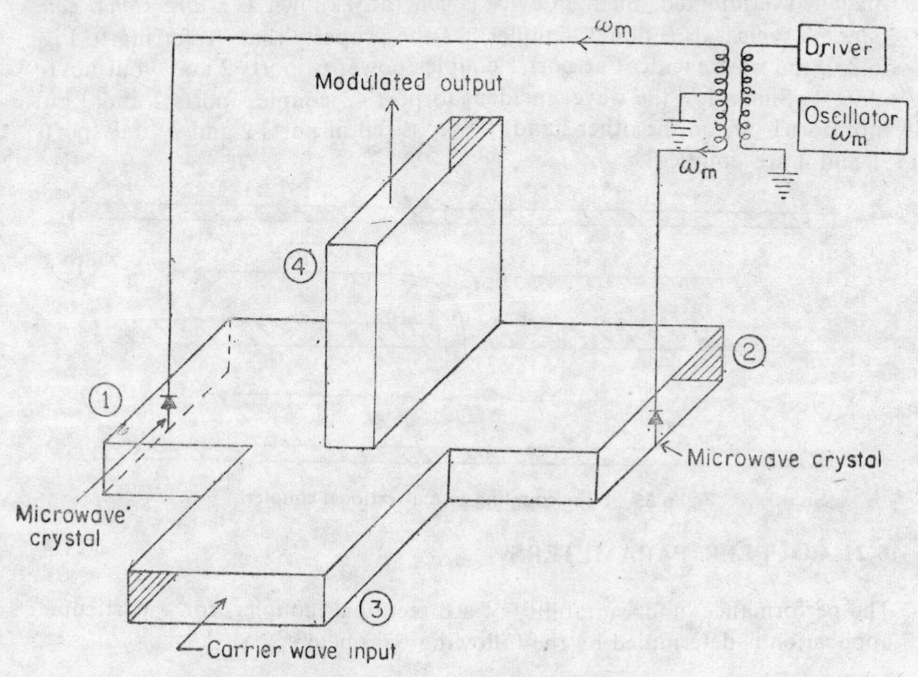


**Fig. 6.24** Magic tee double side band suppressed carrier wave modulator.

Single side band modulation can be obtained by filtering the undesirable side band. This may not be practicable when the side bands are close together. For such cases the arrangement similar to Fig. 6.24 is used. In which two crystal detectors are now driven in phase quadrature and their location is no longer symmetrical with respect to the plane of symmetry. Crystal detector (2) is closer to the wall as compared to crystal detector (1) by a length $\lambda_g/8$. As a result, only the lower side band comes out of port 4. The bands may be interchanged by interchanging the locations of the crystal detectors.

## 6.20 DIRECTIONAL COUPLERS

In many microwave circuits it is often desired that:

(*i*) The power injected at a given point travels in only one direction.

(*ii*) The power at a given point may be divided to travel in the same or in opposite direction but the energy abstracted from each bifurcation appears as a separate output and the abstraction and division of power do not, in any way, affect the power in the original line.

(*iii*) Wherever desired the energy of the two lines are added together.

(*iv*) The signals from the two channels are compared in magnitude as well as in phase.

The device that satisfies the above requirements must be a reciprocal, conservative, four-port junction that is reflectionless with all of its ports matched terminated. Such a device is generally known as a *directional coupler*. A typical directional coupler has the property that (referring to Fig. 6.25), the wave incident at port 1 couples power to ports 2 and 3 but not to port 4. Similarly, the wave incident to port 4, couples ports 2 and 3 but not port 1. If, on the other hand, power is fed in ports 2 and 3, only ports 1 and 4 are coupled.

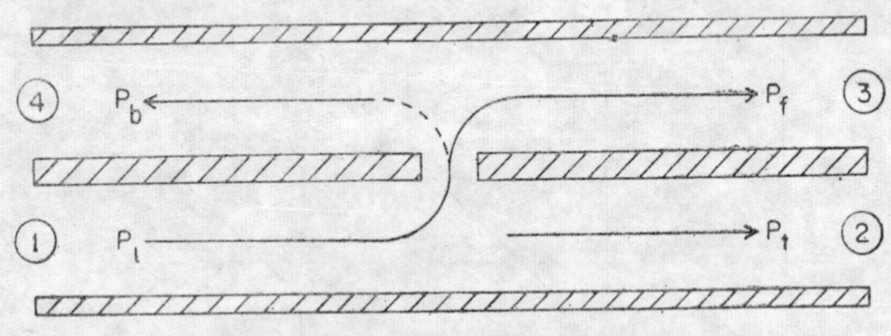


Fig. 6.25 Power coupling in a directional coupler.

## 6.21 COUPLER PARAMETERS

The performance and suitability of a directional coupler for a particular application is determined by the following parameters.

### (*i*) Coupling

The ratio expressed in decibels of the power incident and the power coupled in the auxiliary arm in the forward direction, is known as the coupling coefficient $C$ or simply coupling. It is, however, assumed that the coupler does not couple any power in the backward direction and all the ports except the coupled ports in question are matched terminated. Mathematically,

$$C = 10 \log_{10} (P_i/P_f) \text{ db} \tag{6.99}$$

Parameter $C$ defines the degree of coupling; the smaller is $C$, the greater is the power coupled. It also indicates the amount of attenuation within the coupler.

### (*ii*) Directivity

The power coupled by a directional coupler is measured in terms of directivity. The directivity $D$ of a coupler is defined as the ratio, expressed in decibels, of the power coupled in the forward direction to the power coupled in the backward direction of the auxiliary arm with unused terminals which are matched terminated. Mathematically.

$$D = 10 \log_{10} \frac{P_f}{P_b} \ db \tag{6.100}$$

Parameter $D$ is a measure of the discrimination of a directional coupler beween forward and backward waves in the main waveguide.

*(iii)* **Insertion Loss**

Insertion loss, $L$, measures the power loss arising in a line when a directional coupler is inserted in it. It is defined as the ratio, expressed in decibels, of the power incident to the power transmitted in the main line of the coupler when the auxiliary arms are matched terminated. Mathematically,

$$L = 10 \log_{10} P_i/P_t \ db \tag{6.101}$$

*(iv)* **Isolation**

The capability of a directional coupler to isolate coupled ports is expressed in terms of the parameter, isolation ($I$). It is defined as the ratio, expressed in decibels, of the power incident in the main arm to the backward power coupled in the auxiliary arm, with all other ports matched terminated

$$I = 10 \log_{10} \frac{P_i}{P_b} \ db. \tag{6.102}$$

For an ideal coupler $D$ and $I$ are infinite while $C$ and $L$ are zero. If the coupler is not absolutely symmetric there may be as many parameters as the ports such as $C_1$, $C_2$, $C_3$, $C_4$, $D_1$, $D_2$, $D_3$, $D_4$, etc.

## 6.22 SCATTERING MATRIX OF A DIRECTIONAL COUPLER

Many important properties of a directional coupler may be derived if its scattering matrix is known.

Since a directional coupler is a four-port matched device, its scattering matrix will be a four by four square matrix with diagonal elements zero.

$$[S] = \begin{bmatrix} 0 & S_{12} & S_{13} & S_{14} \\ S_{21} & 0 & S_{23} & S_{24} \\ S_{31} & S_{32} & 0 & S_{34} \\ S_{41} & S_{42} & S_{43} & 0 \end{bmatrix}. \tag{6.103}$$

Moreover, the device is symmetric, hence

$$S_{ij} = S_{ji}; \ i \text{ and } j = 1, 2, 3, 4. \tag{6.104}$$

Using Eq. (6.104) in Eq. (6.103)

$$[S]=\begin{bmatrix} 0 & S_{12} & S_{13} & S_{14} \\ S_{12} & 0 & S_{23} & S_{24} \\ S_{13} & S_{23} & 0 & S_{34} \\ S_{14} & S_{24} & S_{34} & 0 \end{bmatrix}. \tag{6.105}$$

Referring to Fig. 6.25, from the property of the directional coupler,

$$S_{13}=S_{31}=0 \text{ and } S_{24}=S_{42}=0. \tag{6.106}$$

Hence,

$$[S]=\begin{bmatrix} 0 & S_{12} & 0 & S_{14} \\ S_{12} & 0 & S_{23} & 0 \\ 0 & S_{23} & 0 & S_{34} \\ S_{14} & 0 & S_{34} & 0 \end{bmatrix}. \tag{6.107}$$

The device is loss-less, hence from the unitary property $[S]\,[S]^{\dagger}=[1]$, whose

$$(1, 1) \text{ term yields, } |S_{12}|^2+|S_{14}|^2=1 \tag{6.108a}$$

$$(1, 3) \text{ term yields, } S_{12}S_{23}^{*}+S_{14}S_{34}^{*}=0 \tag{6.108b}$$

$$(2, 4) \text{ term yields, } S_{12}S_{14}^{*}+S_{23}S_{34}^{*}=0 \tag{6.108c}$$

From Eqs. (6.108b and c)

$$|S_{12}|\,|S_{23}|=|S_{14}|\,|S_{34}|$$

$$|S_{12}|\,|S_{14}|=|S_{23}|\,|S_{34}|$$

from which we obtain

$$|S_{23}|=|S_{14}|=\beta \text{ (say)} \tag{6.109a}$$

and

$$|S_{12}|=|S_{34}|=\alpha \quad \text{(say)}. \tag{6.109b}$$

Equations (6.109a and b) state that the coupling from arm 2 to 3 is equal to the coupling from arm 1 to 4, and also coupling from arm 1 to 2 is equal to coupling from arm 3 to 4.

Even at this stage the scattering coefficients have a great deal of arbitrariness as regards their phases. However, this can be removed by the proper choice of reference planes on the ports. Let us select these reference planes such that (*a*) $S_{12}$ is positive real, i.e. $\alpha$ is positive real; and ($S_{14}$) is positive imaginary, i.e. $\beta$ is positive imaginary. Then Eq. (6.107) can be written as

$$[S]=\begin{bmatrix} 0 & \alpha & 0 & \beta j \\ \alpha & 0 & j\beta & 0 \\ 0 & j\beta & 0 & \alpha \\ j\beta & 0 & \alpha & 0 \end{bmatrix} \tag{6.110}$$

where $\alpha$ and $\beta$ are related by Eq. (6.108a), that is

$$\alpha^2+\beta^2=1 \tag{6.111}$$

Coupling of such an ideal coupler is given by

$$C=-20 \log \beta \ db. \tag{6.112}$$

If the coupler is not perfect the directivity is given by

$$D=20 \log |\beta/S_{13}| \ db. \tag{6.113}$$

Note that $S_{13}=0$ for an ideal coupler.

Before we switch on to various types of practical couplers we prove an important theorem regarding an ideal directional coupler.

*Theorem*: *The necessary and sufficient condition for a device to be a directional coupler is that all of its ports are perfectly matched.*

*The Condition is Necessary*. Suppose port 2 (Fig. 6.25) is not matched then a part of the power transmitted from port 1 is reflected and another part of this power is coupled to port 3. This is in contradiction to the definition of a directional coupler. Hence port 2 must be matched and the same is true for all other ports.

*The Condition is Sufficient*. The scattering matrix for a perfectly matched four-port junction is

$$[S]=\begin{bmatrix} 0 & S_{12} & S_{13} & S_{14} \\ S_{21} & 0 & S_{23} & S_{24} \\ S_{31} & S_{32} & 0 & S_{34} \\ S_{41} & S_{42} & S_{43} & 0 \end{bmatrix} \tag{6.114}$$

Chosing the reference planes in various ports in such a way that $S_{21}$ and $S_{43}$ are positive real while $S_{41}$ is positive imaginary, and since the juuction is loss-less, $[S]\,[S]^{\dagger}=[I]$ or

$$S_{21}\,S_{23}^* - S_{41}\,S_{43}^* = 0 \tag{6.115a}$$

$$-S_{21}\,S_{41}^* + S_{23}\,S_{43}^* = 0. \tag{6.115b}$$

Multiplying Eq. (6.115 a) by $S_{21}$ and Eq. (6.115 b) by $S_{43}$ and substracting we have

$$S_{23}\,(S_{21}^2 - S_{43}^{*2}) = S_{23}\,(S_{21}^2 - S_{43}^2) = 0.$$

That is, either $S_{23}=0$ or $S_{21}=S_{43}$.
If $S_{23}=0$, we obtain from Eqs. (6.115a and b)

$$S_{23}=S_{41}=S_{14}=0. \tag{6.116}$$

Thus, there is no coupling between ports 2 and 3, and 1 and 4. Hence the device is a directional coupler. Moreover, if $S_{23}\neq 0$, $S_{21}=S_{43}=\alpha$ (real-positive) and $S_{23}=S_{41}=j\beta$ ;($\beta$=real-positive) then substituting these results in Eq. (6.114), gives Eq. (6.110) and hence, the proposed condition.

## 6.23 DIRECTIONAL COUPLERS IN USE

In general, a directional coupler is designed by placing two uniform waveguides side by side (broad or narrow dimension common) with an appropriately located coupling hole or aperture (*s*). The coupling of two guides may be effected in a number of ways and hence there are many types of couplers which are in practice.

### (*i*) Bathe-hole Coupler

In this coupler, the coupling is achieved by means of a circular hole drilled in the centre of the common broader wall of the waveguides inclined at an angle $\theta$ as shown in Fig. 6.26(a). However, the coupling may also be achieved by placing the guides parallel and off-setting the aperture as shown in Fig. 6.26(b).

The $TE_{10}$ mode of amplitude $A$, incident at port 1, produces a normal electric dipole in the aperture plus a tangential magnetic dipole proportional and in the same direction as the magnetic field of the incident wave. In the auxiliary waveguide (upper one) the normal electric dipole and the axial component of the magnetic dipole radiate symmetrically in both directions. The transverse component of the magnetic dipole radiates antisymmetrically, so that by proper adjustment of the angle $\theta$ or the aperture position $d$ [Figs. 6.26(a) and (b)], the radiation in port 4 can be *cancelled* and that in port 3 *enhanced.*

The field radiated in both the directions by an electric dipole is found to have an amplitude

$$B_1 = -\frac{j\omega\epsilon_0}{abY_\omega}\,\frac{2}{3}\,Y_0^3\,A\sin^2\frac{\pi d}{a}$$

where $\gamma_o$ is the radius of the aperture; $a$ and $b$ are waveguide dimensions; $Y_\omega$ is the wave admittance of the guide; and $d$ is the distance of the aperture from the side. The field radiated by the magnetic dipole in the direction of port 4 is

$$B_2 = \frac{j\omega\mu_o Y_\omega}{ab}\,\frac{4}{3}\,\gamma_0^3\,A\left(\sin^2\frac{\pi d}{a} + \frac{\pi^2}{\beta^2 a^2}\cos^2\frac{\pi d}{a}\right)$$

and in the direction of port 3

$$B_3 = \frac{j\omega\mu_o}{ab} Y_\omega \frac{4}{3} \gamma_0^3 A \left( -\sin^2 \frac{\pi d}{a} + \frac{\pi^2}{\beta^2 a^2} \cos^2 \frac{\pi d}{a} \right)$$

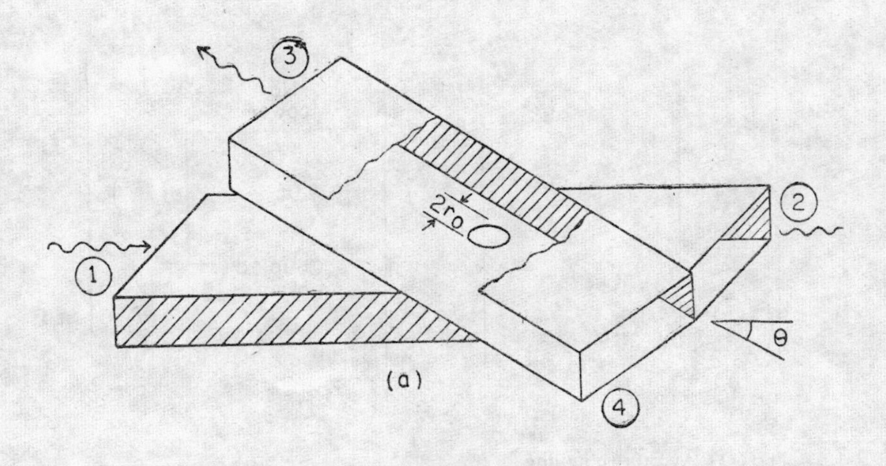


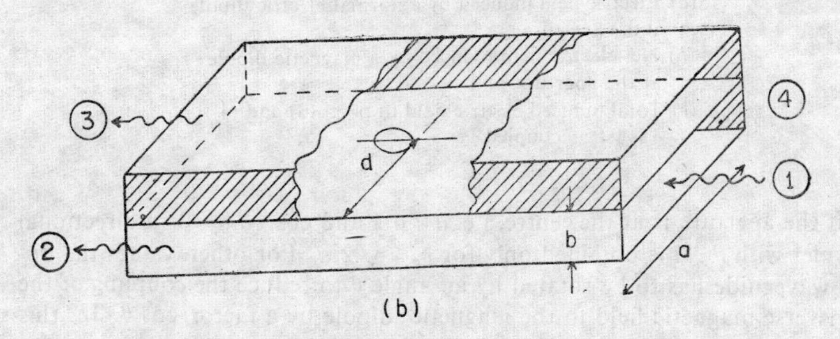


**Fig. 6.26** Bathe hole directional couplers (*a*) Two guides inclind at an angle *θ*. (*b*) Aperture off-setted.

For the device to be a directional coupler, i.e. port 3 is uncoupled, we must have $B_1+B_3=0$ (as illustrated in Fig. 6.27) which yields

$$\sin \frac{\pi d}{a} = \frac{\lambda_o}{\sqrt{6}\, a} \,. \tag{6.117}$$

However, if port 4 is uncoupled, we will have

$$B_1 + B_2 = 0$$

or

$$\sin \frac{\pi d}{a} = \frac{\lambda_o}{\sqrt{2\,\lambda_o^2 - a^2}} \tag{6.118}$$

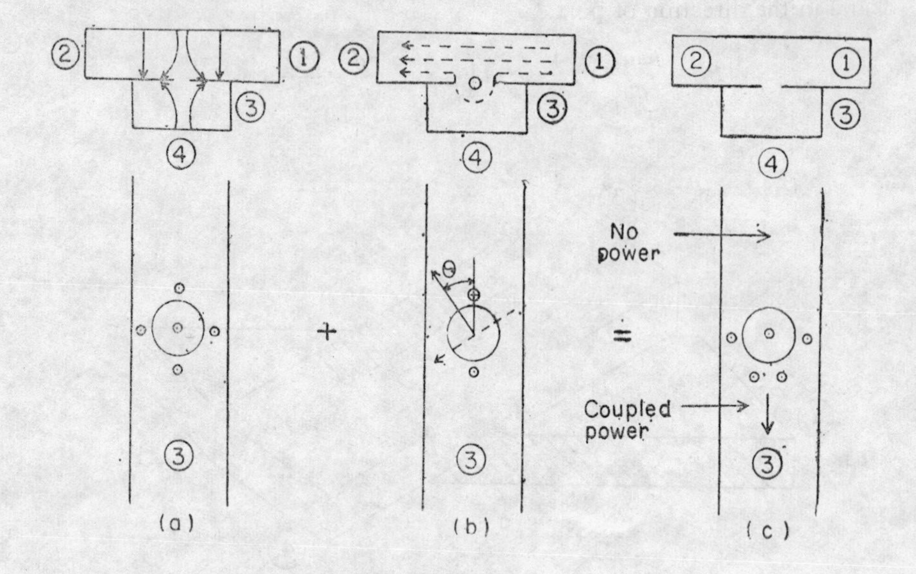


Fig. 6.27 Aperture coupling
(*a*) Electric field induced by a normal electric dipole at the aperture
(*b*) Net electric field induced by a magnetic dipole at the aperture
(*c*) Total induced electric field in ports (3) and (4), 4 is left uncoupled.

If the aperture is at the centre, i.e. $d=a/2$ and $\cos \pi d/a=0$, a directional coupler with $\theta=0$ is obtained only for $\lambda_o=\sqrt{2}\,a$. For other values the upper waveguide has to be rotated by an angle $\theta$ to reduce the coupling of the transverse magnetic field to the magnetic dipole by a factor $\cos\theta$. In this case the device will be a directional coupler when

$$B_2 \cos \theta = -B_1,$$

or

$$\cos\theta = \frac{k_o^2}{2\beta^2} = \frac{1}{2}\,(\lambda_g/\lambda_o)^2. \tag{6.119}$$

For the coupler of Fig. 6.26(a), the coupled wave in port 3 has an amplitude $B_1+B_3\cos\theta$ and hence the coupling is given by

$$C = -20\log\left|\frac{B_1+B_3\cos\theta}{A}\right| = -20\log\frac{4}{3}\frac{\beta r_o^2}{ab}\left(\cos\theta+\frac{k_o}{2\beta^2}\right)$$

$$= -20\log\frac{4}{3}\frac{\beta \gamma_o^2}{ab}\left\{\cos\theta+\frac{1}{2}\left(\frac{\lambda_g}{\lambda_o}\right)^2\right\}. \tag{6.120}$$

The directivity is given by

$$D=20 \log \left|\frac{B_1+B_3 \cos \theta}{B_1+B_2 \cos \theta}\right|=20 \log \frac{2\beta^2 \cos \theta+k_o^2}{2\beta^2 \cos \theta-k_o^2}$$

$$=20 \log \frac{2 \cos \theta+(\lambda_g/\lambda_o)^2}{2 \cos \theta-(\lambda_g/\lambda_0)^2}. \qquad (6.121)$$

The scattering matrix for this type of coupler is*

$$[S]=\begin{bmatrix} 0 & \sqrt{1-\beta^2} & 0 & \pm j\beta \\ \sqrt{1-\beta^2} & 0 & \pm j\beta & 0 \\ 0 & \pm j\beta & 0 & \sqrt{1-\beta^2} \\ \pm j\beta & 0 & \sqrt{1-\beta^2} & 0 \end{bmatrix} \qquad (6.122)$$

$\beta$ is real** and is positive for the top wall (broad wall) coupler and negative for side (narrow) wall coupler. For a 3 *db* coupler, equal power is coupled in both ports 2 and 3. Hence, Eq. (6.122) can be reduced to the following simple form

$$[S]_{3db}=\begin{bmatrix} 0 & 1 & 0 & \pm j \\ 1 & 0 & \pm j & 0 \\ 0 & \pm j & 0 & 1 \\ \pm j & 0 & 1 & 0 \end{bmatrix}. \qquad (6.123)$$

(*ii*) **Two-hole Coupler**

In this type of coupler, two waveguides are coupled by drilling two identical holes or slots on the common wall. The holes are spaced at a distance $s$ and may be on any of the walls, broad [Fig. 6.28(a)] or narrow [Fig. 6.28(b)] but the wall has to be parallel to the electric field of the $TE_{10}$ mode. The operation of such a coupler may be understood with the help of Fig. 6.28(c). Let $B_f$ and $B_b$ be the aperture coupling coefficients of the holes in the forward and backward directions, i e. the amplitudes of the waves coupled in the auxiliary guide in the forward and backward directions when a wave of unit amplitude is incident in the main guide. Assume that a very small power is coupled in the first aperture so the amplitude of the wave arriving at the second hole can also be taken to be unity. Then, the total forward wave in the upper guide at plane *bb* [Fig. 6.28(c)] will be $2B_f\, e^{-j\beta s}$, where $\beta s$ is the change in phase arising due to the propagation of the wave by a dis-

*For the derivation the reader is referred to J.L. Altman, *op. bit*.

**$\beta$ used in the scattering matrix is a constant (scattering coefficient) and should not be confused with the phase constant $\beta$.

tance $s$. However, the total backward wave arriving at the plane $aa$ in the upper waveguide will be $B_b(1-e^{-2j\beta s})$. Since the forward path lengths in

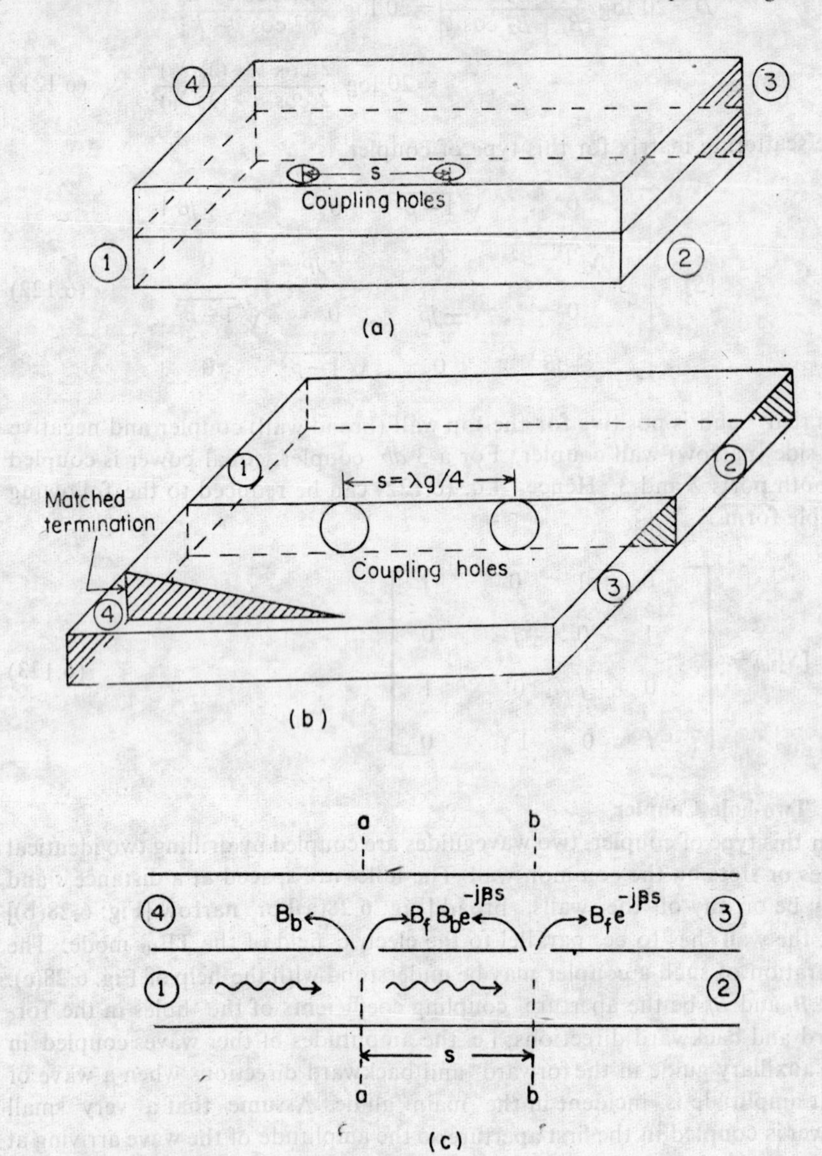


Fig. 6.28 Two hole couplers (*a*) Broad wall or top wall (Electric coupling) (*b*) Narrow wall or side wall (magnetic coupling), (*c*) Principle of two hole coupler

the two waveguides are always the same, the forward waves always add in phase. However, the backward waves may be added out of phase, resulting in the complete cancellation of the wave when $2\beta s = n\pi$; $n = 1, 3, 5 \ldots$ i.e.

the distance $s$ is an odd multiple of $\lambda_g/4$. Under these conditions, the device will behave as a directional coupler, the coupling of which is given by

$$C = -20 \log |2B_f| \tag{6.124}$$

and the directivity is

$$D = 20 \log \frac{2|B_f|}{|B_b|\,|1+e^{-2j\beta s}|} = 20 \log \left|\frac{B_f}{B_b}\right| + 20 \log |\sec \beta s| \tag{6.125}$$

Equation (6.125) reveals that the directivity is the sum of the inherent directivity of the single aperture plus a directivity associated with the two element array. As $B_f$ and $B_b$ are slowly varying functions of frequency, the coupling $C$ is not particularly frequency sensitive. However, directivity $D$ is a sensitive function of frequency because of the array factor sec $\beta s$.

*(iii)* **Schwinger Reversed-phase Coupler***

This coupler is designed to interchange the frequency sensitivity of coupling $C$ and directivity $D$. As shown in Fig. 6.28(d) a reversed phase coupler is designed by drilling two identical slots separated by $\lambda_g/4$ common to the

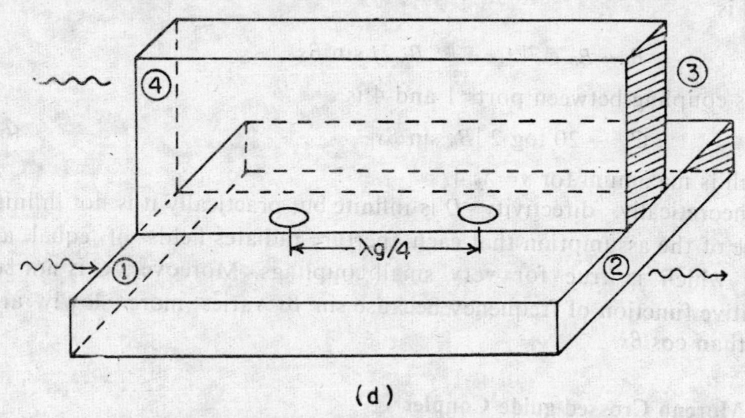


**Fig. 6.28(*d*)** Schwinger (two hole) reversed phase coupler.

top wall of the main waveguide (1-2) and to the side wall of the auxiliary waveguide (3-4). The slots are so oriented that the field radiated by the first aperture is negative to that radiated by the second aperture. This is achieved by properly orienting the slot locations. Referring to Fig 6.28(d) for the first approximation the electric coupling may be disregarded since the slots have the wrong orientation with respect to waveguide (3-4); for the same reason the $x$-component of the magnetic field may be neglected. Now the $H_z^o$ for the TE$_{10}$ mode incident at port 1 can be written as (Eq. 4.62)

$$H_z^o = B \cos \frac{\pi x}{a}.$$

*For design details the reader may refer to T.N. Anderson, "Directional coupler design nomograms", *Microwave J.*, Vol. 2, (May 1959).

Let the value of $H_z^o$ at the first slot be

$$H_{z1}^o = B \cos \frac{\pi x_1}{a}.$$

At the second slot, it becomes

$$H_{z1}^o = B \cos \frac{\pi x_2}{a} e^{-j\pi/2} = jB \cos \frac{\pi x_1}{a}$$

since $x_2 = a - x_1$.

The cancellation of the wave occurs at port 4 because $H_{za}^o$ becomes $-jB \cos \pi x_1/a$ at the second slot and at port 3 there is the addition of the waves because $H_{zb}^o$ becomes $B \cos \pi x_1/a$ at the first slot. If we interpret these results with reference to Fig. 6.28(c), we find that the fields radiated at the first aperture and second aperture are respectively $B_f$, $B_b$ and $-B_f$, $-B_b$. At plane *bb* in the upper waveguide, the total field is now $B_f - B_f = 0$ under all conditions. Hence port 3 is not coupled. At plane *aa* the total field is

$$B_b - B_b e^{-2j\beta s} = e^{-j\beta s} B_b 2j \sin \beta s.$$

Thus coupling between ports 1 and 4 is

$$C = -20 \log 2 |B_b \sin \beta s| \tag{6.126}$$

which is maximum for $s = \lambda_g/4$.

Theoretically, directivity $D$ is infinite but practically it is not infinite because of the assumption that each aperture radiates fields of equal amplitude which is true for very small couplings. Moreover, $C$ is not such a sensitive function of frequency because $\sin \beta s$ varies more slowly around $\pi/2$ than $\cos \beta s$.

### (*iv*) Moreno Crossed-guide Coupler

As shown in Fig. 6.29, this coupler is designed by arranging two waveguide sections at right angles with two coupling holes drilled through the common portion of their top wall. These couplers utilize the polarization properties of the dominant $TE_{10}$ mode in obtaining the directional coupling character. The size and location of the apertures is so chosen that each of the holes 1 and 2 are located at a distance 1/4 from the waveguide sides and are thus offset by approximately $\lambda_g/4$ in the normal operating range of the waveguide.

In this case also only magnetic fields need to be considered. At hole $A$, the magnetic field in waveguide 1-2 is right hand circularly polarized and it couples a right hand circularly polarized magnetic field in waveguide 3-4 which produces propagation towards port 3. However, the phase of this wave will lag by 90° because the wave has made a right hand 90° turn. A similar radiation occurs at hole $B$. The left hand wave in waveguide 1-2 couples a left hand wave that propagates in port 3, of course with a phase lead of 90°. At port 3, both coupled waves will be in phase since the wave from $B$ has lost a total phase of 90° (– 180° through distances and a gain of 90° through the hole), and the wave from $A$ has also lost a phase of 90°. It is, however, not difficult to infer that the waves arriving at port 4 are 180° out of phase and hence cancel out.

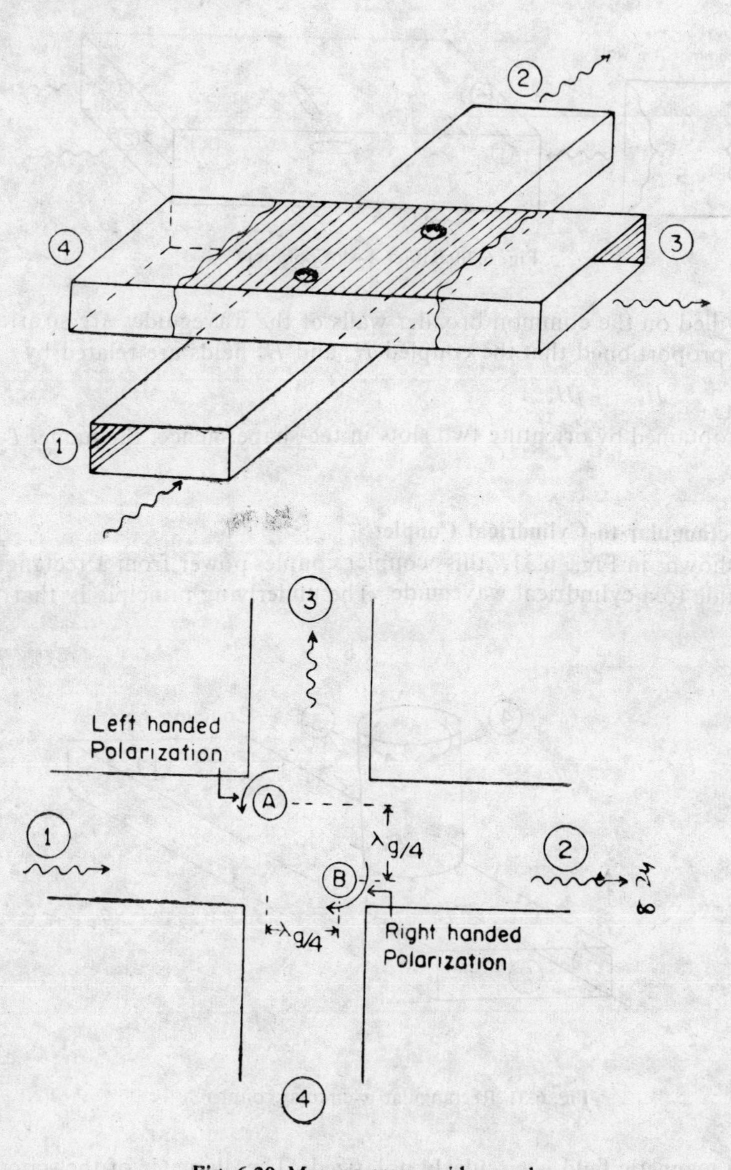


Fig. 6.29 Moreno cross-guide coupler.

The directivity of the coupler is the function of the directivities of each aperture and is enhanced by the $\lambda_g/4$ offset arrangement which not only increases coupling but also tends to cancel reverse coupling.

### (v) Riblet T-slot Coupler

This coupler too utilizes circularly polarized coupling in achieving directional coupling characteristics. As shown in Fig. 6.30, the two coupling

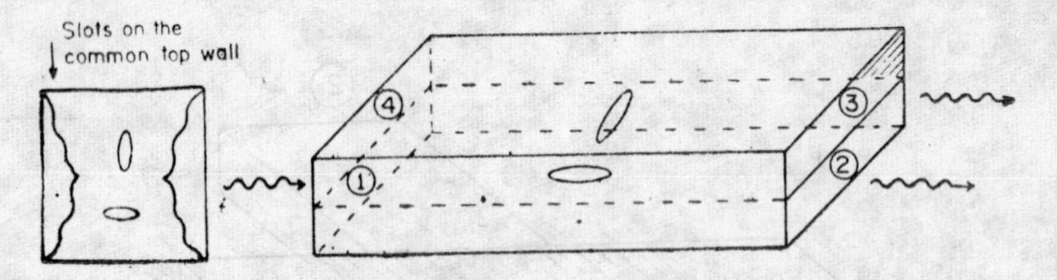


**Fig. 6.30** Riblet T-slot coupler.

slots drilled on the common broader walls of the waveguides are so arranged and proportioned that the coupled $H_x$ and $H_z$ fields are related by

$$H_z = \pm jH_x.$$

This is obtained by orienting two slots in tee shape; hence, the name *T*-slot coupler.

### (vi) Rectangular-to-Cylindrical Coupler

As shown in Fig. 6.31, this coupler couples power from a rectangular waveguide to a cylindrical waveguide. The underlying principle is that the

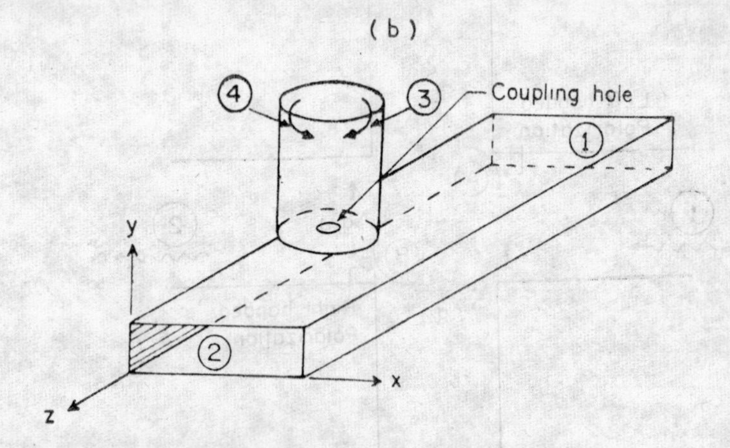


**Fig. 6.31** Rectangular to circular coupler.

coupled magnetic field is circularly polarized. The diameter of the circular waveguide is so selected that the coupling transmits only the dominant $TE_{11}$ mode in the auxiliary guide and so the coupled wave appears to be circularly polarized. Port 3 will correspond to left hand circular polarization if propagation is from port 1 to port 2 (Fig. 6.31). Port 4 will couple (right hand circular polarization) when the propagation is from port 2 to port 1.

### (*vii*) Small Loop Coupler

As shown in Fig. 6.32, this coupler basically couples power from a rectangular waveguide to a coaxial line. It is designed by inserting a loop in

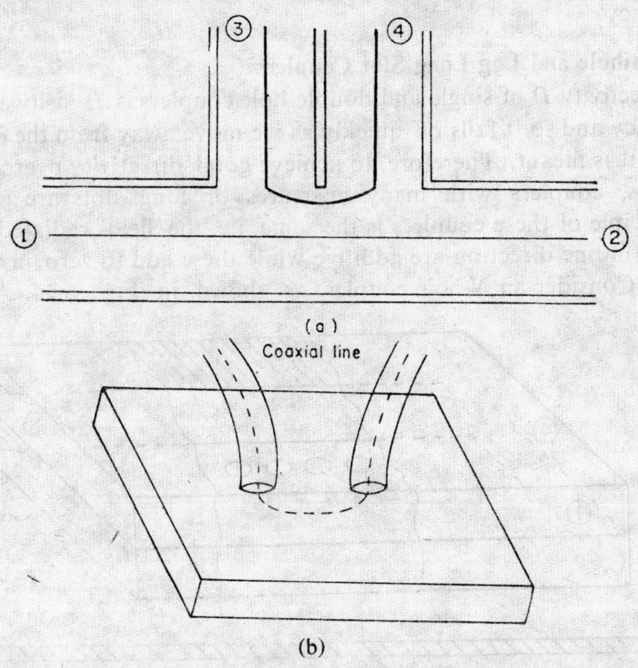


Fig. 6.32 (*a*) Small loop couplers Coaxial type,
(*b*) Waveguide type.

the centre of the broader wall of the rectangular waveguide. The same loop forms the central conductor of the coaxial ports 3 and 4. The loop couples symmetric electric fields at port 3 and port 4 from the electric field originating from port 1 [Fig. 6.33(a)]. The dimensions are so chosen that the magnetic field is also coupled and is such that it induces an antisymmetric electric field equal in magnitude to the symmetrical one and hence fields at port 3 are added while they cancel at port 4, as shown in Fig. 6.33(b) and

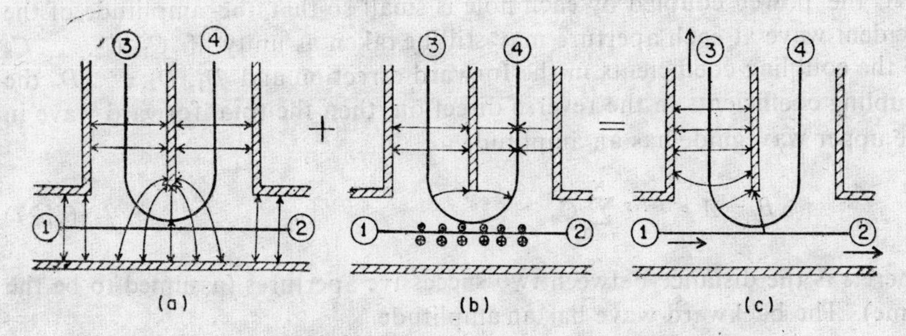


Fig. 6.33 Coupling of fields in a small loop coaxial coupler
(*a*) Electric coupling. (*b*) Magnetic coupling.
(*c*) Net effect total induced electric field.

(c). The additive fields produce additive currents ($I_e+I_h$) in the conductor of port 3 and substractive ($I_e-I_h=0$) currents in port 4. For frequencies remote from the guide cut-off frequency, these couplers are little sensitive to frequency.

**(viii) Multihole and Log Long Slot Couplers**

The directivity $D$ of single and double hole couplers is a sensitive function of frequency and so it falls off quickly as we move away from the frequency for which it is meant. Therefore, to achieve good directivity over a band of frequencies, couplers with many apertures or long slots are used. The basic principle of these couplers is the same, i.e. the fields radiated by the apertures in one direction are additive while these add to zero in the other direction. Consider an $N$-hole coupler as shown in Fig. 6.34. Assuming

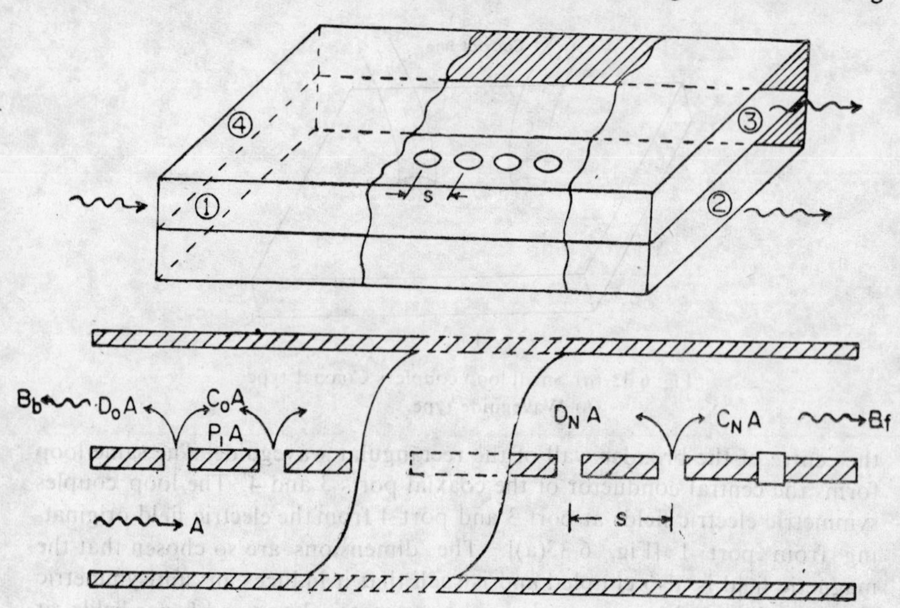


**Fig. 6.34** Multihole directional coupler.

that the power coupled by each hole is small so that the amplitude of the incident wave at each aperture may still be taken as unity. if $C_1, C_2, \ldots C_n$ be the coupling coefficients in the forward direction and $D_1, D_2, \ldots D_n$ the coupling coefficients in the reverse direction, then the total forward wave in the upper waveguide has an amplitude

$$B_f = A\, e^{-j\beta Ns} \sum_{n=0}^{N} C_n \tag{6.127}$$

where $s$ is the distance between two successive apertures (assumed to be the same). The backward wave has an amplitude

$$B_b = A \sum_{n=0}^{n} D_n\, e^{-j\beta 2ns}. \tag{6.128}$$

The coupling and directivity are respectively given by

$$C = -20 \log \left| \sum_{n=0}^{N} C_n \right| \tag{6.129}$$

$$D = -20 \log \frac{\left| \sum_{n=0}^{N} D_n \, e^{-j\beta 2ns} \right|}{\left| \sum_{n=0}^{N} C_n \right|}. \tag{6.130}$$

Assuming that all the apertures are similar so that the coupling coefficients $D_n$ can be expressed as a product of frequency independent amplitude constant $d_n$ and a frequency dependent factor $T_b$,

$$D = -C - 20 \log |T_b| - 20 \log \left| \sum_{n=0}^{N} d_n \, e^{-2j\beta ns} \right|. \tag{6.131}$$

The first two terms in Eq. (6.131) give the directivity associated with the individual apertures, and the last term is the directivity arising from the array. Various types of multihole couplers have been designed depending on the design of the array factor. A discussion on all these is beyond the scope of the present volume and may be found in the literature*.

For coupling, instead of multiholes a long slot may also be used as shown in Fig. 6.35. The action of the coupler may be understood by considering

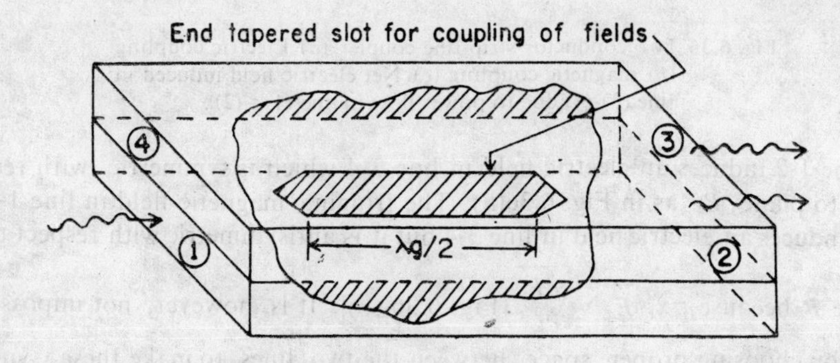


Fig. 6.35 Long slot coupler.

each point of the slot as a coupling aperture. It is not difficult to visualize that if the length of the slot is $\lambda_g/2$. the power coupled in the reverse direction in the auxiliary arm will be zero. However, the flow of power from the main waveguide to the auxiliary waveguide cannot be ruled out but will definitely be small if the slot is narrow. A careful analysis shows that if the ends of the slot are tapered (to avoid reflections at the ends), perfect directivity is possible even with a slot that is short and wide, and there is optimum slot length for which all the power entering the main waveguide is

*R.E. Collin, *Foundations for Microwave Engineering*, McGraw-Hill Book Co. (1966).

transferred to the auxiliary waveguide. For a small coupling, however, the power coupled varies as the square of the slot length and the sixth power to the width. Such a coupler, besides having high directivity and high attainable coupling, is a markedly of frequency sensitive.

### (*ix*) Two Conductor-Strip Line Coupler

Consider a system of two parallel conductor lines located in proximity to each other as strip lines. Assuming that a TEM wave is propagating in the lower line from ports 1 to 2 such that the electric and magnetic fields are maximum at a plane $P_1$ of Fig. 6.36, say at $t=0$. The fringing electric field

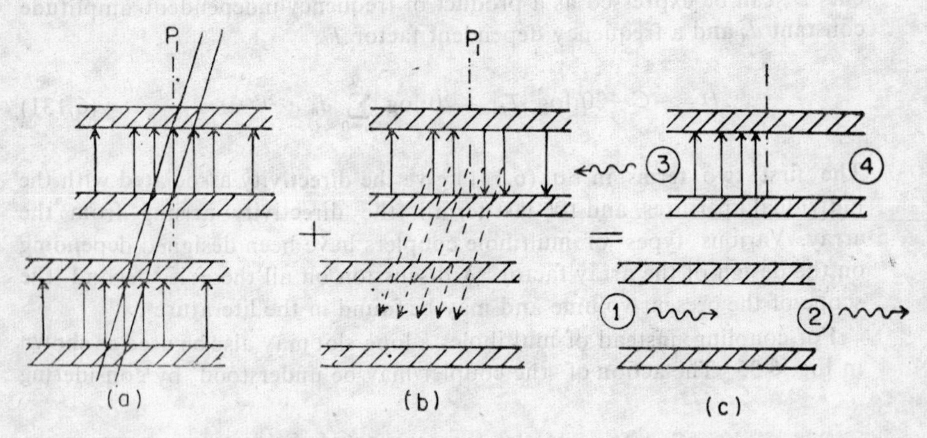


**Fig. 6.36** Two conductor-strip line coupler (*a*) Electric coupling (*b*) magnetic coupling (*c*) Net electric field induced in line (3)-(4) due to power flow in line (1)—(2).

in line 1-2 induces an electric field in line 3-4 which is symmetric with respect to plane $P_1$ as in Fig. 6.36(a). The fringing magnetic field in line 1-2 also induces an electric field in line 3-4 but it is antisymmetric with respect to plane $P$ because $\nabla x \ \vec{E}=-\frac{\partial \vec{B}}{\partial t}$ [Fig. 6.36 (b)]. It is, however, not impossible, by choosing proper space between the two lines, to make these symmetric and antisymmetric fields cancel out. If this happens, there will be no flow of power in port 4 [Fig. 6.36(c)] and the device will work as a directional coupler.

## 6.24 APPLICATIONS OF DIRECTIONAL COUPLERS

A directional coupler is basically a power monitoring device and, is used in measuring equipment and for checking the performance of microwave equipments. In addition, directional couplers are also used to design some of the important circuit elements such as phase shifters, variable impedances and balanced duplexers.

### (i) Power Measurements

Since the coupling of a directional coupler is independent of the power in the main line, a matched calibrated detector placed at the auxiliary line gives an accurate indication of power in the main line without affecting the power flow in the main line. For absolute power measurement, however, the coupler and detector may be suitably calibrated by absolute means such as a calorimeteric-wattmeter.

### (ii) Reflectometer*

Since a directional coupler is a reciprocal four-port matched device, a directional coupler containing matched calibrated detectors in both ports of the auxiliary line may be used to sample or detect the power flowing in the forward and backward directions simultaneously. This type of device is known as a reflectometer. Because it is difficult to match the two detectors simultaneously it is convenient to use two directional couplers with two separate detectors. If the incident (forward) and reflected (backward) outputs from the detectors are fed to a ratiometer one can directly measure the reflection coefficient of the mismatch at port (2). Reflectometers also provide quick and accurate measurement of VSWR in a given frequency band and can also be used to measure low and high power. A typical reflectometer arrangement is depicted in Fig. 6.37. An obvious advantage of this

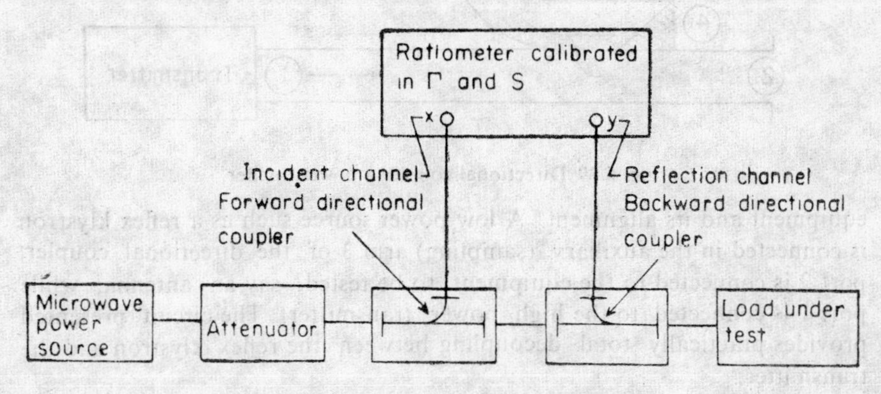


**Fig. 6.37** Simplified reflectometer system.

method is that, compared to ordinary slotted line measurements, it provides quick and more accurate measurement of reflected power by which we can match a load to the line without loss of much time.

### (iii) Fixed Attenuators and Directional Power Deviders

Figure 6.38 depicts a fixed attenuator designed using a directional coupler. When ports 2 and 4 are terminated in matched loads, the power coming out of port 3 is the coupling of the directional coupler. This type of device

*Sisodia, M.L. and Raghuvanshi G.S., Basic Microwave Techniques and Laboratory, Manual Wiley Eastern Ltd., New Delhi (1987).

provides 10 to 30 *db* attenuation and is used in designing wide band decoupling attenuators.

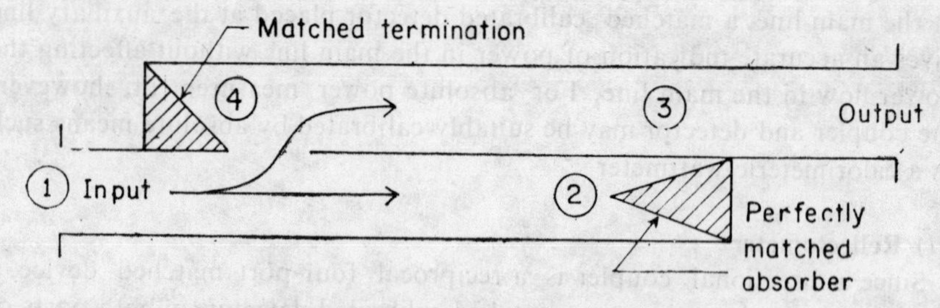


**Fig 6.38** Directional coupler as fixed attenuator.

Directional power dividers are the equipments which can simultaneously sample the forward and backward power travelling in a line independently. A typical directional power divider is depicted in Fig. 6.39 used to test radar

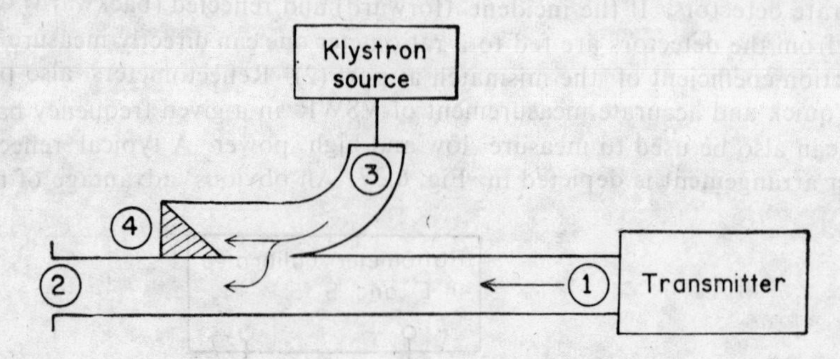


**Fig. 6.39** Directional coupler a power divider.

equipment and its alignment. A low power source such as a reflex klystron is connected in the auxiliary (sampling) arm 3 of the directional coupler; port 2 is connected to the equipment to be tested, say an antenna, while port 1 is connected to the high power transmitter. The circuit presented provides practically total decoupling between the reflex klystron and the transmitter.

A directional coupler used as a waveguide stretcher is depicted in Fig. 6.40. Arms 2 and 4 are provided with movable short circuits each of which is at an electrical distance of $\phi$ with respect to a given reference plane (towards the junction) at the ports. The scattering matrix of the 3 *db* side coupler used here is then given by (Eq. 6.123). The normalized output wave is given by

$$[S]_{3db} = \begin{bmatrix} 0 & 1 & 0 & j \\ 1 & 0 & j & 0 \\ 0 & j & 0 & 1 \\ j & 0 & 1 & 0 \end{bmatrix}.$$

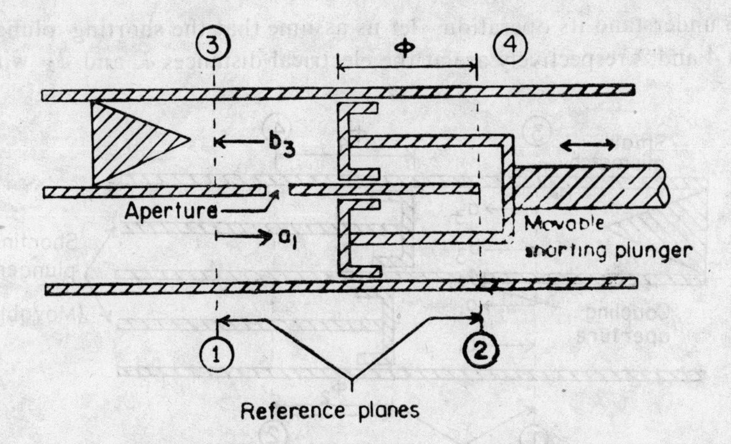

**Fig. 6.40** Waveguide stretcher (phase shifter) using a directional coupler.

The normalized output is thus given by

$$\begin{bmatrix} b_1 \\ b_2 \\ b_3 \\ b_4 \end{bmatrix} = \frac{1}{\sqrt{2}} \begin{bmatrix} 0 & 1 & 0 & j \\ 1 & 0 & j & 0 \\ 0 & j & 0 & 1 \\ j & 0 & 1 & 0 \end{bmatrix} \begin{bmatrix} a_1 \\ -b_2\, e^{j2\phi} \\ 0 \\ -b_4\, e^{j2\phi} \end{bmatrix}$$

$$= \frac{1}{\sqrt{2}} \begin{bmatrix} 0 \\ a_1 \\ -\sqrt{2}\, ja_1\, e^{j2\phi} \\ ja_1 \end{bmatrix}$$

because $a_2 = -b_2\, e^{j2\phi}$ and $a_4 = -b_2\, e^{j2\phi}$.
Thus, the output at port 3 is

$$b_3 = -ja_1\, e^{j2\phi} = a_1\, e^{j(2\phi - \pi/2)}.$$

The above relation suggests that by varying $\phi$ and hence the positions of the shorting plungers in ports 2 and 4, we can vary the phase of the outgoing wave. Port 3 which is uncoupled has to be terminated in a matched load while designing such a waveguide stretcher or a phase shifter.

### (*iv*) Variable Impedance or Matching Device

If the positions of two shorting plungers that terminate ports 2 and 4 of the 3 *db* directional coupler in Fig. 6.41 are varied independently, what we have is a variable impedance.

To understand its operation, let us assume that the shorting plungers in arms 4 and 2 respectively are at the electrical distances $\phi_4$ and $\phi_2$ with res-

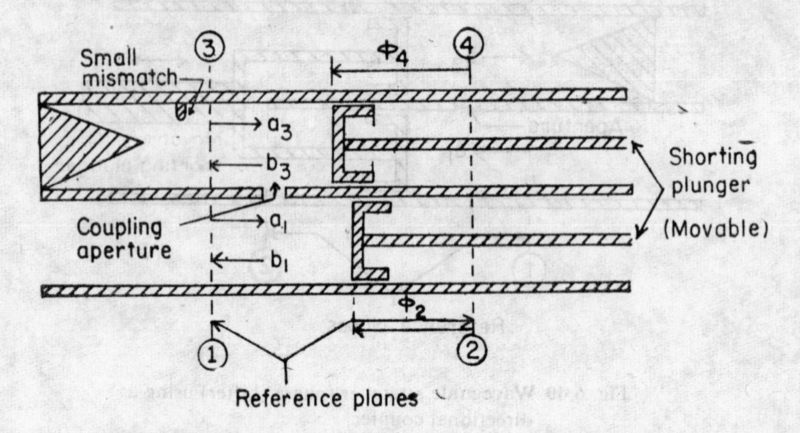


Fig. 6.41 Variable impedance or impedance matching device.

pect to the given reference planes. Further, suppose $a_3 \neq 0$, i.e. there is some coupling or mismatch. Then we have,

$$\begin{bmatrix} b_1 \\ b_2 \\ b_3 \\ b_4 \end{bmatrix} = \frac{1}{\sqrt{2}} \begin{bmatrix} 0 & 1 & 0 & j \\ 1 & 0 & j & 0 \\ 0 & j & 0 & 1 \\ j & 0 & 1 & 0 \end{bmatrix} \begin{bmatrix} a_1 \\ -e^{j2\phi_2} b_2 \\ a_3 \\ -e^{j2\phi_4} b_4 \end{bmatrix}$$

or

$$b_1 = \tfrac{1}{2} \left[a_1 \left(e^{j2\phi_2} - e^{j2\phi_4}\right) + ja_3 \left(e^{j2\phi_2} + e^{j2\phi_4}\right)\right]$$

$$b_2 = \frac{1}{\sqrt{2}} (a_1 + ja_3)$$

$$b_3 = -\tfrac{1}{2} \left[ja_1 \left(e^{j2\phi_2} + e^{j2\phi_4}\right) - a_3 \left(e^{j2\phi_2} - e^{j2\phi_4}\right)\right]$$

$$b_4 = \frac{1}{\sqrt{2}} (ja_1 + a_3).$$

For port 1 be matched, we must have $b_1 = 0$ or

$$\frac{e^{j2\phi_2} + e^{j2\phi_4}}{e^{j2\phi_2} - e^{j2\phi_4}} = j \frac{a_1}{a_3}$$

or

$$e^{j2(\phi_2 - \phi_4)} = -\frac{1 + j\dfrac{a_1}{a_3}}{1 - j\dfrac{a_1}{a_3}}.$$

which has a magnitude 1. We, therefore, conclude that the system depicted in Fig. 6.41 can provide matching to any impedance connected at port 2.

(*vi*) **Balanced Duplexer**

A typical balanced duplexer designed using two directional couplers is shown in Fig. 6.42. When the magnetron connected at port 1 of the first

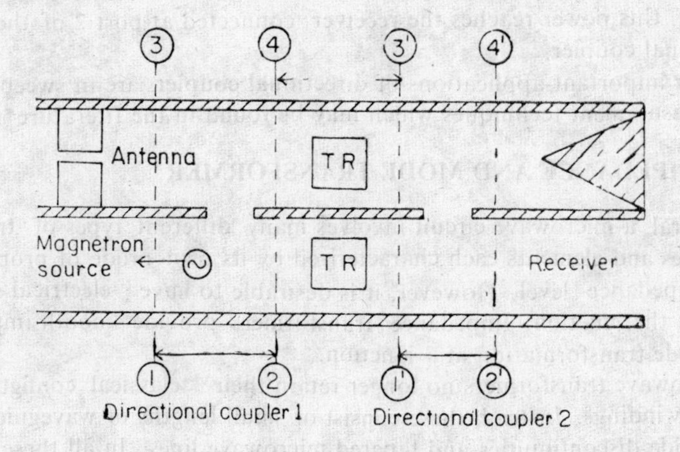


**Fig. 6.42** Balanced duplexer system employing two directional couplers.

directional coupler fires a pulse, the two TR tubes connected at ports 2 and 4 of the first directional coupler produce effective short circuits at some symmetrical electrical distance $\phi$ from the reference planes in ports 2 and 4. The outgoing signal proceeds to port 3, which is terminated in a matched antenna. In between pulses (when the magnetron does not transmit any power) TR tubes act as transmission lines and join the two directional couplers, so the power received by the antenna (also known as the antenna echo signal) is applied to the first coupler. Output from the first coupler, $[b]$ is given by

$$\begin{bmatrix} b_1 \\ b_2 \\ b_3 \\ b_4 \end{bmatrix} = \frac{1}{\sqrt{2}} \begin{bmatrix} 0 \\ ja_3 \\ 0 \\ a_3 \end{bmatrix}.$$

This is applied to the input of the second coupler. Then, the output from the second coupler $[b']$ is given by

$$\begin{bmatrix} b_1' \\ b_2' \\ b_3' \\ b_4' \end{bmatrix} = \frac{1}{\sqrt{2}} \begin{bmatrix} 0 & 1 & 0 & j \\ 1 & 0 & j & 0 \\ 0 & j & 0 & 1 \\ j & 0 & 1 & 0 \end{bmatrix} \frac{1}{\sqrt{2}} \begin{bmatrix} 0 \\ ja_3 e^{-j\phi_0} \\ 0 \\ a_3 e^{-j\phi_0} \end{bmatrix}$$

$$=\frac{1}{2}\begin{bmatrix} 0 \\ 2ja_3e^{-j\phi_0} \\ 0 \\ 0 \end{bmatrix}.$$

Clearly, this power reaches the receiver connected at port 2 of the second directional coupler.

Other important applications of directional couplers are in sweep frequency measurement techniques which may be found in the literature*.

## 6.25 IMPEDANCE AND MODE TRANSFORMER

In general, a microwave circuit involves many different types of transmission lines and elements each characterized by its own mode of propagation and impedance level. However, it is desirable to have "electrical smoothness" at the junction. Impedance transformers provide smooth impedance and mode transformation at a junction.

Microwave transformers no longer retain their "classical configuration" of two windings. Instead, these consist of small lengths of waveguide lines, waveguide discontinuities and tapered microwave lines. In all these devices the electrical smoothless is effected by cancelling the intrinsic reflections with the reflections produced due to the purposely produced discontinuity. In the following sections we shall discuss some useful microwave transformers.

## 6.26 TUNING SCREWS AND STUBS

In Chapter 2 we discussed the matching of lines using studs which are shorted lines of certain tunable lengths. A single screw tuner, shown in Fig. 6.43,

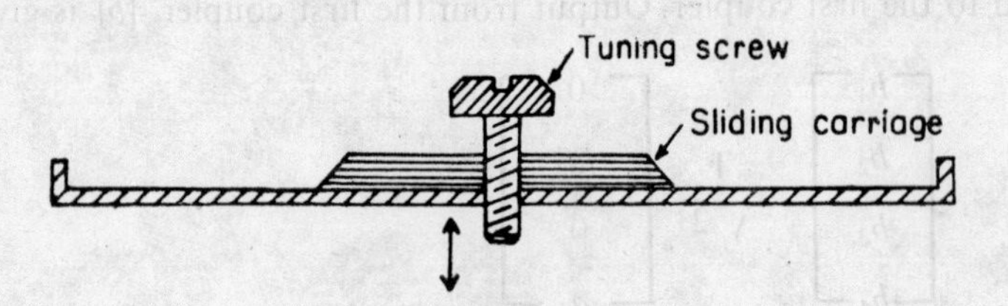


Fig. 6.43 Single slide-screw tuner.

is equivalent to a single stub tuner. It consists of a variable depth screw mounted on a sliding carriage free to move longitudinally along the guide over a distance more than $\lambda_g/2$. The screw penetrates into the guide through a centred narrow slot in the broad wall of the guide. The screw presents a variable shunt capacitive susceptance when penetration is small and a match

*T.S Leverghetta, *Microwave Measurements and Techniques*, Artech House Inc., Washington Street Massahuseetts (1976).

can therefore be obtained in all the cases because the screw can move over a distance more than $\lambda_g/2$. Practically, the tunable range of susceptances for a single screw tuner is small so sometimes a triple screw tuner shown in Fig. 6.44 is used. The separation between the screws is $3\lambda_g/8$. It is equivalent to

Fig. 6.44 Triple screw tuner.

a triple stub tuner. The waveguide version of a double stub tuner is the *E-H* tuner shown in Fig 6.45. It consists of a magic tee whose *E*- and *H*-arms are shorted by movable shorting plungers. Since both *E*- and *H*-arms are tunable, this tuner can be used for a wide range of impedances.

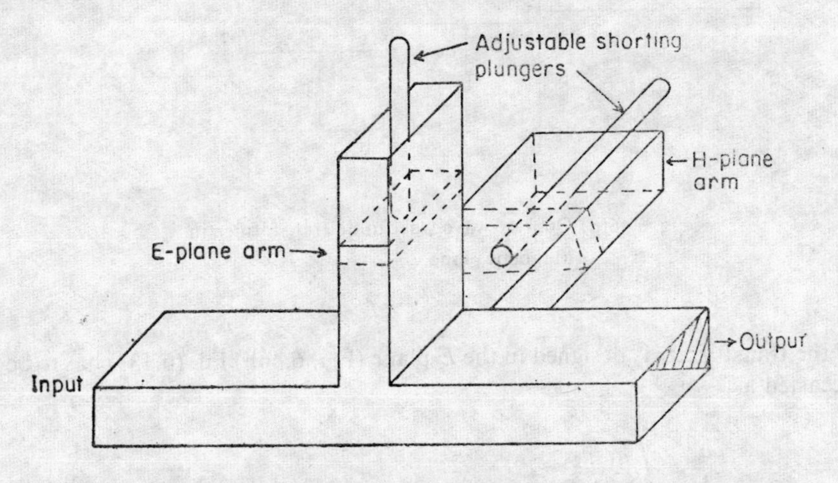


Fig. 6.45 *E-H* tuner (hybrid tuner).

## 6.27 QUARTER-WAVE TRANSFORMER

This transformer makes use of the impedance transformation properties of a quarter-wave line. As shown in sec. 2.15 the input impedances that appear when a quarter-wave line is terminated in an impedance $Z_2$, is given by

$$Z_1 = Z_{in} = Z^2/Z_2 \text{ or } Z = \sqrt{Z_1 Z_2} \tag{6.132}$$

where $Z$ is the characteristic impedance of the line. It follows from Eq. (6.132) that two lines having characteristic impedances can be joined by a quarter-wave line (transformer) if the transformer has a characteristic impedance which is the geometric mean of the characteristic impedances of two lines.

A typical waveguide transformer (Fig. 6.46) consists of two waveguides (differing in their $b$-dimensions) joined by a third waveguide such that

$$b=\sqrt{b_1 b_2}. \tag{6.133}$$

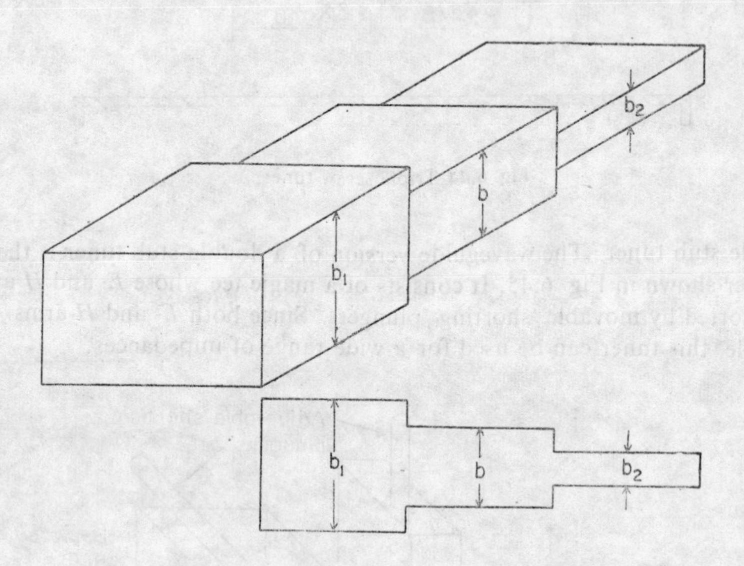


**Fig. 6.46(*a*) Quarter-wave waveguide transformer in Magnetic plane.**

If the transformer is designed in the $E$-plane (Fig. 6.46b) Eq. (6.133) has to be recasted as

$$\frac{a}{\lambda_g}=\sqrt{\frac{a_1}{\lambda_{g_1}}\frac{a_2}{\lambda_{g_2}}} \tag{6.134}$$

where $\lambda_g$, $\lambda_{g_1}$ and $\lambda_{g_2}$ are the guide wavelengths respectively of the transformer and the lines to be matched.

A quarter-wave transformer provides perfectly smooth transformation at a particular frequency at which it is a quarter-wave long and so it has a narrow bandwidth. To obtain an expression for the bandwidth, we recall Eq. (2.51) and say that for perfect match

$$Z_1=Z_{in}=Z\,\frac{Z_2+j\,Z\tan\beta l}{Z+j\,Z_2\tan\beta l}$$

where $l$ is the length of the transformer and $\bar{\beta}=2\pi/\lambda$.

The corresponding reflection coefficient is given by (Eq. 2.79)

$$\Gamma=\frac{Z_{in}-Z_1}{Z_{in}+Z_1}=\frac{Z(Z_2-Z_1)+j(Z^2-Z_1Z_2)\tan\bar{\beta}l}{Z(Z_2+Z_1)+j(Z^2+Z_1Z_2)\tan\bar{\beta}l}$$

$$=\frac{Z_2-Z_1}{Z_2+Z_1+2j\sqrt{Z_1Z_2}\tan\bar{\beta}l}. \quad (6.135)$$

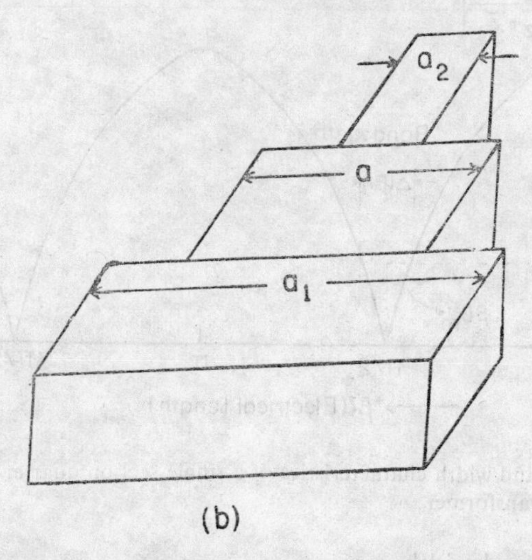


Fig. 6.46(*b*) Quarter-wave waveguide transformer in electric plane.

The magnitude of the reflection coefficient which is a measure of mismatch can readily be obtained

$$|\Gamma|=\frac{|Z_2-Z_1|}{\{(Z_2+Z_1)^2+4Z_1Z_2\tan^2\bar{\beta}l\}^{1/2}}$$

$$=\frac{1}{\left\{1+\left(\frac{2\sqrt{Z_1Z_2}}{Z_2-Z_1}\sec\bar{\beta}l\right)^2\right\}^{1/2}}. \quad (6.136)$$

For a quarter-wave transformer $\bar{\beta}l=\pi/2$

$$|\Gamma|=\frac{Z_2-Z_1}{2\sqrt{Z_1Z_2}}|\cos\bar{\beta}l|. \quad (6.137)$$

The plot of Eq. (6.137) [Fig. 6.47] basically shows that the variation of mismatch with frequency is periodic. The periodicity arises due to the periodicity of $Z_{in}$. Moreover, $|\Gamma|\propto\cos\bar{\beta}l$, hence $|\Gamma|$ rapidly increases as we move away from $\pi/2$. If $|\Gamma_m|$ be the maximum tolerable mismatch then the band width of a quarter-wave transformer is given by

$$(\bar{\beta}l)_m=\cos^{-1}\left\{\frac{2|\Gamma_m|\sqrt{Z_1Z_2}}{(Z_2-Z_1)\sqrt{1-|\Gamma_m|^2}}\right\}. \quad (6.138)$$

For TEM waves, $\beta l = f/f_o \cdot \pi/2$, where $f_o$ is the frequency where $\beta l = \pi/2$ (transformer is quarterwave long). Therefore, the band width is given by

$$\Delta f = 2\,(f_o - f_m) = 2\left[f_o - \frac{2f_o}{\pi}(\beta l)_m\right]$$

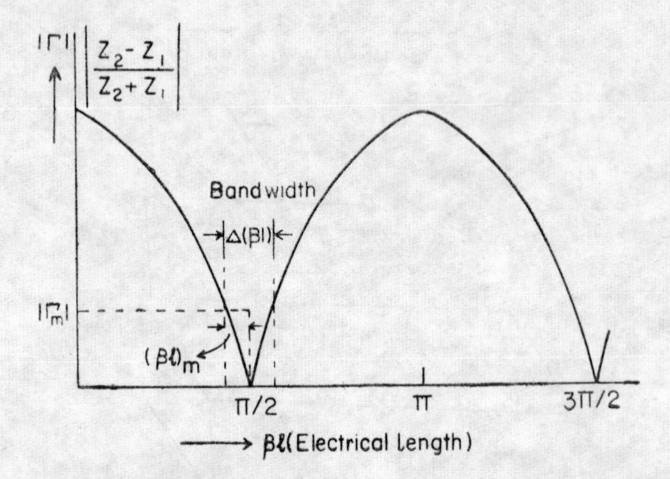


**Fig. 6.47** Band width characteristics of a single section quarter-wave transformer.

or the fractional band width is

$$\frac{\Delta f}{f_o} = 2 - \frac{4}{\pi}\cos^{-1}\left|\frac{2\,|\Gamma_m|\sqrt{Z_1 Z_2}}{(Z_2 - Z_1)\sqrt{1 - |\Gamma_m|^2}}\right|. \tag{6.139}$$

This band width may be adequate in a number of microwave systems. However, in many cases broad band operation is desired which can be obtained by using many quarter-wave sections in cascade.

## 6.28 MULTISECTION QUARTER-WAVE TRANSFORMER*

A wide band multisection transformer is shown in Fig. 6.48. In this transformer electrical smoothness is achieved by cancelling reflections arising from various steps. The band width of such a transformer may be maximized by varying the heights in accordance with certain rules. Some of the common rules are:

(*i*) Linear law, i.e. all impedance steps are equal. The normalized impedance of the $n^{\text{th}}$ section is $z_n = (1 + an)$, $a$ is some constant.

*For details the reader may refer to:

1. S.B. Cohn, "Optimum design of stepped transmission line transformers", *IRE Trans* (on MTT), (April 1955).
2. R.E. Collin, "Theory and design of wide band multisection quarter-wave transformers", *Proc. IRE*, Vol. 43, (Feb. 1955).
3. J.L. Altman, *Microwave Circuits*, New York: Van Nostrand (1964).

(*ii*) Exponential law, i.e. percentage change at each step is constant. In this case, $z_n=(1+a)^{n-1}$.

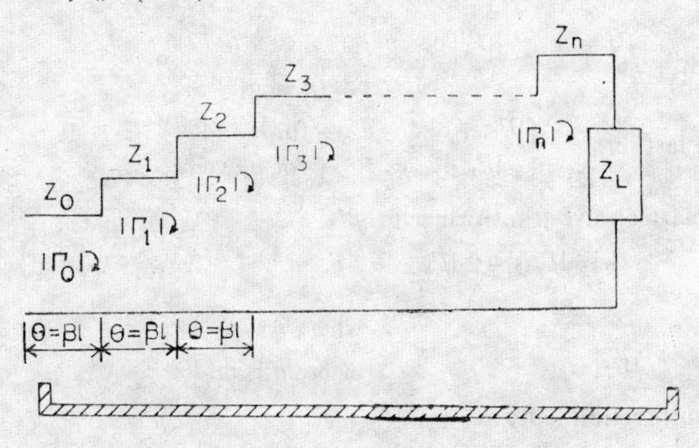


Fig. 6.48 Multisection quarter wave transformer.

(*iii*) Binomial law, i.e. steps are proportional to coefficients of binomial distribution. In this case

$$\ln\left(\frac{z_n}{z_{n-1}}\right)=B\,(1+a)^{n-1}$$

$B$ is also a constant.

(*iv*) Chebyshev law, i.e. steps are proportional to Chopyshev polynomial coefficients.

In this case

$$\ln\left(\frac{z_n}{z_{n-1}}\right)\propto T_n(x)$$

where $T_1(x)=x$, $T_2(x)=2x^2-1$, $T_3(x)=4x^3-3x$ etc. with

$$T_n(x)=2T_{n-1}-T_{n-2}.$$

Of these only binomial and Chebyshev transformers are most commonly used. Before we move on to these transformers let us consider a general case of an $n$-section quarter-wave transformer shown in Fig. 6.48. Let the characteristic impedances (wave impedances in case of waveguides) of various sections be $Z_1Z_2 \ldots Z_n$ respectively. Also let the electrical lengths of all the steps be the same, $\beta l=\theta=\lambda_g/4$ (actual or corrected at a given central frequency $f_o$. The reflection coefficient at the $n$th section is given by

$$|\Gamma_n|=\frac{Z_{n+1}-Z_n}{Z_{n+1}+Z_n}. \tag{6.140}$$

Assuming small reflections, the total reflection coefficient at the input of the transformer is given by

$$|\Gamma_T|=|\Gamma_0|+|\Gamma_1|\,e^{-2j\theta}+|\Gamma_2|\,e^{-4j\theta}+\ldots|\Gamma_n|\,e^{-2jn\theta}. \tag{6.141}$$

Assuming further that the discontinuities are symmetrical, i.e. $|\Gamma_o| = |\Gamma_n|$. $|\Gamma_1|=|\Gamma_{n-1}|$ . . . etc. Equation (6.141) can be written as

$$\Gamma_T=2e^{-jn\theta} \{|\Gamma_0| \cos n\theta+|\Gamma_n| \cos (n-2)\theta+ \ldots |\Gamma_p| \cos (n-2p)\, \theta \ldots\}$$

$$\text{with the last term}= \begin{cases} |\Gamma_{(n-1)/2}| \cos \theta, & \rightarrow \text{for } n \text{ odd} \\ \frac{1}{2}\, |\Gamma_{n/2}| & , \rightarrow \text{for } n \text{ even.} \end{cases} \tag{6.142}$$

For a quarter-wave transformer $\theta=\beta l=\pi/2$

$$|\Gamma_T|=|\Gamma_{n/2}|-2\, |\Gamma_{\frac{n-2}{2}}|+2|\Gamma_{\frac{n-4}{2}}| \ldots \pm 2|\Gamma_0| \quad \text{where } n \text{ is even}$$

$$|\Gamma_T|=0 \quad \text{where } n \text{ is odd.}$$

For a multisection halfwave transformer

$$\phi=\beta l=\pi \text{ and so}$$

$$T= \begin{cases} 2\, (|\Gamma_1|+|\Gamma_2|+ \ldots |\Gamma_T|)+|\Gamma_{n/2}|, & n \text{ is even} \\ -2\, (|\Gamma_1|+|\Gamma_2|+ \ldots |\Gamma_{\frac{(n-1)}{2}}|), & n \text{ is odd.} \end{cases} \tag{6.143}$$

## 6.29 BINOMIAL TRANSFORMER

If we select the impedances in accordance with binomial distribution and use Eq. (6.140), we find that various reflection coefficients may be expressed in terms of $|\Gamma_0|$, i.e.

$$|\Gamma_0|=|\Gamma_0|,\ |\Gamma_1|=n\, |\Gamma_0|,\ |\Gamma_2|=\frac{n(n-1)}{2\,!}\,|\Gamma_0|.$$

$$|\Gamma_3|=\frac{n(n-1)\,(n-2)}{3\,!}\, |\Gamma_0|, \ldots$$

Substituting these values in Eq. (6.142) and summing up we have

$$|\Gamma_T|=2\, |\Gamma_0|\, e^{-jn\theta}\, A_n \cos^n \theta. \tag{6.144}$$

Equation (6.144) indicates that all derivatives of $|\Gamma_T|$ at $f=f_0$ are zero upto the $(n-1)^{\text{th}}$ term and hence a binomial transformer is maximally flat as shown in Fig. 6.49. The fractional band width is given by

$$\frac{\Delta f}{f}=\frac{2\,(f_0-f_m)}{f_0}=2-\frac{4}{\pi}\, \cos^{-1} \left\{\frac{2\,|\Gamma_m|}{\ln\,(Z_L/Z_0)}\right\}^n \tag{6.145}$$

where $Z_L$ is the impedance to be matched.

To design a typical binomial transformer, we need to obtain the impedances $Z_n$ at various steps. These may be evaluated under small reflection conditions as follows. We can write

$$\ln\left(\frac{Z_{n+1}}{Z_n}\right)=2\,\frac{Z_{n+1}-Z_n}{Z_{n+1}+Z_n}=2\, |\Gamma_n|. \tag{6.146}$$

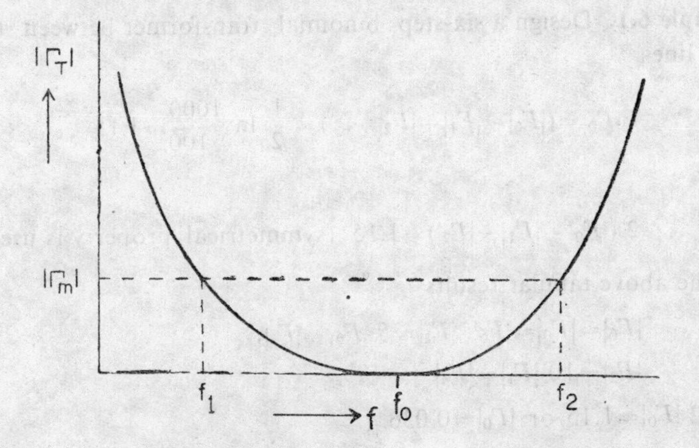


Fig. 6.49 Response curve for a binomial transformer.

Using Eq. (6.146), we can write

$$|\Gamma_0|+|\Gamma_1| \ldots = \frac{1}{2}(\ln Z_n - \ln Z_0) = \frac{1}{2} \ln \left|\frac{Z_n}{Z_0}\right|$$

or

$$\frac{1}{2} \ln \left|\frac{Z_n}{Z_0}\right| = f\,|\Gamma_0|. \tag{6.147}$$

Equation (6.147) is a means to select impedances $Z_n$. $f\,|\Gamma_0|$ may be evaluated from distribution coefficients. However, an easy way of obtaining these impedances is by offsetting each row by one place to right and adding up to get the next term as illustrated below:

| | | | | | | | |
|---|---|---|---|---|---|---|---|
| $n=1$ | 1 | | | | | | |
| | | 1 | | | | | |
| $n=2$ | 1 | 1 | | | | | |
| | | 1 | 1 | | | | |
| $n=3$ | 1 | 2 | 1 | | | | |
| | | 1 | 2 | 1 | | | |
| $n=4$ | 1 | 3 | 3 | 1 | | | |
| | | 1 | 3 | 3 | 1 | | |
| $n=5$ | 1 | 4 | 6 | 4 | 1 | | |
| | | 1 | 4 | 6 | 4 | 1 | |
| $n=6$ | 1 | 5 | 10 | 10 | 5 | 1 | |
| | | 1 | 5 | 10 | 10 | 5 | 1 |
| $n=7$ | 1 | 6 | 15 | 20 | 15 | 6 | 1 |

. . . . .

. . . . .

**Example 6.1.** Design a six-step binomial transformer between 100 and 1000 Ω lines.

$$f\,|\Gamma_0|=(|\Gamma_0|+|\Gamma_1|+|\Gamma_2|\ldots)=\frac{1}{2}\ln\frac{1000}{100}=1.15$$

or

$$2\,(|\Gamma_0|+|\Gamma_1|+|\Gamma_2|)=1.15 \text{ (symmetrical property is used here).}$$

Using the above tabular results

$$|\Gamma_0|=|\Gamma_0|=|\Gamma_6|,\ |\Gamma_1|=5\,|\Gamma_0|=|\Gamma_5|,$$
$$|\Gamma_2|=10\,|\Gamma_1|=|\Gamma_4|$$

Thus, $32\,|\Gamma_0|=1.15$ or $|\Gamma_0|=0.036$.

Where $|\Gamma_0|=0.036$, $|\Gamma_1|=0.180$, $|\Gamma_3|=0.360$.

The impedances are

$$z_1=(\text{antilog } 2\times 0.036)\times 100=107.5\ \Omega$$
$$z_2=(\text{antilog } 2\times 0.180)\times 107.5=154\ \Omega$$
$$z_3=(\text{antilog } 2\times 0.360)\times 154=316\ \Omega$$
$$z_4=(\text{antilog } 2\times 0.360)\times 316=648\ \Omega$$
$$z_5=(\text{antilog } 2\times 0.180)\times 648=930\ \Omega.$$

## 6.30 CHEBYSHEV TRANSFORMER

In this transformer reflection coefficients are selected to be proportional to the Chebyshev polynomials. As the polynomial oscillates between $\pm 1$ for $x \leqslant 1$ and increases indefinitely for $x > 1$, the pass band characteristics of this type of transformer are such that $|\Gamma_T|$ oscillates between 0 and $|\Gamma_m|$, where $|\Gamma_m|$ is the maximum tolerable mismatch, as shown in Fig. 6.50. For this transformer, Eq. (6.142) may be written as*

$$|\Gamma_T|=A\,e^{-jn\theta}\,T_n\,(\sec\theta_m\cos\theta),$$

where $A$ is a constant to be determined and $\theta_m$ is the maximum tolerable electrical length.
When $\theta=0$,

$$|\Gamma_T|=\frac{Z_L-Z_0}{Z_L+Z_0}=A\,T_n\,(\sec\theta_m)$$

or

$$A=\frac{Z_L-Z_0}{(Z_L+Z_0)\,T_n\,(\sec\theta_m)}$$

*R.E. Collin, *Foundations for Microwave Engineering*, McGraw-Hill Book Co. (1966).

Hence, $$|\Gamma_T| = e^{-jn\theta} \frac{Z_L - Z_0}{Z_L + Z_0} \frac{T_n(\sec\theta_m\cos\theta)}{T_n(\sec\theta_m)}.$$

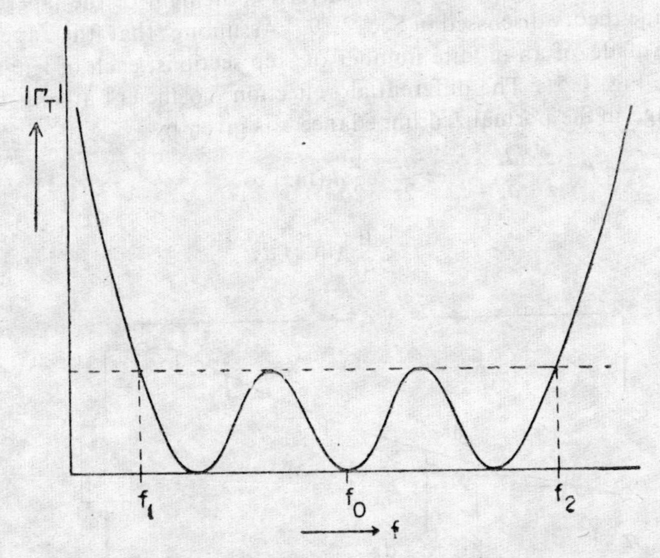


Fig. 6.50 Passband characteristics of a Chebyshev transformer.

In a passband $T_n(\sec\theta_m\cos\theta)$ has its maximum value unity. Therefore,

$$|\Gamma_m| = \frac{Z_L - Z_0}{(Z_L + Z_0)\,T_n(\sec\theta_m)}$$

or

$$T_n(\sec\theta_m) = \frac{Z_L - Z_0}{Z_L + Z_0}(|\Gamma_m|)^{-1}.$$

Since,

$$T_n\left(\frac{\cos\theta}{\cos\theta_m}\right) = \cos n\left(\cos^{-1}\frac{\cos\theta}{\cos\theta_m}\right)$$

(property of $T_n$ polynomials).

Therefore,

$$\sec(\theta_m) = \cos\left(\frac{1}{n}\cos^{-1}\frac{Z_L - Z_0}{Z_L + Z_0}(\Gamma_m|)^{-1}\right). \tag{6.148}$$

Equation (6.148) gives $\theta_m$ in terms of passband tolerance $|\Gamma_m|$. Once $\theta_m$ is specified, it is easy to design the transformer because all $T_n$'s are known.

## 6.31 TAPERED SECTION TRANSFORMER

The two waveguides differing in dimensions they may be connected by a smoothly tapered waveguide section. No doubt, such a tapered transformer provides electrical smoothness when the wave impedances of the two waveguides are the same. However, if the wave impedances of the guides differ, careful design of the transformer provides electrical smoothness.

A tapered transformer may be analyzed in the light of the tapered transmission line theory discussed in Sec. 2.19. Assuming that the tapered section to be made of an infinite number of step sections; each of length $dz$, as shown in Fig. 6.51. The differential reflection coefficient arising from the step change in the normalized impedance $z$ is given by

$$d\Gamma = \frac{z+dz-z}{z+dz+z} \simeq \frac{dz}{2z} = \frac{1}{2} d\,(\ln z)$$

$$= \frac{1}{2}\frac{d}{ds}(\ln z)\,ds$$

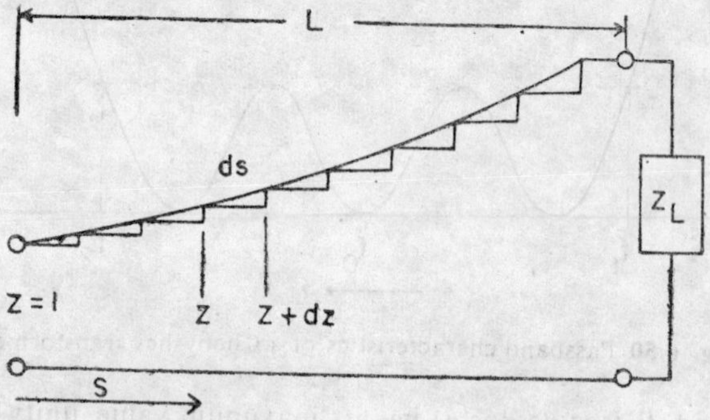

**Fig. 6.51** Tapered section transformer.

The contribution to the reflection coefficient from this step at the input of the taper is given by

$$d\Gamma_i = e^{-2\bar{\beta}s}\,\frac{1}{2}\,\frac{d}{ds}(\ln z)\,ds.$$

The total input reflection coefficient may be supposed to be made of such contributions from each small step of length $ds$. Then the total reflection coefficient is given by

$$|\Gamma_i| = \frac{1}{2}\int_0^L e^{-2\bar{\beta}s}\,\frac{d}{ds}(\ln z)\,ds. \tag{6.149}$$

Equation (6.149) forms the basis of various types of tapered transformers. Exponential, triangular and Chebyshev tapers are some common type of tapers.

*Exponential Taper*. In this transformer normalized impedance varies exponentially with the distance from unity to a desired value $z_L$. Consequently, ln ($z$) varies linearly with $s$ and Eq. (6.149) can be integrated using

$$\ln(z) = \frac{s}{L}\ln(z_L)$$

$$|\Gamma_i| = \frac{1}{2}\int_0^L \frac{\ln(z_L)}{L}\,e^{-2j\bar{\beta}s}\,ds = \frac{1}{2}\,e^{-j\bar{\beta}L}\ln z_L\,\frac{\sin\bar{\beta}L}{\bar{\beta}L} \tag{6.150}$$

The response curve (plot of Eq. 6.150) of such an exponential taper is shown in Fig. 6.52. It is seen that when the length of the taper is less than $\lambda/2$ the mismatch is quite small.

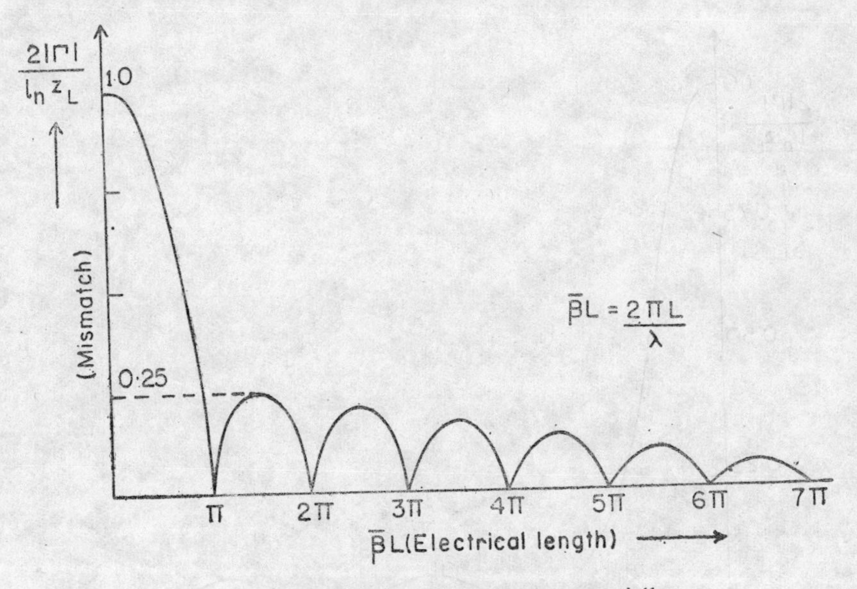


**Fig. 6.52** Response curve of an exponentially tapered transformer.

*Triangular Taper.* If the taper is triangular, i.e.

$$\frac{d}{ds}(\ln z) = \begin{cases} \frac{4s}{L^2} \ln (z_L), & 0 \leqslant s \leqslant L_2 \\ \frac{4}{L^2}(L-s) \ln (z_L), & L/2 \leqslant s \leqslant L. \end{cases}$$

we have a triangular tapered transformer which has more favourable properties. For such a taper total input reflection coefficient is given by*

$$|\Gamma_i| = \frac{1}{2} e^{-j\beta \bar{L}} \ln (z_L) \left\{ \frac{\sin \beta L/2}{\beta L/2} \right\}^2. \tag{6.151}$$

The response curve of a triangular tapered transformer is shown in Fig. 6.53. It is noted that the transformer action is smoother but the minimum required length is 1.63 $\lambda/2$ as compared to $\lambda/2$ in exponential taper.

*Chebyshev Taper.* A Chebyshev tapered transformer consists of an infinite number of Chebychev transformers discussed in Sec. 6.28. However, this transformer has a length 27 percent shorter than the triangulas taper for the same passband tolerance and a lower cut-off frequency.

*J.L. Altman, *Microwave Circuits* New York: Van Nostrand (1964).

If the number of sections in a binomial transformer is increased to infinity what we have is a Gaussian transformer. The study of these transformers is beyond the scope of present volume.

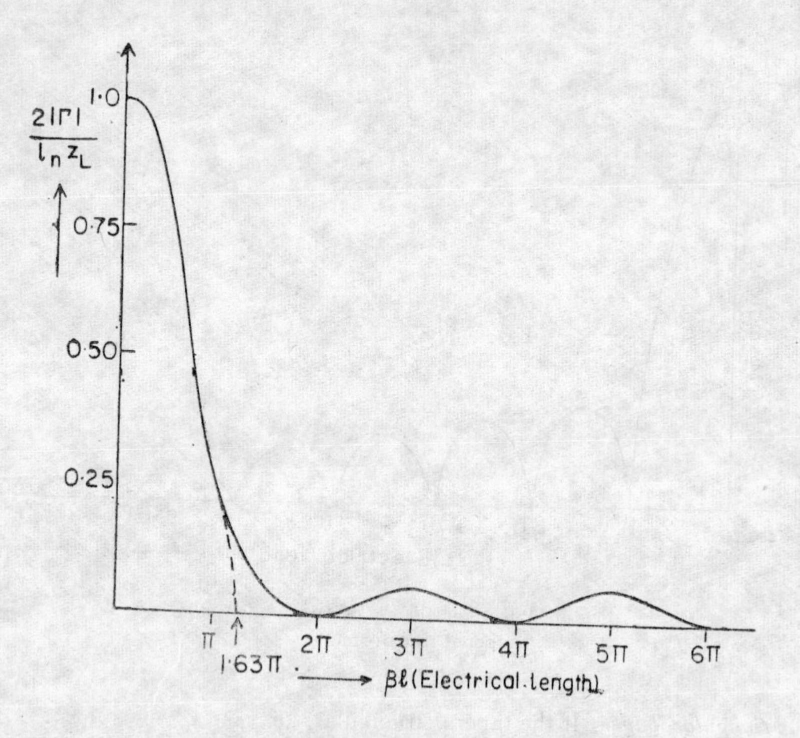


**Fig. 6.53** Response of a triangular taper transformer.

## 6.32 DIELECTRIC TRANSFORMERS

Two waveguides of the same cross-section but of different characteristics may be matched by filling them with different dielectrics provided Eq. (6.132) is satisfied. Typical dielectric transformers are depicted in Figs. 6.54(a) through (e). The dielectrics commonly used have values ranging from $\epsilon_r=2.5$ to 20. A typical dielectric is obtained by mixing 8 percent by weight of ethyl cellulose to ethyl acitamide or simply lead chloride mixed with polystyrene. Referring to Fig. 6.54(b), the quarter-wave section is, in this case, the region over which an overlap of media prevails. The width of the slot determines the relative magnitude of the component reflections arising from the two discontinuities while the depth of the slot determines their phase difference. Or in transmission line language, the slot provides the characteristics impedance $Z$ and the width provides the length of the quarter-wave dielectric transformer. In cylindrical geometry (Fig. 6.54(e)), two fins are provided which are parallel to the electric field of the dominant mode and are of length $\lambda_g/4$.

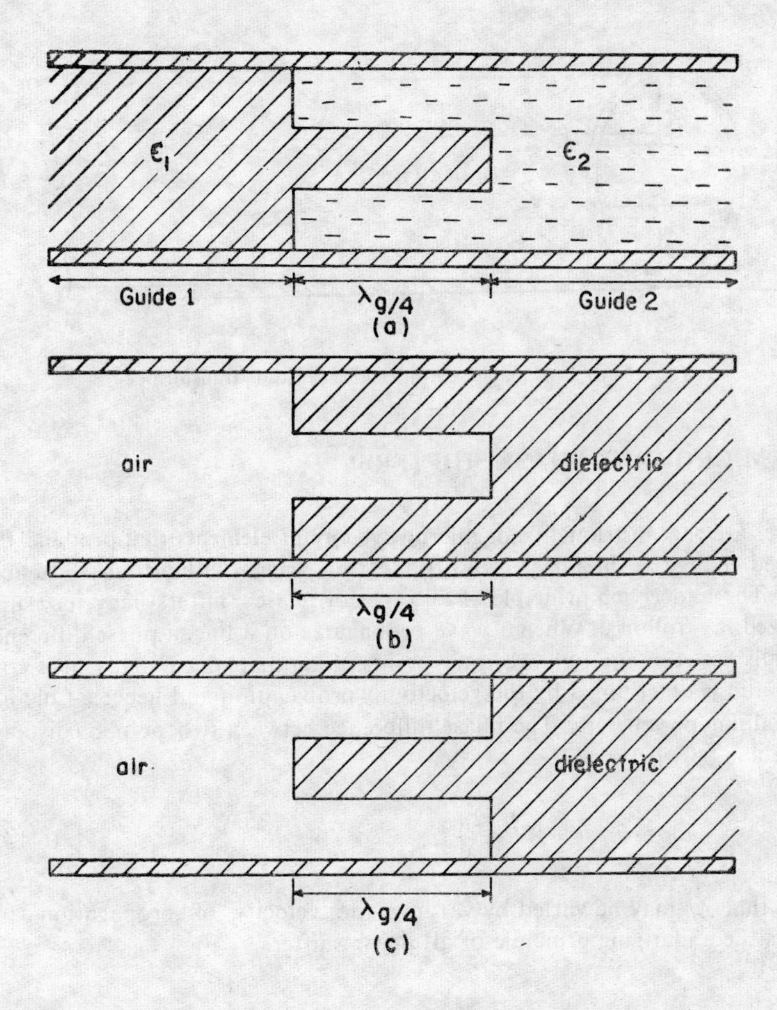


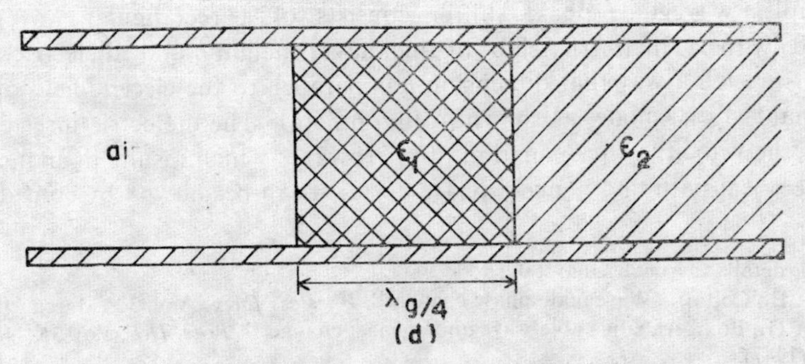


**Fig. 6.54** Dielectric transformers (*a*) General form (*b*) and (*c*) Rectangular waveguide type with dielectric as quarter wave transformer (*d*) Loaded Waveguide type.

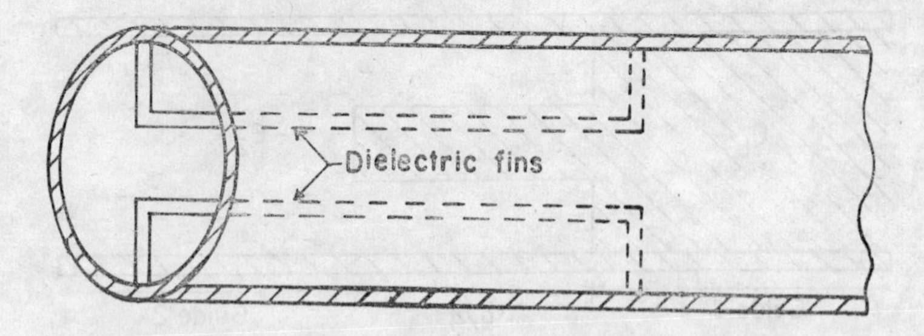


**Fig. 6.54(*e*)** Fins loaded cylindrical waveguide transformer.

## 6.33 MICROWAVE PHASE SHIFTERS*

These are the instruments or microwave circuit elements that produce the desired change in the phase of a propagating wave, without any attenuation. The underlying principle of all types of phase shifters may be summarized as follows. When a wave propagates on a line, a phase difference prevails between any two arbitrary points along its path. Usually, it is constant and is determined by the velocity of propagation and hence a function of medium parameters. The phase difference between two points, $l$ distance apart can be written as

$$\Delta\phi=\phi_2-\phi_1=\beta l=\frac{2\pi}{\lambda}\, l. \tag{6.152}$$

Note that $\Delta\phi$ may be varied by varying the velocity of propagation and this is the underlying principle of all phase shifters.

## 6.34 DIELECTRIC PHASE SHIFTER

A simplest waveguide phase shifter consists of a rectangular waveguide loaded with a dielectric slab of thickness $t$, height $h$ and dielectric constant $\epsilon$ in such a way that dimension $h$ is parallel to the electric field of the dominant $TE_{10}$ mode as shown in Fig. 6.55(a). The dielectric inserted reduces the velocity of propagation of microwaves which results in an increased electrical path and hence a phase delay. It can be shown* by considering

*For details the reader may refer to:

1. R.E., Collin, "Waveguide phase changer", *Wireless Engg.*, Vol. 32 (March 1955).
2. A.G., Fox "An adjustable waveguide phase changer" *Proc. IRE*, Vol. 35, No. 12 Dec. (1947).
3. J. Helszain, "*Passive and Active Microwave Circuits*," Wiley-Interscience, New York (1978).

*J.L. Altman, "*Microwave Circuits*", Van Nostrand New York (1964)

the waveguide to be perturbed due to the interaction of the dielectric that the propagation constant of the perturbed waveguide $\beta$ is given by

$$\frac{\omega}{v}=\beta=\beta_0+\epsilon_0(\epsilon'-1)\frac{\int_{\Delta s} E_0'^* E' \, ds}{4P} \tag{6.153}$$

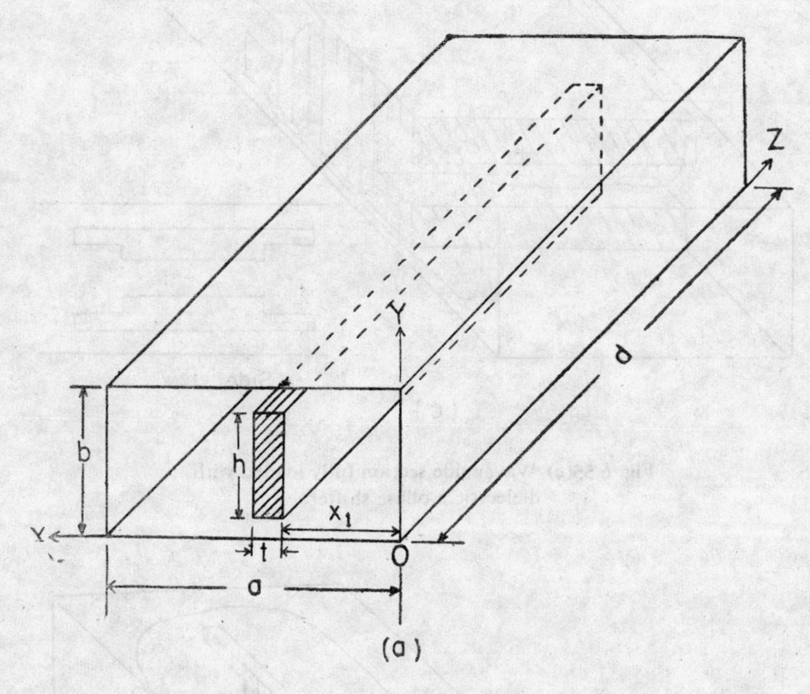


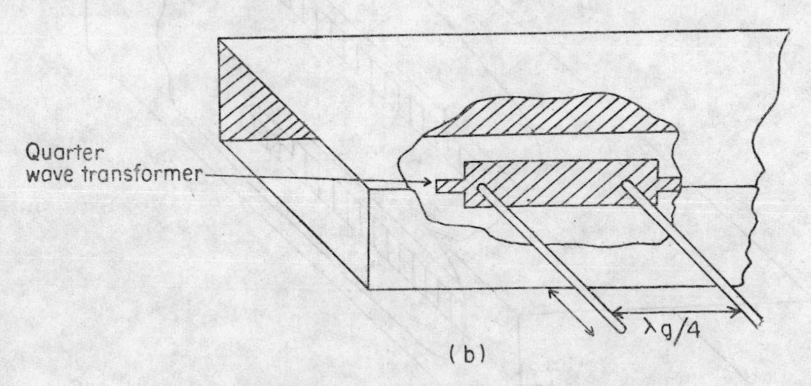


**Fig. 6.55** Waveguide dielectric phase shifter. (*a*) principle (*b*) practical form.

$\beta_0$ is the unperturbed propagation constant. $\epsilon_0$ is the freespace permittivity, $E_0$ and $E_0'$ are the electric fields in the $TE_{10}$ mode without and with the dielectric respectively. Using the electric field of the dominant $TE_{10}$ mode

$$E_y'=E_y=E_m \sin \pi x/a. \tag{6.154}$$

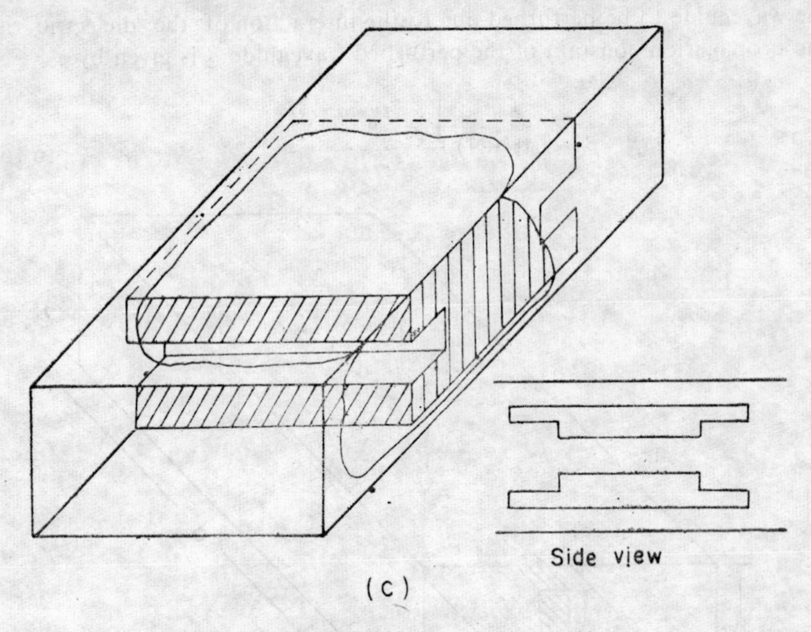


**Fig. 6.55**(*c*) **Waveguide section fully loaded with dielectric a phase shifter.**

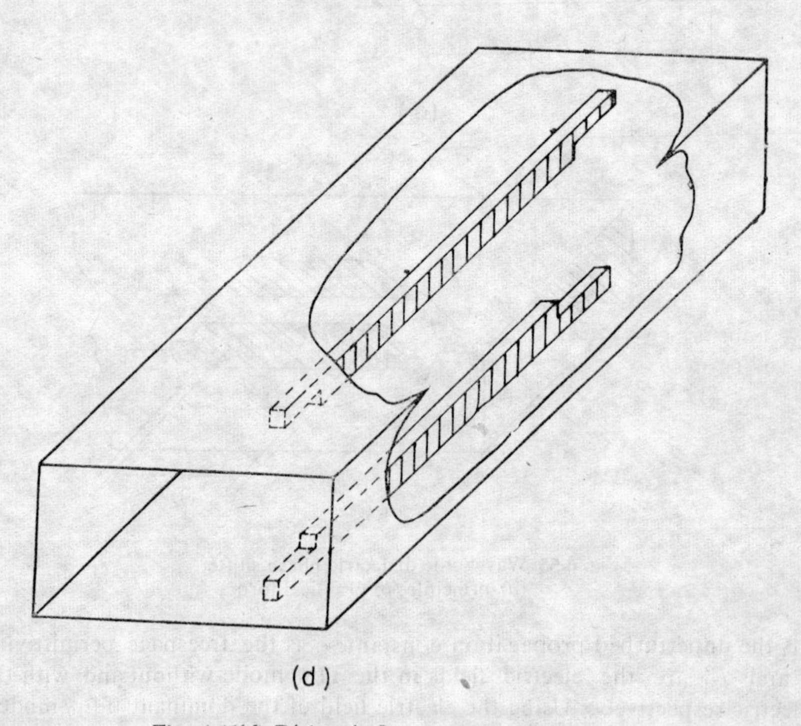


**Fig. 6.55**(*d*) **Dielectric fins waveguide phase shifter.**

where $E_m$ is the peak electric field at $x=a/2$; the output power flowing through the guide is given by

$$P=\frac{1}{2Z_{TE}}\int_s |E_t|^2\, ds=\frac{E_m^2}{2Z_{TE}}\int_0^b\int_0^a \sin^2\frac{\pi x}{a}\,dxdy$$

$$=\frac{E_m^2}{4Z_{TE}}\,a\,b=\frac{E_m^2}{4Z_{TE}}\,S \tag{6.155}$$

where $Z_{TE}=\sqrt{\frac{\mu_0}{\epsilon_0}\left(\frac{\lambda_{g0}}{\lambda}\right)^2}$ and $S$ is cross-section.

Using Eq. (6.155) in Eq. (6.153), we obtain,

$$\beta=\beta_0+(\epsilon'-1)\left(\omega\epsilon_0\sqrt{\frac{\mu_0}{\epsilon_0}}\right)\frac{\lambda_{g0}}{\lambda}\sin^2\frac{\pi x_1}{a}\left(\frac{\Delta S}{S}\right) \tag{6.156}$$

or in wavelengths

$$\frac{1}{\lambda_g}=\frac{1}{\lambda_{g0}}+(\epsilon'-1)\left(\frac{\Delta S}{S}\right)\frac{\lambda_{g0}}{\lambda_g^2}\sin^2\frac{\pi x_1}{a}.$$

It is noted from Eq. (6.156) that propagation characteristics of a perturbed waveguide depend upon $\epsilon'$, $t$, $h$ and the location of the slab $x_1$. Clearly, $(\beta-\beta_0)=\Delta\beta$ is maximum when the slab is at $x=a/2$ which is expected because the electric field is maximum at $x=a/2$. Thus, if the slab is moved laterally we can introduce the desired phase change and the system will become a variable phase changer as shown in Fig. 6.55(b). It may, however, be mentioned that Eq. (6.153) is subjected to perturbation assumptions and hence theory fails if $h \gg t$ besides other factors.

In practice, the ends of the dielectric slab are tapered so as to avoid reflections at the ends. For narrow band operation one quarter-wave step transformer is made in the dielectric slab at each of its ends. Sometimes, two fins are also used in place of a slab. Typical practical dielectric phase shifters are shown in Figs. 6.55(c) and (d). The supporting pins of a variable attenuator (Fig. 6.55b) are $\lambda_g/4$ distance apart to minimize reflections arising from these pins.

## 6.35 WAVEGUIDE STRETCHER

If a waveguide having a longitudinal slot on the broad face is subjected to a pressure on the other faces by means of a clamp, its broader dimension is reduced, which in turn results in decreased electrical length and hence the system works as a waveguide phase changer or *line stretcher*. The phase change produced in the outgoing wave is proportional to the change in the broader dimension of the waveguide and so the desired phase shift may be obtained. However, the slot is made sufficiently long so that when the line stretcher is compressed to its smallest width, the change in dimension is still so gradual that serious reflections do not occur. A typical line stretcher is shown in Fig. 6.56.

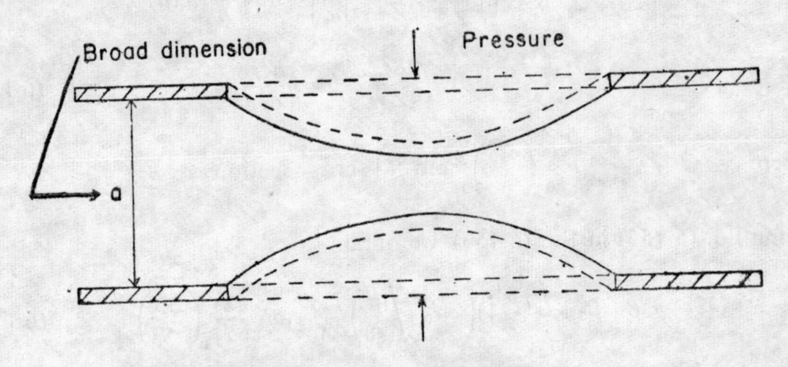


Fig. 6.56 Waveguide stretcher.

## 6.36 LINEAR PHASE SHIFTER

A typical linear dielectric phase shifter (Fig. 6.57) consists of three dielectric slabs placed in a rectangular waveguide such that the central slab is

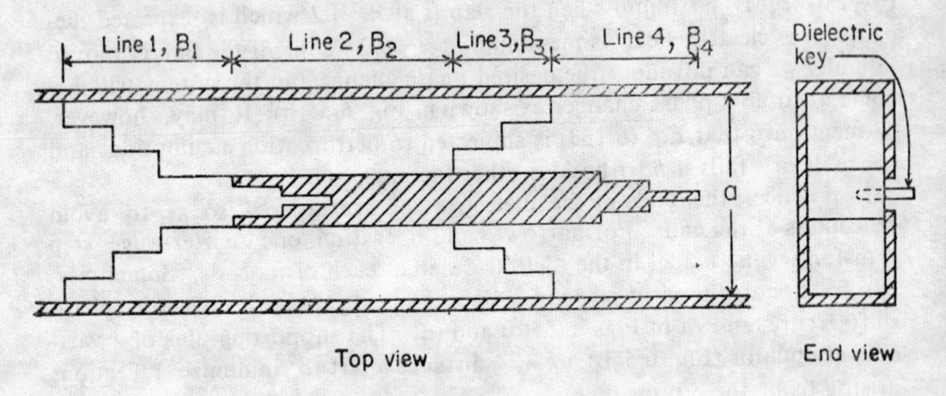


Fig. 6.57 Linear dielectric phase shifter.

movable longitudinally by a suitable drive mechanism. A dielectric key protruded through a long centred slot cut in one broad face of the guide is attached to the central slab. The ends of the slabs are so cut that these constitute multisection broad band quarter-wave transformers that provide a match between unloaded and dielectric loaded waveguides. It is found that when the central slab is about 0.3 a times wide, $\beta_1+\beta_2>\beta_2+\beta_4$. Therefore, referring to Fig. 6.57, the displacement of the central slab towards the right by a distance $l$ amounts to an increase in the length of the lines

1 and 3 by an amount $l$ and to decrease in the length of the lines 2 and 4 by the same amount. Consequently, the wave propagating through the phase changer undergoes a phase change

$$\Delta\phi=\{(\beta_1+\beta_3)-(\beta_2+\beta_4)\}\ l. \quad (6.157)$$

Thus varying $l$, the desired phase change may be obtained. However, there is an upper limit to the phase change determined by the values of $\beta_1$, $\beta_2$, $\beta_3$ and $\beta_4$. The central slab has to be displaced about 16 cm to obtain a 360° phase shift for a guide 0.9 inch loaded with a dielectric $\epsilon=2.56$.

## 6.37 DIFFERENTIAL PHASE SHIFTER: QUARTER-AND HALF-WAVE PLATES

Consider a circular waveguide diametrically loaded with a thin dielectric slab as shown in Fig. 6.58. Clearly the dominant $TE_{11}$ mode will suffer

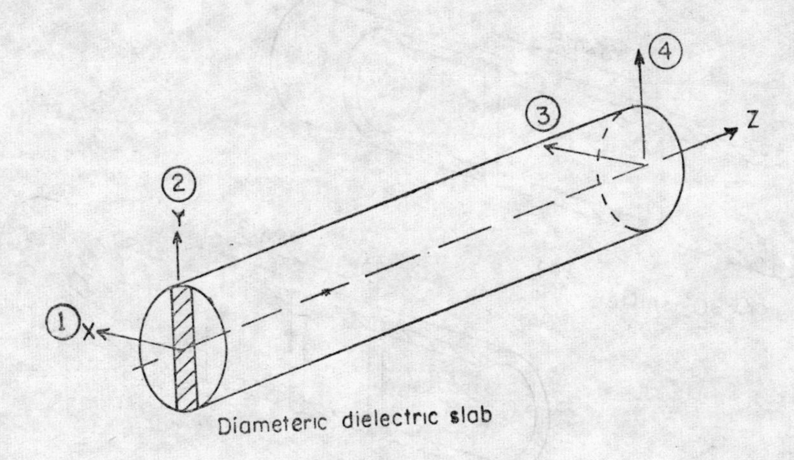


**Fig. 6.58** Quarter-half-wave plate using dielectric slab in a circular waveguide.

maximum phase delay when its electric field lies in the principal dielectric plane and minimum (almost zero) when it lies perpendicular to it. Now a wave approaching the slab, polarized at some intermediate angle may be considered to be made up of two components, one lying in the plane of the slab and another perpendicular to it. Further, if both these degenerate $TE_{11}$ modes are matched (by suitably tapering the slab), then while passing through such a section the mode polarized || to the slab will be retarded more than the one ⊥ to it. After passing through the dielectric region the two degenerate modes combine but the wave suffers a net differential phase shift. Such a device is called a differential phase shifter.

The differential phase shift obtainable from the differential phase shifter depends upon the length, the dielectric constant and the thickness of the slab. If these parameters are suitably adjusted, it is not difficult to obtain a differential phase shift of $\pi/2$ or $\pi$ and accordingly we can have quarter- or half-wave sections. These are respectively designated as $\Delta 90°$ and $\Delta 180°$.

Quarter- and half-wave sections can also be designed utilizing other means, e.g. by having an elliptical cross-section, fins or ridges or pins into the waveguides, etc. However, the desirable feature is that these are essentially anisotropic in the transverse plane, i.e. they must "favour" one direction over the direction perpendicular to them. In iris, fins and ridge type of differential phase shifters the susceptance arising from the discontinuity introduces a relative phase difference between two polarizations. The inductive iris has the effect of retarding the phase while the capacitive iris advances it. Typical quarter-half-wave plates designed using metal fins are shown in Fig. 6.59 (a and b).

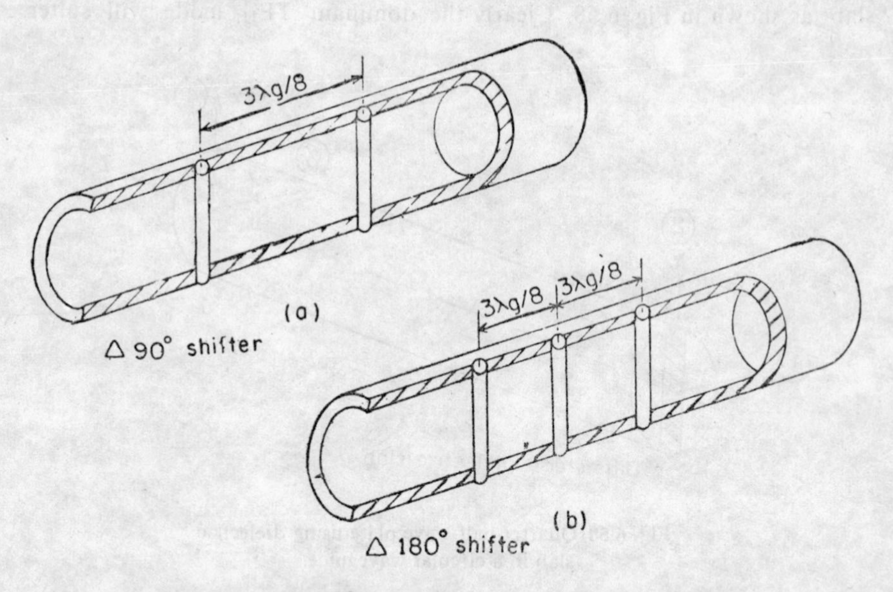


**Fig. 6.59** Metal fin type quarter (*a*) and half (*b*) wave plate.

The properties of quarter-and half-wave sections can best be understood using a scattering matrix.

## 6.38 SCATTERING MATRIX OF A QUARTER-WAVE SECTIONS

In physical appearance, a quarter-wave section is a two-port, one input and one output network, but electrically it has to be considered as a four-port network because each of the degenerate $TE_{10}$ polarizations at the input and output forms a separate port. Numbering these parts as shown in Fig. 6.58 such that ports 1 and 2 form a right handed system with direction of polarization along $z$-axis, any incident wave may be considered to be the vectorial sum of the waves entering at ports 1 and 2.

Suppose $a_1$ and $a_2$ be the normalized waves incident at ports 1 and 2 respectively. Then the outgoing normalized waves at ports 3 and 4 respectively are given by

$$b_3=a_1\, e^{-j\phi_1},\ b_4=a_2\, e^{-j\phi_1}\, e^{-j\pi/2}=-ja_2\, e^{-j\phi_1}. \tag{6.158}$$

The polarization travelling parallel to the dielectric slab suffers an additional phase delay of $\pi/2$ to $\phi_1$, the phase delay being produced due to wave propagation in the waveguide of length $l$.

From the reciprocity of ports 1, 3 and 2, 4 we have,

$$b_1=a_3\, e^{-j\phi_1},\ b_2=a_4\, e^{-j\phi_1}\, e^{j\pi/2}=-ja_2\, e^{-j\phi_1}. \tag{6.159}$$

From Eqs. (6.158) and (6.159), it follows that

$$\begin{bmatrix} b_1 \\ b_2 \\ b_3 \\ b_4 \end{bmatrix} = e^{-j\phi_1} \begin{bmatrix} 0 & 0 & 1 & 0 \\ 0 & 0 & 0 & -j \\ 1 & 0 & 0 & 0 \\ 0 & -j & 0 & 0 \end{bmatrix} \begin{bmatrix} a_1 \\ a_2 \\ a_3 \\ a_4 \end{bmatrix}. \tag{6.160}$$

Hence, the scattering matrix of a quarter-wave section is

$$[S]_{\Delta_{90}^\circ}=e^{j\phi_1} \begin{bmatrix} 0 & 0 & 1 & 0 \\ 0 & 0 & 0 & -j \\ 1 & 0 & 0 & 0 \\ 0 & -j & 0 & 0 \end{bmatrix}. \tag{6.161}$$

Some important properties may now be derived from Eq. (6.160).

**1. A wave linearly polarized in a plane oriented at +45° with respect to port 1, is transformed into a right-handed circularly polarized wave.**

In this case $a_1=a_2=a\, e^{j\phi_0}$ ($\phi_0$ is the arbitrary phase angle). So the output is

$$\begin{bmatrix} b_1 \\ b_2 \\ b_3 \\ b_4 \end{bmatrix} = e^{-j\phi_1} \begin{bmatrix} 0 & 0 & 1 & 0 \\ 0 & 0 & 0 & -j \\ 1 & 0 & 0 & 0 \\ 0 & -j & 0 & 0 \end{bmatrix} e^{j\phi_0} \begin{bmatrix} a_1 \\ a_2 \\ 0 \\ 0 \end{bmatrix}$$

$$= e^{-j(\phi_1-\phi_0)} \begin{bmatrix} 0 \\ 0 \\ 0 \\ -ja \end{bmatrix}. \tag{6.162}$$

Equation (6.162) represents a right-handed circularly polarized wave;

$$b_3^2 + b_4^2 = a^2.$$

**2. A wave linearly polarized in a plane oriented at $-45°$ with respect to port 1, is transformed into a left-handed circularly polarized wave.**

In this case we have

$$a_1 = -a_2 = a\, e^{j\phi_0}. \text{ So,}$$

$$\begin{bmatrix} b_1 \\ b_2 \\ b_3 \\ b_4 \end{bmatrix} = e^{-j\phi_1} \begin{bmatrix} 0 & 0 & 1 & 0 \\ 0 & 0 & 0 & -j \\ 1 & 0 & 0 & 0 \\ 0 & -j & 0 & 0 \end{bmatrix} e^{j\phi_0} \begin{bmatrix} a \\ a \\ 0 \\ 0 \end{bmatrix} = e^{-j(\phi_1 - \phi_0)} \begin{bmatrix} 0 \\ 0 \\ a \\ ja \end{bmatrix}. \tag{6.163}$$

That is outgoing wave is left-handed circularly polarized.

**3. A right-handed circularly polarized wave is transformed into a wave linearly polarized 45° with respect to port 1.**

In this case we have

$$a_1 = ae^{j\phi_0}$$

$$a_2 = ae^{j\phi_0}\, e^{-j\pi/2}$$

$$= -ja\, e^{j\phi_0}.$$

So,

$$\begin{bmatrix} b_1 \\ b_2 \\ b_3 \\ b_4 \end{bmatrix} = e^{-j\phi_1} \begin{bmatrix} 0 & 0 & 1 & 0 \\ 0 & 0 & 0 & -j \\ 1 & 0 & 0 & 0 \\ 1 & -j & 0 & 0 \end{bmatrix} e^{j\phi_0} \begin{bmatrix} a \\ -ja \\ 0 \\ 0 \end{bmatrix} = e^{-j(\phi_1 - \phi_0)} \begin{bmatrix} 0 \\ 0 \\ a \\ -a \end{bmatrix}. \tag{6.164}$$

**4. A left-handed circularly polarized wave is transformed into a linearly polarized wave with respect to Port 1.**

$$\begin{bmatrix} b_1 \\ b_2 \\ b_3 \\ b_4 \end{bmatrix} = e^{-j\phi_1} \begin{bmatrix} 0 & 0 & 1 & 0 \\ 0 & 0 & 0 & -j \\ 1 & 0 & 0 & 0 \\ 0 & -j & 0 & 0 \end{bmatrix} e^{j\phi_0} \begin{bmatrix} a \\ ja \\ 0 \\ 0 \end{bmatrix} = e^{-j(\phi_1 - \phi_0)} \begin{bmatrix} 0 \\ 0 \\ a \\ a \end{bmatrix}. \tag{6.165}$$

Thus, in brief, a quarter-wave section is a device that will produce a circularly polarized wave when a linearly polarized wave is incident upon it and vice versa.

## 6.39 SCATTERING MATRIX OF A HALF-WAVE SECTION

In a half-wave section unlike a quarter-wave section, the dielectric slab is oriented at an arbitrary angle $\theta$ to the right hand positive with respect to port 1 as shown in Fig. 6.60. Following Fig. 6.60, if a normalized wave $a_1$

Fig. 6.60 Ports in a half-wave plate for scattering matrix consideration.

is incident at port 1, instead of going through port 3, the wave will be divided into ports 3 and 4. The incident wave $a_1$ may be considered to be the vectorial sum of two orthogonal components parallel and perpendicular to the slab, say, $a_1^{\parallel}$ and $a_1^{\perp}$. As they pass through the device $a_1^{\perp}$ loses an arbitrary phase $\phi_2$ but $a_1^{\parallel}$ loses a phase $\phi_2+\pi$. These components may be written as

$$a_1^{\perp}=a_1 \sin\theta$$
$$a_1^{\parallel}=a_1 \cos\theta.$$

The corresponding outputs are

$$a_{1\,out}^{\perp}=a_1 \sin\theta\, e^{-j\phi_2}$$
$$a_{1\,out}^{\parallel}=a_1 \cos\theta\, e^{-j\phi_2}\, e^{-j\pi}=-a_1 \cos\theta\, e^{-j\phi_2}$$

Therefore, the output at port 3 will be

$$\begin{aligned} b_3 &= a_{1\,out}^{\perp} \sin\theta + a_{1\,out}^{\parallel} \cos\theta \\ &= a_1 (\sin^2\theta - \cos^2\theta)\, e^{j\phi_2} \\ &= -a_1 \cos 2\theta\, e^{-j\phi_2}. \end{aligned} \tag{6.166}$$

Similarly, $$\begin{aligned} b_4 &= -a_{1\,out}^{\perp} \cos\theta + a_{1\,out}^{\parallel} \sin\theta \\ &= -a_1 (\sin\theta \cos\theta + \sin\theta \cos\theta)\, e^{-j\phi_2} \end{aligned}$$

or $$b_4 = -a_1 \sin 2\theta\, e^{-j\phi_2}. \tag{6.167}$$

Thus, the first column and hence the first row (due to symmetry) of the scattering matrix has elements 0, 0, $-\cos 2\theta$, $-\sin 2\theta$.

Similarly, considering $a_2$ to be the vectorial sum of two orthogonal components, it may be proved that the second column and hence the second row of elements of the scattering matrix are given by 0, 0, $-\sin 2\theta$, $\cos 2\theta$. The complete scattering matrix is

$$[S]_{180^\circ}=e^{-j\phi_2}\begin{bmatrix} 0 & 0 & -\cos 2\theta & -\sin 2\theta \\ 0 & 0 & -\sin 2\theta & \cos 2\theta \\ -\cos 2\theta & -\sin 2\theta & 0 & 0 \\ -\sin 2\theta & \cos 2\theta & 0 & 0 \end{bmatrix}. \tag{6.168}$$

It is noted that if $\theta$ is varied, the scattering coefficient also varies.

Some of the important properties of a half-wave plate may be derived from Eq. (6.168).

1. **A right-handed circularly polarized wave applied to a half-wave plate is transformed into a left-handed circularly polarized wave and vice versa. Furthermore, each component of the circularly polarized wave suffers a phase delay twice the space angle $\theta$ in case of right-handed polarization and a phase advance of twice the space angle $\theta$ in case of left-handed polarization, in addition to a fixed phase delay.**

Consider a circularly polarized incident wave

$$\begin{bmatrix} a_1 \\ a_2 \\ a_3 \\ a_4 \end{bmatrix} = e^{j\phi_0}\begin{bmatrix} a \\ \mp ja \\ 0 \\ 0 \end{bmatrix}; \quad \begin{matrix} - \rightarrow \text{RH polarization} \\ + \rightarrow \text{LH polarization} \end{matrix}$$

Then, the output is

$$\begin{bmatrix} b_1 \\ b_2 \\ b_3 \\ b_4 \end{bmatrix} = e^{-j\phi_2}\begin{bmatrix} 0 & 0 & -\cos 2\theta & -\sin 2\theta \\ 0 & 0 & -\sin 2\theta & \cos 2\theta \\ -\cos 2\theta & -\sin 2\theta & 0 & 0 \\ -\sin 2\theta & \cos \theta\theta & 0 & 0 \end{bmatrix} e^{j\phi_0}\begin{bmatrix} a \\ \mp ja \\ 0 \\ 0 \end{bmatrix}$$

$$= a\, e^{-j(\phi_2-\phi_0)}\begin{bmatrix} 0 \\ 0 \\ -\cos 2\theta \pm j\ \sin 2\theta \\ -\sin 2\theta \pm j\ \cos 2\theta \end{bmatrix}$$

$$= a\, e^{-j(\phi_2-\phi_0)} \begin{bmatrix} 0 \\ 0 \\ -e^{\mp j 2\theta} \\ \mp j\, e^{\mp j2\theta} \end{bmatrix} = a\, e^{-j(\psi \pm 2\theta)} \begin{bmatrix} 0 \\ 0 \\ a \\ \pm ja \end{bmatrix} \tag{6.169}$$

where $\psi = \phi_2 - \phi_0 + \pi$ does not depend upon $\theta$.

Equation (6.169) verifies the property stated above.

2. **When a wave linearly polarized at an angle $\theta$ is applied to a half-wave plate, it is transformed into a wave linearly polarized at an angle $(2\theta - \theta_1)$.**

Consider a linearly polarized incident wave polarized at an angle $\theta$, then

$$\begin{bmatrix} a_1 \\ a_2 \\ a_3 \\ a_4 \end{bmatrix} = e^{j\phi_0} \begin{bmatrix} a\cos\theta_1 \\ a\sin\theta_1 \\ 0 \\ 0 \end{bmatrix}$$

The output is given by

$$\begin{bmatrix} b_1 \\ b_2 \\ b_3 \\ b_4 \end{bmatrix} = e^{-j\phi_2} \begin{bmatrix} 0 & 0 & -\cos 2\theta & -\sin 2\theta \\ 0 & 0 & -\sin 2\theta & \cos 2\theta \\ -\cos 2\theta & -\sin 2\theta & 0 & 0 \\ -\sin 2\theta & \cos 2\theta & 0 & 0 \end{bmatrix}$$

$$a\, e^{j\phi_0} \begin{bmatrix} \cos\theta_1 \\ \sin\theta_1 \\ 0 \\ 0 \end{bmatrix} = a e^{-j(\phi_2 - \phi_0)} \begin{bmatrix} 0 \\ 0 \\ -(\cos 2\theta \,\cos\theta_1 + \sin 2\theta \,\sin\theta_1) \\ -(\sin 2\theta \,\cos\theta_1 + \cos 2\theta \,\sin\theta_1) \end{bmatrix}$$

$$= a e^{-j\psi} \begin{bmatrix} 0 \\ 0 \\ \cos(2\theta - \theta_1) \\ \sin(2\theta - \theta_1) \end{bmatrix} \tag{6.170}$$

where $\quad \psi=\phi_2-\phi_0+\pi.$

Thus the output is polarized at an angle $(2\theta-\theta_1)$. However, if $\theta=\theta_1$, there is no change in the plane of polarization. On the other hand, if $\theta-\theta_1=45°$, i.e. the incident wave is polarized at $-45°$ with respect to the half-wave section, then,

$$(2\,\theta-\theta_1)=2(\theta-\theta_1)+\theta_1=90°+\theta_1.$$

That is, the orientation angle is rotated by 90°.

## 6.40 PRECESSION ROTARY PHASE CHANGER

A precession rotary phase changer is shown in Fig. 6.61(a). The instrument consists of two rectangular to circular waveguide tapered transitions, together with two quarter-wave sections on both the sides of the free rotatable central half-wave section. The quarter-wave sections are oriented at an angle 45° relative to the broad wall of the rectangular waveguide. The incoming linearly polarized. $TE_{11}$ mode is decomposed into two modes polarized parallel and perpendicular to the quarter-wave section. When a half-wave section happens to be in its zero set position, the outgoing wave suffers a total phase of $90°+180°=270°$. Consequently, the wave going out of the second quarter-wave section suffers a total phase $270°+90°=360°$, i.e. no phase change under ideal conditions. However, when the central half-wave section is rotated by an angle $\theta$, the outgoing wave suffers a phase delay of $2\theta$.

The basic operation of the device can be understood with reference to Fig. 6.61(b), which explicitly shows all the components except transitions. Of course, instead of orienting the quarter-wave section at 45° to the rectangular waveguide, the latter is oriented at an angle 45° to the first.

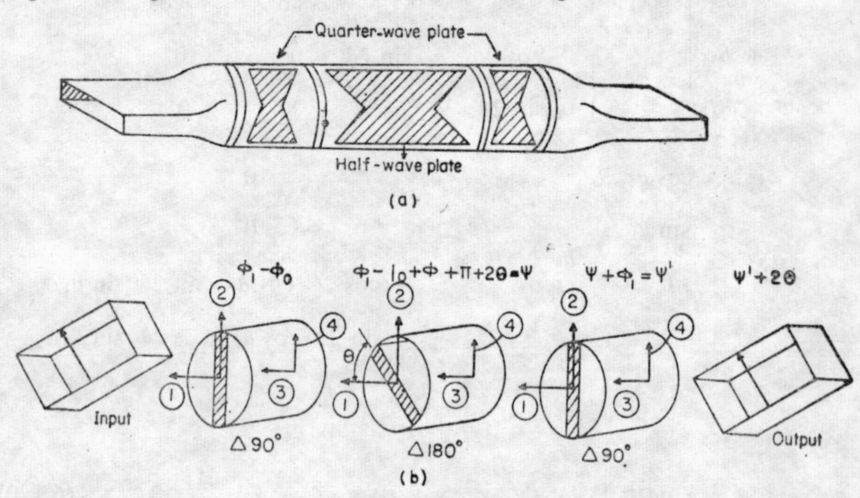


**Fig. 6.61** Precision rotary phase changer (*a*) Compact form (*b*) component form explaining working principle.

Taking* the normalized input wave to be $\sqrt{2}\ a\ e^{j\phi_0}$, the normalized waves entering at ports 1 and 2 of the first quarterwave section can be written as $a_1=a_2=a\ e^{j\phi_0}$. The outgoing normalized wave is then given by

$$[b]=[S]_{\Delta 90^\circ}\ [a]$$

$$\begin{bmatrix} b_1 \\ b_2 \\ b_3 \\ b_4 \end{bmatrix} = e^{-j(\phi_1-\phi_0)} \begin{bmatrix} 0 \\ 0 \\ a \\ ja \end{bmatrix} \tag{6.171}$$

which is a right-handed circularly polarized wave.

Since the wave entering at port 1 of the half-wave section is the wave emerging out of port 3 of the first quarter-wave section and that the wave entering port 2 is the one coming out of port 4 of this quarter-wave section, the wave being applied to the input of the half-wave section is

$$[a']=e^{-j(\ 1-\phi_0)} \begin{bmatrix} a \\ -ja \\ 0 \\ 0 \end{bmatrix}.$$

The output from the half-wave section is

$$[b']=[S]\Delta_{180^\circ}[a']$$

where $[S]\Delta_{180}{}^\circ$ is given by Eq. (6.168). That is,

$$\begin{bmatrix} b_1' \\ b_2' \\ b_3' \\ b_4' \end{bmatrix} = e^{-j\phi_2} \begin{bmatrix} 0 & 0 & -\cos 2\theta & \sin 2\theta \\ 0 & 0 & -\sin 2\theta & \cos 2\theta \\ -\cos 2\theta & -\sin 2\theta & 0 & 0 \\ -\sin 2\theta & \cos 2\theta & 0 & 0 \end{bmatrix} \times$$

$$e^{-j(\phi_1-\phi_0)} \begin{bmatrix} a \\ -ja \\ 0 \\ 0 \end{bmatrix} = a\ e^{-j(\phi_1+\phi_2-\phi_0)} \begin{bmatrix} 0 \\ 0 \\ -\cos 2\theta+j\ \sin 2\theta \\ -\sin 2\theta-j\ \cos 2\theta \end{bmatrix}$$

*The particular form is taken for mathematical Convenience.

$$= a\, e^{-j(\phi_1+\phi_2-\phi_0)} \begin{bmatrix} 0 \\ 0 \\ -e^{-2j\theta} \\ -je^{-2j\theta} \end{bmatrix} = e^{-j(\psi+2\theta)} \begin{bmatrix} 0 \\ 0 \\ 0 \\ ja \end{bmatrix} \tag{6.172}$$

where $\psi = \phi_1 + \phi_2 - \phi_0 + \pi$.

The output of the half-wave section is the input of the second quarter-wave section. Hence the output of the second quarter-wave section is

$$\begin{bmatrix} b_1'' \\ b_2'' \\ b_3'' \\ b'' \end{bmatrix} = e^{-j\phi_1} \begin{bmatrix} 0 & 0 & 1 & 0 \\ 0 & 0 & 0 & -j \\ 1 & 0 & 0 & 0 \\ 0 & -j & 0 & 0 \end{bmatrix} e^{-j(\psi+2\theta)} \begin{bmatrix} 0 \\ 0 \\ a \\ ja \end{bmatrix} = e^{-j(\psi'+2\theta)} \begin{bmatrix} 0 \\ 0 \\ a \\ a \end{bmatrix} \tag{6.173}$$

where $\psi' = \psi + \phi_1$.

Therefore, the output issuing out of the rectangular waveguide is

$$\sqrt{2}\, a\, e^{-j(\psi'+2\theta)} \tag{6.174}$$

$\psi'$ is fixed and does not depend upon spatial orientation of the half-wave section slab. Thus, by varying $\theta$ by $\Delta\theta$, the phase of the output is differentially delayed by $2\Delta\theta$.

There may be several halfwave sections instead of one, in that case the output will be

$$\sqrt{2}\, a\, e^{-j}(\psi' + 2\theta_1 - 2\,\theta_2 + 2\theta_3 + \ldots (-1)^{n-1}\, 2\theta_n) \tag{6.175}$$

where $\theta_i$'s may be positive, negative or zero.

The simple dependence of the phase change on a mechanical rotation may be used to calibrate the phase changer precisely and is the chief advantage of the rotary phase changer.

## 6.41 MICROWAVE ATTENUATORS

The passive elements used to control the amount of microwave power transferred from one point to another on a microwave transmission line are called microwave attenuators. Generally, these elements control the flow of microwave power either by reflecting and/or absorbing it in some dissipative elements. Attenuators may be fixed or variable depending on the requirements. The fixed types are used as isolators in various microwave circuits such as padding in microwave oscillators. The variable type are commonly used in bridge setup networks to control signal levels. An ideal attenuator when placed in a transmission line, must present a good impedance match at both the terminals, i.e. it should be a *well-matched reciprocal device.* The attenuation is a function of frequency and so due care should be taken if standard attenuation is desired.

In the following sections, we shall discuss some of the common waveguide attenuators. However, a brief reference to coaxial attenuators will also be made.

## 6.42 WAVEGUIDE ATTENUATORS

### (*i*) Flap or Card Attenuators

Typical fixed and variable type waveguide pad attenuators are shown in Figs. 6.62(a) and (b). The fixed type [Fig. 6.62(a)] consists of a dissipative element (pad) placed in a section of the waveguide with its plane parallel to the electric field and at the position where the electric field of the dominant $TE_{10}$ mode is maximum. The dissipating element is generally in the form of card or pad. One class of such pads consist of thin metallic films coated on a glass plate. A baked-on metallic film combining platinum and

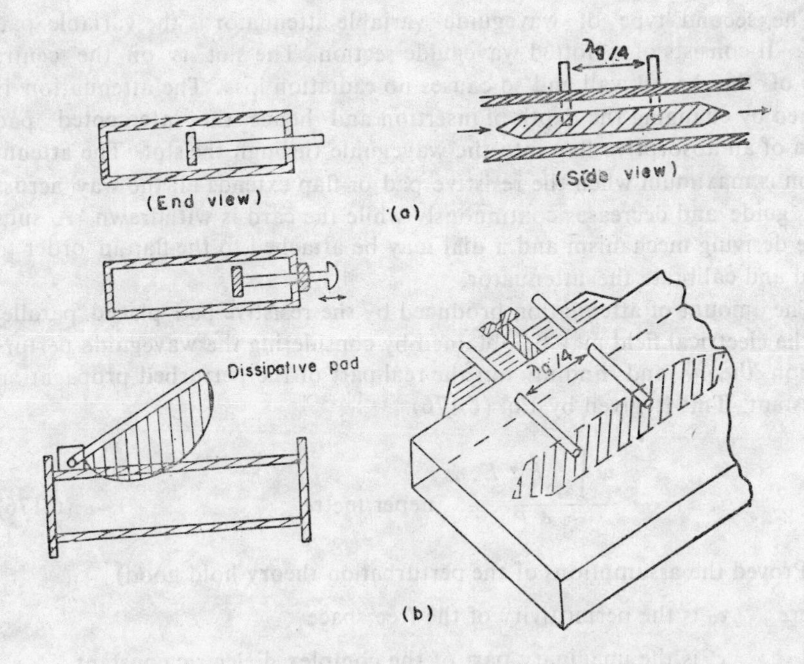


Fig. 6.62 Flap or card type waveguide attenuators (*a*) Fixed type (*b*) Variable type: variable pad area and variable position type.

palladium and also an evoporated film of chromium or nichrome with a protective film of magnesium fluoride have been satisfactorily used. The base material, glass, is chosen because it does not react with the film, its surface is smooth, and it can maintain its original shape at the temperatures we come across. The card is held parallel to the electric field by means of two thin metal rods.

To minimize reflections, the rods are held normal to the electric field and spaced quarter-wavelength apart, and the ends of the pad are tapered. When a dominant $TE_{10}$ mode enters the waveguide attenuator, the electric field tengential to the pad causes a dissipative current and is thus absorbed. The amount of microwave power thus dissipated or the attenuation produced depends upon the strength of the electric field, i.e. location of the pad within the waveguide, area of the pad intercepted by the electric field and the frequency. Obviously, such a fixed attenuator may be made *variable* either by changing the pad location or the pad area intercepted by the propagating electric field. The waveguide variable attenuators are shown in Fig. 6.62(b). A nob and gears (not shown in figure) control the movement of the card from the wall to the centre. The amount of attenuation introduced is controlled by changing the position occupied by the absorbing plate inside the waveguide, moving the plate from the narrow wall towards the waveguide axis thus producing an increase in attenuation, which becomes maximum when the plate is moved in the region of maximum electric field intensity, i.e, along the axis.

The second type of waveguide variable attenuator is the variable pad area. It consists of a slotted waveguide section. The slot is on the centre line of the broad wall and so causes no radiation loss. The attenuation is varied by changing the depth of insertion and hence the intercepted pad area of an absorptive flap into the waveguide through the slot. The attenuation is maximum when the resistive pad or flap extends all the way across the guide and decreases continuously while the card is withdrawn. A suitable deriving mechanism and a dial may be attached to the flap in order to read and calibrate the attenuator.

The amount of attenuation produced by the resistive pad placed parallel to the electrical field may be obtained by considering the waveguide perturbation theory and finding out the real part of the perturbed propagation constant. This is given by Eq. (6.176).

$$\alpha = \epsilon_0 \epsilon'' \frac{\omega \int_{\Delta s} E_0'^* E' \, ds}{4P} \text{ neper/metre} \qquad (6.176)$$

(Proved the assumptions of the perturbation theory hold good).

where $\epsilon_0$ is the permittivity of the free space

$\epsilon''$ is the imaginary part of the complex dielectric constant, $\epsilon = \epsilon_0 (\epsilon' - j\epsilon'')$

$\omega$ is the angular frequency of the waves

$s$ is the intercepted area

$e_0'$ is the electric field of the empty waveguide without $e^{j(\omega t-\gamma_0 z)}$ dependence

$E'$ is the electric field of the perturbed waveguide without $e^{j(\omega t-\gamma z)}$ dependence.

Taking the electric field in the TE10 mode as

$$E = E_0' = E_m \sin \frac{\pi x}{a}$$

we can obtain the values of the numerator and denominator which yield

$$\alpha = \epsilon'' \epsilon_0 \frac{\omega\, t\, h \sin^2 \frac{\pi x_1}{a} Z_{TE10}}{ab}$$

$$= \frac{h}{R_s} \sqrt{\frac{\mu_0}{\epsilon_0}} \frac{\lambda_{g0}}{\lambda} \frac{1}{ab} \sin^2 \frac{\pi x_1}{a} \text{ neper/metre} \tag{6.177}$$

where $t$ is the thickness to which the field penetrates in the resistive pad and $h$ is the height of the resistive pad in the waveguide, $R_s$ is the surface resistance of the lossy card measured in $\Omega/m^2$ and is related to the conductivity $\sigma$ as

$$R_s = \frac{1}{\sigma t}$$

provided $t$ is less than $\delta$, the skin depth, $x_1$ is the distance of the pad from the guide wall. It is clear from Eq. (6.177) that attenuation can be varied by either changing $x_1$, i.e. pad location or $h$, the depth of penetration.

### (*ii*) Cut-off Attenuators

It was pointed out in Chapter 3 that if the dimensions of the waveguide are such that $\lambda \gg \lambda_c$ ($\lambda_c$ is the cut-off wavelength) the wave is exponentially attenuated. Therefore, a waveguide of cross-section sufficiently small to ensure $\lambda \gg \lambda_c$ can be used as an attenuator. The principle of a cut-off attenuator is illustrated in Fig. 6.63. A circular waveguide section operating below cut-off is excited by a coaxial input, the output is similarly coupled to a coaxial line. The guide is below cut-off and hence the wave is attenuated with the constant of attenuation,

$$\alpha = \frac{2\pi}{\lambda_c} \sqrt{1-(\lambda_c/\lambda)^2}. \tag{6.178}$$

The attenuation produced is therefore given by

$$L=20 \log_{10}\left|\frac{E_{in}}{E_{out}}\right|=20 \log_{10} e^{\alpha l}\simeq 8.68\ \alpha l\ db, \tag{6.179}$$

where $E_{in}$ and $E_{out}$ are respectively the input and output electric fields and $l$ is the length of the cut-off waveguide. We note from Eq. (6.179) that attenuation is a linear function of $l$ and does not depend upon frequency, which is an advantage. However, the major disadvantage is that the attenuation is achieved by reducing the coupling between input and output due to intrinsic reflections in the cut-off waveguide. Consequently, a high degree of attenuation implies a reflection coefficient equal to unity which is often undesirable because reflected power may damage set-ups. Moreover, in general, there is considerable mismatch inevitably present at the device input and output, i.e. almost a purely reactive input and output impedance. Owing to these reasons, cut-off attenuators are generally used in conjunction with dissipative attenuators.

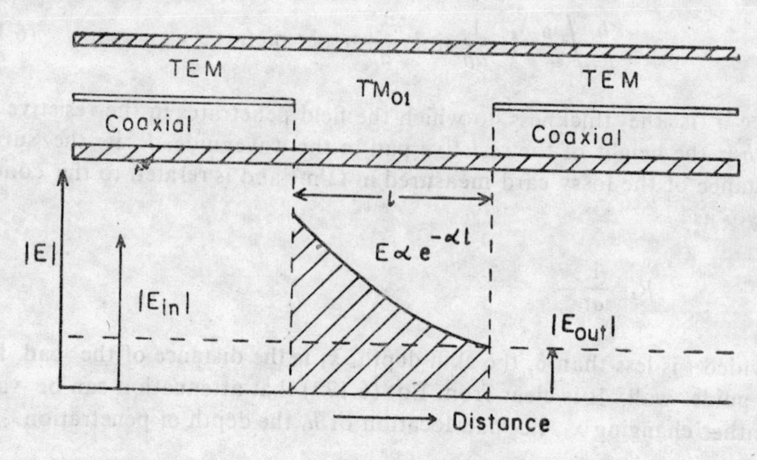


Fig. 6.63 Coaxial cut-off attenuator and its working principle.

### (*iii*) Lossy-Wall Attenuators

From Eq. (4.108) it is noted that a waveguide section may be used as an attenuator if $R_s$ is made sufficiently large. These attenuators are used in high power attenuation. However, these attenuators suffer the demerits of having low frequency sensitivity, fixed value of attenuation needed for a large value of $R_s$ or large length of the attenuator for appreciable attenuation.

## 6.43 PRECISION VARIABLE ATTENUATORS

The most satisfactory precision attenuator is the variable rotary attenuator depicted in Fig. 6.64(a). It comprises the following components:

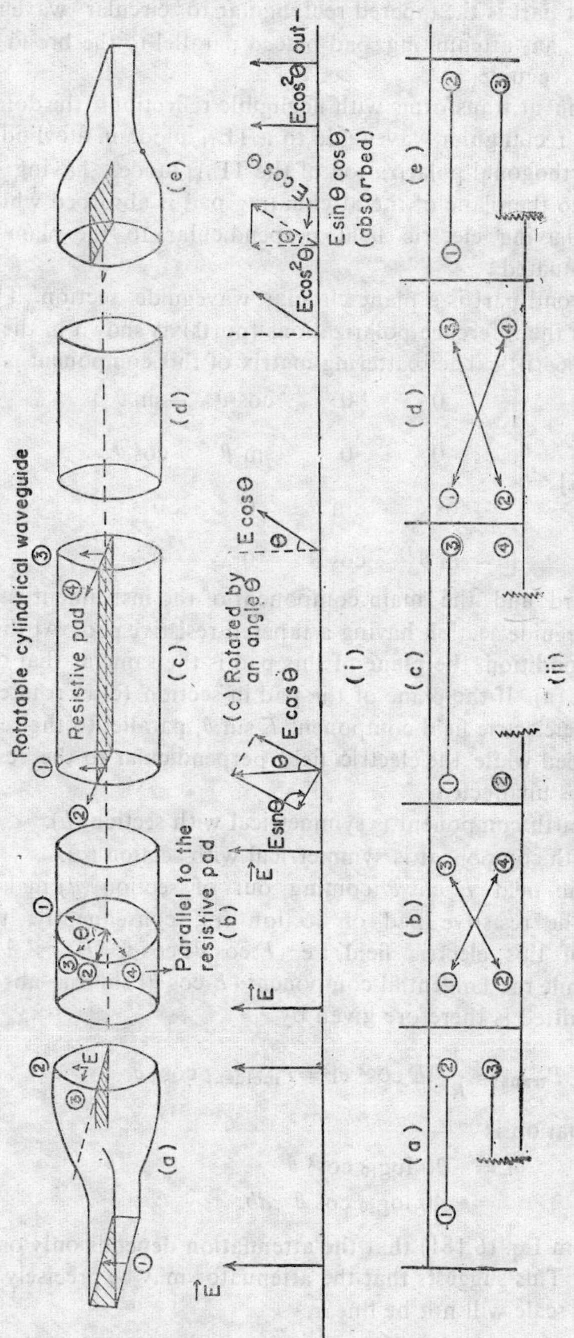


Fig. 6.64 Precision variable attenuator (i) Component form with working principle (ii) Equivalent circuit.

(*a*) The first part is the tapered rectangular to circular waveguide transition having an attenuating pad placed parallel to the broad wall of the rectangular waveguide.

This component transforms with negligible reflections, the dominant $TE_{10}$ mode in the rectangular waveguide to a $TE_{11}$ mode in a cylindrical waveguide. The orthogonal polarization of the $TE_{11}$ mode, having its electric field parallel to the plane of the attenuating pad is absorbed while the other polarization having electric field perpendicular to the plane of the pad passes unattenuated.

(*b*) The second part is a plane circular waveguide section. This section only changes the reference polarizations (ports) as shown in the equivalent circuit (Fig. 6.64(ii)). The scattering matrix of this component is

$$[S] = \begin{bmatrix} 0 & 0 & \cos\theta & -\sin\theta \\ 0 & 0 & \sin\theta & \cos\theta \\ \cos\theta & \sin\theta & 0 & 0 \\ -\sin\theta & \cos\theta & 0 & 0 \end{bmatrix} \tag{6.180}$$

(*c*) The third and the main component of the instrument is a rotatable circular waveguide section having a tapered resistive pad. At the minimum attenuation condition, the plane of this pad is the same as that of the pad in component (a). If the plane of the pad in section (c) is rotated through an angle $\theta$, the electric field component $E \sin\theta$, parallel to the resistive pad will be absorbed while the electric field perpendicular to the resistive pad, $E \cos\theta$ passes unaffected.

(*d*) The fourth component is symmetrical with section (*b*).

(*e*) The fifth component is symmetrical with section (*a*).

The electric field $E \cos\theta$ coming out of section (*d*) makes an angle $(90-\theta)$ with the resistive pad of section (*e*); consequently, the normal component of this electric field, i.e. $(E\cos\theta)\cos\theta = E\cos^2\theta$ passes unattenuated while the tangential component $(E\cos\theta)\sin\theta$ is absorbed. The power transmitted is therefore given by

$$P_{\text{trans}} = \frac{1}{R}\,|E\cos^2\theta|^2 = P_{\text{incident}}\cos^2\theta$$

Or the attenuation is

$$\begin{aligned} L &= -20\log_{10}\cos^2\theta \\ &= -40\log_{10}\cos\theta \quad db. \end{aligned} \tag{6.181}$$

It is clear from Eq. (6.181) that the attenuation depends only on the angle of rotation $\theta$. This suggests that the attenuator may be precisely calibrated. Note that the scale will not be linear.

## 6.44 HIGH POWER ATTENUATORS

The attenuators discussed so far cannot be used at high powers because of high heat production when power is dissipated in the attenuator pads. A typical high power microwave attenuator consisting of a waveguide hybrid junction is shown in Fig. 6.65. One port of the hybrid is terminated in a variable short circuit in series with one of the two connecting lines of a cascade arrangement of two hybrids. The high power signal is dissipated in a high power load such as circulating air or water.

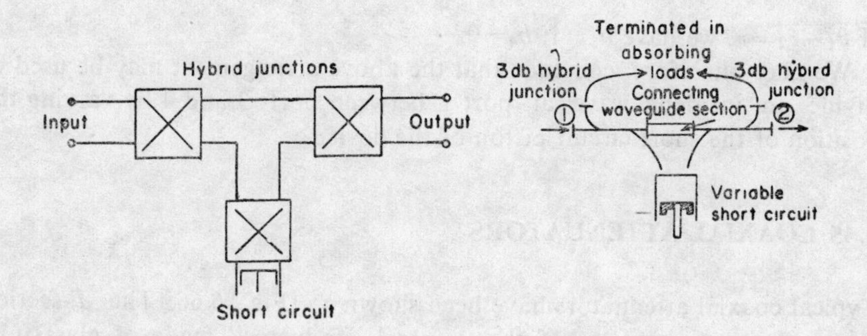


Fig. 6.65 High power microwave attenuator using 3*db* hybrids.

Following the current terminology, the normalized outgoing waves for the first hybrid for a normalized unit input wave at port 1, is given by

$$b_1=0,\ b_2=0,\ b_3=\frac{1}{\sqrt{2}},\quad b_4=\frac{j}{\sqrt{2}}.$$

$b_3$ is now phase shifted by the series hybrid through a phase angle $\theta_1$ which is determined by the position of the variable short circuit, whereas $b_4$ is phase shifted through a fixed phase angle $\theta_2$ determined by the series waveguide section connecting the two hybrids. The incident waves at the input terminals of the output hybrid are therefore given by

$$a_1=\frac{1}{\sqrt{2}}\,e^{-j\theta_1},\quad a_2=\frac{j}{\sqrt{2}}\,e^{-j\theta_2}$$

Using the scattering matrix Eq. (6.123) we can write

$$\begin{bmatrix} b_1 \\ b_2 \\ b_3 \\ b_4 \end{bmatrix}=\frac{1}{\sqrt{2}}\begin{bmatrix} 0 & 0 & 1 & j \\ 0 & 0 & j & 1 \\ 1 & j & 0 & 0 \\ j & 1 & 0 & 0 \end{bmatrix}\begin{bmatrix} \frac{1}{\sqrt{2}}e^{-j\theta_1} \\ \frac{j}{\sqrt{2}}e^{-j\theta_2} \\ 0 \\ 0 \end{bmatrix} \tag{6.182}$$

Thus the outgoing waves at the last hybrid are

$$b_1=0,\ b_2=0,\ b_3=\frac{e^{-j\theta_1}-e^{-j\theta_2}}{2}$$

$$b_4=\frac{j(e^{-j\theta_1}+e^{-j\theta_2})}{2}\ \text{Setting } \theta_1=\theta_2 \text{ gives}$$

$$b_3=0,\ b_4=j.$$

If $\theta_1-\theta_2=\pi$, we have $b_3=1$, $b_4=0$.

We may, therefore, conclude that the above arrangement may be used to divide an incident wave at port 1 between ports 3 and 4 by varying the position of the short circuit piston of the device.

## 6.45 COAXIAL ATTENUATORS

Typical coaxial attenuators have been shown in Fig. 6.66. The *T*-section type attenuator consists of the "central conductor" made of glass tube having two resistive metalized film sections on either side of a metalized circular disc. The distributed type attenuator consists of higher resistive main attenuating section in the centre and lower resistive matching sections on either side of the central conductor. The high resistivity of the inner conductor is achieved by maintaining the thickness of the resistive film less than the skin depth of the film metal.

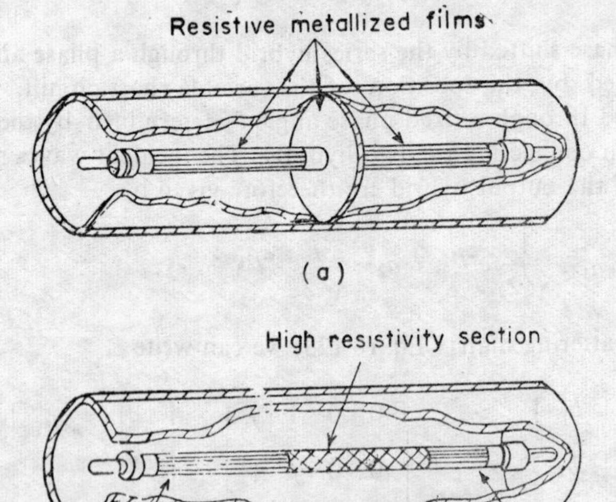


**Fig. 6.66** Coaxial attenuators (*a*) T-section type. (*b*) Distributed type.

Variable type coaxial attenuators utilize cut-off phenomena as discussed in Sec. 6.38*ii*.

## 6.46 MICROWAVE TERMINATIONS

Microwave terminations are the precise loads required in microwave measurement and calibration systems. Most common among them are matched terminations and variable precession short circuits. A matched load or a matched termination absorbs all the power incident on it without any reflection. In this respect it is equivalent to an absolutely black body in microwave range. On the other hand, a microwave line terminated in a matched load does not have any reflection and hence is equivalent to terminating the line with its characteristic impedance. The short circuit termination produces an adjustable reactive load at the desired point on a microwave line.

## 6.47 MATCHED TERMINATIONS

Matched waveguide terminations are constructed by mounting a power-absorbing card or pad in the space near the closed end of a waveguide section as shown in Fig. 6.67. A typical attenuator card consists of powdered iron or carbon mixed with a binder and deposited on a dielectric strip,

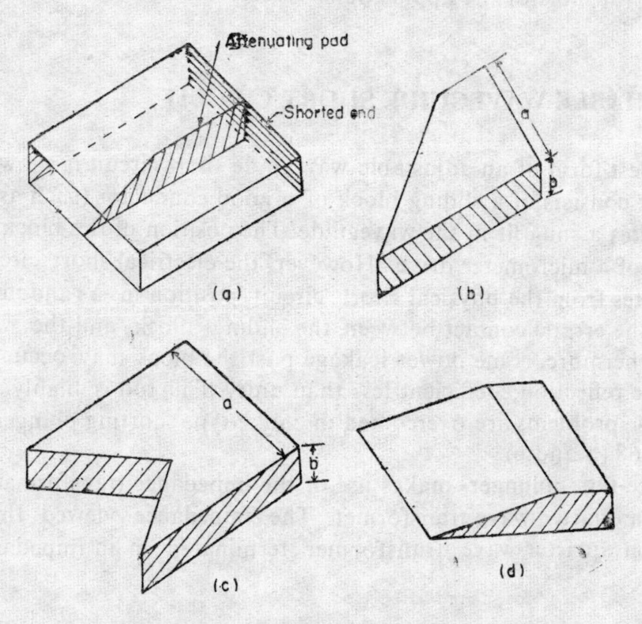


**Fig. 6.67** Waveguide matched terminations (*a*) pad type (*b*) (*c*) and (*d*) lossy type.

or porcelain containing silicon carbide. The reflections arising from the end are minimized by tapering the card. The card is placed parallel to the dominant $TE_{10}$ mode at a place where the electric field is maximum to have maximum attenuation. As the card has finite thickness, the reflections arising from it cannot be ruled out. Moreover, it is located at $E_{max}$ which will give maximum reflected power. To avoid this the pad is kept closer to the side wall and its length is increased.

Matched loads may also be designed by loading the whole of the waveguide with lossy materials such as powdered metal, lossy dielectric, wood, flowing water, sand, etc. However, due care should be taken to avoid reflections at the front end. In high power matched terminations graphite mixed with cement, and aquadag-coated sand have been used as these can withstand high temperatures. Moreover, to remove the heat from the waveguide, radiating copper fins are added to the waveguide casing. For additional cooling, if required, forced air may be blown against the fins.

In fact, a matched termination is an attenuator terminated in a short circuit. If $d$ be the length of the attenuator then the power reflected back and appearing at the end of the matched termination will be

$$P_{inc}\, e^{-2\alpha d} \cdot e^{-2\alpha d} = P_{inc}\, e^{-4\alpha d}.$$

Making $\alpha$ very large it is not difficult to obtain very low VSWR in front of the matched termination. A typical attenuation pad of 20 *db* ($\alpha d$) causes an attenuation of 40 *db* in total and results in a VSWR of 1.02, which is negligible for most of the applications.

## 6.48 VARIABLE WAVEGUIDE SHORT CIRCUIT

The simplest form of an adjustable waveguide short circuit is shown in Fig. 6.68(a). It consists of a sliding block of a good conductor (such as copper) which makes a snug fit in the waveguide. The position of the block is varied by means of a micrometer drive. However, the electrical short circuit position deviates from the physical short circuit position in a random manner owing to the erratic contact between the sliding block and the waveguide wall. Furthermore, some power leakage past the block may occur thereby making the reflection coefficient less than unity. This too is highly undesirable. These problems are overcomed in choke-type shorting plungers shown in Figs. 6.68 (b and c).

A choke-type plunger makes use of the impedance transformation properties of a quarter-wave transformer. The impedance viewed through a two section quarter-wave transformer terminated in an impedance $Z_s$ is given by

$$Z_s' = \left(\frac{Z_1}{Z_2}\right)^2 Z_s \tag{6.183}$$

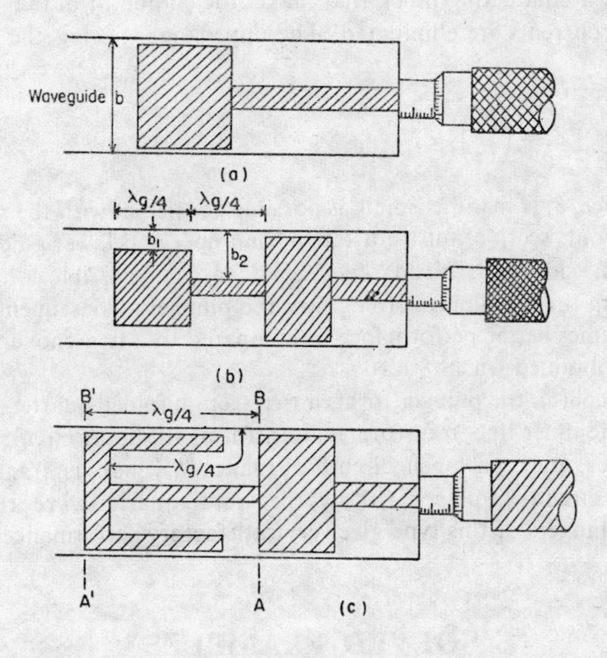


Fig. 6.68 Waveguide short-circuits (*a*) adjustable type (*b*) choke type (*c*) folded choke type.

where $Z_1$ and $Z_2$ are the impedances of the quarter-wave transformers. Thus, if in Eq. (6.183), $Z_2$ is chosen to be much greater than $Z_1$, $Z_s'$ will approximately be a short circuit, better than $Z_s$ by a factor $(Z_1/Z_2)^2$. This constitutes the under lying principle of choke-type shorting plungers. However, the performance of such short circuit terminations deteriorates when the frequency is changed because the transformer sections no more remain quarter-wave long. No doubt, by proper design a band width of 10 percent may be achieved.

To understand how the designs of Figs. 6.68(b and c) overcome the erratic contact behaviour and increase band width, consider the $TE_{10}$ mode incident on a rectangular waveguide, constituting the short circuit. The field pattern causes an axial current to flow across the gap between the upper and lower waveguide walls and vertically across the front face of the sliding block. The currents flowing along the side walls flow in a vertical direction and do not cross the gap between the waveguide walls and the front face of the plunger. Therefore, the erratic performance arises from the axial current flowing on the upper and lower walls. Thus, if the plunger height is made non-uniform as in Fig. 6.68(b), i.e. the front section is a quarter guide wavelength long and less than the guide height by an amount $2b_1$, and is followed by another quarter-wave section having height less than the guide height by $2b_2$ and a conducting block that makes the sliding fit in the guide, then the erratic currents are eliminated. The impedance viewing the plunger is given by

$$Z_s' = \left(\frac{b_1}{b_2}\right)^2 Z_s \tag{6.184}$$

In practice, $b_1$ is made as small as possible consistent with the requirement that the front section must not touch the upper and lower guide walls. (It may touch side walls.) The $b_2$ is made as large as possible consistent with maintaining the mechanical strength of the plunger. Consequently, a more than 100 times better performance as compared to a non-choke-type plunger can be obtained when $b_2 = 10b_1$.

In Fig. 6.68(c), the plunger forms a two section folded quarter-wave transformer. The inner line transformer transforms the short circuit due to the sliding block into an open circuit at plane $AB$ which is transformed to the short circuit at plane $A'B'$ by the outer quarter-wave transformer. Shorting plungers of this type give very satisfactory performance.

## SOLVED EXAMPLES

**Example 6.1.** A component of normalized admittance $Y$ is placed across a microwave line. Obtain the scattering matrix referred to the plane of the shunt admittance.

*Solution*: The scattering matrix can be written as

$$[S] = \begin{bmatrix} a_{11} & a_{12} \\ a_{21} & a_{22} \end{bmatrix}.$$

The junction is symmetrical so

$$a_{11} = a_{22} \quad \text{and} \quad a_{12} = a_{21}$$

Clearly, $a_{11}$ is the reflection coefficient while $a_{21}$ is the transmission coefficient

$$a_{11} = \Gamma_{11} = \frac{Y_0 - Y_L}{Y_0 + Y_L} = \frac{Y_0 - (y\,Y_0 + Y_0)}{Y_0 + (y\,Y_0 + Y_0)}$$

$$= \frac{-y}{2+y}$$

and

$$a_{21} = T_{21} = \frac{2\,Y_0}{(Y_0 + Y_0\,y)} = \frac{2}{2+y}$$

$$\therefore \quad [S]=\begin{bmatrix} \dfrac{-y}{2+y} & \dfrac{2}{2+y} \\ \dfrac{2}{2+y} & \dfrac{-y}{2+y} \end{bmatrix}$$

**Example 6.2.** A junction is made between the transmission lines of characteristic impedances $Z_{01}$ and $Z_{02}$. Find the scattering matrix with both reference planes in the plane of the junction.

*Solution*: Clearly,

$$a_{11}=\Gamma_{11}=\frac{Z_{02}-Z_{01}}{Z_{02}+Z_{01}}$$

$$a_{22}=\Gamma_{22}=\frac{Z_{01}-Z_{02}}{Z_{02}+Z_{01}}$$

$$a_{21}=\Gamma_{21}=\frac{2(Z_{01}\ Z_{02})^{1/2}}{Z_{01}+Z_{02}}=a_{12}$$

$$\therefore \quad [S]=(Z_{02}+Z_{01})^{-1}\begin{bmatrix} Z_{02}-Z_{01} & 2(Z_{01}\ Z_{02})^{1/2} \\ 2(Z_{02}Z_{01})^{1/2} & Z_{01}-Z_{02} \end{bmatrix}.$$

**Example 6.3.** A waveguide WG 16, feeding an antenna of frequency 9.29 GHz shows a VSWR 1.7 with a maximum electric field at 6.2 cm from the commencement of the corner in the waveguide. Determine the size and location of an inductive iris which will bring the VSWR unity.

*Solution*: The problem may be solved using the Smith Chart of Fig. Ex. 6.3.

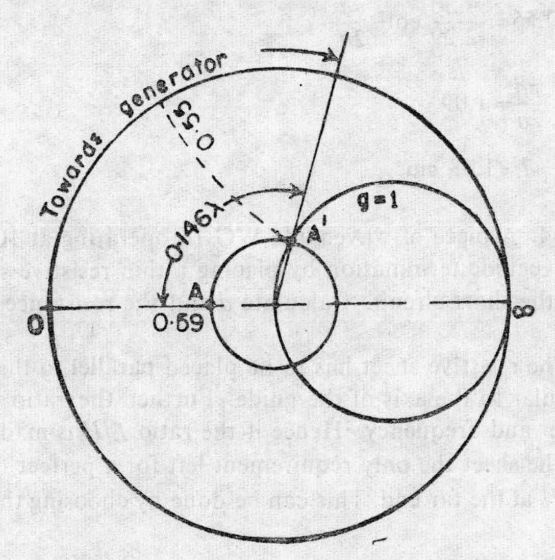


Fig. Ex. 6.3.

For WG 16, $a=2.29$ cm, $b=1.02$ cm.

Thus, $f_{01}=c/2a=6.55$ GHz.

The guide wavelength is given by

$$\lambda_g=2\,\pi/\beta=c(f^2-f_{10}^2)^{1/2}=4.56 \text{ cm.}$$

The maximum, therefore, occurs at

$$=\frac{6.2}{4.56}=1.35\ \lambda_g$$

before the reference. This point in terms of admittance $Y/Y_0=1/1.7=0.59$ is $A$ in the Smith Chart. Now VSWR=1 requires $g=1$. Thus point $A$ is to be transformed to $A'$ by moving through a distance $0.146\ \lambda_g$ towards the generator. Point $A$ is characterized by the admittance $1+j\,0.55$. To reduce this admittance to unity, the addition of an inductive susceptance $-j\,0.55$ is needed.

Note that the same point $A$ can also be reached by going towards the load (antenna) a distance $(1-0.146)\ \lambda_g=0.854_g=3.9$ cm. This point will be feasible since it is far enough away from the bend for the field to be a virtually pure dominant mode.

The susceptance of a symmetrical iris is given by

$$\frac{B}{Y_0}=\frac{\lambda_g}{a}\cot^2\frac{\pi d}{2a}$$

which gives,

$$0.55=\frac{4.56}{2.29}\cot^2\frac{\pi d}{2a}$$

or

$$\frac{\pi d}{2a}=1.09.$$

Hence, $d=1.58$ cm.

**Example 6.4.** A piece of waveguide WG 16 operating at 10 GHz is used to make a waveguide termination by placing a thin resistive sheet at a distance $d$ from the short circuit. Calculate $d$ and the resistance per unit area of the sheet.

*Solution*: The resistive sheet has to be placed parallel to the electric field, i.e. perpendicular to the axis of the guide. Further the ratio $E/H$ depends on the mode and frequency. Hence if the ratio $E/H$ is made equal to the resistance of the sheet the only requirement left for a perfect match is the vanishing of $H$ at the far end. This can be done by choosing the guide length to be $\lambda_g/4$

Thus, $d=\lambda_g/4\simeq 10$ cm

and resistance per unit area

$$\frac{E}{H}=\frac{\omega\mu_0}{\beta}=\frac{\omega\mu_0}{\{\omega^2/c^2-(\pi/a)^2\}^{1/2}}.$$

Substituting $a=2.29$ cm, $\omega=2\pi\times10^{10}$,

$$=4\times10^{-7}.$$

We have $\frac{\omega\mu_0}{\beta}=499\ \Omega.$

**Example 6.5.** It is required to make an attenuator for measuring attenuation upto 100 db with an accuracy of 1 db. The internal dimensions of the guide are 5 cm×2.5 cm. What is the attenuation of such a guide operating at very low frequencies? Assuming that the coupling is perfect, what is the maximum frequency at which the attenuator can be used without recalibration.

*Solution*: The waveguide will be an attenuator when the propagation constant is a real quantity, i.e. (for the $TE_{10}$ mode)

$$\gamma=\alpha=\{(\pi/a)^2-\omega^2/c^2\}^{1/2}.$$

At low frequencies

$$\alpha\simeq\pi/a$$

$$=\frac{\pi}{0.05}$$

$$=20\ \pi$$

$$=546\text{ db m}^{-1}.$$

As frequency is increased $\alpha$ will decrease. It will be 1 percent in error when

$$\left(\frac{\omega}{c}\right)^2=0.02\ (\pi/a)^2$$

if

$$f=\frac{c}{2a}\sqrt{0.02}=424\text{ MHz}.$$

**Example 6.6.** Show that any lossless network, two of whose parts are matched and decoupled, must be matched and decoupled as the other two parts. (Such a system is called a biconjugate network.)

*Solution*: The scattering matrix of the given network can be written as

$$[S]=\begin{bmatrix} 0 & 0 & S_{13} & S_{14} \\ 0 & 0 & S_{23} & S_{24} \\ S_{31} & S_{32} & S_{33} & S_{34} \\ S_{41} & S_{42} & S_{43} & S_{44} \end{bmatrix}.$$

Applying the unitary property to $[S]$

$$i=j=1 : |S_{31}|^2+|S_{41}|^2=1$$
$$i=j=2 : |S_{32}|^2+|S_{42}|^2=1$$
$$i=j=3 : |S_{31}|^2+|S_{32}|^2+|S_{33}|^2+|S_{34}|^2=1$$
$$i=j=4 : |S_{41}|^2+|S_{42}|^2+|S_{43}|^2+|S_{44}|^2=1.$$

Adding the last two equations and substituting the first two equations, we obtain

$$|S_{33}|^2+|S_{43}|^2+|S_{34}|^2+|S_{44}|^2=0.$$

The resultant scattering matrix is thus

$$[S]=\begin{bmatrix} 0 & 0 & S_{13} & S_{14} \\ 0 & 0 & S_{23} & S_{24} \\ S_{31} & S_{32} & 0 & 0 \\ S_{41} & S_{42} & 0 & 0 \end{bmatrix}$$

which proves the proposal.

## PROBLEMS

1. A rectangular waveguide with width $a=2.5$. cm faces a sudden transition to a long waveguide of width $a=1.2$ cm. Derive the scattering matrix for the junction neglecting the lengths of the guide. How is the scattering matrix modified if both the guides have equal length, say, 10 cm.

2. Derive scattering matrices for the following junctions shown in Fig. Prob. 6.2.

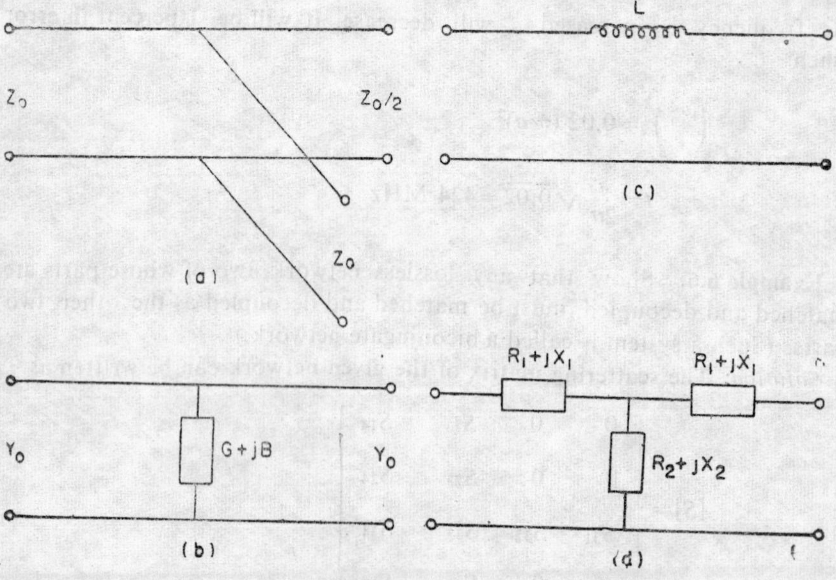


Fig. Prob. 6.2

3. Measurements on a microwave transistor gave the following scattering parameters, referred to 50Ω characteristic impedance

$$S_{11}=0.76\angle 145°, \qquad S_{12}=0.013\angle 55°$$

$$S_{21}=1.12\angle 25°, \qquad S_{22}=0.26\angle -164°,$$

Estimate the corresponding impedance and admittance parameters.

4. A two-port network with given scattering parameters is terminated in a load $Z_L$. Obtain the scattering parameters of the new system.

5. A *Y*-junction is terminated as shown in Fig. Prob. 6.5. Determine the scattering matrices connecting ports (3) and (1), (1) and 2, and (2) and (3). Can there be matching of any of the two lines?

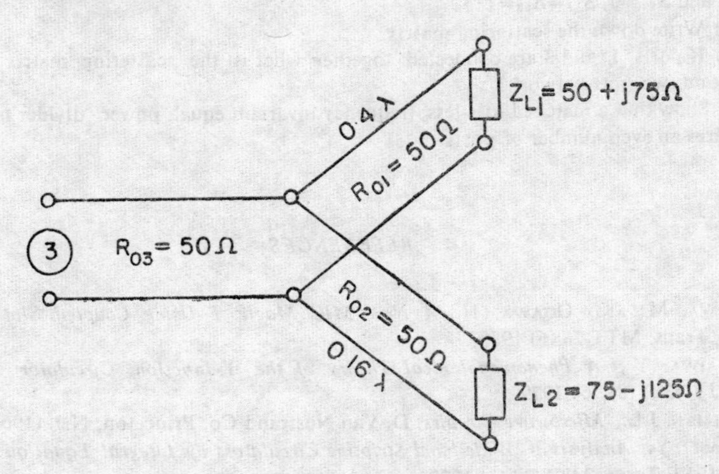


**Fig. Prob. 6.5**

6. Can there be a perfectly matched three-port junction?

7. A three-port microwave network is described, at a particular frequency, by the following port admittance matrix:

$$\begin{bmatrix} 3+j & 8+j3 & 27-j10 \\ 1+j10 & 7-j5 & 12+j4 \\ 5+j2 & 1+j0 & 13-j24 \end{bmatrix}.$$

If port 3 is short-circuited and rendered inaccessible, determine the port admittance of the remaining two ports (units are immaterial).

8. "A loss-less, matched, reciprocal 5-port junction is physically realizable at one frequency". Justify.

9. "A frequency invariant loss-less, reciprocal matched *n*-port junction is physically realizable if *n* is even". Comment.

10. Prove that in a byconjugate network the magnitude of the coupling coefficients must be equal.

11. Prove that a loss-less reciprocal 4-port junction, with resistive port terminations, has ports 1 and 2 matched, with 1 only decoupled from 3, and 2 only decoupled from 4 and matched at all four ports.

12. Design a four section maximally flat transformer of all pass lattice sections for a normalized band $\omega=2$, $\omega_c=1$ and load ratio $R_2/R_1=3$.

13. Given the unitary reciprocal scattering matrix ($r_1=r_2=r_3=1$), determine whether *Z* and/or *Y* exists

$$[S]=\begin{bmatrix} \frac{3-a}{6} & \frac{3+a}{3\sqrt{2}} & \frac{1-a}{2\sqrt{3}} \\ \frac{3+a}{3\sqrt{2}} & -a/3 & \frac{1-a}{\sqrt{6}} \\ \frac{1-a}{2\sqrt{3}} & \frac{1-a}{\sqrt{6}} & \frac{1+a}{\sqrt{2}} \end{bmatrix}, \quad a=e^{j45}$$

Find the network representation by the direct transformation of $[S]$.

14. A four-port reciprocal loss-less junction has all the non-zero elements real equal to $C$, and $S_{13}=0$, $S_{12}=S_{23}=C$.

(*i*) Write down the scattering matrix.

(*ii*) If ports 1 and 3 are connected together, what is the scattering matrix of the resultant two-port network?

15. Show that a matched loss-less, frequency invariant equal power divider network requires an even number of ports.

## *REFERENCES*


AIKAWA, M. AND OGAWA, H., *A New MIC Magic T Using Coupled Slot Lines* IEEE Trans. MTT-28 (6) 1980.

AKAIWA. Y., *A Phenomenological Theory of the Y-Junction Circulator* Trans. IECEJ J 62-B (8) 764 1979.

ALTMAN, J.L., *Microwave Circuits*, D. Van Nostrand Co. Princeton, N.J. (1964).

AYASLI, Y., *Analysis of Wide-band Stripline Circulators by Integral Equation Technique* IEEE Trans. MTT-28 (3) 1980.

ATWATER, H.A., *Introduction to Microwave Theory*, McGraw-Hill Book Co. (1962).

BAHL, I.J. AND GUPTA, K.C., *Design of Loaded Line p-i-n Diode Phase Shifter Circuits* IEEE Trans. MTT-28 (3) 1980.

BITTAR, G. AND VESZELY, G., *A General Equivalent Network of the Input Impedance of Symmetric Three-Port Circulators*, IEEE Trans. MTT-28 (7) 1980.

BRONWELL, A.B. AND BEAM, R.E., *Theory and Applications of Microwaves*, McGraw-Hill Book Co. (1947).

CARLIN, H.J. AND GIORTANO, A.B., *Network Theory*, Prentice-Hall Inc. 1964

CARROLL, J.E. AND RIGG, P.R., *Matrix Theory for n-line Microwave Coupler Design*, IEE Proc. 127H (6) 1980.

CHAKRABORTY, A. AND SANYAL, G.S. *Transmission Matrix for a Linear Double Taper in Rectangular Waveguides*, IEEE Trans. MTT 28 (6) 1980.

COLLIN, R.E. *Foundations for Microwave Engineering*, McGraw-Hill Book Co. (1966).

CRISTAL, E.G., *A Continuously Variable Coaxial Line Attenuator* ,IEEE Trans. MTT-28 (3) 1980.

DE RONDE, F.C., *A Simple Full-band, Matched 180° E-Plane Waveguide Band*, IEEE Trans. MTT-28 (4) 1980.

GHOSE, R N., *Microwave Circuit Theory and Analysis*, McGraw-Hill Book Co. (1963).

GULDBRANDSEN, T. AND GULDBRANDSON, B., *Novel 4-port Microwave Rotary Phase Shifter and Method for its Precision Calibration*, IEE Proc. 128H (1) 1981.

GULDBRANDSEN, T. AND GULDBRANDSON, B., *Novel Rotary Van Attenuater and Variable Directional Coupler Including Methods for Their Precision Calibration*, IEE Proc. 128H (1)1981.

HARVEY, A.F., *Microwave Engineering*', Academic Press (1963).

HELSZAJAIN, J., *Measurement of Symmetrical Waveguide Discontinuities Using Eigenvalue Approach*, IEE Proc. 127H (2) 1980.

HEWLETT PACKARD, S, *Parameter Design Application Note 154, Hewlett Packard*, Palo Alto Calif., (1972).

HOPFER, S, '*Analog Phase Shifter for 8-18 GHz, Microwave J.* 22 (3) 1979.

ISHII, T.K., *Microwave Engineering*, Ronald Press New York (1966).

KERNS, D.M., *Basis of Application of Network Equations to Waveguide Problems*, J. Res. Natl. Bur. Std. 42 1949.

LEVINE, P., *Installation and R.F. Matching of Power Terminations for Microstrip and Stripline Applications*, Microwave J. 9 (1980).

LEVY, R., Improved Single and Multiaperture Waveguide Coupling Theory, Including Explanation of Mutual Interactions IEEE Trans. MTT-28 (4) 1980.

LIAO, S.Y., *Microwave Devices and Circuits*, Prentice-Hall Inc. N.J. (1980).

MIYOSHI, T. AND MIYAUCHI, S., *The Design of Planar Circulators for Wide Band Operation*, IEEE Trans. MTT-28 (3) 1980.

MONTGOMERY, G.C. *et. al.*, *Principles of Microwave Circuits*', McGraw-Hill Book Co. (1948).

NYSTROM, G.L., *Analysis and Synthesis of Broad Band Symmetric Power Dividing Tees*, IEEE Trans. MTT-28 (11) 1980.

PANNENBORG, A.E., *On the Scattering Matrix of Symmetrical Waveguide Junctions*, Philips Res. Rept. 7 131-157 (1952).

PROC. 7th Eutopean Microwave Conference, Denmark, Sept. 1977. (Secs PC-1 and PC-2).

REICH, H.J., *et. al.*, *Microwave Principles*, D. Von Nostrand Co., Princeton, J.N.J. (1966).

RIGG, P.R. AND CARROLL, J.E., *Three Line Broad Band Codirectional* Microwave Couplers Using Planar Comb and Herrigbone Microstrip-Lines IEE, Proc. 127H (6) (1980).

RUBIN, D. AND SAUL, D.S, *The Microstrip Duplixer—A New Tool for Millimeter Waves*, Microwave J. 6 1980.

SANDER K.F. AND G.A.L., *Transmission and Propagation of Electromagnetic Waves*, Cambridge University Press, London (1978).

SALEH, A.A.M., *Theorems on Match and Isolation in Multiport Networks*, IEEE Trans. MTT-28 (4) 1980.

SLEEMART W., *Coupled Line Directional Couplers and All Pass Networks Synthesis*, IEEE Trans. MTT-16 (8) 1968.

SOUTHWORTH, G.C., *Principles and Applications of Waveguide Transmission*, D. Van Nostrand Co., N.J. (1980).

TANAKA T. '*Ridge-shaped Narrow Wall Directional Coupler Using $TE_{10}$ $TE_{20}$ and $TE_{30}$ Modes*, IEEE Trans. MTT-28 (3) 1980.

# 7

# MICROWAVE FERRITES

## 7.1 INTRODUCTION

Passive components used in microwave circuitry are generally obtained by introducing materials in waveguides and coaxial lines so as to change their propagation characteristics in a desired manner. The devices discussed in Chapter 6 are examples of this. The variation in these devices (variation in attenuation or phase) is achieved by mechanically displacing the coated or uncoated dielectric. Since the method is mechanical, there is a limit to the rate at which these changes can be made. To overcome this, ferromagnetic materials, whose properties can be varied electrically, are used to control the propagation characteristics so as to obtain the desired properties. These materials are known as *ferrites*.

Ferrite is a generic name given to ferromagnetic or ferrimagnetic insulators with resistivity as much as $10^{14}$ times greater than that of metals, dielectric constants around 10 to 15 or greater, and relative permeabilities of several thousands. These are ceramic-like materials having general chemical composition $MO \cdot Fe_2O_3$, where $M$ is a divalent metal such as magnesium, manganese, nickel, zinc, cobalt, iron, etc. or a mixture of these and are made by sintering a mixture of metallic oxides.

The magnetic properties that make ferrites useful in microwave applications, arise from the interaction of the dipole moment of an electron associated with its spin with the propagating electromagnetic wave. The high values of resistivity and dielectric constant play a dominant role in making these interactions strong. The high resistivity enables an electromagnetic wave to penetrate the material and the magnetic field component of the wave thus interacts with the magnetic moment of the ferrite. The interaction is revealed in the remarkable behaviour of the microwave permeability of ferrites. The permeability exhibits a pronounced resonance at a frequency which is simply related to the strength of the applied magnetic field within the ferrite. This resonance absorption of an electromagnetic wave by a magnetized ferrite within a waveguide is used in many important microwave devices.

Ferrites are also characterized by their special property of exhibiting different permeabilities in different directions. In other words, unlike an isotropic medium, the permeability $B/H$ of a ferrite is not a scalar quantity but a tensor. This tensor nature of microwave permeability of a ferrite makes it very suitable in designing non-reciprocal circuit elements. Moreover, the microwave permeability of a magnetized ferrite causes cross-coupling of linear and circularly polarized waves which would remain uncoupled in an isotropic medium. This results in irreversible Faraday rotation which is used in many important microwave devices.

In this chapter we aim to study in brief the microwave propagation in ferrites and microwave devices utilizing ferrites. However, for further details and for the preparation and properties of ferrites, refer to the literature.*

## 7.2 MICROWAVE PROPAGATION IN FERRITES

Ferrites are essentially ferrimagnetic** substances. These substances have *domains* of uniform magnetization in which uncancelled spins are all mutually parallel. Thus the macroscopic magnetization vector of a domain behaves as one unit under the influence of applied magnetic field. Therefore, to investigate the response of ferrites to microwaves, electron-spin magnetization may be treated as a classical unit endowed with angular momentum and magnetic moment, and classical mechanics may be applied to the system of electron spins.

### Permeability and Susceptibility Matrices

Consider an assembly of electron spins, represented by a spin angular momentum $\vec{J}$, placed in a steady magnetic field $\vec{H_0}$. This magnetic field will exert a torque on the magnetic moment $\vec{M}$ of the spin and will try to rotate it into alignment with the field, obeying the equation of angular motion

$$\vec{T} = \frac{d\vec{J}}{dt}. \tag{7.1}$$

Due to angular momentum, spin does not change its angle with respect to the applied magnetic field but precesses gyroscopically around $\vec{H_0}$, as

*1. A.E. Robinson, "The preparation of magnesium-manganese ferrite for microwave applications", *Proc. IEE,* Part B Supp. 5 (1957).

2. *Proc. I.R.E.*, Ferrite Issue, Vol. 44 (Oct. 1956).

3. B. Lax, and K.J. Button, *Microwave Ferrites and Ferrimagnetics*, McGraw-Hill Book Co. (1962).

4. P.J.B. Clarricoats, *Microwave Ferrites*, London: Chapman and Hall Ltd. (1961).

5. R.F. Soohoo, *Theory and Applications of Ferrites*, Englewood Cliffs N.J.: Prentice-Hall (1960).

6, H.A. Atwater, *Introduction to Microwave Theory*, McGraw-Hill Book Co. (1962).

**For details the reader may refer to C. Kittel, *Introduction to Solid State Physics*, New Delhi: Wiley Eastern Ltd. (III Edn.) (1971).

shown in Fig. 7.1. It is this tendency to precession which is responsible for the magnetic properties of ferrites.

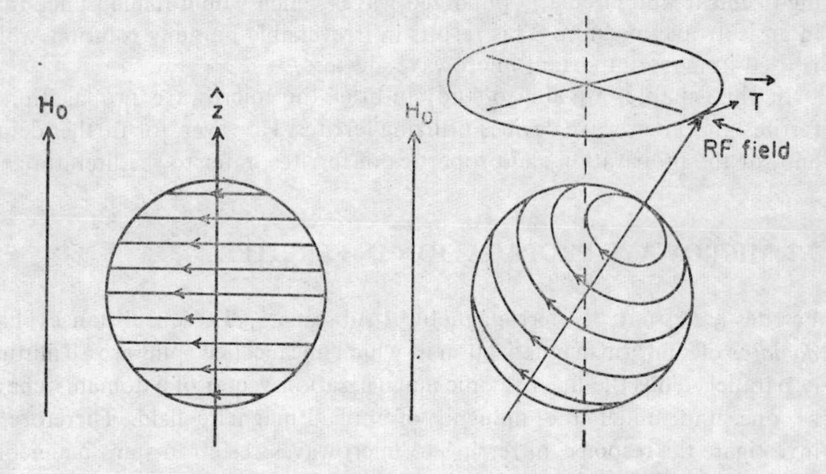


**Fig. 7.1** The electron spin tries align and as a result presses around $\vec{H}$ itself with the applied magnetic field.

The ratio of magnetization (magnetic moment per unit volume) to the angular momentum is called the gyromagnetic ratio, $\gamma$,* that is

$$\gamma = \frac{M}{J} \quad \text{or} \quad \vec{M} = \gamma \vec{J}. \tag{7.2}$$

Moreover, the torque $\vec{T}$ exerted on the electron spin by the magnetizing field $\vec{H}$ is given by

$$\vec{T} = \mu_0(\vec{M} \times \vec{H}). \tag{7.3}$$

Substituting Eqs. (7.1) and (7.2) in Eq. (7.3) yields,

$$\frac{1}{\gamma}\frac{d\vec{M}}{dt} = \mu_0(\vec{M} \times \vec{H}). \tag{7.4}$$

Assuming the time variation of $M$ as $e^{j\omega t}$ where $\omega$ is the angular frequency of the source, Eq. (7.4) can be reduced to

$$\vec{M} = \frac{\mu_0 \gamma}{j\omega}(\vec{M} \times \vec{H}). \tag{7.5}$$

Equation (7.5) is an interrelation between $\vec{M}$ and $\vec{H}$ and can be used to evaluate susceptibility constants $\left(\chi = \frac{M}{H}\right)$ of the assembly of atoms in our model.

*Note that $\gamma$ there is not the propagation constant.

For simplicity, let us consider the case of a small signal when a small high frequency vector field $\vec{H_1}$ is superimposed on a large steady magnetic field $\vec{H_0}$ in the $z$-direction. This results in a large constant magnetization $\vec{M_0}$ in the $z$-direction and small time varying components of magnetization $\hat{x}M_x$ and $\hat{y}M_y$ in the $x$- and $y$-directions respectively.

Thus we can write

$$\vec{H}=(\hat{x}H_x+\hat{y}H_y)\,e^{j\omega t}+\hat{z}H_0 \tag{7.6}$$

$$\vec{M}=(\hat{x}M_x+\hat{y}M_y)\,e^{j\omega t}+\hat{z}M_0 \tag{7.7}$$

where z components of $\vec{M}$ and $\vec{H}$ have been neglected because of the above assumptions, Eq. (7.5) can be broken into three equations in phasor component form as

$$j\omega M_x=\mu_0\gamma(M_yH_z-M_zH_y)$$

$$j\omega M_y=\mu_0\gamma(M_zH_x-M_xH_z)$$

$$j\omega M_z=\mu_0\gamma(M_xH_y-M_yH_x)$$

which on using Eqs. (7.6) and (7.7) and the assumptions implied therein that the product of two high frequency terms is negligible, can be written as

$$j\omega M_x=\mu_0\gamma(M_yH_0-M_0H_y)$$

$$j\omega M_y=\mu_0\gamma(M_0H_x-M_xH_0)$$

$$j\omega M_z=0.$$

The solution of these equations for $M_x$, $M_y$ and $M_z$ gives

$$M_x=\frac{\omega_0\omega_M}{\omega_0^2-\omega^2}\,H_x+j\,\frac{\omega\,\omega_M}{\omega_0^2-\omega^2}\,H_y \tag{7.8a}$$

$$M_y=-\,j\frac{\omega\,\omega_M}{\omega_0^2-\omega^2}\,H_x+\frac{\omega_0\,\omega_M}{\omega_0^2-\omega^2}\,H_y \tag{7.8b}$$

$$M_z=0 \tag{7.8c}$$

where

$$\omega_0=-\,\mu_0\gamma H_0 \tag{7.8d}$$

$$\omega_M=-\,\mu_0\gamma M_0. \tag{7.8e}$$

Equations (7.8a) through (7.8e) establish relations between components of magnetization and the magnetizing field. In the scalar case the ratio of $M/H$ expresses magnetic susceptibility $\chi$ of the medium. However, in our present case magnetization is not the same in all the directions and so $\chi$ can be said to be a tensor. The components of $\chi$ can be obtained by transforming the above equations in the following matrix form

$$\begin{bmatrix} M_x \\ M_y \\ M_z \end{bmatrix} = \begin{bmatrix} \chi_{xx} & \chi_{xy} & 0 \\ \chi_{yx} & \chi_{yy} & 0 \\ 0 & 0 & 0 \end{bmatrix} \begin{bmatrix} H_x \\ H_y \\ H_z \end{bmatrix} \tag{7.9a}$$

where

$$\chi_{xx} = \chi_{yy} = \frac{\omega_0 \, \omega_M}{\omega_0^2 - \omega^2}$$

$$\chi_{xy} = -\chi_{yx} = j \frac{\omega \, \omega_M}{\omega_0^2 - \omega^2}.$$

Again magnetic induction $\vec{B}$ is related to the magnetization $\vec{M}$ as

$$\vec{B} = \mu \vec{H} = \mu_0 (\vec{H} + \vec{M}) \tag{7.10a}$$

or

$$\begin{bmatrix} B_x \\ B_y \\ B_z \end{bmatrix} = \mu_0 \begin{bmatrix} 1+\chi_{xx} & \chi_{xy} & 0 \\ \chi_{yx} & 1+\chi_{yy} & 0 \\ 0 & 0 & 1 \end{bmatrix} \begin{bmatrix} H_x \\ H_y \\ H_z \end{bmatrix} \tag{7.10b}$$

Or, in matrix notation

$$[B] = [\mu][H] \tag{7.11a}$$

where $\mu$ is the permeability tensor for the ferrite and can be written as

$$[\mu] = \begin{bmatrix} \mu & -jk & 0 \\ jk & \mu & 0 \\ 0 & 0 & \mu_0 \end{bmatrix} = \begin{bmatrix} \mu_{11} & \mu_{12} & 0 \\ \mu_{21} & \mu_{22} & 0 \\ 0 & 0 & \mu_0 \end{bmatrix} \tag{7.11b}$$

where
$$\mu = \mu_0 (1 + \chi_{xx}) = \mu_0 \left( 1 + \frac{\omega_0 \, \omega_M}{\omega_0^2 - \omega^2} \right) \tag{7.11c}$$

$$k = \mu_0 (\chi_{yx}) = -\mu_0 \left( \frac{\omega \omega_M}{\omega_0^2 - \omega^2} \right).$$

$\omega_0$ is generally known as the gyromagnetic resonance frequency because at $\omega = \omega_0$, $\mu = \mu_0$ and permeability is no more a tensor but becomes a scalar. $\omega_M$ is a constant of a material and depends upon the strength of the saturation magnetization $M_0$ in the ferrimagnetic domains and has values of the order of ($2\pi \times 400$ to $2\pi \times 1000$) MHz.

It is noted from Eq. (7.11) that the application of a small high-frequency field say, $H_x$, in the $x$-direction perturbs the precession of the dipole around the $z$-axis. In the course of its perturbed precession, the projection of the magnetization vector normal to the $z$-axis rotates through all directions in the $xy$ plane, including a component of magnetic field $B_y$ in the $y$-direction, in phase quadrature with the applied field $H_x$, as given by

$$B_y = +jkH_x$$

This property of the saturated magnetic ferrite that alternating magnetic field components in two orthogonal directions are coupled, is utilized in waveguide coupling junctions such as a tetrahedral junction*.

So far we have assumed the ferrite to be loss-less; however, if losses are present these may be accounted for by introducing a damping term in Eq. (7.1). The damping tries to reduce the angle of precession $\phi$. It is found** that the susceptibility coefficients become complex:

$$\chi_{xx} = \chi_{yy} = \chi' - j\chi'' = \chi \tag{7.12}$$

$$\chi_{xy} = -\chi_{yx} = j(K' - jK'') = jK. \tag{7.13}$$

where

$$\chi' = \frac{\omega_0\, \omega_M(\omega_0^2 - \omega^2) + \omega_M\, \omega_0\, \omega^2\, \alpha^2}{[\omega_0^2 - \omega^2(1+\alpha^2)]^2 + 4\omega_0^2\, \omega^2\, \alpha^2} \tag{7.14}$$

$$\chi'' = \frac{\omega\, \omega_M\, \alpha[\omega_0^2 + \omega^2(1+\alpha^2)]}{[\omega_0^2 - \omega^2(1+\alpha^2)]^2 + 4\, \omega_0^2\, \omega^2\, \alpha^2} \tag{7.15}$$

$$K' = \frac{\omega\, \omega_M[\omega_0^2 - \omega^2(1+\alpha^2)]}{[\omega_0^2 - \omega^2(1+\alpha^2)]^2 + 4\, \omega_0^2\, \omega^2\, \alpha^2} \tag{7.16}$$

$$K'' = \frac{2\, \omega_0^2\, \omega_0\, \omega_M\, \alpha}{[\omega_0^2 - \omega^2(1+\alpha^2)]^2 + 4\, \omega_0^2\, \omega^2\, \alpha^2} \tag{7.17}$$

and $\alpha$ is the dimensionless damping constant.

**Propagation Characteristics**

Having evaluated the permeability and susceptibility matrices we now proceed to analyze the propagation of microwaves in a saturated (magnetized) ferrite medium.

In our discussions we shall consider microwave propagation in an infinite magnetized ferrite medium. Although only ferrites of finite size are employed in variour microwave components, there is justification for considering an infinite ferrite medium. A number of features of infinite medium theory are common to waveguides containing ferrites and this simplifies the mathematics involved in the analysis. Consider first the general case when a wave

*For details the reader may refer to J.A. Weiss, 'Tetrahedral Junction . . .' *J. Appl. Phys. Suppl.*, to Vol. 31, No. 5 (1960).

**R.F. Soohoo, *op. cit.*

is propagating in a direction ($\vec{r}$) that makes an angle $\theta$ with the direction of magnetization, i.e. $z$, as shown in Fig. 7.2. Without loss of generality it may be assumed that $\vec{r}$ lies in the $x$-$z$ plane and all field vectors possess a dependence $e^{j\omega t - \vec{\gamma} \cdot \vec{r}}$. Then field vectors $\vec{E}$, $\vec{H}$, $\vec{B}$ and $\vec{D}$ are related by Maxwell's equations

$$\nabla X \vec{E} = -j\omega \vec{B} \tag{7.18}$$

$$\nabla X \vec{H} = j\omega \vec{D}. \tag{7.19}$$

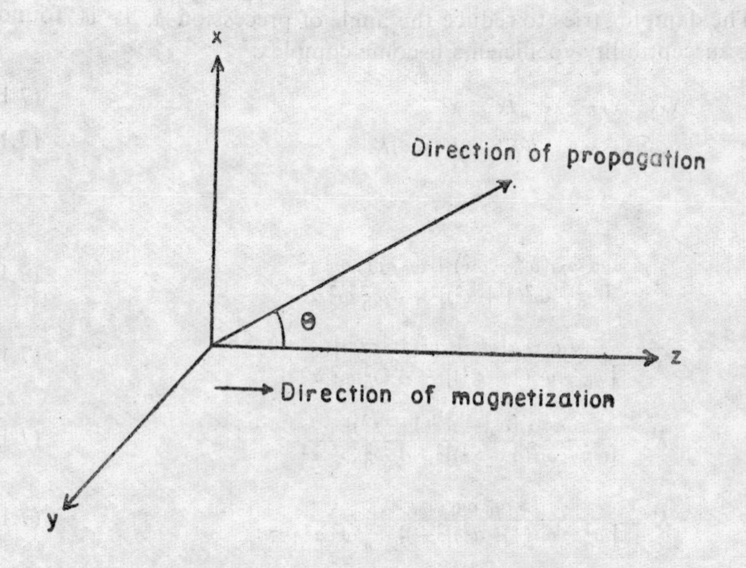


Fig. 7.2 Coordinates in infinite ferrite medium.

The ferrite medium may be assumed to possess isotropic scalar permittivity $\epsilon$, thus

$$\vec{D} = \epsilon \vec{E}. \tag{7.20}$$

In a ferrite $\vec{B}$ and $\vec{H}$ are related by Eq. (7.11)

$$\begin{bmatrix} B_x \\ B_y \\ B_z \end{bmatrix} = \begin{bmatrix} \mu & -jk & 0 \\ +jk & \mu & 0 \\ 0 & 0 & \mu_0 \end{bmatrix} \begin{bmatrix} H_x \\ H_y \\ H_z \end{bmatrix}. \tag{7.21}$$

Substitution of the values of components $B_x$, $B_y$, $B_z$ in Eq. (7.21) from Eqs. (7.18) and (7.19), gives

$$\gamma \cos\theta\, E_y = -j\omega(\mu H_x - jkH_y) \tag{7.22a}$$

$$-\gamma \cos\theta\, E_x + \gamma \sin\theta\, E_z = -j\omega(jkH_x + \mu H_y) \tag{7.22b}$$

$$-\gamma \sin\theta\, E_y = -j\omega\mu_0 H_z \tag{7.22c}$$

$$\gamma \cos\theta\, H_y = j\omega\epsilon E_x \tag{7.22d}$$

$$\gamma \sin\theta\, H_z - \gamma \cos\theta\, H_x = j\omega\epsilon E_y \tag{7.22e}$$

$$\gamma \sin\theta\, H_y = j\omega\epsilon E_z. \tag{7.22f}$$

From Eqs. (7.22a) and (7.22c), (7.22c) and (7.22e), (7.22b) and (7.22d), and (7.22f), three equations each involving only $H_x$, $H_y$ and $H_z$ are obtained.

$$j\omega\mu \sin\theta\, H_x + \omega k \sin\theta\, H_y + j\omega\mu_0 \cos\theta\, H_z = 0$$

$$-\gamma^2 \sin\theta \cos\theta\, H_x + 0 + (\gamma^2 \sin^2\theta + \omega^2\mu_0\epsilon)\, H_z = 0$$

$$+j\omega^2\epsilon k\, H_x + (\omega^2\mu\epsilon + \gamma^2)\, H_y + 0 = 0.$$

For a non-trivial solution of the above equations the determinant of the coefficients must vanish.

$$\begin{vmatrix} j\omega\mu \sin\theta & \omega k \sin\theta & j\omega\mu_0 \cos\theta \\ -\gamma^2 \sin\theta \cos\theta & 0 & \gamma^2 \sin^2\theta + \omega^2\mu_0\epsilon \\ j\omega^2\epsilon k & \omega^2\mu\epsilon + \gamma^2 & 0 \end{vmatrix} = 0 \tag{7.23}$$

Equation (7.23) may be expanded and solved for the propagation constant $\gamma$ as done by Polder*. However, the general case is of little importance. Instead we will confine our attention to the two important cases, viz. (*i*) $\theta = 0$, i.e. when propagation is along the direction of magnetization which will yield *microwave Faraday rotation*, a non-reciprocal phenomenon; and (*ii*) $\theta = 90°$, i e. when propagation is normal to the magnetization. This will correspond to the *birefringence effect*. These will be discussed in next sections.

**Propagation in the Direction of Magnetization ($\theta = 0°$)**

When the propagation is in the direction of magnetization, factors $\sin\theta$ and $\cos\theta$ may be removed in Eq. (7.23) and the determinant then reduces to

$$\begin{vmatrix} \mu & -jk & \mu_0 \\ -\gamma^2 & 0 & \omega^2\mu_0\epsilon \\ j\omega^2\epsilon k & \gamma^2 + \omega^2\mu\epsilon & 0 \end{vmatrix} = 0. \tag{7.24}$$

Expanding the determinant and factorizing it, we have the propagation equation

*D. Polder, "On the Theory of Ferromagnetic Resonance", *Philosophical Magazine*, 40 (1949).

$$\mu\{-\omega^2\mu_0\epsilon(\gamma^2+\omega^2\mu\epsilon)\}-jk\{\omega^2\mu_0\epsilon(+j\omega^2\epsilon k)\}$$
$$+\mu_0\{-\gamma^2(\gamma^2+\omega^2\mu\epsilon)\}=0$$

or $$-\omega^2\mu\epsilon(\gamma^2+\omega^2\mu\epsilon)+\omega^2k\epsilon\cdot\omega^2k\epsilon-\gamma^2(\gamma^2+\omega^2\mu\epsilon)=0$$

or $$-(\gamma^2+\omega^2\mu\epsilon)(\gamma^2+\omega^2\mu\epsilon)+\omega^2k\epsilon\cdot\omega^2k\epsilon=0$$

or $$[\gamma^2+\omega^2\mu\epsilon+\omega^2k\epsilon][\gamma^2+\omega^2\mu\epsilon-\omega^2k\epsilon]=0$$

or $$[\gamma^2+\omega^2\epsilon(\mu+k)][\gamma^2+\omega^2\epsilon(\mu-k)]=0. \tag{7.25}$$

One solution is given by

$$\gamma=\pm j\omega\sqrt{\epsilon(\mu+k)}. \tag{7.26}$$

and another solution is

$$\gamma=\pm j\omega\sqrt{\epsilon(\mu-k)}. \tag{7.27}$$

The propagation constant in Eq. (2.26) is generally denoted by $\gamma_-$ while that in Eq. (2.27) is by $\gamma_+$ (the reason for this choice will shortly become evident). Hence,

$$\gamma_\pm^2=-\omega^2[\epsilon(\mu+k)]. \tag{7.28}$$

Substituting the value of $E_x$ from Eq. (7.22b) in Eq. (7.22d) we have

$$-\gamma^2H_y=\omega^2\epsilon(\mu H_y+jkH_x). \tag{7.29}$$

Substituting Eq. (7.28) in Eq. (7.29) gives

$$H_y=jH_x \tag{7.30}$$

and $$H_y=-jH_x \tag{7.31}$$

respectively for $\gamma_-$ and $\gamma_\pm$.

In the $z=0$ plane, the field component $H_{x(\gamma\pm)}$ has dependence $He^{j\omega t}$ while $H_{y(\gamma\pm)}$ has dependence $\pm jHe^{j\omega t}$ respectively. Taking the real parts of the complex fields we have

$$H_x=H\cos\omega t,\ H_y=-H\sin\omega t \tag{7.32}$$

and $$H_x=H\cos\omega t,\ H_y=+H\sin\omega t. \tag{7.33}$$

Equations (7.32) and (7.33) clearly indicate that the fields are circularly polarized in the said plane. As shown in Fig. 7.3, Eq. (7.32) defines a field vector that rotates anticlockwise when viewed in the direction of the steady magnetic field. Equation (7.33) defines a field vector which rotates clockwise when similarly viewed. As $k$ changes sign when magnetization is reversed, it becomes evident that the sense of circular polarization must be defined with respect to that direction. When the *rf* magnetic field rotates clockwise, when viewed in the direction of static magnetization, the sense of circular polarization is said to be positive. On the other hand, the field is said to be negatively polarized when the *rf* magnetic field rotates in the opposite direction, i.e. counter-clockwise. This justifies the $\pm$ signs with the propagation constant.

It thus turns out that circularly polarized waves constitute a natural mode of propagation in magnetized ferrites. However, if the direction of propa-

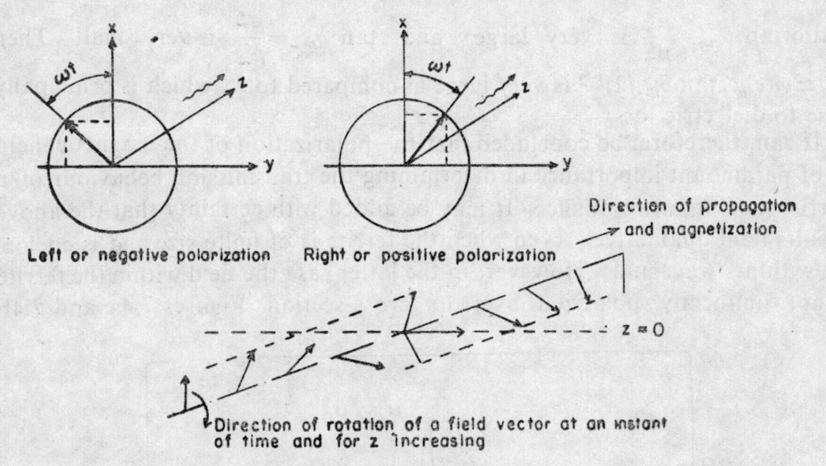


Fig. 7.3 Diagrams showing that a wave, defined to be negative circularly polarized, rotates clockwise as $z$ increases positively, at an instant of time.

gation other than the magnetization is considered, it will be found that there are, again, two modes of propagation but these are no longer circularly polarized TEM waves.

Further, it becomes clear from Eq. (7.28) that the ferrite presents a scalar permeability $\mu_-=\mu+k$ to negatively circularly polarized waves and a scalar permeability $\mu_+=\mu-k$ to positively circularly polarized waves. If losses are assumed, each $\mu_+$ and $\mu_-$ has real and imaginary parts.

$$\mu_\pm=\mu'_\pm-j\mu''_\pm. \tag{7.34}$$

The propagation constants $\gamma_\pm$ may also be written as

$$\gamma_\pm=\alpha_\pm+j\beta_\pm \tag{7.35}$$

and

$$\epsilon=\epsilon'-j\epsilon''. \tag{7.36}$$

Then substitution of Eqs. (7.34) through (7.36) in Eq. (7.28) and its separation into real and imaginary parts yields explicit expressions for the attenuation and phase constants for a circularly polarized wave propagating in a ferrite.

$$\alpha_\pm=\frac{\omega}{\sqrt{2}}[\{(\mu'_\pm\epsilon'-\mu''_\pm\epsilon'')^2+(\mu'_\pm\epsilon''+\mu''_\pm\epsilon')^2\}^{1/2} -(\mu'_\pm\epsilon'-\mu''_\pm\epsilon'')]^{1/2} \tag{7.37}$$

$$\beta_\pm=\frac{\omega}{\sqrt{2}}[\{(\mu'_\pm\epsilon'-\mu''_\pm\epsilon'')^2+(\mu'_\pm\epsilon''+\mu''_\pm\epsilon')^2\}^{1/2} +(\mu'_\pm\epsilon'-\mu''_\pm\epsilon'')]^{1/2}. \tag{7.38}$$

Moreover, it is noted that $\mu_-$ remains real at resonance even when the components $\mu$ and $k$ contain large imaginary parts. In contrast, the permeability for positive circular polarization passes through resonance and the ratio $\tan \delta_{m+} = \frac{\mu''_+}{\mu'_+}$ is very large* and $\tan \delta_{m-} = \frac{\mu''_-}{\mu'_-}$ is very small. Then $\alpha_+ \simeq \omega(\epsilon'\mu' \tan \delta_{m+}/2)^{1/2}$ is very large as compared to $\alpha_-$ which is principally due to dielectric loss.

It can, therefore, be concluded that the polarization of the magnetic field is of paramount importance in determining the transmission behaviour of a ferrite medium at resonance. It may be added with certainty that the above results hold qualitatively even when the ferrite is of finite size and is enclosed within a waveguide. However, in the latter case the field within the ferrite is not uniformly polarized over its cross-section. Figures 7.4a and 7.4b

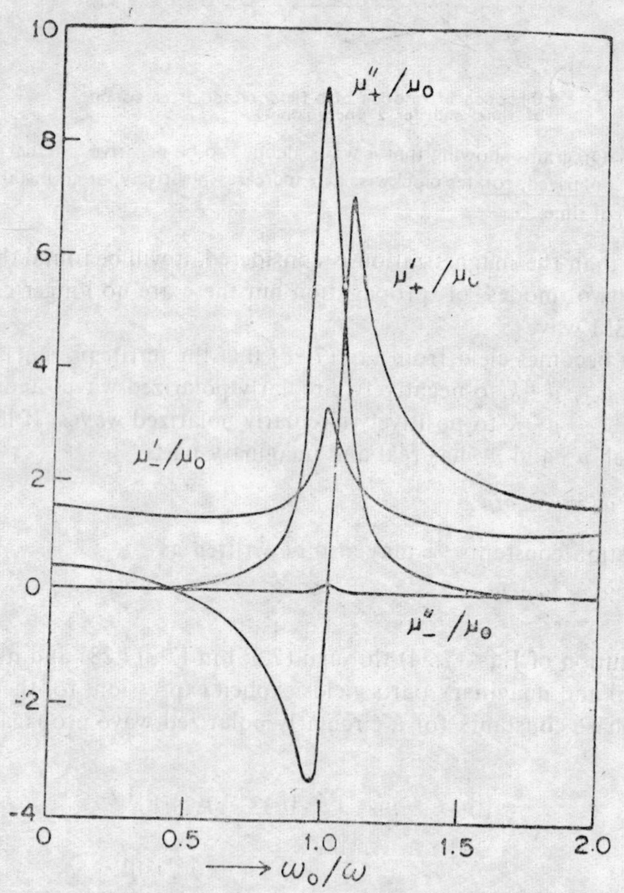


**Fig. 7.4(*a*)** Variation of the real and imaginary components of permeability for circularly polarized waves in a ferrite with $\omega_0/\omega$ for $f = \frac{\omega}{2\pi} = 10$ GHz $\omega_m/2\pi = 5.6$ GHz and $\alpha = 0.05$

*There is a condition for this that line width $\ll$ saturation magnetization. For details the reader is referred to Poder, *op. cit*.

respectively depict the behaviour of ferrite parameters and propagation characteristics with the change in frequency for circularly polarized modes.

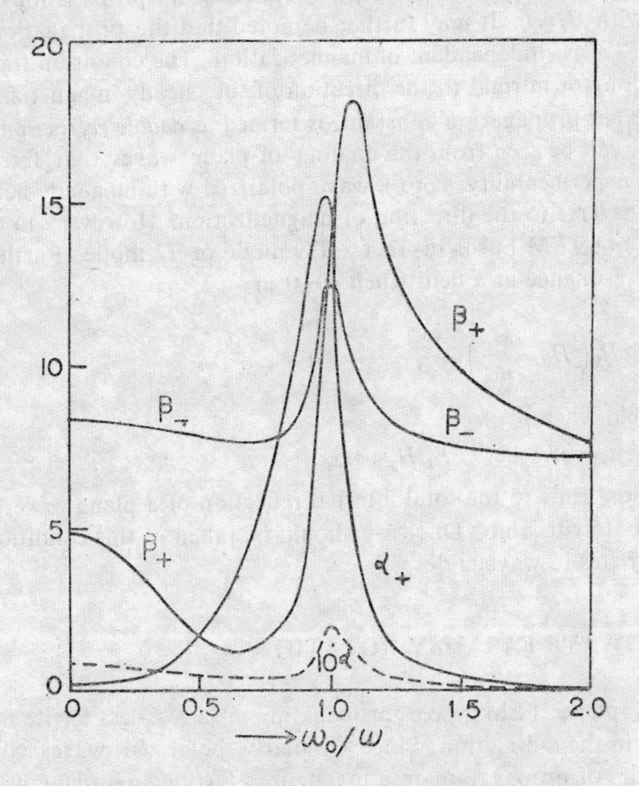


Fig. 7.4(*b*) Propagation and attenuation constants for circularly polarized waves in a ferrite versus $\omega_0/\omega$ and $\epsilon=10\epsilon_0$. The dotted curve is $10\alpha_-$ versus $\omega_0/\omega$ since $\alpha_-$ is very small.

**Propagation Normal to the Direction of Polarization ($\theta=90°$)**

In this case the determinant in Eq. (7.23) simplifies to

$$\begin{vmatrix} j\omega\mu & \omega k & 0 \\ 0 & 0 & \gamma^2+\omega^2\mu_0\epsilon \\ j\omega^2\epsilon k & \gamma^2+\omega^2\mu\epsilon & 0 \end{vmatrix} = 0 \tag{7.39}$$

Expanding and factorizing, we find two solutions for $\gamma$

$$\gamma = \pm j\omega(\epsilon\mu_0)^{1/2} \tag{7.40}$$

and

$$\gamma = \pm j\omega\left[\epsilon\left(\frac{\mu^2-k^2}{\mu}\right)\right]^{1/2} = \pm j\omega(\epsilon\mu_e)^{1/2}. \tag{7.41}$$

It can easily be visualized that Eq. (7.40) defines the propagation constant of a wave with $H_x=H_y=0$ while Eq. (7.41) defines the propagation constant of a wave with $H_z=0$. It may further be noted that the propagation constant in Eq. (7.40) is independent of magnetization. The condition that waves polarized along or normal to the direction of the steady magnetizing field have a different propagation constant, is termed as *double refraction or birefringence.* It can be seen from the analogy of plane waves that ferrite presents a scalar permeability $\mu_e$ to a wave polarized with magnetic field components transverse to the direction of magnetization. However, in contrast the wave is not TEM but is in fact a TE mode or $H$ mode. Furthermore, $\mu_e$ exhibits resonance at a field when $\mu=0$ or

$$H_0\left(H_0+\frac{M_0}{\mu_0}\right)=\frac{2}{\gamma^2}.$$

Also at a field when $\mu=k$

$$\mu_e=0, \quad \text{i.e.} \quad E_z/H_y=0.$$

This condition leads to the total internal reflection of a plane wave from a semi-infinite ferrite slab. Due care should be taken of this condition while installing ferrites in waveguides.

## 7.3 MICROWAVE FARADAY ROTATION

Consider a plane TEM wave propagating in a loss-less ferrite medium, magnetized in the $z$-direction. Since circularly polarized waves constitute normal mode of propagation in a magnetized ferrite, so a plane wave may be decomposed into the sum of left- and right-handed circularly polarized waves having propagation constants $\beta_-$ and $\beta_+$. Since $\beta_-\neq\beta_+$, therefore, the plane of polarization of the linearly polarized wave rotates as the wave propagates through a magnetized medium. This rotation is known as *microwave Faraday rotation.* Moreover, as $\beta_-$ and $\beta_+$ are functions of direction and magnitude of magnetization and not of direction of propagation, the rotation will be non-reciprocal.

To evaluate the magnitude and direction of rotation a linearly polarized wave may be decomposed into the sum of a left- and right-circularly polarized wave as

$$\vec{E}=\hat{x}E_0=(\hat{x}+j\hat{y})\frac{E_0}{2}+(\hat{x}-j\hat{y})\frac{E_0}{2} \tag{7.42}$$

The propagation constant for left and right polarizations are respectively $\gamma_-=j\beta_-$ and $\gamma_+=j\beta_+$, so the wave at $z=l$ becomes

$$\vec{E}=(\hat{x}+j\hat{y})\frac{E_0}{2}e^{-j\beta_-l}+(\hat{x}-j\hat{y})\frac{E_0}{2}e^{-j\beta_+l}$$

$$=\hat{x}\frac{E_0}{2}(e^{-j\beta_- l}+e^{-j\beta_+ l})+j\hat{y}\frac{E_0}{2}(e^{-j\beta_- l}-e^{-j\beta_+ l})$$

$$=\frac{E_0}{2}e^{-j(\beta_-+\beta_+)l/2}[\hat{x}(e^{-j(\beta_--\beta_+)l/2}+e^{j(\beta_--\beta_+)l/2})$$

$$+j\hat{y}(e^{-j(\beta_--\beta_+)l/2}-e^{j(\beta_--\beta_+)l/2})]$$

$$=E_0\,e^{-j(\beta_-+\beta_+)l/2}[\hat{x}\cos(\beta_+-\beta_-)l/2-\hat{y}\sin(\beta_+-\beta_-)l/2] \quad (7.43)$$

Equation (7.43) indicates that the resultant wave is a linearly polarized wave that has undergone a phase delay of $(\beta_-+\beta_+)\,l/2$. The angle that the new plane of polarization makes with the old, i.e. $x$-axis is given by

$$\theta=\tan^{-1}\frac{E_y}{E_x}=\tan^{-1}[-\tan(\beta_+-\beta_-)\,l/2]=-(\beta_+-\beta_-)l/2 \quad (7.44)$$

(minus sign indicates phase delay).

If the same travelling wave propagates in the negative direction, it may be seen that the plane of polarization continues to rotate in the same direction. Thus, if the wave travels back to $z=0$, i.e. in the negative $z$-direction a distance $l$ then the original direction of polarization is not restored; instead, the wave will arrive back at $z=0$ polarized at an angle $2\theta$ relative to the $x$-axis. This result can be easily derived by noting that $\beta_-$ and $\beta_+$ are independent of the direction of propagation so the resultant wave reaching back to $z=0$ will be

$$E=(\hat{x}+j\hat{y})\frac{E_0}{2}e^{-2j\beta_- l}+(\hat{x}-j\hat{y})\frac{E_0}{2}e^{-2j\beta_+ l}. \quad (7.45)$$

Equation (7.45) clearly indicates that the new polarization makes an angle

$$2\theta=-(\beta_+-\beta_-)l$$

with the axis of $x$ (initial polarization).

Hence, *regardless of the direction of propagation*, the phase of the propagating wave is delayed by an amount $(\beta_-+\beta_+)/2$ rad/metre where $\beta_\pm$ is given by Eq. (7.38). These results are valid so long as losses are negligible, i.e. $\omega \gg \omega_0$ which is true throughout the microwave frequency spectrum if $H_0$ is merely sufficient to saturate the ferrite. However, if $-\frac{\gamma M_0}{\omega\mu_0} \ll 1$ (which is true for frequencies>35 GHz), the rotation per metre can be written as

$$\theta/l \simeq \tfrac{1}{2}(\epsilon\mu_0)^{1/2}\left(\gamma\frac{M_0}{\mu_0}\right). \quad (7.46)$$

Although Eq. (7.46) is valid above 35 GHz frequency, the dependence of $\theta$ on $M_0$ and $\epsilon$ remains unquestioned, i.e. rotation per unit length is proportional to the saturation magnetization of the ferrite and to its permittivity. Also rotation is independent of frequency in the microwave range of frequencies. This conclusion is slightly modified when ferrite is confined in a waveguide, owing to bounded structure.

## 7.4 WAVEGUIDES CONTAINING MAGNETIZED FERRITES

Most of the ferrite microwave devices have a magnetized ferrite of desired dimension placed in a waveguide in a desired configuration. Two types of configurations are most commonly used and these have been shown in Fig. 7.5.

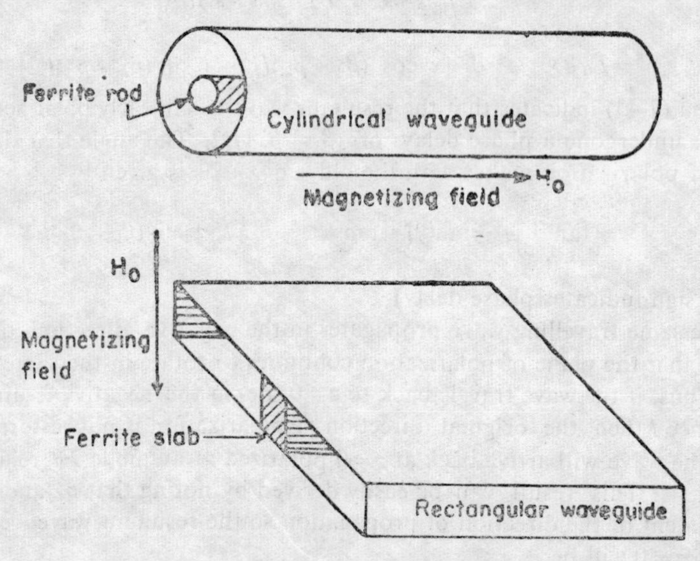


Fig. 7.5 Waveguides containing magnetized ferrite.

The problem of finding the propagation constants can be tackled in two ways. In the first method, the fields may be found by directly solving the boundary value problem*. Solutions with arbitrary constants are obtained for the ferrite and free space regions and, by matching methods across the boundaries, the determinental equation for the propagation constants, is obtained. The second method is used when the ferrite inserted is small as compared to the cross-section of the waveguide and constitutes finding the propagation constants as perturbations of those for the waveguides without the ferrite**.

No doubt a few boundary value problems have been solved and these lead to complicated transcenedental relations for the propagation constants and fields. Our aim in this section is not to include all this in our discussion but to outline the major steps of analysis to indicate the factors which distinguish this problem from that of the isotropic guide. We will examine thecase of a cylindrical guide with D.C. magnetic field along the axis. Here, a complete set of modes may be found. The principal distinction from the isotropic (empty guide) case is that the natural modes cannot be divided into TE and TM modes, rather they have both $E_z$ and $H_z$ components of fields.

*1. B. Lax and K.J. Button, *op cit*.
2. R.F. Soohoo, *op cit*.
** B. Lax and K.J. Button, *op cit*.

Maxwell's equations for the case under consideration may be written as

$$\nabla\times\vec{H}=j\omega\epsilon\vec{E},\quad \nabla\times\vec{E}=-j\omega[\mu]\,\vec{H} \tag{7.47}$$

where $[\mu]$ is the magnetic permeability tensor. Assuming $e^{-\gamma z}$ variation in fields, the above equations may be written in the following component form

$$\frac{1}{\rho}\,\frac{\partial H_z}{\partial\phi}+\gamma H_\phi=j\omega\epsilon E_\rho \tag{7.48}$$

$$-\gamma H_\rho-\frac{\partial H_z}{\partial\rho}=j\omega\epsilon E_\phi \tag{7.49}$$

$$\frac{1}{\rho}\,\frac{\partial(\rho H_\phi)}{\partial\rho}-\frac{1}{\rho}\,\frac{\partial H_\rho}{\partial\phi}=j\omega\epsilon E_z \tag{7.50}$$

$$\frac{1}{\rho}\,\frac{\partial E_z}{\partial\phi}+\gamma E_\phi=-j\omega\,(\mu_{11}H_\rho+\mu_{12}H_\phi) \tag{7.51}$$

$$-\gamma E_\rho-\frac{\partial E_z}{\partial\rho}=-j\omega\,(\mu_{21}H_\rho+\mu_{22}H_\phi) \tag{7.52}$$

$$\frac{1}{\rho}\,\frac{\partial}{\partial\rho}(\rho E_\phi)-\frac{1}{\rho}\,\frac{\partial E_\rho}{\partial\phi}=-j\omega\mu_0 H_z. \tag{7.53}$$

The RHS of Eqs. (7.51) and (7.52) follows from Eq. (7.11b).

Equations (7.48), (7.49), (7.51) and (7.52) may be solved to find $E_\rho$, $E_\phi$, $H_\rho$ and $H_\phi$ in terms of the derivatives of $E_z$ and $H_z$. These are of the following exemplified form:

$$E_\rho=a'\frac{\partial E_z}{\partial\rho}+b'\,\frac{1}{\rho}\,\frac{\partial E_z}{\partial\rho}+c'\,\frac{\partial H_z}{\partial\phi}+d'\,\frac{1}{\rho}\,\frac{\partial H_z}{\partial\phi}\,. \tag{7.54}$$

The substitution of these values for transverse components into Eqs. (7.50) and (7.53) yields the coupled equations of the form:

$$\Delta_T^2\,E_z+aE_z+bH_z=0 \tag{7.55}$$

$$\Delta_T^2\,H_z+cH_z+dE_z=0 \tag{7.56}$$

where subscript $T$ signifies that the derivatives are with respect to transverse coordinates $\rho$ and $\phi$. Coefficients $a$, $b$, $c$ and $d$ in Eqs. (7.55) and (7.56) may be found in the literature*.

Since Eqs. (7.55) and (7.56) are coupled, there cannot be a possible mode of the form $E_z=0$ or $H_z=0$, i.e. TE or TM mode. However, these may be classified by their nature at the cut-off. Further, it can be shown that at cut-off $\gamma=0$ and the coupling coefficients $b$ and $d$ vanish. The modes are; therefore, generally called *quasi-TE* or *quasi-TM* modes depending on whether they have $H_z$ only or $E_z$ only at cut-off frequency.

*R.F., Soohoo, *op. cit.*

The solutions for Eqs. (7.55) and (7.56) are found in terms of Bessel's functions:

$$E_z = [A_n J_n (K_1 \rho) + B_n J_n (K_2 \rho)] \, e^{jn\phi} \tag{7.57}$$

$$H_z = [a_1 A_n J_n (K_1 \rho) + b_1 B_n J_n (K_2 \rho)] \, e^{jn\phi} \tag{7.58}$$

where $K_1$, $K_2$, $a_1$ and $b_1$ are rather complicated expressions involving $\epsilon$, $\omega$, $\gamma$ and coefficients for the permeability tensor. Now, in the ferrite-filled, perfectly conducting guide $E_z = 0$ and $\frac{\partial H_z}{\partial \rho} = 0$ at the guide radius. The application of these two conditions leads to the evaluation of the ratio $A_n/B_n$ and a transcendental determinental equation for $\gamma$. The transverse fields may then be found using Eq. (7.54). For a partially filled guide, field expressions for TE and TM waves must be written for the region outside the ferrite. Then the $z$ and $\phi$ components of the ferrite fields are equated to those of the outside fields including both TE and TM waves at the ferrite boundary. The four resulting expressions, with two conditions at the guide wall constitute a set of six linear homogeneous equations in the six coefficients of Bessel functions in field expressions. This leads to the ratio of five coefficients to the sixth and also the propagation coefficient. Figure 7.6(a)

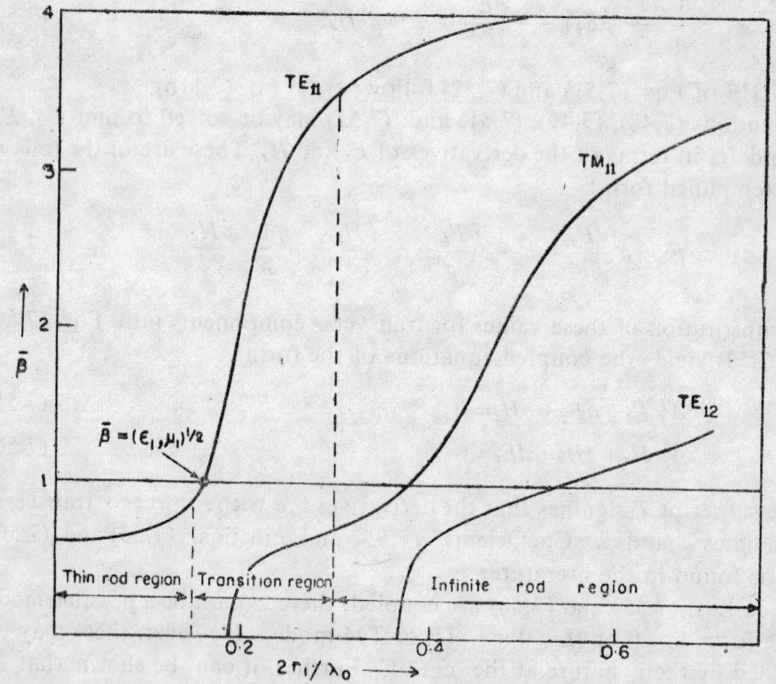


**Fig. 7.6**(*a*) Variation in the normalized phase constant $\bar{\beta}$ with respect to rod diameter/free space wavelength, $\frac{2\gamma_0}{\lambda_0}$ for ferrimic $R_1$ ferrite magnetized to saturation, for negative circularly polarized modes.

depicts the variation of normalized phase constant $\bar{\beta}$ as a function of rod diameter/free space wavelength.

The propagation characteristics for a rectangular waveguide loaded with magnetized ferrite are depicted in Fig. 7.6(b).

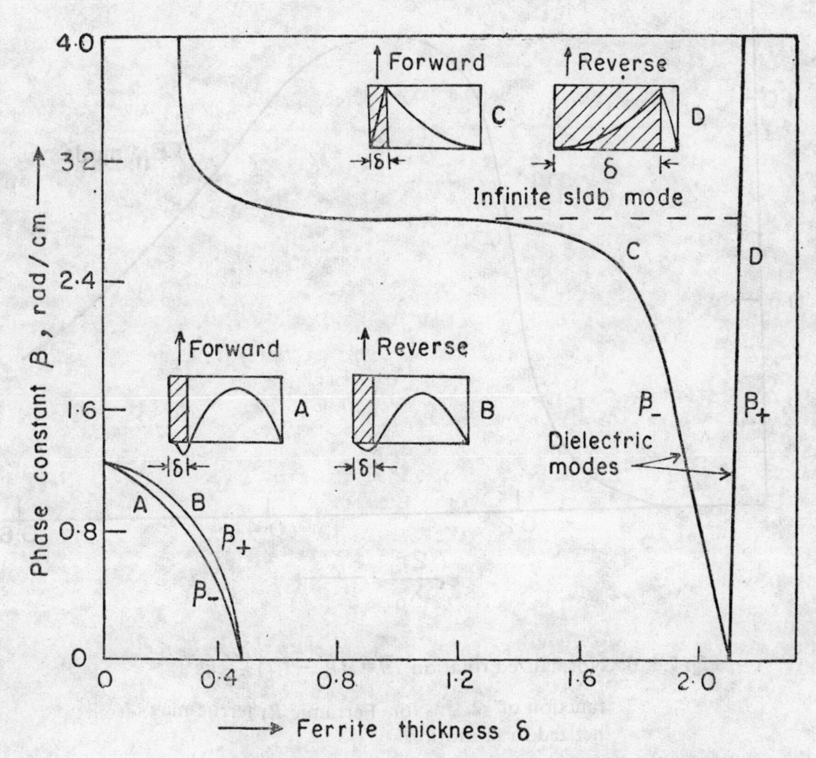


Fig. 7.6(*b*) Propagation constant versus thickness of ferrite slab in rectangular waveguide 0.9″×0.4″ at 1 GH $H_0$=3,000 Gauss.

The analysis shows that the dimensions of the ferrite effect the rate of Faraday rotation. It is found that decreasing the guide radius decreases the rate of rotation. Also, decreasing the ratio of the radius of the ferrite to the radius of the guide decreases the rotation. Figure 7.6(c) depicts the variation in normalized rotation $\bar{\theta}=(\bar{\beta}_{-}-\bar{\beta}_{+})\frac{1}{2}=\theta\,\frac{\lambda_0}{2\pi}$ as a function of rod diameter/ free space wavelength $2\gamma_1/\lambda_0$ for a typical magnetized (upto saturation) ferrite.

Having analyzed the properties of ferrites and the propagation of microwaves in ferrites, we now proceed to study microwave components utilizing ferrites. Many of the ferrite microwave devices are non-reciprocal, i.e. propagation characteristics are functions of the direction of propagation or from the scattering matrix point of view

$$S_{ij}\neq S_{ji}. \tag{7.59}$$

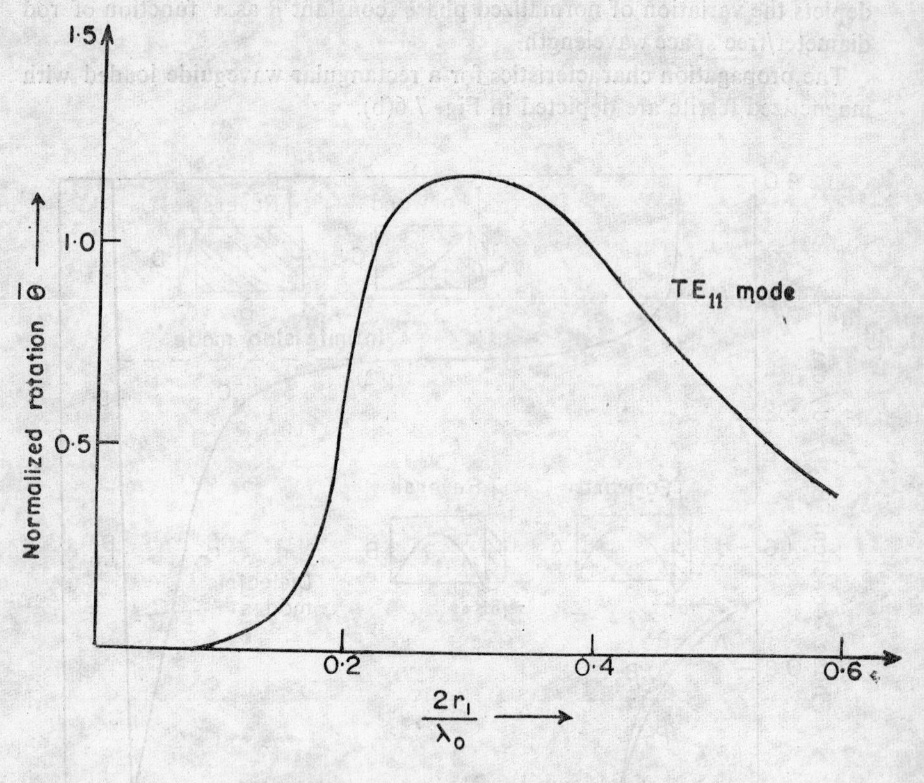


Fig. 7.6(*c*) Normalized rotation $\bar{\theta}=\frac{1}{2}(\bar{\beta}_- - \bar{\beta}_\pm)\frac{\theta\lambda_0}{2\pi}$ as a function of $2\gamma_0/\lambda_0$ for Ferramic $R_1$ ferrite magnetized to saturation.

However, if the device is loss-less

$$[S]^\dagger [S]=[I]. \tag{7.60}$$

([*I*] is the unit matrix) which holds whether the device is reciprocal or not.

## 7.5 GYRATOR

A gyrator, as introduced by Hogan*, is essentially a two-port non-reciprocal phase shifter with a phase difference of 180° between forward and backward directions of propagation. A gyrator may be designed by employing the non-reciprocal property of Faraday rotation. Figure 7.7 illustrates a typical microwave gyrator. It consists of a rectangular guide with a 90° twist connected to a circular guide, which in turn is connected to another rectangular guide at the other end. The two rectangular guides have the

*C.L. Hogan, *Revs. Modern Phys.*, Vol. 25 (1953).

same orientation at ports (1) and (2). Central circular guide contains an axially magnetized thin cylindrical ferrite rod with ends tapered (to avoid reflections) that produces a 90° Faraday rotation of the $TE_{11}$ dominant mode in a circular waveguide. Consider a wave propagating from left to right. By virtue of the twist of the waveguide the plane of polarization of the wave coming from left is rotated counter clockwise by 90° before reaching the ferrite. The ferrite produces an additional 90° rotation as shown in Fig. 7.7. Thus, the wave progressing towards the right is inverted, or equivalently, shifted in phase by 180°. For a wave passing from right to left the Faraday rotation is still 90° *in the same sense* and this is cancelled by the rotation in the twisted section which is 90° in clock wise sense. Thus, the non-reciprocal behaviour of the gyrator increases the electrical length of the line in a given direction by $\lambda/2$ as compared to the other direction.

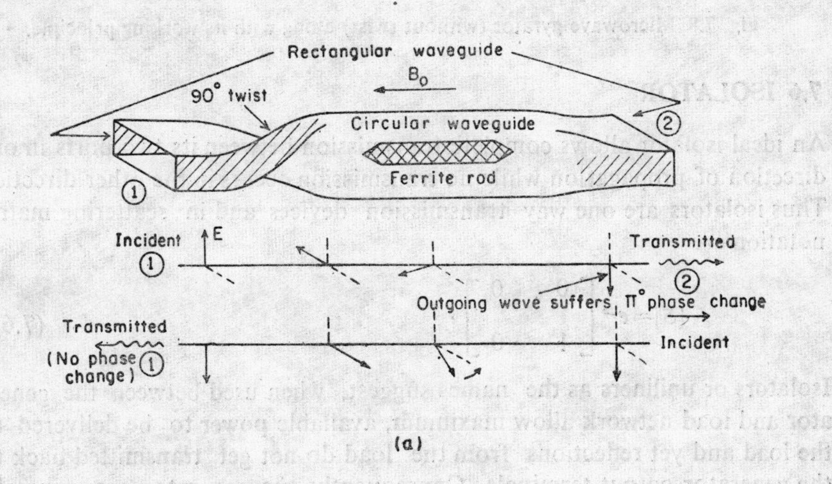


**Fig. 7.7** *(a)* Microwave gyrator (with 90° twist) along with its working principle.

A gyrator may also be designed without twist if the inconvenience of having the input and output rectangular guides oriented at 90° can be tolerated. A typical gyrator without twist is shown in Fig. 7.8. A wave travelling from left to right suffers a 90° rotation in its plane of polarization by the magnetized ferrite. As output guide has been rotated by 90° relative to the input so the output wave will have the right polarization to propagate in the output guide. However, when the propagation is from right to left, the wave arriving in guide 1 will have its polarization changed by 180° as illustrated in Fig. 7.8. Thus it works as a gyrator.

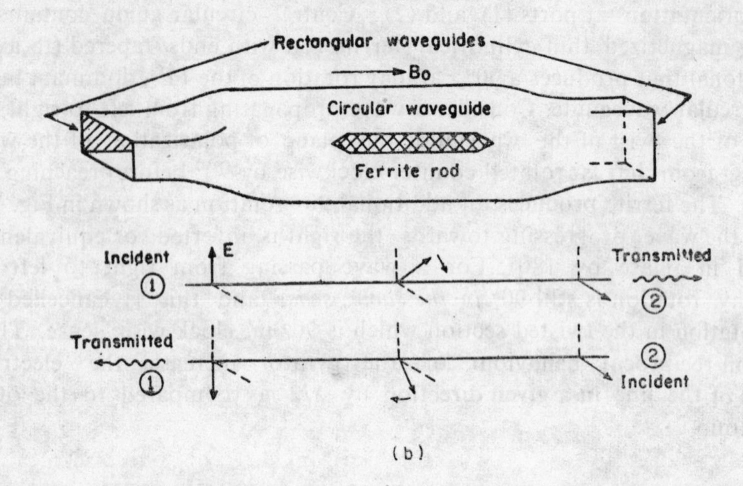


Fig. 7.8 Microwave gyrator (without twist) along with its working principle.

## 7.6 ISOLATOR

An ideal isolator allows complete transmission between its two ports in one direction of propagation while no transmission occurs in the other direction Thus isolators are one way transmission devices and in scattering matrix notation

$$[S]=e^{j\phi}\begin{bmatrix} 0 & 0 \\ 1 & 0 \end{bmatrix}. \tag{7.61}$$

Isolators or uniliners as the names suggest, when used between the generator and load network allow maximum, available power to be delivered to the load and yet reflections from the load do not get transmitted back to the generator output terminals. Consequently, the generator *sees* a matched load and effects such as power output variation and frequency pulling, with variation in load impedance are avoided.

Several realizations of isolators are possible, based on different principles.

### (*i*) Faraday Rotation Isolator

As shown in Fig. 7.9 the Faraday rotation isolator is similar to a gyrator in construction except that it employs a 45° twist section and a 45° Faraday rotation magnetized ferrite; in addition, it has thin resistive (absorbing) tapered vanes inserted in the input and output guides to absorb the field that is polarized, with the electric field parallel to the wide dimension of the guide. The operation of the device is as follows: a $TE_{10}$ dominant mode in a rectangular waveguide excites a $TE_{11}$ linearly polarized mode in a circular waveguide and vice versa. A wave propagating from left to right has its polarization rotated 45° in the counter clockwise direction by the 45° twist

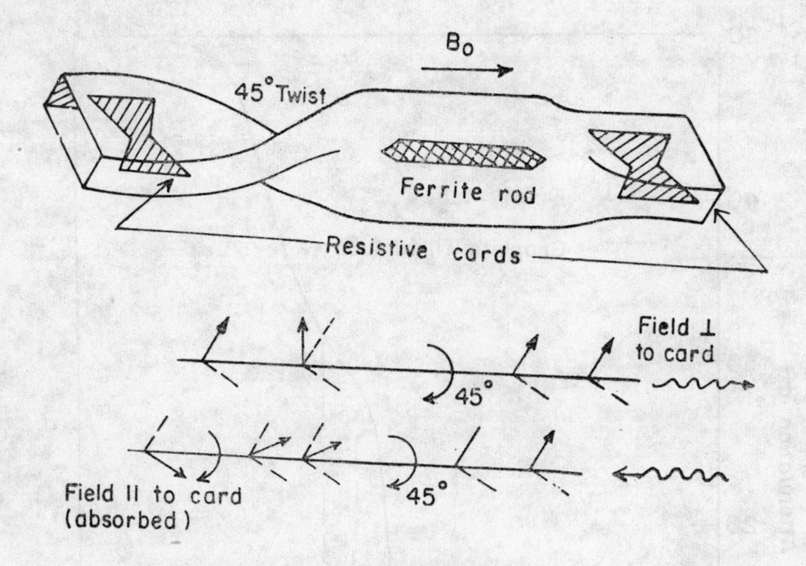


Fig. 7.9 Faraday rotation isolator along with its working principle.

section, and 45° clockwise by the Faraday rotator (the ferrite is magnetized in such a way); consequently, the wave will emerge with the correct polarization at the output port. However, if there are higher modes or waves with cross (incorrect) polarizations, they will be absorbed in the attenuating vane at the output. Now a wave travelling from right to left will suffer 45° rotation in the clockwise direction due to the waveguide twist and a 45° clockwise rotation due to the Faraday rotator. Consequently, the wave will have electric field parallel to the attenuating pad and consequently will be absorbed.

The reversal in the direction merely reverses the direction of transmission and absorption.

(*ii*) **Circularly Polarized Isolator**

A Faraday rotation isolator discussed above suffers one drawback. Since isolation greatly depends on the Faraday rotation and twist design, broad band isolators are difficult to design. An isolator of broad band operation with a wide range of variable attenuation was suggested by Clarricoats*. He used a circular waveguide containing ferrite with dielectric lossy material such as graphite power. Figure 7.10(a) shows attenuation as a function of the applied field for a ferrite tube with and without graphite powder packed in the bore. It is noted that the peak in attenuation is observed for negative circular polarization. Moreover, if a field of about 300 oersteds is applied, an isolation in excess of 30 db can be obtained using an arrangement as

*P.J.B., Clarricoats, "Some properties of circular waveguides containing ferrites", *Proc. I.E.E.*, Part B **104** (1957).

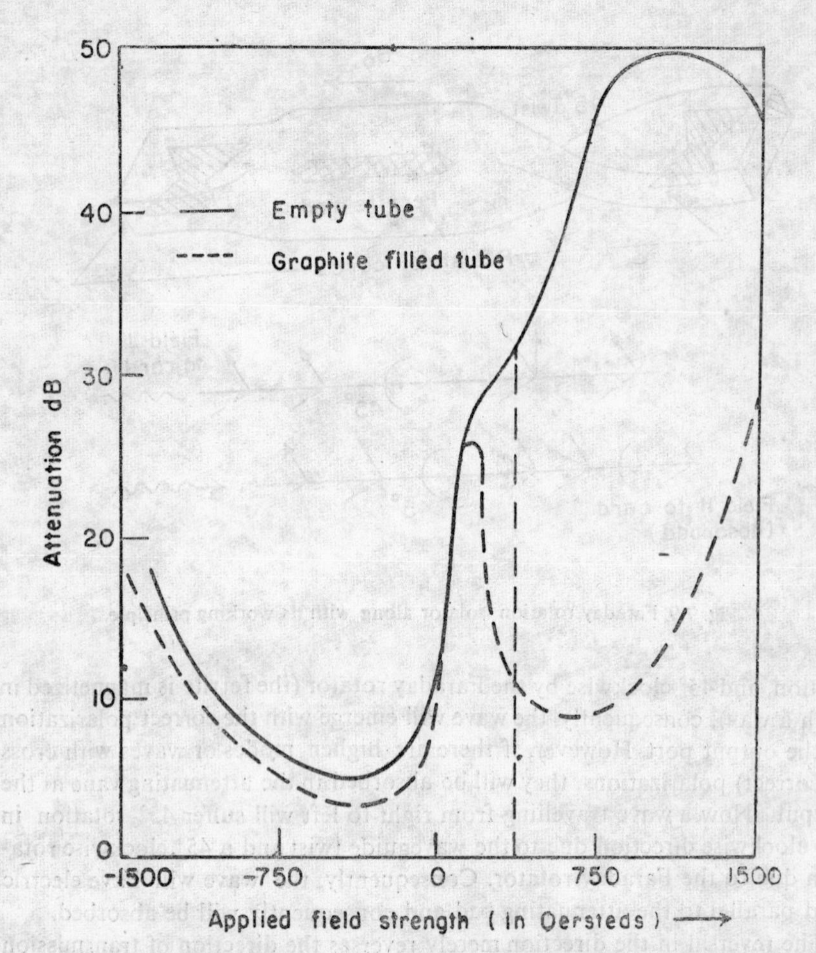


Fig. 7.10(*a*) Attenuation as a function of applied magnetic field for foam supported ferrite tube, external diameter 0.42″, internal diameter 0.22″ and length 1.75″ Frequency 9.52 GHz and waveguide diameter 0.575″.

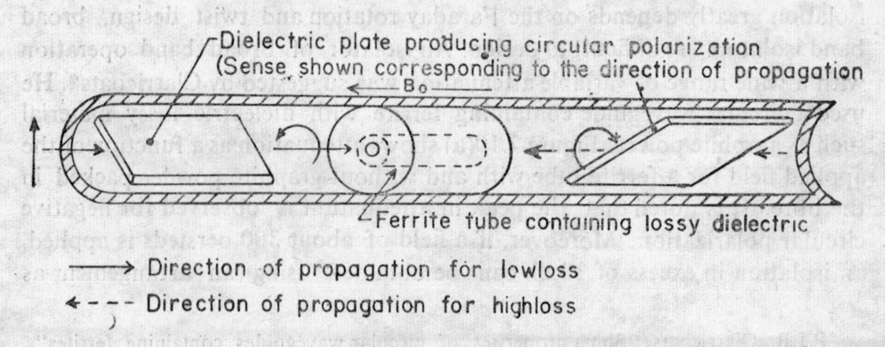


Fig. 7.10(*b*) Circularly polarized wave isolator.

depicted in Fig. 7.10(b). Propagation in the forward direction is by means of positive circular polarization and so does not suffer any attenuation, but in the reverse direction propagation is by means of negative circular polarization and so it suffers attenuation and the arrangement of Fig. 7.10(b) acts as broad band isolator. It should, however, be noted that attenuation characteristics vary with different ferrite shapes and with different types of lossy materials contained within and applied to the surface of the ferrite.

### (*iii*) Resonance Isolator

From the curves of Figs. 7.4(a and b) it is clear that the attenuation constant for negative circular polarization is always very small whereas that for positive circular polarization it is very large in the vicinity of ferromagnetic resonance $\omega \simeq \omega_0$. If, however, a transversely magnetized ferrite slab of vanishing thickness is located in a rectangular waveguide at a distance nearly equal to one quarter of the waveguide width, similar resonance effects are observed. This phenomena is used to design *resonance* isolators. Typical resonators utilizing one and two ferrite slabs are shown in Fig. 7.11.

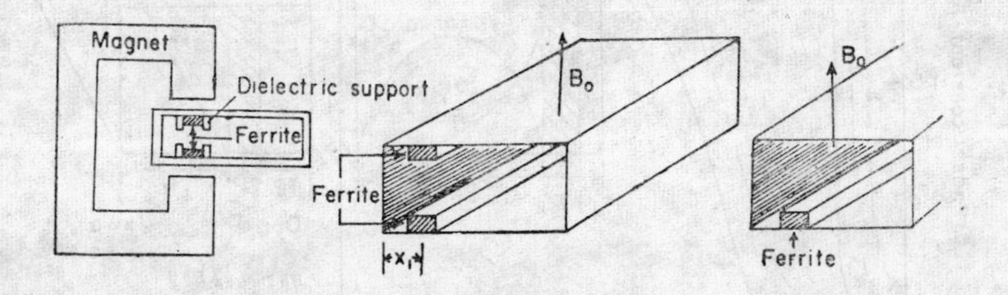


Fig. 7.11 Rectangular-waveguide ferrite resonance isolator.

The resonance isolators may also be designed utilizing circular waveguides but the principal advantages of the rectangular structures are their compactness, simplicity of construction and relatively high power handling capacity.

The location(s) of the ferrite rod(s) is such that at the slab, the field is circularly polarized. As the sense of polarization depends upon the direction of propagation, so the propagation in one direction, the magnetic field is negative circularly polarized and suffers little attenuation, whereas in the reverse direction the field is positive circularly polarized and rapidly attenuated.

Though resonance isolators are easy to design, one of the disadvantages is the need for a relatively high static field to produce resonance. At high microwave frequencies, fields of the order $10^4$ $G$ are required. Moreover, the ratio of backward attenuation to forward attenuation is smaller than that in the Faraday rotation isolators.

### (*iv*) Field Displacement Isolator

From outward appearance, the field displacement isolator resembles the resonance isolator but it differs in principle. The device relies on the non-reciprocal field pattern of the dominant mode in a ferrite-loaded rectangular waveguide. A typical isolator of this kind consists of a transversely magnetized ferrite rod bounded with an attenuative strip and placed near one of the waveguide walls as shown in Fig. 7.12(a). The basic operation may be under-

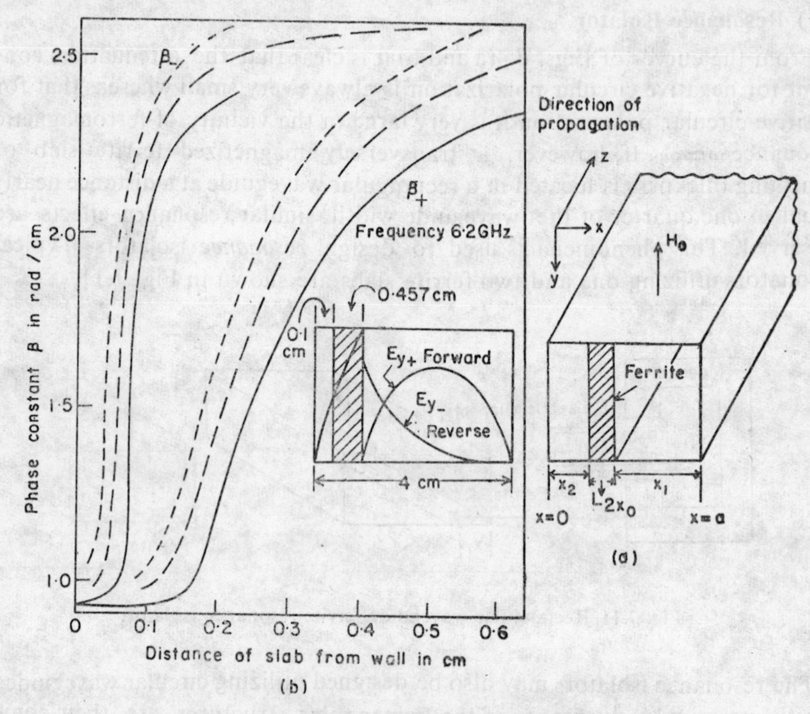


Fig. 7.12(*a*) Waveguide loacied with a ferretic in field displacement isolater
(*b*) Phase constants $\beta_+$ and $\beta_-$ as functions slab position in the waveguide in the field displacement isolater configuration.

stood with reference to Fig. 7.12(b) which depicts $\beta_+$ and $\beta_-$ as a function of slab position and for different values of internal static magnetic field.* It is noted that when moderately thick slab is located so that $x_2 \simeq 0.1$ cm $\overline{\beta}_- > 1$ while $\overline{\beta}_+ < 1$, the wave number $K_{0\pm}$ is given by

$$K_{0\pm} = \sqrt{\omega^2 \epsilon_0 \mu_0 - \beta_\pm^2}$$
$$= \omega \sqrt{\epsilon_0 \mu_0} \sqrt{1 - \overline{\beta}_\pm} = \beta_0 \sqrt{1 - \overline{\beta}_\pm} \tag{7.62}$$

*K.J. Button, "Theoretical analysis of the operation of the field displacement ferrite isolator", *Trans IRE*, MTT-6 (1958).

where $K_{0-}$ is the wave number in the empty waveguide region corresponding to the propagation in the negative $z$-direction and $K_{0+}$ corresponds to propagation in the positive $z$-direction

$$\bar{\beta}_{\pm}=\frac{\beta_{\pm}}{\beta_0}\,.$$

It therefore follows that $K_{0-}$ is imaginary while $K_{0+}$ is real for the said slab position.

The fields in the empty waveguide region are given by

$$E_{y\pm}=A_1 \sin K_{0\pm}\,(x_2+x_0+x);\; x_0 > x > (x_2+x_0)$$

Region $A$

$$E_{y\pm}=A_2 \sin K_{0\pm}\,(x_1+x_0-x);\; (x_1+x_0) > x > x_0$$

Region $B$

If $K_{0-}$ is imaginary, the field in region $B$ has an exponential dependence on $x$ and a maximum value at the ferrite air interface. Also, if $K_{0+}$ is real, the field in region $B$ has a sinusoidal dependence on $x$ and when $K_{0+}\,x_1\simeq\pi$, the field at the air interface is nearly zero. Thus by placing a thin resistive film at the interface, differential attenuation may be achieved. It has, however, been assumed in above discussion that the presence of the resistive thin film does not modify the fields in the waveguide. Thus, in brief, at the ferrite position $\mu_{eff}$ is small and so also $\vec{B}$; consequently, $\vec{E}$ is also small as required by Maxwell's first curl equation. However, for the propagation in reverse direction $\mu_{eff}$ is large, $\vec{B}$ is large and so $\vec{E}$ is also large and consequently there may be sizeable attenuation due to the attenuation pad bounded by the ferrite.

Field displacement isolators have been designed for operation between frequencies 4 and 58 GHz. These are characterized by a high isolation ratio, a very low attenuation in the forward direction and quite a large inherent band width. Moreover, the magnetic field strength required is also smaller than that for resonance isolators but the major disadvantage is that it is a low power level device. To achieve a high power field displacement isolator one has to design a three-port circulator with one arm terminated in a high power load.

## 7.7 CIRCULATOR

A circulator allows complete transmission between adjacent arms taken in one sense of circulation only. For example a, signal incident in arm 1 is transmitted unattenuated to arm 2, a signal incident in arm 2 is transmitted to arm 3 and so on. Circulators of 3 and 4 arms or ports are most common, however, a circulator with any number of arms can be built as a principle. Various types of circulators are discussed below.

*(i)* **Faraday Rotation Circulator**

As shown in Fig. 7.13(a) a four-port Faraday rotation circulator consists of a rectangular waveguide (port 1), a circular waveguide through suitable transition containing a 45° Faraday ferrite rotator, a 45° twist and a rectangular waveguide (port 2). Port 3 is a rectangular waveguide fastened perpendicular to the circular waveguide near to port 1 such that the narrow dimension of port 3 is parallel to the broad dimension of the waveguide of port 1. Port 4 is again a rectangular waveguide perpendicular to port 2 with the narrow dimension parallel to the broader dimension of the waveguide of port 2 and located at the side of port 2 relative to the 45° ferrite rotator.

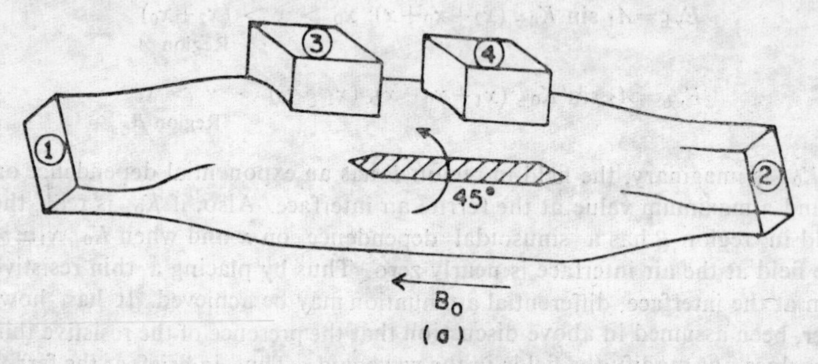


**Fig. 7.13**(*a*) Faraday rotation circulator.

The operation is as follows:

The $TE_{10}$ mode incident in port 1 excites a linearly polarized $TE_{11}$ mode in a circular waveguide. This linearly polarized wave is rotated by 45° in passing through the ferrite rotator and thus excites a $TE_{10}$ mode in arm 2. Thus the wave entering port 1 has correct polarization to be issued out of

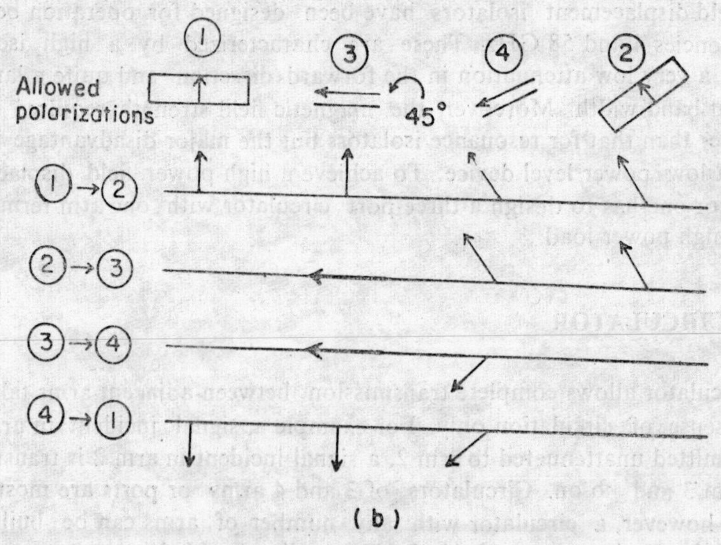


**Fig. 7.13**(*b*) Action of Faraday rotation circulator.

port 2 but not ports 3 and 4. Similarly, the wave incident at port 2 has correct polarization to be issued out of port 3 but not from ports 2 and 4, and so on as shown in Fig. 7.13(b).

An ideal circulator is a matched device; however, its performance may be impaired by many factors. A four-port circulator has a scattering matrix of the form

$$[S]=\begin{bmatrix} 0 & 0 & 0 & 1 \\ 1 & 0 & 0 & 0 \\ 0 & 1 & 0 & 0 \\ 0 & 0 & 0 & 1 \end{bmatrix}. \tag{7.63}$$

(*ii*) **Gyrator Circulator**

Figure 7.14(a) depicts a four-port circulator utilizing two identical magic tees and a gyrator. The gyrator produces an additional phase shift of 180° for propagation in the direction from $a$ to $b$. The propagation electrical lengths $b$ to $a$, $c$ to $d$ and $d$ to $c$ are equal. The wave incident in port 1 splits into two equal amplitudes in-phase waves propagating in the side arms of the magic tee. These waves arrive at $a$ and $c$ in phase, and hence emerge out of the adder arm 2. A wave incident in port 2 will split in two in-phase equiamplitude waves. These waves arrive at $b$ and $d$ at a phase difference of $\pi$ due to the presence of the gyrator and consequently emerge out of the subtractor arm of the tee, i.e. from port 3. Similarly, the wave incident in port 3 will emerge out of port 4 and the wave incident in port 4 in port 1.

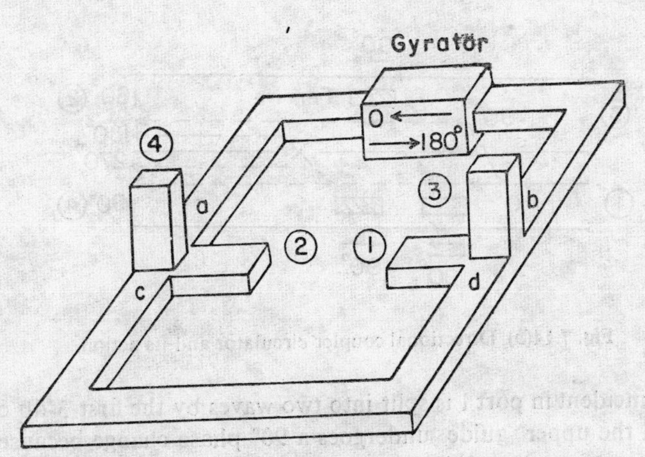


**Fig. 7.14(*a*)** Circulator using two magic tees and gyrator.

(*iii*) **Directional-Coupler Circulator**

A compact four-port circulator may be designed utilizing 3 db side-hole directional couplers and rectangular waveguide *non-reciprocal phase shifters*.

The non-reciprocal phase shifter consists of a rectangular waveguide having a thin magnetized ferrite slab placed at a position where the $TE_{10}$ mode is circularly polarized. Since the phase constants for negative and positive polarizations are different, consequently, chosing the proper length of the ferrite and the biasing field, a non-reciprocal differential phase change,

$$\phi=(\beta_+-\beta_-)\, l \tag{7.64}$$

can be obtained.

In our present case $\phi=90°$. A four-port circulator utilizing two 90 non-reciprocal phase shifters is shown in Fig. 7.14(b). The phase shifters are oppositely biased. One of the guides constituting the directional coupler is also loaded with a 90° reciprocal dielectric phase shifter.

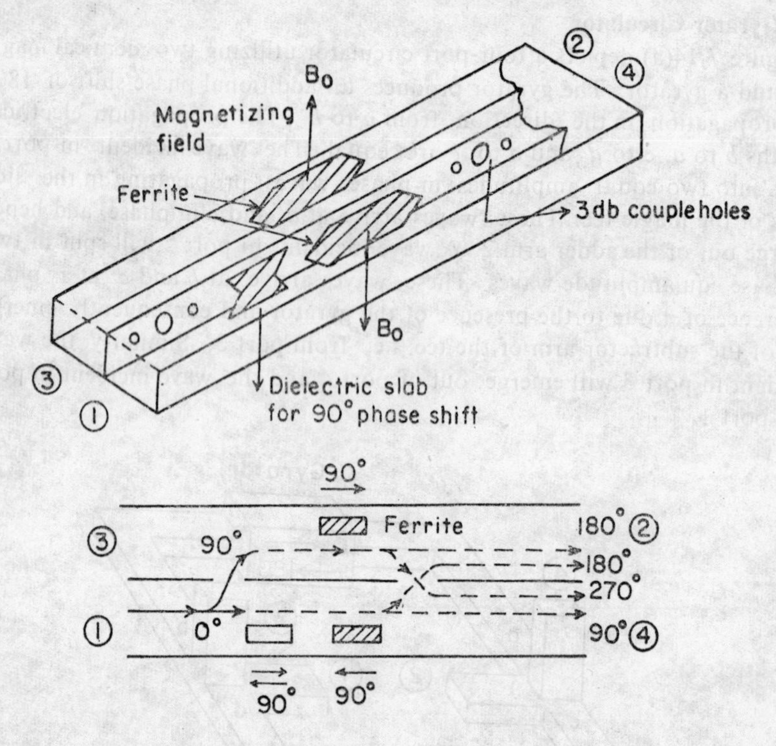


Fig. 7.14(*b*) Directional coupler circulator and its action.

The wave incident in port 1 is split into two waves by the first 3 db coupler, the wave in the upper guide undergoes a 90° phase change because of the transmission properties of the coupling aperture. The wave in the upper waveguide arrives at the second aperture with a relative phase of 180°, and the wave in the lower guide with a relative phase of 90°. The second aperture again splits the wave similarly. As illustrated in Fig. 7.14(b), the waves arrive in phase at port 2 and out of phase in port 4. Similarly, it can be shown that the ports are coupled in the sequence 1→2→3→4→1 and not in the reverse sequence.

Rectangular waveguide circulators have been built for operation at frequencies between 1.2 and 35 GHz with transmission loss-less than 0.5 db and isolation exceeding 20 db over band widths between 5 and 10 percent. Compared with rotational circulators, these circulators have a better power handling capacity and are more compact.

### (*iv*) Ferrite Y-Junction Circulator

This type of waveguide circulator employs a 120° *H*-plane *Y*-junction with a ferrite post placed in the centre of the junction. The ferrite post is magnetized normal to the plane of the junction.

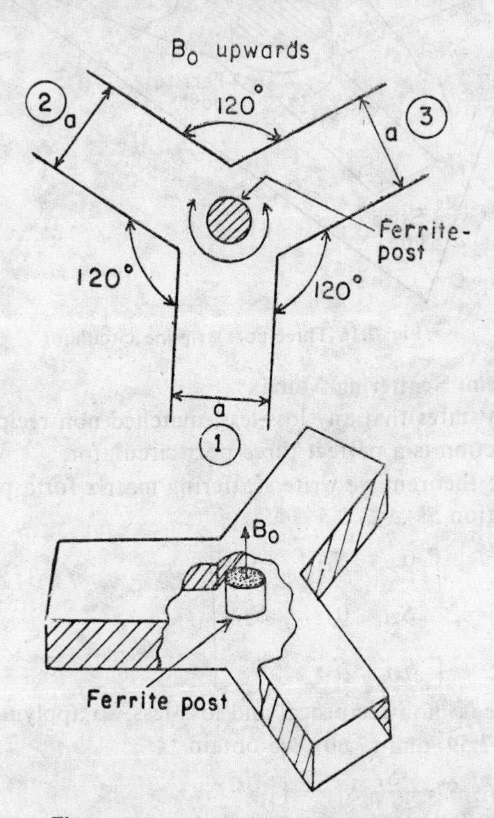


Fig. 7.15 Ferrite *Y*-junction circulator.

Referring to Fig. 7.15, the $TE_{10}$ mode wave incident in port 1 is diffracted around the ferrite post forming two surface waves propagating in opposite directions around the magnetized ferrite port. The magnetizing field and the diameter of the ferrite post are so chosen that the axis of port 2 coincides with the antinode of the electric field intensity and a node appears simultaneously at port 3. Consequently, a signal incident in arm 1 will issue out of arm 2 and no energy comes out of arm 3. The non-reciprocal behaviour of the device arrives from the fact that the phase velocities for the waves propagating clockwise and counter clockwise around the ferrite post are different.

At frequencies 3 GHz and lower a microstrip version of these circulators is used instead of waveguides which are more compact and simple to design. A typical microwave strip line three-port circulator utilizing a ferrite port is shown in Fig. 7.16.

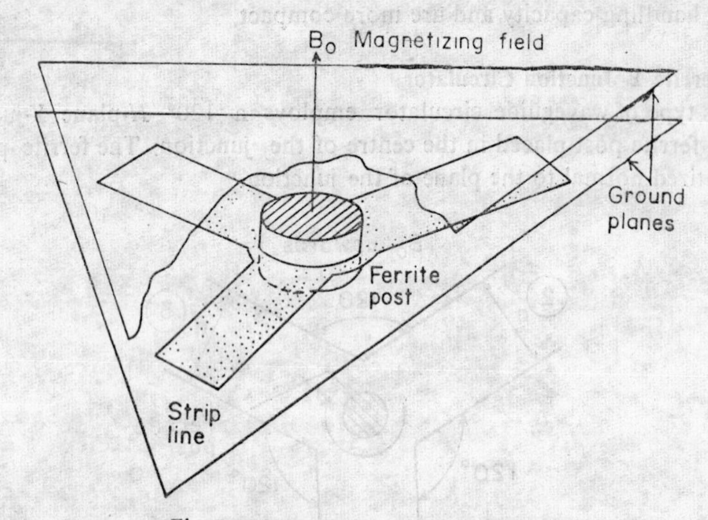


**Fig. 7.16** Three-port stripline circulator.

***Carline Theorem: Scattering Matrix**

This theorem states that any loss-less, matched non-reciprocal three-port microwave junction is a perfect three-port circulator.

To prove the theorem we write scattering matrix for a perfectly matched three-port junction as

$$[S] = \begin{bmatrix} 0 & S_{12} & S_{13} \\ S_{21} & 0 & S_{23} \\ S_{31} & S_{32} & 0 \end{bmatrix}. \tag{7.65}$$

Since the device is non-reciprocal and loss-less, so applying the conditions stated in Eqs. (7.59) and (7.60), we obtain

$$S_{12} S_{12}^* + S_{13} S_{13}^* = 1 \tag{7.66}$$

$$S_{21} S_{21}^* + S_{23} S_{23}^* = 1 \tag{7.67}$$

$$S_{31} S_{31}^* + S_{32} S_{32}^* = 1 \tag{7.68}$$

$$S_{13} S_{23}^* = S_{12} S_{32}^* = S_{21} S_{31}^* = 0 \tag{7.69}$$

*H.J. Carline, "Principles of gyrator networks", *Microwave Res. Inst. Symp. Ser.* Polytech. Inst. Brooklyn, Vol. 4 (1955).

Taking $S_{21} \neq 0$, then from Eq. (7.69)

$$|S_{31}|=0$$

or from Eq. (7.68) $\quad |S_{32}|=1$

or from Eq. (7.69) $\quad |S_{12}|=0$

or from Eq. (7.66) $\quad |S_{13}|=1 \qquad (7.70)$

or from Eq. (7.69) $\quad |S_{23}|=0$

or from Eq. (7.67) $\quad |S_{21}|=1$

These equations imply that there is perfect transmission from port 1 into port 2, from port 2 into port 3 and from port 3 into port 1, $|S_{21}|=|S_{32}|=|S_{13}|=1$, and there is zero transmission in the opposite direction $|S_{12}|=|S_{23}|=|S_{31}|=0$.

Therefore, the resultant scattering matrix for any matched *loss-less*, non-reciprocal three-port junction must have the form

$$[S]=\begin{bmatrix} 0 & 0 & S_{13} \\ S_{21} & 0 & 0 \\ 0 & S_{32} & 0 \end{bmatrix}. \tag{7.71}$$

If the locations of terminal planes are so chosen that the phase angles of $S_{13}$, $S_{21}$ and $S_{32}$ are zero then $S_{13}=S_{21}=S_{32}=1$ and consequently the scattering matrix becomes

$$[S]=\begin{bmatrix} 0 & 0 & 1 \\ 1 & 0 & 0 \\ 0 & 1 & 0 \end{bmatrix}. \tag{7.72}$$

The most common application of a circulator as mentioned earlier, is as a decoupling isolator at high power levels where a common isolator becomes unsuitable [Fig. 7.17(a)]. However, a more interesting and practically more important application is the use of circulators in so-called reflex amplifier circuits which include, among others, such devices as paramagnetic amplifiers of the quantum mechanical type and semiconductor diode amplifiers. Such amplifiers are essentially two terminal devices of the regenerative type in which the amplified signal appears at the same port as the input signal. The amplified signal is separated by means of a circulator and directed to the load, for instance, a receiver. A typical arrangement is shown in Fig. 7.17(b). Circulators are also used in parametric amplifiers and duplexer circuits. Figure 7.17(c) depicts a typical ferrite duplexer circuit.

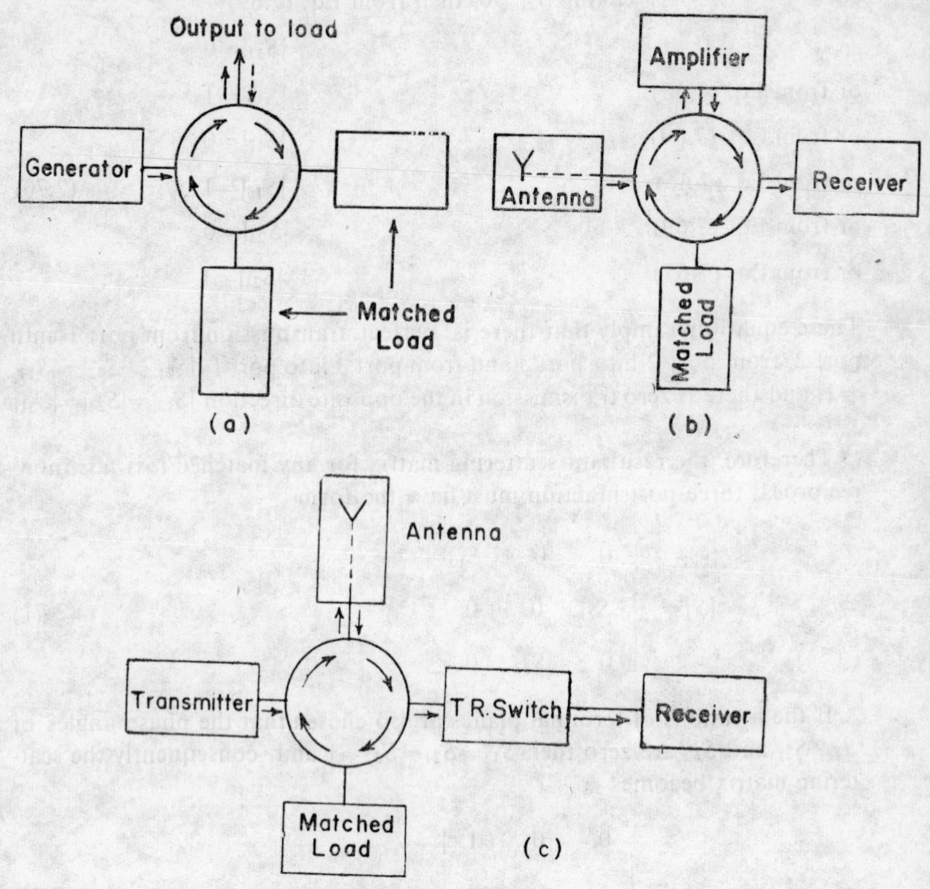


Fig. 7.17(*a*) Circulator as decoupling isolator.
(*b*) Circulator in reflex amplifier circuit.
(*c*) Circulator in a duplexer circuit.

## 7.8 VARIABLE ATTENUATOR AND SWITCHES

Variable attenuators and switches commonly used in modulation circuits can be designed using ferrites. The simplest Faraday rotation variable attenuator and switch comprises a section of circular waveguide containing a longitudinally magnetized ferrite rod in which a linear polarized wave can be rotated by an angle between 0° and 90° depending on the strength of the applied magnetic field. Two attenuating pads are aligned at right angles to each other and placed at either end of the ferrite section as shown in Fig. 7.18. Thus a wave rotated by 90° passes unattenuated while a wave not suffering any rotation is completely absorbed. Consequently, the device can function as a switch or as a variable attenuator. The amount of attenuation is given by

$$-10 \log_{10} (\sin^2 \theta + \epsilon^2 \cos^2 \theta) \text{ db}$$

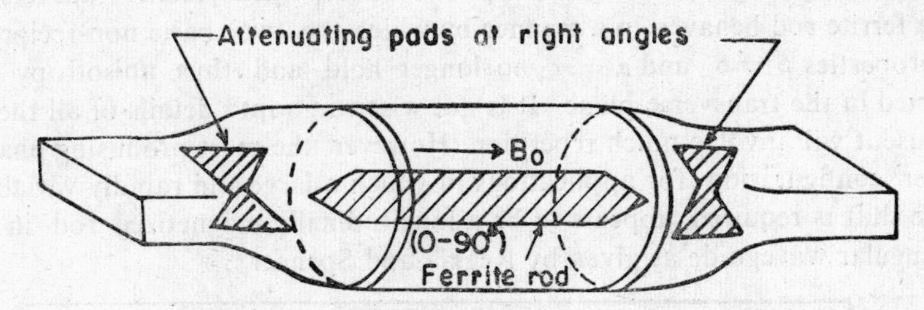


Fig. 7.18 Variable attenuator using a ferrite rotator.

where $\theta$ is the angle of rotation and $\epsilon$ is the ellipticity of the transmitted wave. It is, however, assumed that the ferrite itself does not introduce any attenuation and $\epsilon \ll 1$. Moreover, if the two attenuating pads are aligned to be parallel, the device functions as a switch corresponding to the zero magnetizing field.

The performance of such attenuators and switches may detoriate due to many factors such as hysteresis losses, rise in temperature and eddy currents. Hysteresis effects were examined by Sullivan and LeCraw*, and they loaded the cavity instead of the waveguide with the ferrite to improve performance.

In general, a circularly polarized isolator may be used as a variable attenuator when a varying magnetic field is applied to the ferrite but such an attenuator requires large changes in the field to produce corresponding changes in attenuation and so is suitable when slow variations in the attenuation are desired. Similarly, since resonance and field displacement attenuators too require considerably larger magnetic fields, these type of attenuators are suitable only when slow changes are specified.

Soohoo** has investigated an attenuator that utilizes cut-off conditions of the waveguide loaded with ferrite. A single transversely magnetized ferrite slab is inserted and the cut-off behaviour of the $TE_{10}$ mode is exploited. It is observed that as magnetic field increases the effective permeability $\mu_{eff}$ decreases, as does the propagation constant $\gamma$ of the mode $TE_{10}$. Finally, at a value of the field which depends inversely on slab thickness, the $TE_{10}$ mode is cut-off. As the field strength increases beyond cut-off, the attenuation increases rapidly.

## 7.9 VARIABLE PHASE SHIFTER

The circularly polarized isolator of Fig. 7.10(b) becomes a variable phase shifter if its lossy dielectrics are taken away. Similarly, the gyrator configuration becomes a reversible phase shifter if the ferrite rod inside it (Fig. 7.7)

*R.F. Sullivan and R.C. LeCraw "A new type of ferrite microwave switch", *Jour. Appl. Phys.*, 26 (1955).

**R.F. Soohoo, "A ferrite cut-off switch", *Trans. IRE*, MTT-7 (1959).

is transversely magnetized. This happens because, when biased transversely, a ferrite rod behaves in a manner by which the two basic non-reciprocal properties $\beta_+ \neq \beta_-$ and $\alpha_+ \neq \alpha_-$ no longer hold and thus anisotropy is reverted in the transverse plane. It is not wise to go into details of all these because it will involve much repetition. However, the most promising phase shifter configuration for applications in which a large and rapidly variable phase shift is required, appears to be a longitudinally magnetized rod in a rectangular waveguide as given by Reggia and Spencer*.

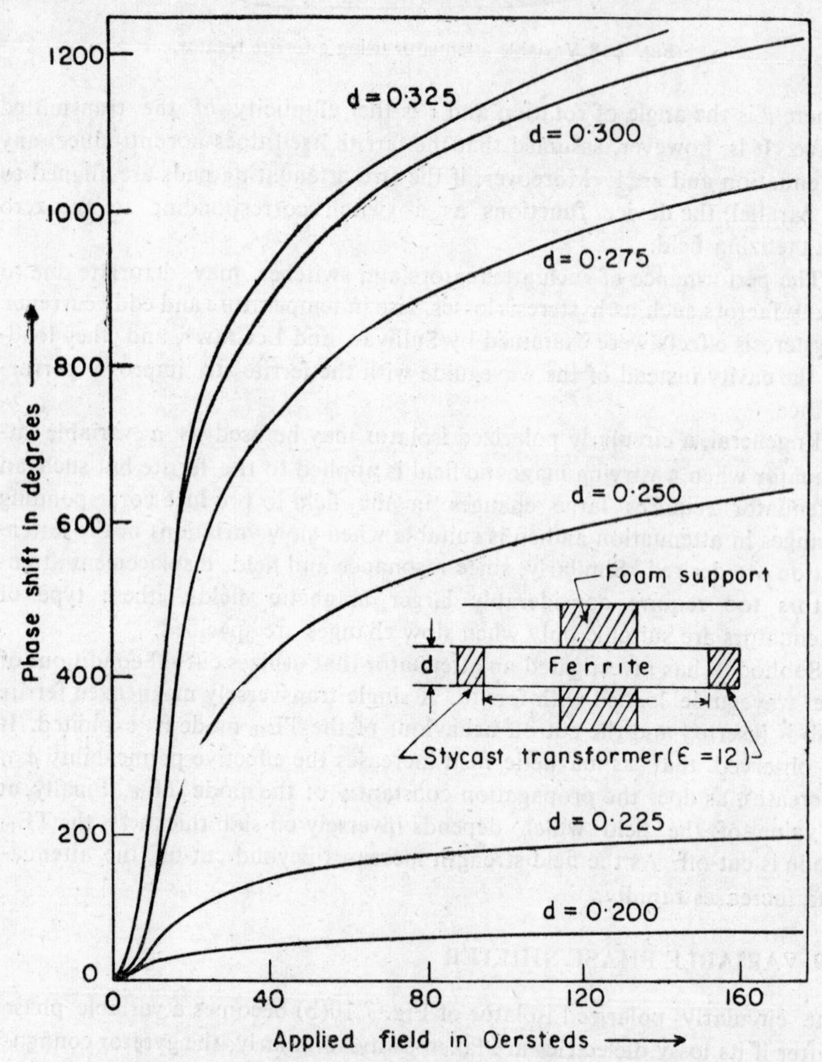


Fig. 7.19 Phase shift produced by longitudinally magnetized R1 Ferrite rod in a rectangular waveguide 0.9″×0.4″ at frequency 9.1 GHz for various rod diameters.

*F. Reggia and E.G. Spencer "A new technique in ferrite phase shifting for beam scanning in microwave antennas", *Proc. IRE,* 45 (1957).

The basic principle underlying the device may be summarized with reference to Fig. 7.19. For a sufficiently small ferrite rod diameter, the ferrite only perturbs the empty waveguide fields slightly and under these conditions the change in phase shift with magnetic field arises from the change in the diagonal component of the permeability $\mu$. In the unmagnetized state $\mu$ has a value 0.7 for a typical ferrite (Ferramio $R$ 1) while at saturation $\mu=0.9$. When the rod diameter is increased beyond about 0.2 inches, the field is more strongly perturbed and the phase shift, which is still reciprocal, increases rapidly with radius. It was observed that the phase constant was similar to that of a circular waveguide containing an axial ferrite rod supporting a $TE_{11}$ limit mode with negative circular polarization. Hence, it appears that the wave excited in the rectangular waveguide with ferrite structure, by the incident $TE_{10}$ mode has a field which is negative circularly polarized within the ferrite. Consequently, the change in waveguide dimensions does not appreciably affect the configuration and so the phase shifter may also be designed in a circular waveguide configuration.

A ferrite variable phase shifter may also be designed without moving parts by varying a resultant magnetic field produced in space by four coils as in Fig. 7.20.

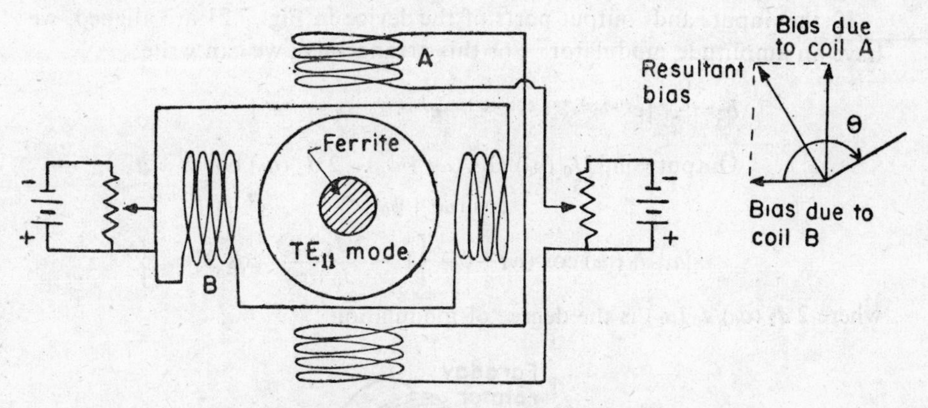


**Fig. 7.20** Ferrite electronic variable phase shifter

## 7.10 FERRITE MODULATORS

Amplitude and frequency modulators may be designed using Faraday rotation and ferrite reciprocal phase shifters.

A double side band, suppressed carrier modulator may be constructed by connecting two rectangular waveguides at right angles through a Faraday rotator biased with a microwave magnetic field, with suitable transitions as shown in Fig. 7.21. The wave outgoing Faraday rotator can be written as

$$b_2=\frac{|a_1|}{2}\, j\,[\underbrace{e^{j(\omega t+\psi_0-\theta)}}_{\text{(positive polarization)}} - \underbrace{e^{j(\omega t+\psi_0+\theta)}}_{\text{(negative polarization)}}] \tag{7.73}$$

where $\theta$ is the Faraday rotation, dependent on the applied magnetic field, $\psi_0$ is the common phase shift due to wave propagation. Suppose the peak value of $\theta$ is $\alpha$* and the applied magnetic field varies with microwave frequency $\omega_m$, then

$$\theta = \theta_0 \cos \omega_m t \tag{7.74}$$

Substituting Eq. (7.74) in Eq. (7.73) gives

$$b_2 = \frac{|a_1|}{2} j[e^{j(\omega t+\psi_0-\theta_0 \cos \omega_m t)} - e^{j(\omega t+\psi_0+\theta_0 \cos \omega_m t)}$$

which can be simplified to yield

$$\text{Output} = |a_1| \,[2\, J_1\,(a_0) \cos \omega_m t \sin (\omega t + \psi_0') + 2\, J_3\,(a_0) \cos 3\, \omega_m t \sin (\omega t + \psi_0')]$$

where $\psi_0'$ is a new constant phase angle.

It may be noted that output has the same form as the reflected wave from either crystal in a single side band modulator. Hence, if a microwave signal is split, with each half fed to the device of Fig. 7.21 but with each device driven in quadrature and then recombined in a magic tee (but with one arm longer by $\lambda_g/4$), then result will be equivalent to a single side band modulator.

If the input and output ports of the device in Fig. 7.21 are aligned, we have an amplitude modulator. For this arrangement we can write

$$b_2 = \frac{|a_1|}{2} [e^{j(\omega t+\psi_0-\alpha_0 \cos \omega_m t)} + e^{j(\omega t+\psi_0+\alpha_0 \cos \omega_m t)}]$$

$$\text{Output} = |a_1| \,[J_0\,(a_0) \cos (\omega t + \psi_0') - 2\, J_2\,(a_0) \cos 2\, \omega_m t \cos (\omega t + \psi_0)]$$

$$= |a_1|\, J_0\,(a_0) \cos (\omega t + \psi_0') \left\{1 - \frac{2\, J_2\,(a_0)}{J_0\,(a_0)} \cos 2\, \omega_m t\right\}$$

where $2\, J_2\,(a_0)/J_0\,(a_0)$ is the degree of modulation.

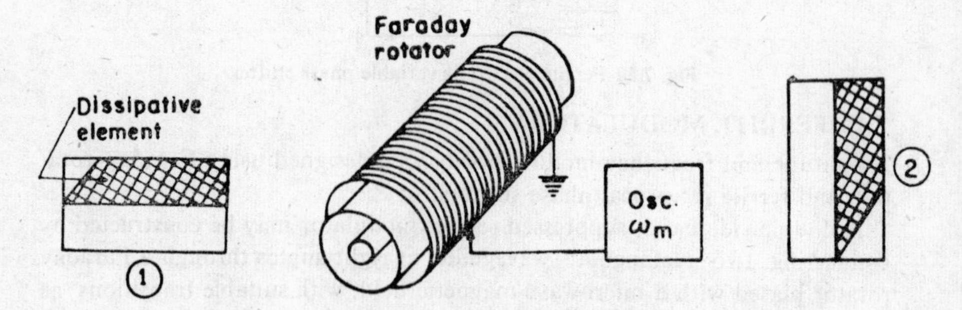


Fig. 7.21 Faraday double side band modulator.

It is possible to produce various types of modulation by combinations of the schemes described above. For example, from amplitude modulation, it is possible to obtain double side band suppressed-carrier modulation by rejecting the carrier out of phase; or, it is possible to obtain single side band modulation by combining amplitude modulation and frequency modulation, provided that the indices of modulation are small; or from double side band suppressed-carrier modulation, it is possible to obtain either amplitude modulation or frequency modulation by rejecting the carrier in a proper phase relationship.

**Frequency Modulation**

A typical ferrite frequency modulator consists of a reciprocal Faraday rotation phase shifter whose permanent magnet is replaced by a solenoid as shown in Fig. 7.22. The attenuating pads serve as the eliminator of any

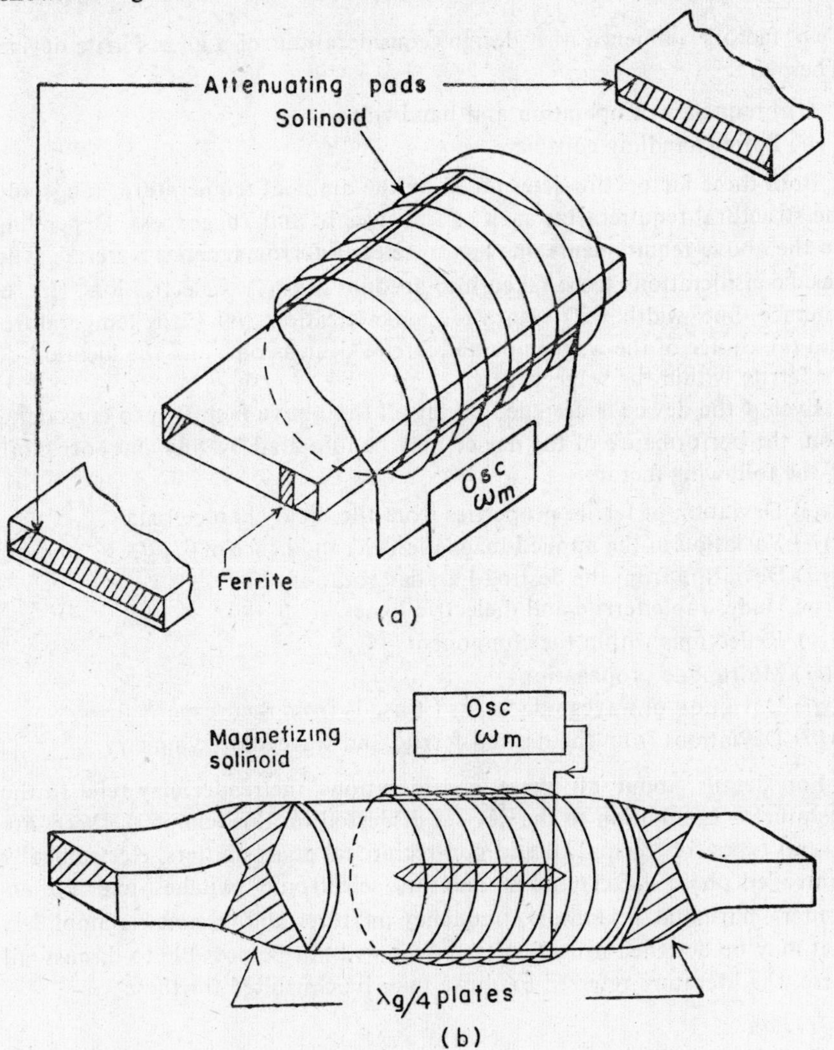


Fig. 7·22 Ferrite frequency modulators.

cross polarizations. Since the phase shift is a function of the applied current hence desired frequency modulation may be achieved. Frequency modulation may also be achieved by inserting a Faraday rotator (solenoid-controlled) between two quarter-wave plates. Since the phase of the resulting wave varies as $\theta = \alpha_0 \sin \omega_m t$, hence the resultant spectrum contains an infinite number of side bands, located symmetrically with respect to $\omega$, and all separated by $\omega_m$.

## 7.11 FACTORS INFLUENCING DESIGN AND PERFORMANCE OF MICROWAVE FERRITE COMPONENTS

Two factors influence the design considerations of a given ferrite device. These are:

(*i*) Frequency of operation and bandwidth
(*ii*) Power handling capacity

Both these factors are determined by the ambient temperature range and the structural requirements such as size, weight and ruggedness. Depending on the above requirements one has to select a ferromagnetic material. The basic considerations to be taken into account are: (*i*) dielectric loss; (*ii*) resonance line width; (*iii*) saturation magnetization; (*iv*) Curie temperature; and (*v*) choice of the waveguide and ferrite dimensions, and the location of the ferrite within the waveguide.

Even if the device is designed taking all the above factors into consideration, the performance of the device may be impaired by any one or more of the following factors:

(*i*) Deviation of ferrite properties from the ideal characteristics.
(*ii*) Variation in the applied magnetic field and hence in $\beta_+$, $\beta_-$, $\alpha_+$ and $\alpha_-$.
(*iii*) Deviation from the desired Faraday rotation.
(*iv*) Undesirable ferrite and dielectric losses.
(*v*) Reflections within the component.
(*vi*) Multimode propagation.
(*vii*) Deviation of waveguide twists (if used) from their exact values.
(*viii*) Deviations from the desired ferrite and waveguide symmetry.

For details about all these considerations, the reader may refer to the literature*. In addition to the devices discussed in this chapter, there are several types of reciprocal and non-reciprocal phase shifters, electronically controlled phase shifters and modulators, electronic switches and power-limiters, harmonic generators, frequency mixtures and parametric amplifiers that may be designed using ferrites. As it will not be possible to discuss all these, the literature referred to above may be consulted for these.

*1. B. Lax and K.J. Button, *op. cit.*
2. P.J.B. Clarricoats, *op. cit.*

# SOLVED EXAMPLES

**Example 7.1.** Using the Landu-Lifshitz form of the equation of damped motion

$$\frac{d\vec{M}}{dt}=\gamma(\vec{M}\times\vec{H})-\lambda/|M|^2\,\vec{M}\times(\vec{M}\times\vec{H}).$$

Evaluate the permeability matrix.

*Solution*: Assuming small signal conditions we can write

$$\vec{H}=H_0\hat{z}+\vec{H}_x+\vec{H}_y$$

and $$\vec{M}=M_0\hat{z}+\vec{M}_x+\vec{M}_y.$$

Letting $$-\frac{\gamma M_0}{\mu_0}=\omega_M,\quad -\gamma H_0=\omega_0$$

$$\chi_0=\frac{M_0}{H}=\frac{\omega_M}{\omega_0}\mu_0,\quad \bar{\chi}_0=\frac{\chi_0}{\mu_0}\ (\chi_0 \text{ is static susceptibility}).$$

We can write

$$\frac{d\vec{M}}{dt}=\gamma(\vec{M}\times\vec{H})+\lambda[\vec{H}-(\vec{H}\cdot\vec{M})/|\vec{M}|^2\vec{M}]$$

which on using time dependence $e^{j\omega t}$ can be reduced to

$$j\omega M_x=\omega_M\mu_0H_y-\omega_0M_y+\lambda H_x-\frac{\lambda}{\chi_0}M_x$$

or $$\left(j\omega+\frac{\lambda}{\chi_0}\right)M_x=\omega_M\mu_0H_y+\lambda H_x-\omega_0M_y. \tag{i}$$

Similarly,

$$\left(j\omega+\frac{\lambda}{\chi_0}\right)M_y=-\omega_M\mu_0H_x+\lambda H_y+\omega_0M_x. \tag{ii}$$

Solving these for $M_x$ and $M_y$

$$M_x=\frac{\mu_0[\omega_0\omega_M+j\omega\lambda/\mu_0+(\omega_0/\chi_0)(\lambda/\mu_0)^2]\,H_x+j\omega\omega_MH_y}{\omega_0^2+\omega^2+2\,j\omega/\bar{\chi}_0\cdot\lambda/\mu_0+(\omega/\omega_M\cdot\lambda/\mu_0)^2}$$

$$=\chi H_x-jkH_y\ \text{(say)}.$$

Similarly,

$$M_y=jkH_x+\chi H_y$$

Then, using

$$\vec{B}=\mu_0\vec{H}+\vec{M}$$

and $$\vec{B}=[\mu]\vec{H}$$

we have

$$\vec{B} = \begin{bmatrix} \mu & -jk & 0 \\ jk & \mu & 0 \\ 0 & 0 & \mu_0 \end{bmatrix} \vec{H}$$

where

$$\mu = \mu_0 + \chi = \mu_0 + \frac{\mu_0\left[\omega_0\omega_M + j\omega\lambda/\mu_0 + \left(\frac{\mu_0}{\chi_0}\right)\left(\frac{\lambda}{\mu_0}\right)^2\right]}{(\omega_0^2 - \omega^2) + 2j\frac{\omega}{\chi_0}\frac{\lambda}{\mu_0} + \left(\frac{\omega_0}{\omega_M}\frac{\lambda}{\mu_0}\right)^2}$$

$$k = \frac{-\omega\omega_M\mu_0}{(\omega_0^2 - \omega^2) + 2j\frac{\omega}{\chi_0} + \left(\frac{\omega_0}{\omega_M}\frac{\lambda}{\mu_0}\right)^2}.$$

**Example 7.2.** The expression for the change in the propagation constant, $\delta\gamma$, arising from a transversely magnetized ferrite is given by (for the $TE_{no}$ mode)

$$\delta\gamma = \frac{j\omega \int_{S'} \{(\mu - \mu_0)\, H_T \cdot H_{T_n}^* - jk(\hat{y} \times H_T) \cdot H_{T_n}^* + (\epsilon - \epsilon_0) E_T \cdot E_{T_n}^*\} ds}{2 \int_S E_{tn} \times H_{tn} \cdot \hat{z}\, ds}$$

where the wave is propagating in the $z$-direction. Then, referring to Fig. Ex. 7.2 show that a differential phase shift

$$\frac{2\pi k\, x_0}{\mu a^2} \sin 2K_0 x_2 \text{ is obtained.}$$

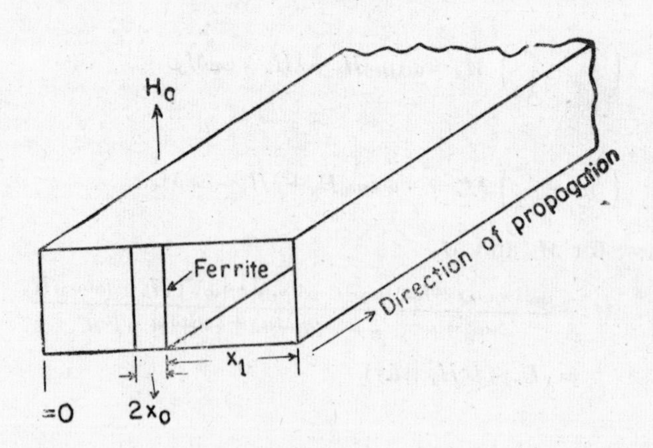


**Fig. Ex. 7.2**

*Solution*: In general,

$$\vec{H}_T = \vec{H}_x + \vec{H}_z,\ \vec{E}_T = \vec{E}_x + \vec{E}_z$$

$$\vec{H}_{tn} = \vec{H}_{xn} + \vec{H}_{yn},\ \vec{E}_{tn} = \vec{E}_{xn} + \vec{E}_{yn}$$

where $H_T$ and $E_T$ represent fields in a perturbed waveguide while $\overrightarrow{H_{tn}}$ and $\overrightarrow{E_{tn}}$ represent the normal mode fields in an unperturbed waveguide. ($T$ refers to the fields transverse to the direction of magnetization while $t$ denotes fields tanqential to the direction of propagation.)

Assuming that an infinitely long slab is immersed in the field of unperturbed mode,

$$H_x = H_{x_n} - N_x \frac{M_x}{\mu_0}.$$

Since fields are uniform in the $y$-direction

$$N_y = 0 \text{ and } N_x = 1 \text{ then}$$

$$H_x = H_{x_n} - \frac{M_x}{\mu_0}.$$

We can writte

$$M_x = H_x - jkH_x$$

or $$H_x = \frac{\mu_0 H_{x_n} + jkH_{zn}}{\mu} \text{ (because } N_z = 0,\ H_z = H_{z_n}).$$

Moreover, $E_y = E_{yn}$.

Now fields in the unperturbed waveguide are given by

$$H_{x_n} = -\frac{\beta_0}{\omega\mu_0} \sin K_0 x$$

$$H_{z_n} = j\frac{K_0}{\omega\mu_0} \cos K_0 x$$

$$E_{y_n} = \sin K_0 x$$

where $$K_0 = \frac{n\pi}{a} \text{ and } K_0^2 = \omega^2\epsilon_0\mu_0 - \beta_0^2.$$

Assuming the ferrite to be loss-less, $\delta\gamma = j\delta\beta$ and substituting values in the given expression for $\delta\beta$, we have

$$\delta\beta = \frac{2x_0\beta_0}{a}\left\{\frac{\chi}{\mu_0}\sin^2 K_0 x_2 + \frac{K_0^2}{\beta_0^2}\left(\frac{\chi}{\mu_0} - \frac{k^2}{\mu\mu_0}\right)\cos^2 K_0 x_2 - \frac{k\pi}{\mu a\beta_0}\sin 2K_0 x_2 + \frac{\overline{\chi}_e \omega^2 \epsilon_0\mu_0}{\beta_0^2}\sin^2 K_0 x_2\right\}.$$

The differential phase shift is given by

$$|\delta\beta_+ - \delta\beta_-| = \frac{4\pi k x_0}{\mu a^2} \sin 2K_0 x_2.$$

(In the expression for $\delta\beta$, the only reciprocal term is

$$\frac{k\pi}{\mu a\beta_0} \times \frac{2x_0\beta_0}{a} \sin 2K_0 x_2\Big).$$

**Example 7.3.** Obtain an expression for effective permeability when a transversely magnetized ferrite slab is placed in an arbitrary position in Fig. Ex. 7.3.

*Solution*: The permeability tensor for the given configuration is

$$[\mu]=\begin{bmatrix} \mu & 0 & -jk \\ 0 & \mu_0 & 0 \\ jk & 0 & \mu \end{bmatrix}.$$

From Maxwell's equations

$$\nabla\times\vec{E}=-j\omega\mu\,\vec{H}$$

$$\nabla\times\vec{H}=-j\omega\epsilon\,\vec{E}.$$

Taking the field dependence of the form $e^{j\omega t-\gamma z}$, Maxwell's equations become

$$\gamma E_y=j\omega(\mu H_x-jk\,H_z) \qquad (i)$$

$$\frac{\partial E_y}{\partial x}=-j\omega(jkH_x+\mu H_z) \qquad (ii)$$

$$\gamma H_x+\frac{\partial H_z}{\partial x}=-j\omega\epsilon E_y \qquad (iii)$$

and

$$\gamma E_x+\frac{\partial E_z}{\partial x}=j\omega\mu_0 H_y \qquad (iv)$$

$$\gamma H_y=j\omega\epsilon E_x \qquad (v)$$

$$\partial H_y/\partial x=j\omega\epsilon E_z. \qquad (vi)$$

The first three equations determine the properties of $TE_{no}$ modes, while the last three determine the properties of $TM_{no}$ modes. Since only $TE_{no}$ modes are influenced by the magnetic conduction of the saturated ferrite and the dominant mode is $TE_{10}$ we shall confine ourselves to $TE_{no}$ modes. Eliminating $H_z$ and objaining $H_x$ from the first two equations, we have

$$\gamma E_y+j\frac{k}{\mu}\frac{\partial E_y}{\partial x}=-j\omega\,\frac{\mu^2-k^2}{\mu}\,k_x$$

$$-j\,\frac{k}{\mu}\,\gamma E_y+\frac{\partial E_y}{\partial x}=-j\omega\,\frac{\mu^2-k^2}{\mu}\,H_z.$$

Using the third of the set of equations

$$\frac{\partial^2 E_y}{\partial x^2}\left\{+\gamma^2+\omega^2\epsilon\left(\frac{\mu^2-k^2}{\mu}\right)\right\}E_y=0$$

or

$$\frac{\partial^2 E_y}{\partial x^2}+K_f^2\,E_y=0$$

where $\quad K_f^2=\gamma^2+\omega^2\epsilon\,\mu_{eff}$ and $\mu_{eff}=\dfrac{\mu^2-k^2}{\mu}.$

It may be noted that $\mu_{eff}=0$ when $\mu=k$. This corresponds to resonance condition.

**Example 7.4.** Show that it is impossible to construct a perfectly matched, loss-less, reciprocal three-port junction.

*Solution*: A circulator is a non-reciprocal perfectly matched three-port junction and has a scattering matrix

$$[S]=\begin{bmatrix} 0 & 0 & S_{13} \\ S_{21} & 0 & 0 \\ 0 & S_{32} & 0 \end{bmatrix}.$$

If the junction is reciprocal in addition to being perfectly matched and lossless, then from reciprocity we have

$$S_{21}=S_{12}=0,\ S_{13}=S_{31}=0,\ S_{32}=S_{23}=0.$$

Hence $[S]$ vanishes. This proves the proposal.

**Example 7.5.** Show that a turnstile junction shown in Fig. Ex. 7.5 behaves as a four-port circulator. Take the scattering matrix as

$$[S]=\frac{1}{2}\begin{bmatrix} 0 & 1 & 0 & 1 & \sqrt{2} & 0 \\ 1 & 0 & 1 & 0 & 0 & \sqrt{2} \\ 0 & 1 & 0 & 1 & -\sqrt{2} & 0 \\ 1 & 0 & 1 & 0 & 0 & -\sqrt{2} \\ \sqrt{2} & 0 & -\sqrt{2} & 0 & 0 & 0 \\ 0 & \sqrt{2} & 0 & -\sqrt{2} & 0 & 0 \end{bmatrix}$$

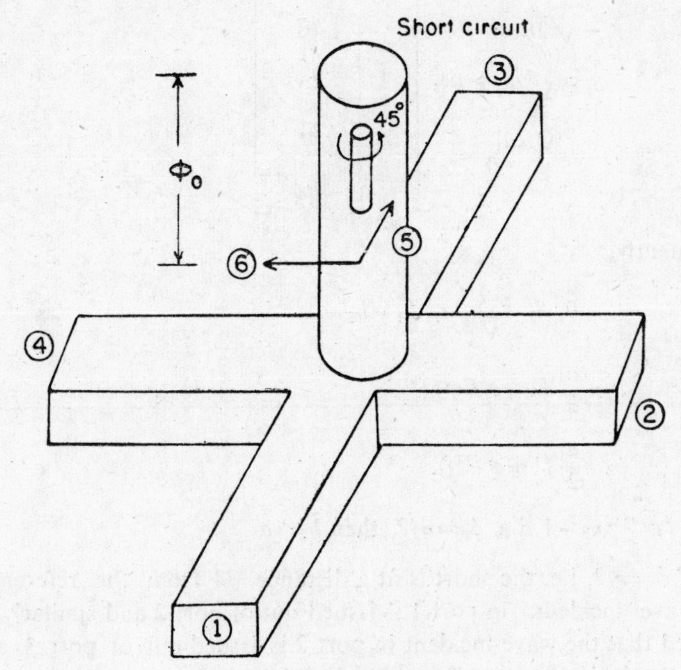


Fig. Ex 7.5

*Solution*: The normalized input wave $a_5$ results from $-b_6$, changed in sign by the short circuit and delayed in phase by $2\phi_0$, where $\phi_0$ is the phase shift from the reference planes to the short circuit. Since the 46° Faraday rotator shifts the polarization plane by a total of 90° counter clockwise, hence

$$a_5 = b_6\, e^{-j2\phi_0}$$

Similarly, $a_6 = -b_5\, e^{-j2\phi_0}$.

Therefore, the input wave $a_1$ is transformed as

$$\frac{1}{2}\begin{bmatrix} 0 & 1 & 0 & 1 & \sqrt{2} & 0 \\ 1 & 0 & 1 & 0 & 0 & \sqrt{2} \\ 0 & 1 & 0 & 1 & -\sqrt{2} & 0 \\ 1 & 0 & 1 & 0 & 0 & -\sqrt{2} \\ \sqrt{2} & 0 & -\sqrt{2} & 0 & 0 & 0 \\ 0 & \sqrt{2} & 0 & -\sqrt{2} & 0 & 0 \end{bmatrix} \begin{bmatrix} a_1 \\ 0 \\ 0 \\ 0 \\ b_6\, e^{-j2\phi_0} \\ -b_5\, e^{-j2\phi_0} \end{bmatrix} = \begin{bmatrix} b_1 \\ b_2 \\ b_3 \\ b_4 \\ b_5 \\ b_6 \end{bmatrix}$$

or

$$\frac{1}{2}\begin{bmatrix} \sqrt{2} b_6\, e^{-j2\phi_0} \\ a_1 - \sqrt{2} b_5\, e^{-j2\phi_0} \\ -\sqrt{2} b_6\, e^{-j2\phi_0} \\ a_1 + \sqrt{2} b_5\, e^{-j2\phi_0} \\ \sqrt{2} a_1 \\ 0 \end{bmatrix} = \begin{bmatrix} b_1 \\ b_2 \\ b_3 \\ b_4 \\ b_5 \\ b_6 \end{bmatrix}.$$

Consequently,

$$b_6 = 0,\ b_5 = \frac{1}{\sqrt{2}}\, a_1,\ b_1 = b_3 = 0$$

$$b_2 = \frac{a_1}{2}\,(1 - e^{-j2\phi_0})$$

$$b_4 = \frac{a_1}{2}\,(1 + e^{-j2\phi_0}).$$

Taking $e^{-j2\phi_0} = -1$, i.e. $\phi_0 = \pi/2$, then $b_2 = a_1$.

Hence if $\phi = \pi/2$, i.e. the short is at a distance $\lambda/4$ from the reference the plane wave incident, in port 1 is issued out of port 2 and similarly it can be proved that the wave incident in port 2 is issued out of port 3 and so on, and the device acts as a four-port circulator.

**Example 7.6.** Two gyrators are connected in cascade, neglecting the phase and assuming gyration resistance $R$ in each case. Draw the equivalent four terminal network made up of classical $R$, $C$ and $L$ elements and the same transfer function.

*Solution*: The gyrator is a four terminal network having impedance matrix

$$[Z] = \begin{bmatrix} 0 & -R \\ R & 0 \end{bmatrix}.$$

The input impedance is thus related to the load impedance as

$$Z_{in} = R/Z_L.$$

Now, the second gyrator has infinite load impedance and therefore, a zero input impedance, i.e. a short circuit. If output is short-circuited so that the input impedance of the second gyrator is infinite, the input will be given by

$$Z_{in} = R^2 j\omega C = j\omega\ (R^2 C).$$

The system thus behaves as an *a* series inductor circuit with series inductance $L = R^2C$ as shown in Fig. Ex. 7.6.

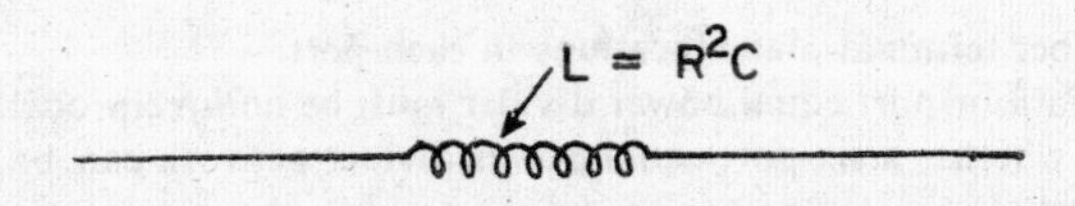


**Fig. Ex. 7.6**

## PROBLEMS

1. Give an account of the electrical properties of ferrites which differentiate them from other magnetic materials.
2. Obtain the permeability matrix for a transversely magnetized ferrite.
3. What is Faraday rotation? Show that it is a non-reciprocal phenomenon.
4. What is a gyrator? Give some of its applications.
5. How does an isolator differ from an attenuator? Explain the various types of ferrite isolators with suitable diagrams.
6. List the basic characteristics of a circulator. Discuss any one type with uses.
7. "Any loss-less matched non-reciprocal three-port microwave junction is a circulator". Discuss.
8. How does a ferrite variable attenuator differ from other attenuators?
9. How are frequency and amplitude modulations achieved by variable ferrite phase shifters?

10. Obtain the field configuration that exists in a rectangular waveguide (operating in the $TE_{10}$ mode) when a thin transversely magnetized ferrite slab parallel to its length is placed inside it.

11. Show that for a TEM wave propagation in a direction perpendicular to $B_0$ in an infinite ferrite medium, the two solutions are linearly polarized waves with propagation constants

$$\gamma_+ = j\omega\sqrt{\epsilon_0}\left[\frac{(1+\chi)^2 - K^2}{1+\chi}\right]^{1/2}$$

$$\gamma_- = j\omega\sqrt{\mu_0\epsilon}.$$

12. Show that if $\omega K\sqrt{\epsilon/\mu}\, z/2 \ll 1$, the angle of rotation of polarization in a ferrite is proportional to the distance $z$ of wave travel.

13. Show that the scattering matrix for an ideal loss-less $N$-port circulator can be put into the form

$$[S] = \begin{bmatrix} 0 & 0 & 0 & \dots & 0 & 1 \\ 1 & 0 & 0 & \dots & 0 & 0 \\ 0 & 1 & 0 & \dots & 0 & 0 \\ \dots & \dots & \dots & \dots & \dots & \dots \\ \dots & \dots & \dots & \dots & \dots & \dots \\ 0 & 0 & 0 & \dots & 1 & 0 \end{bmatrix}$$

by choosing proper terminal-plane locations in each port.

14. Prove that a four-port equal power divider must be non-reciprocal.

15. Determine whether a six-port equal power divider network can be realized by a reciprocal network.

Deduce the real scattering matrix which represents the six-port equal power divider network.

16. Show that an $n$-port circulator with a minimum number of gyrators can be realized by the inter-connection of three-and four-port circulators.

17. A three-port circulator designed for unit resistance terminations (1→2→3) has port 2 terminated in 1 Ω, port 3 *in* −1 Ω. Show that port 1 must satisfy the condition $v=i=0$. (This is called a *nullator*.)

If two resistances at ports 1 and 2 are interchanged, show that $v$ and $i$ at port 1 are independent. (This is the *norator*.)

Show that the norator is a one-port non-reciprocal junction.

18. A one way line is formed by terminating port 3 of a three-port circulator (1→2→3) in a matched load. Suppose the port 3 termination is not matched and has a reflection coefficient $0.1e^{-j30°}$, what is the scattering matrix of the resultant approximation to a one-way line?

19. (*i*) Can a four-port circulator be obtained by combining two three-port circulators?

(*ii*) Can a four-port circulator be used as a three-port circulator?

## *REFERENCES*

Adams, S.F., 'Microwave Theory and Applications', Prentice-Hall Inc. (1969).

Atwater, H.A., 'Introduction to Microwave Theory', McGraw-Hill Book Co., New York (1962).

BERK, A.D. and STRUNEWASSES, E., 'Ferrite Directional Coupler' Proc. IRE 1956.

BOWNESS, C., 'Microwave Ferrites and Their Applications' Microwave J. 1 13-21 July-Aug 1958.

CASTILLO, J.B., 'Loss-less Three Port H-Plane Waveguide Junction Loaded Coaxially with Inhomogeneous Ferrite Cylinders: Performance Evaluation' IEEE Proc. MTT-18 (1) 1970.

CLARRICOATS, P., 'Microwave Ferrites', John Wiley & Sons Inc., New York 1961.

CLAVIN, A., 'High Power Ferrite Load Isolators Trans. Inst. Radio Enggrs. MTT-3 (1955).

COOPER, B.R. and VOGT, O., 'Phys. Rev. B1 1218 1970.

COPPLETONE, R.J., 'Cycle H-Plane Junction: Exact Three Dimensional Field Theory' IEEE Proc. MTT-27 (1) 1979.

DE SANTIS, P., 'Edge-Guided-Wave Circulators for 8-12 GH Band: MIC Realization' IEEE Proc. MTT-23 (6) 1975.

FOURNET, F.C. '50 KW C Differential Phase Shift Circulator at 2450 MH' IEEE Proc. MTT-26 (5) 1978.

GUREVICH, A.G.; 'Ferrites at Microwave Frequencies', Heywood and Co. good bibliography 469 ref. 1960.

HELSZAJN, J., 'Principles of Microwave Ferrite Engineering', Wiley Interscience 1969.

HELSAJN, J., 'High Peak Power Differential Phase Shift Circulators: Biasing at Direct Magnetic Field Between Subsidiary and Main Resonances' IEEE Proc. MTT-19 (1) 1971.

HELSZAJN, J., 'E-Plane Junction Circulators: Mode Charts' IEEE Proc. MTT-20 (2) 1972.

HELSZAJN, J., 'Characteristics of Circulators Using Triangular and Disc Resonators Symmetrically Loaded with Magnetic Ridges' IEEE Trans. MTT-28 (6) 1980.

KHILLA,, A.M., 'Cylindrical Cavity Containing Full Height Triangular Ferrite Post: Point Matching Solution: Magnetic Tuning Characteristics' IEEE Proc. MTT-27 (6) 1979.

KNERR, R.H., 'Lumped Element Circulator for 4 GH Region: Compatible with Integrated Circuits' IEEE Proc. MTT-21 (3) 1973.

KUMAR, R.C., 'Bibliography of Ferrite Junction Circulators from Jan. 56 to Sept. 1969' IEEE Proc. MTT-18 (9) 1970.

LAX, B. and BUTTON K.J., 'Microwave Ferrites and Ferrimagnetics', McGraw-Hill Book Co. 1962.

MIYOSHI, T., 'Microwave Ferrite Planar Circuits with Ferrite Substrate Magnetized Perpendicular to Ground Plane', IEEE Proc. MTT-25 (7) 1979.

MUELLER, R.S., 'Microwave Devices Latching Properties' IEEE Proc. MTT-24 (8) 1976.

NANDA, V.P., 'Millimeter-Wave Dielectric Image Line IC' IEEE Proc. MTT-24 (11) 1976.

NARAYANAN, K.G., 'Isolation Ratings of Ferrite Components at High Pulse Powers' IEEE Proc. MTT-18 (6) 1970.

OKAMOTO, N., 'H-Plane Waveguide Junctions with Full Height Ferrites of Arbitrary Shape: Computer Aided Design' IEEE Proc. MTT-27 (4) 1979.

PIOTROWSKI, W.S., 'Millimeter-Wave Waveguide Circulator for 27-34 GH and 31-38 GH Bands' IEEE Proc. MTT-24 (11) 1976.

POIRIER, A.L., 'Integrated Microstrip Circulator for 1.75 to 2.23 GH Range' IEEE Proc. MTT-19 (7) 1971.

PROC. IEEE INTERMAG Conference London, England, April 1975.

PROC. 7th European Microwave Conference, Denmark, Sept. 1977, Tample House, London 1977.

ROBERTS, J., 'High Frequency Application of Ferrites', D. Van Nostrand Co., N.J. (1960).

ROBINSON, G.H., 'Miniature Use of Slot Lines' IEEE Proc. MTT-17 (12) 1969.

SALAY, S.J., 'Input Impedance of Stripline Circulator' IEEE Proc. MTT-19 (1) 1971.

SCHLOMANN, E., 'Theoretical Analysis of Twin-Slab Phase Shifters in Rectangular Waveguide', Proc. G-MTT Symposium Cleauvater, Florida, May 7, 1965.

SCHOTT, F.W., 'Electromagnetic Waves in Ferrite Rods: Classification Scheme', IEEE Proc. MTT-16 (11) 1968.

SHANDWILY, E. and EZZAT, M., 'E-Plane Waveguide Junction Circulators Having Full-Height Ferrite Configuration: Exact Field Theory Treatment' IEEE Proc. MTT-25 (9) 1977.

SHANDWILY, E. and EZZAT M., 'E-Plane Waveguide Junction Circulator Having Two Disk Ferrite Configuration: Exact Field Theory Treatment' Proc. IEEE MTT-25 (9) 1977.

SILBER L.M., 'Ferromagnetic Resonance in Plannar Ferrites', IEEE Proc. MTT-18 (1) 1970.

SMIT, J. and WIJN, H.P.J., 'Physical Properties of Ferrites', Advances in Electronics and Electron Physics, New York (1954).

SOOHOO, R.F., 'Theory and Applications of Ferrites', Prentice-Hall Inc. N.J. (1960).

TAFT, D.R. (*et al.*), 'Ferrite Digital Phase-shifter', G-MTT Symp. Clearwater, May 6 (1965).

UZDY, Z., 'Computer-Aided Design of Stripline Ferrite Junction Circulators' IEEE Trans. MTT-28 (10) 1980.

VARTANIAN, P.H., 'A Broad Band Ferrite Microwave Isolator' Trans. Inst. Radio Engs. MTT-4 1956.

WEISS, M.T., 'Improved Rectangular Waveguide Resonance Isolator' MTT-4 (1956).

WU, Y.S., 'Microstrip-Inserted Pluck Circulator Using Arc Plasma Sprayed Ferrite' IEEE Proc. MTT-23 (6) 1975.

WU, Y.S., 'Microstrip Y-junction Circulators: Octave Bandwidth Operation' IEEE Proc. MTT-22 (10) 1974.

YOSHIHIKO, A., 'Millimeter-Wave Y-Circulators: Bandwidth Enlargement by Loading Resonators in Each Terminal' IEEE Proc. MTT-22 (12) 1974.

# 8

# SLOW WAVE STRUCTURES AND FILTERS

## 8.1 INTRODUCTION

There are many microwave systems in which periodically loaded transmission lines are used. A periodically loaded line may be an ordinary waveguide loaded at periodic intervals with identical obstacles such as irises or a coaxial line loaded with short-or open-circuited resonant quarter-wave stubs giving rise to periodic capacitive or inductive loading, or the structure may consist of strip lines suitably loaded, or it may be a more complicated device such as a series of identical cavities, each with two waveguide outputs, one on each side, the output of one cavity forming the input to the next. The characteristics of these structures which make them available for different uses are: (*i*) passband—stopband characteristics that make them useful in filter design; and (*ii*) support of waves with phase velocities much less than the velocity of light. As a result of these characteristics efficient interaction between the electrons travelling inside the structure and the electromagnetic field existing in the structure, is obtained. This forms the basis of the working of a travelling wave tube (TWT).

Our aim in this chapter is to present the basic aspects of slow wave structures and the properties of the waves travelling in these structures. We shall also introduce the basic aspects of microwave filter theory and filter design.

## 8.2 CAPACITIVELY LOADED COAXIAL LINES

The phase velocity of the waves travelling in a coaxial line is given by

$$v_p=(LC)^{-1/2}=(\mu_o\epsilon)^{-1/2}=(\mu_o k\epsilon_o)^{-1/2} \tag{8.1}$$

where $k$ is the dielectric constant*, and $\mu_o$ and $\epsilon_o$ are the magnetic permeability and electric permittivity respectively of free space. It can be seen from Eq. (8.1) that a significant reduction in $v_p$ can be obtained by increasing either $k$ or any one of $L$ and $C$. Increase in $k$ results in the higher modes. To avoid higher modes, cross-sectional dimensions have to be reduced, which is disadvantageous. Moreover, dielectrics with high $k$ are

*It should not be confused with wavevector.

difficult to obtain. For an electrically smooth line, $v_p$ cannot be decreased by increasing $C$ or $L$ because of the relation $LC=\mu_o\epsilon$. However, if the smoothness restriction is removed an effective increase in $C$ may be achieved without affecting $L$ by adding a lumped shunt capacitance at periodic intervals with periodicity less than the free space wavelength $\lambda_o$. A typical capacitively loaded coaxial line, obtained introducing thin circular diaphragms at regular intervals (which in fact form an integral part of the central conductor) is shown in Fig. 8.1. The fringing electric field in the vicinity of the diaphragm increases the local storage of electrical energy and hence may be accounted for by an additional shunt capacitance $C_o/d$ per unit length, $d$ being the periodicity of the structure ($d<\lambda_o$). It is not difficult to see that the line will be smooth with a phase velocity

$$v_p=\{(C+C_o/d)\,L\}^{-1/2} \tag{8.2}$$

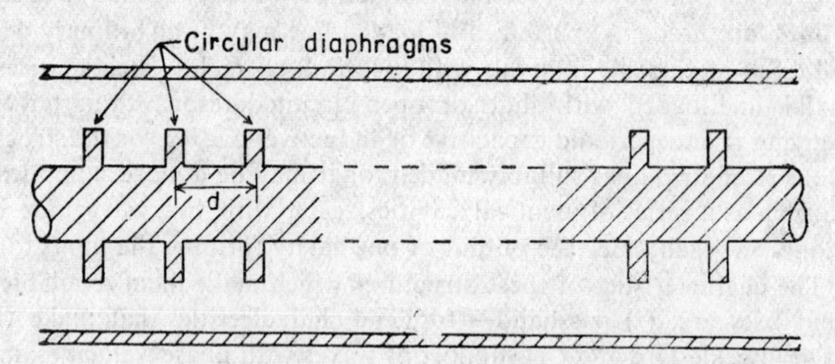


Fig. 8.1 Capacitively loaded coaxial line.

The corresponding phase constant will be

$$\beta=\omega\sqrt{L\,(C+C_o/d)} \tag{8.3}$$

The higher the value of $C_o/d$, the slower is the propagating wave. The normalized characteristic impedance at the input of the "unit cell" of the loaded line is given by

$$z_b=\sqrt{\left\{\frac{L}{C+C_o/d}\right\}}, \qquad \text{for } (\lambda\gg d) \tag{8.4}$$

$z_b$ is not unique for a loaded line because it depends on the choice of terminal planes for a unit cell. Moreover, for unsymmetrical structures with $z_b^+$ (forward wave) and $z_b^-$ (backward wave) characteristics impedances differ. If an unsymmetrical structure is terminated in a normalized load $z_L$, the normalized input impedance at the $n$th section is given by

$$z_n-\zeta=z\frac{z_L-\zeta+jz\tan(N-n)\,\beta d}{Z-j\,(z_L-\zeta)\tan(N-n)\,\beta d} \tag{8.5}$$

where $\quad Z=\dfrac{z_b^+-z_b^-}{2},\; z_b^+=g+Z, \quad$ and $\quad z_b^-=\zeta-Z$

are Bloch-wave characteristic.

$\zeta$ is an impedance parameter defining asymmetry. For a symmetrical structure $\zeta=0$. $N$ is the total number of sections. For waves propagating in the $+z$ direction, the periodic structure must be terminated in a load $z_L=z_b^+$ $=\zeta+z$ to avoid reflection. Similarly, the normalized matched load for waves travelling in the $-z$-direction is $z_L=-z_b^-=z-\zeta$. It may, however, be noted that $z_b$ provides a match at a particular frequency and so for a broad band matching a sophisticated matching network has to be used.

## 8.3 GENERAL CHARACTERISTICS OF WAVE PROPAGATION IN PERIODICALLY LOADED MICROWAVE LINES

We have seen in Section 8.2 that slow waves do exist if a coaxial line or say a microwave line is loaded periodically. In this section, some basic characteristics, common to all such structures are discussed.

**(*i*) General Properties of Slow Waves: Field Distribution, Spatial Harmonics**

The field distribution inside a general slow wave structure is determined by the scalar wave equation

$$\nabla^2 L+k^2 L=0 \tag{8.6}$$

where components of the microwave electric and magnetic fields (i.e., *Ex*, *Ey*, *Ez*, *Hx*, *Hy* or *Hz*) in the guide in question are designated at $L$. The general solution of Eq. (8.6) in the absence of dissipative losses may be written as

$$L=D \begin{matrix}\cos\\ \sin\end{matrix} (\xi x) \begin{matrix}\cos\\ \sin\end{matrix} (\eta y) \exp [j(\omega t-\beta z)]. \tag{8.7}$$

The phase constant $\beta$ is related to the transverse constants $\xi$ and $\eta$ of a wave propagating in the $z$-direction as

$$\beta^2=k^2-(\xi^2+\eta^2). \tag{8.8}$$

Since $v_p=\omega/\beta$, a slow wave can only be expected when $(\xi^2+\eta^2)<1$, i.e. at least one of the transverse constants (or both) are imaginary,

$$\xi=j\xi_1, \quad \eta=j\eta_1$$

where $\xi_1$ and $\eta_1$ are real quantities with dimensions reciprocal to length. Under these conditions Eq. (8.7) can be rewritten as

$$L=D \begin{matrix}\cosh\\ \sinh\end{matrix} \xi_1 x) \begin{matrix}\cosh\\ \sinh\end{matrix} (\eta_1 y) \exp [j(\omega t-\beta z)]. \tag{8.9}$$

Since any one of the transverse constants $\xi$ or $\eta$ can remain real, the field components can also be described as

$$L=D \begin{matrix}\cosh\\ \sinh\end{matrix} (\xi x) \begin{matrix}\cosh\\ \sinh\end{matrix} (\eta y) \exp [j(\omega t-\beta z)]$$

$$L=D \begin{matrix}\cosh\\ \sinh\end{matrix} (\xi_1 x) \begin{matrix}\cos\\ \sin\end{matrix} (\eta y) \exp [j(\omega t-\beta z)]. \tag{8.10}$$

It follows from Eq. (8.10) that the distribution of electric and magnetic fields along at least one of the axes of the transverse cross-section should follow a hyperbolic sine or cosine law as illustrated in Fig. 8.2. Moreover, the tangential electric field cannot be zero at all boundary surfaces at once, because the hyperbolic functions do not have more than one zero point. It may, therefore, be concluded that a slow wave cannot be excited inside a pipe of perfectly conducting metal. In other words, to excite slow waves at least one of the walls of the structure must be discontinuities or a slow wave structure must have an impedance surface; it may be a distributed capacitance or an inductance or a resistance (if losses are tolerable).

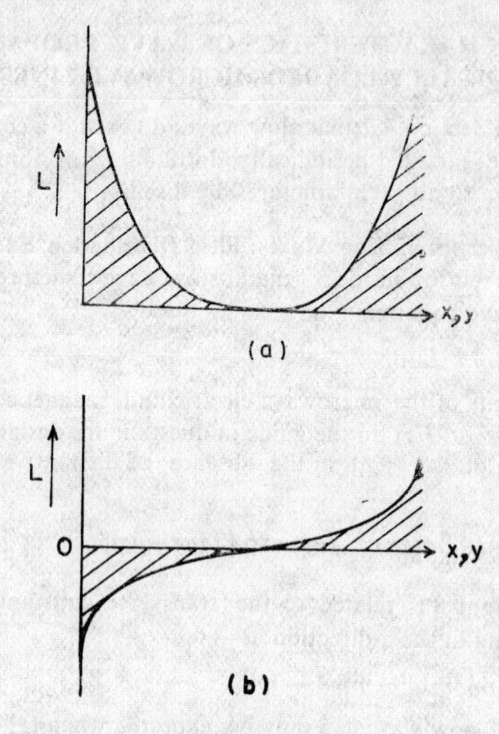


**Fig. 8.2** Field distribution in the transverse cross-section of a slow-wave structure (a) hyperbolic cosine law (b) hyperbolic sin law.

Furthermore, Eq. (8.10) indicates that field intensity decreases exponentially (cosh or sinh) as we move away from the impedance surface. The field is thus forced against the system and the larger the constants $\xi_1$ and $\eta_1$ the stronger the effect.

As a general case let us assume a slow wave structure to be composed of sections of a uniform microwave guide divided by equidistant identical discontinuities as shown in Fig. 8.3. The line may be either a rectangular or circular waveguide, coaxial line or stripline, etc. For simplicity let us assume the guide to be loss-less and supporting only one mode. The solution of Eq. (8.6) may now be taken as an ordinary travelling wave

$$E = E_m e^{j(\omega t - \beta z)} \tag{8.11}$$

Assume further that the slow wave structure is terminated in a matched load so that there is no reflected wave. Now the periodicity of the structure can be accounted for by introducing a periodic factor $F(z)$ in Eq. (8.11)

$$E_m = E_{m1} F(z) \tag{8.12}$$

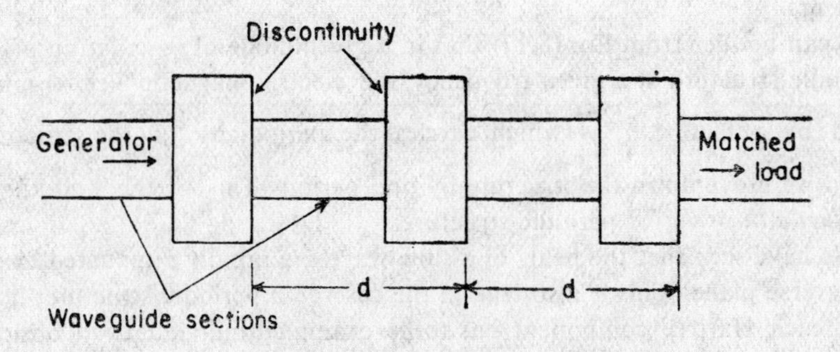


**Fig. 8.3** Waveguide sections equivalent circuit of a periodic slow wave structure.

Function $F(z)$ features spatial periodicity $\alpha$ and so can be expanded into a harmonic series (Fourier expansion).

$$F(z) = \sum_{n=-\infty}^{+\infty} C_n e^{(-j2\pi nz/d)} \tag{8.13}$$

where $n$ is an integer including zero ($n=0, \pm 1, \pm 2, \ldots$) $C_n$ is the expansion coefficient corresponding to a specific harmonic and depends upon the nature of the structure.

Using Eq. (6.12) and (8.13) in Eq. (8.11), we have

$$E = E_{m1} e^{(j\omega t - \beta z)} \sum_{n=-\infty}^{+\infty} C_n e^{-j2\pi nz/d}$$

or

$$E = \sum_{n=-\infty}^{+\infty} E_n e^{j(\omega t - \beta_n z)} \tag{8.14}$$

where,

$$\beta_n = \beta + \frac{2\pi n}{d}. \tag{8.15}$$

The waves* represented by Eq. (8.14) have the same angular frequency $\omega$ but have different propagation constants $\beta_n$ which are periodic in $d$. Such waves are generally referred to as *spatial* or *Hartree harmonics*.

The phase velocity of the $n$th harmonic is

$$(v_p)_n = \omega / \beta_n. \tag{8.16}$$

Thus the higher the harmonic number $n$, the higher will be $\beta_n$ and consequently the lower will be the value of $v_p$ (or the wave will be slower). The harmonic with highest phase velocity is conventionally called a *fundamental*

*Waves in a periodic structure are known as *Bloch waves*.

*Hartree component*; which ordinarily, corresponds to the case $n=0$. Equation (8.16) can then be written as

$$\beta_n = \beta_o + \frac{2\pi n}{d} \tag{8.17}$$

$\beta_0$ is the phase constant corresponding to the fundamental hartree components.

It can be seen from Eq. (8.17) that in a given mode of oscillation of the periodic structure at a given frequency, the electric/magnetic field is multiplied by a factor* $j\frac{2\pi}{d}$ (which carries the periodicity of the structure) when we move down the structure by one period. This is the well-known *Floquet's theorem* for periodic structures.

We have seen that the fields of a slow wave are rapidly attenuated in the transverse plane. This is also true in the case of a periodic structure; however, each Hartree component has to be examined individually. Consider, for instance, the $n$th component, $(E_z)_n$ in rectangular coordinates,

$$(E_z)_n = D^{\cosh}_{\sinh}[(\xi_1)_n\, x]^{\cosh}_{\sinh}\, [(\eta_1)_n\, y]\; e^{j(\omega t - \beta_n z)} \tag{8.18}$$

where $$\beta_n \simeq \{(\xi_1)^2_n + (\eta_1)^2_n\}^{1/2} \tag{8.19}$$

considering the fields to exist only along one axis, say along $\xi$ then $(\eta_1)_n = 0$

or, $$(E_z)_n \simeq D^{\text{Cosh}}_{\sinh}\, (\beta_n x)\; e^{j(\omega t - \beta_n z)}. \tag{8.20}$$

It follows from Eq. (8.20) that the higher the value of $|n|$, the greater will be the wave concentrated at the surface of the slow wave structure. Furthermore, the amplitude of the wave at the surface decreases as $|n|$ is increased. This conclusion can be confirmed by considering the structure of a folded line shown in Fig. 8.4. Clearly, the $z$-component of the electric field is given by

$$E_z = \sum_{n=-\infty}^{+\infty} (E_{zm})_n\, e^{j(\omega t - \beta_n z)}$$

where $(E_{zm})_n$ is the following Fourier expansion constant

$$(E_{zm})_n = \frac{E_o}{d/2}\int_{-b/2}^{b/2} e^{j\beta_n z}\, dz = \frac{2E_o b}{d}\,\frac{\sin \beta_n b/2}{\beta_n b/2} \tag{8.21}$$

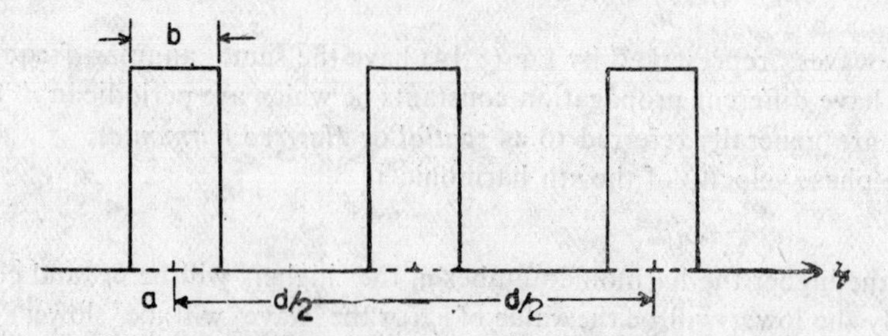


Fig. 8.4 Periodically loaded line (folded line).

*This factor is purely imaginary under the assumed conditions. In general, it is complex.

Figure 8.5 depicts the variation of $(E_{zm})_n$ with $|n|$ and confirms the above statement.

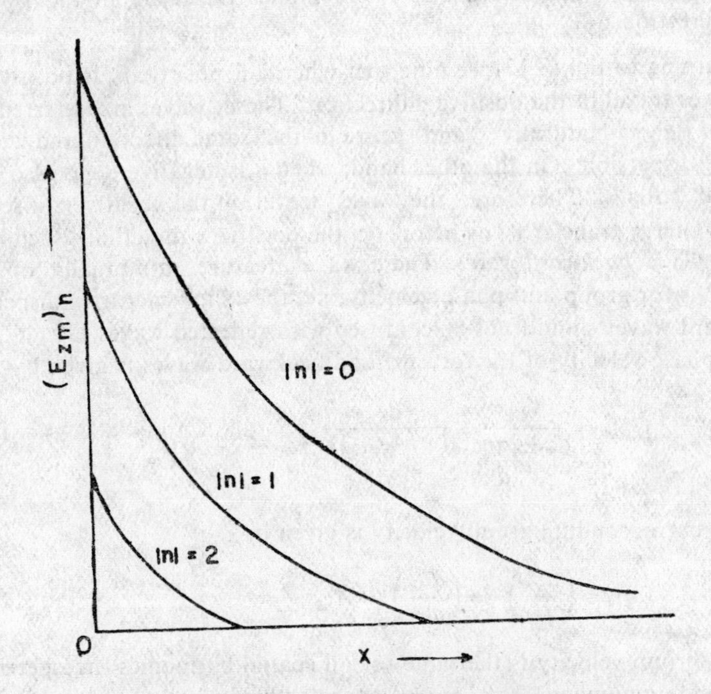


**Fig. 8.5** Variation of the electric field with transverse cross-section of a periodic slow wave structure for various Hartree harmonics

It may be remarked at this stage that $(E_{zm})_n$ is maximum for $n=1$ and so the strong interaction of an electron beam, travelling inside the slow wave structure, with the microwave field existing in the structure, is expected. This concept is important in the design and use of travelling wave tape (TWT) and backward-wave oscillators (BWOs). In practice, a spatial harmonic with $n=0$, $n\pm1$ is used. $n=0$ corresponds to the case of a helical slow wave structure discussed in Section 8.5.

(*ii*) **Dispersion Characteristics**

The concept of dispersion, i.e. dependence of phase velocity on frequency, is of special interest when slow wave structures are used in microwave oscillator and amplifier tubes in which synchronization of the wave and the electron beam forms the basis of tube performance. Four types of dispersion can be observed.

(*a*) *Normal dispersion* implies fall of $|v_p|$ with increase in operating frequency, $\omega$.

(*b*) *Abnormal dispersion* implies increase in $|v_p|$ with the increase in $\omega$.

(c) *Positive or forward dispersion* implies that $v_p$ and $v_g$ are in the same direction.

(d) *Negative or backward dispersion* implies that $v_p$ is in the opposite direction to $v_g$.

Returning to Eq. (8.17) we note that when $n$ is positive $\beta_n$ is positive and the waves travel in the positive $z$-direction. These waves are referred to as *forward waves.* Naturally $v_p$ and $v_g$ are in the same direction and we have positive dispersion. On the other hand, when $n$ is negative $\beta_n$ is also negative and so is $v_p$. Therefore, the waves travel in the negative $z$-direction, though energy transfer is, as before, in the positive $z$-direction. Such waves are known as *backward waves.* These waves feature in mutually opposite directions for group and phase velocity, i.e. they show negative dispersion. Backward waves should not be confused with reflected waves.

The phase velocity of the forward and backward waves is given by

$$(v_p)_n = \frac{\omega}{\beta_o + \frac{2\pi n}{d}} = \frac{d}{d + n\,(\lambda_{\text{slow}})_o} \tag{8.22}$$

and the corresponding group velocity is given by

$$(v_g)_n = \left(\frac{d\beta_n}{d\omega}\right)^{-1} = \left(\frac{d\beta_o}{d\omega}\right)^{-1} = (v_g)_o \tag{8.23}$$

Thus, group velocity is the same for all spatial harmonics irrespective of the harmonic number. This is an expected result because in a slow wave structure the boundary conditions cannot be satisfied by any one wave and the question of transmitting all the energy by any one spatial harmonic is physically not possible.

From Eq. (8.22) some very general and important conclusions can be drawn. Firstly, the curve of $\omega$ as a function of $\beta_o$ is an even function, i.e. change in the sign of $\beta_o$ leaves $\omega$ unchanged. This follows from the symmetry of the structure. Secondly, the $\omega - \beta_o$ curve is a periodic function with period $2\pi/d$, i.e. when $\beta_o$ increases by $2\pi/d$, no change in the physical situation occurs except for the change in the nomenclature, i.e. $\beta_{-1}$ will be $\beta_o$, $\beta_n$ will be $\beta_n + 1$ and so on. We must find the same coefficient for each of the $\beta$'s in this new situation. As a consequence of these conclusions the $\omega$ vs $\beta_o$ curve will have a shape as shown in Fig. 8.6. The slope of the radius vector from the origin to any point of the curve is $\omega\sqrt{\mu\epsilon}/\beta_o = v_p/c$. If this slope is greater than 1 (i.e., 45° line) we have $v_p > c$ and if it is less than 1 the $v_p < c$. If we draw radius vectors at various points of the curve with $\beta$ spaced by $2\pi/d$, we note that the slopes are different. There are many values of $v_p$ or it may be said that there exists many spatial harmonics. Moreover, the higher $n$ components have smaller phase velocity as expected. So both positive and negative dispersions exist in a periodic slow wave structure. Further Eq. (8.23) follows from the curve because the group velocity which is the slope of the curve $\frac{(d\omega)}{(d\beta)}$ is same for all points spaced by $2\pi/d$ i.e. independent of $n$. The group velocity $v_g$ is zero at

$$\beta_o = 0, \frac{\pi}{d}, \frac{2\pi}{d}, \frac{3\pi}{d}, \frac{4\pi}{d} \quad \ldots \text{etc.}$$

Figure 8.6 also depicts various spatial harmonics ($\beta_n$'s). It may be noted that there exists alternately pass and stop bands corresponding to $v_g \neq 0$ and $v_g = 0$. Thus, a periodically loaded line behaves as a bandpass filter.

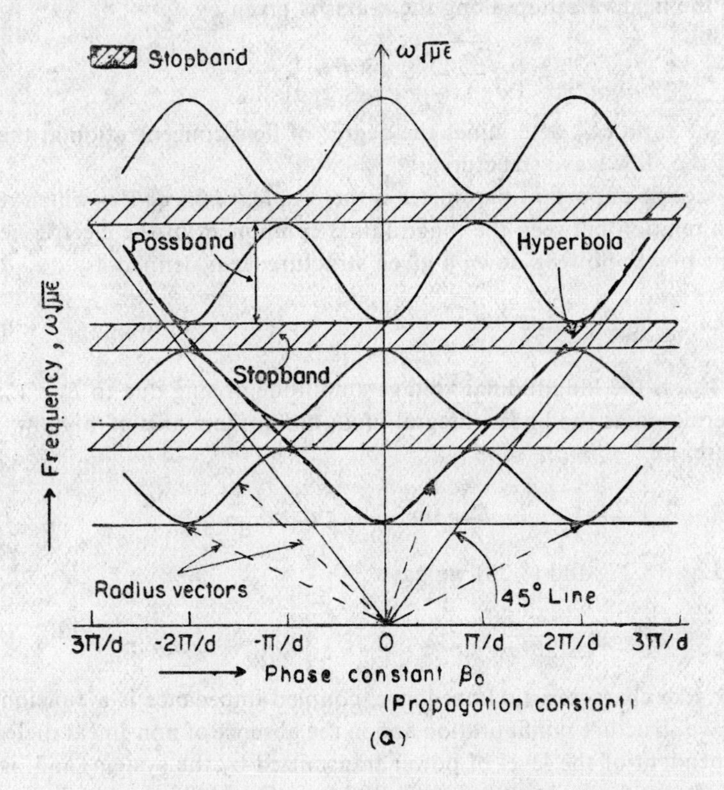


**Fig. 8.6** Frequency as a function of propagation constant for periodically loaded microwave line (Waveguide).

### (*iii*) Slow Wave Structure Parameters

The most important parameter of a slow wave structure is the ratio $c/v_p$ which is the same as the ratio of wavelength in free space to wavelength in a slow wave structure,

$$\frac{c}{v_p} = \frac{\lambda}{\lambda_{slow}} \tag{8.24}$$

where, $$\lambda_{slow} = \frac{v_p}{f} = 2\pi \frac{v_p}{\omega} = \frac{2\pi}{\beta} \tag{8.25}$$

For $\frac{c}{v_p} \gg 1$, $\beta \gg k$ and it follows from Eq. (8.8) that

$$\beta_n = \frac{2\pi}{\lambda}\frac{c}{v_p} \simeq \sqrt{\xi_1^2 + \eta_1^2}.$$

When there is no field variation in one direction, say $\eta_1 = 0$ then,

$$\xi_1 \simeq \sqrt{\frac{2\pi}{\lambda}\frac{c}{v_p}}.$$

Hence the field variation along the $x$-axis is given by

$$L = D_{\sinh}^{\cosh}\left\{\sqrt{\frac{2\pi}{\lambda}\frac{c}{v_p}} \times e^{j(\omega t - \beta z)}\right\}. \tag{8.26}$$

Thus, the ratio $c/v_p$ determines the degree of field concentration at the surface of the slow wave structure.

The second important parameter is the *coupled impedance* which establishes a relation between the longitudinal component of the electric field $E_z$ with the power flowing down a given structure. It is defined as

$$R_{coup} = \frac{V_{zm}^2}{2p} \tag{8.27}$$

where $V_{zm}$ is the longitudinal voltage amplitude arising due to $E_{zm}$. $V_{zm}$ can be determined as the linear integral of $E_z$ along the $z$-axis of a slow wave structure, i.e.

$$V_{zm} = \int_0^{\lambda_{slow}/4} E_{zm} \sin \frac{2\pi}{\lambda_{slow}} z \, dz = \frac{E_{zm}}{\beta}. \tag{8.28}$$

Using Eqs. (8.27) and (8.28) we have

$$R_{coup} = \frac{E_{zm}^2}{2\beta^2 p}. \tag{8.29}$$

Though like characteristic impedance, coupled impedance is a funcion of a slow wave structure configuration and in the absence of non-linear dielectrics is independent of the level of power transmitted by the system and is entirely different from characteristic impedance. Coupled impedance may also be compared with the resonant cavity conductance. The two resemble as far as their physical meaning is concerned—both are independent of the origin of the coordinate system.

To relate coupled impedance to group velocity, we recall the general relation $P = v_g \times W_1$ where $W_1$ is the energy stored per unit length of the slow wave structure. Then Eq. (8.29) reduces to

$$R_{coup} = \frac{E_{zm}^2}{2\beta^2 W_1 v_g} \tag{8.30}$$

Thus, the greater coupling between the slow wave structure (slow waves) and microwave field can be achieved by reducing $W_1$ or $v_g$.

The third equally important parameter is *band width* which is the pass band of the slow wave structure.

These parameters, particularly coupled impedance, play a dominant role in the design and understanding of the basic principle of microwave tubes, e.g. in a travelling wave tube (TWT) in which strong interaction between the electron beam and slow waves is desired.

## 8.4 PERIODICALLY LOADED WAVEGUIDE

Consider a waveguide with periodically spaced irises as shown in Fig. 8.7. Recalling the impedance matrix representation of Chapter 6, it follows that input and output voltages at the $n^{\text{th}}$ and $(n+1)^{\text{th}}$ discontinuities are respectively given by

$$V_n = Z_{11} I_n - Z_{12} I_{n+1}$$

$$V_{n+1} = Z_{21} I_n - Z_{22} I_{n+1} \tag{8.31}$$

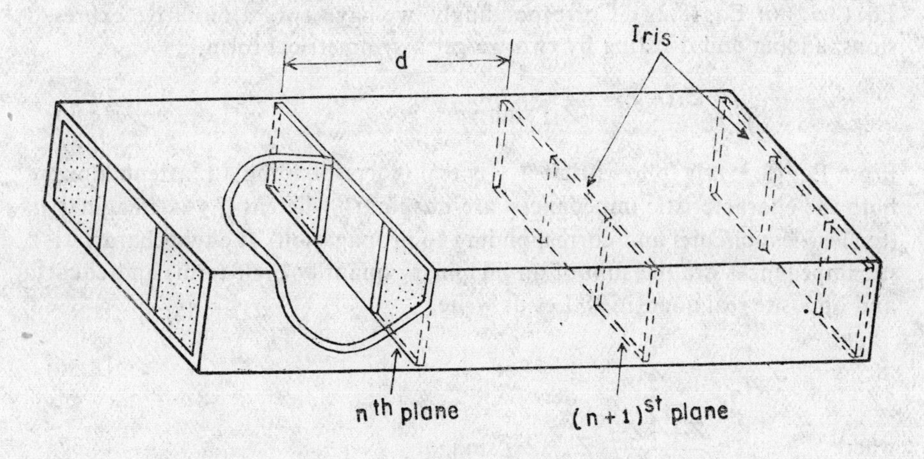


Fig. 8.7 Waveguide loaded with periodically spaced iris.

For a loss-less junction $Z_{12}=Z_{21}$ all $z$'s are purely imaginary. The negative sign to $i_{n+1}$ is assigned, assuming that positive direction is looking into each of the outputs of a 'unit cell', i.e. increasing $z$.

Assuming the solutions as

$$V_n = V_o\, e^{md},\ I_n = I_o\, e^{\gamma nd} \tag{8.32}$$

where $d$ is periodicity (length of one unit cell) of the structure and $\gamma$ is propagation constant. Substitution of Eq. (8.32) in Eq. (8.31) immediately yields,

$$V_o = (Z_{11} - Z_{12}\, e^{\gamma d})\, I_o = (Z_{12}\, e^{-\gamma d} - Z_{22})\, I_o \tag{8.33}$$

or, $$\cosh(\gamma d) = \frac{Z_{11} - Z_{12}}{2Z_{12}} \tag{8.34}$$

Thus $\gamma$ is determined in terms of the constants of the unit cell. Since cosh $\gamma d$ is an even function both negative and positive values of $\gamma$ are permissible. For a loss-less structure $z$'s are all imaginary and so is $\gamma$ (say $\gamma=j\beta$). Then cosh $\gamma d$ goes to cosh $j\beta d=\cos\beta d$ which is real and is $<1$. Therefore, the permissible frequency bands are those for which the RHS of Eq. (8.34) is less than unity (it may be noted that $z$'s are functions of frequency). As the frequency changes, there, will be in general some bands of frequency, in which cosh $i\beta d>1$, having attenuation without propagation, interspersed by other regions in which cosh $j\beta d<1$, where propagation without attenuation occurs. Or the system behaves as a filter. Moreover, $\beta$ is determined only upto an additive constant of $2\pi n/d$; in other words, the frequency is a periodic function of $\beta$, with period $2\pi/d$, just as in Section 8.3. We thus understand., why the curves of Fig. 8.7 are applicable to the system in question.

If a loaded guide is terminated in a matched load the ratio of voltage to current at each unit cell will be the same and will be the characteristic impedance of the system. It can be obtained by dividing $V_n$ by $I_n$, using either Eq. (8.32) or Eq. (8.33). Correspondingly, we have two alternative expressions, adding and dividing by two we get a symmetrical form

$$\frac{V_n}{I_n}=\frac{Z_{11}-Z_{22}}{2}\mp Z_{12}\sinh\gamma d. \tag{8.35}$$

From Eq. (8.35), it follows that if $\gamma$ is real (corresponding to attenuation), both the characteristic impedances are imaginary whereas if $\gamma$ is imaginary (for loss-less circuits and corresponding to propagation), the two characteristic impedances are the sum of an imaginary quantity (first term) and equal and opposite real quantities. Let us write,

$$\frac{V_n}{I_n}=\zeta\mp Z_0 \tag{8.36}$$

where $$\zeta=\frac{Z_{11}-Z_{22}}{2},\ Z_0=Z_{12}\sinh\gamma d \tag{8.37}$$

where the upper sign corresponds to the solution $e^{+\lambda d}$, i.e. the propagation towards the left and the lower $e^{-\gamma d}$ i.e. propagation towards right. Consequently, our solutions are of the form

$$V_n=I\,(\zeta+Z_0)\,e^{-md},\ I_n=I'\,e^{-\gamma nd}$$

and

$$V_n=I\,(\zeta-Z_0)\,e^{\gamma nd},\ I_n=I'\,e^{\gamma nd}$$

Where $I$ and $I'$ are constants having dimensions of current.

Suppose now that the guide is terminated in an impedance $Z_L$ then the voltage at the input of the $n$th unit cell will be the sum of incident and reflected voltage, i.e.,

$$V_n=I\,e^{j\omega t-\gamma nd}\,\{(\zeta+Z_0)-\Gamma_n\,(\zeta-Z_0)\},$$

$$I_n=I'\,e^{j\omega t-\gamma nd}\,\{1-\Gamma_n\}, \tag{8.38}$$

where $\Gamma_n$ is the reflection coefficient still to be determined. Letting $Z_L = V_L / I_L$, it follows from Eq. (8.38) that,

$$Z_L - \zeta = Z_0 \frac{1+\Gamma_L}{1-\Gamma_L}, \quad \Gamma_L = \frac{Z_L - \zeta - Z_0}{Z_L - \zeta + Z_0}. \tag{8.39}$$

The reflection coefficient is given by

$$\Gamma_L = \Gamma_0 \, e^{2\gamma d L}$$

where termination is at the $L$th unit cell. The reflection coefficient at the $n$th section will therefore be,

$$\Gamma_n = \Gamma_0 \, e^{-\gamma d (L-n)}.$$

Hence $Z_n$, the input impedance at the $n$th unit cell is given by

$$Z_n - \zeta = Z_0 \frac{Z_0 \sinh \gamma d\,(L-n) + (Z_L - \zeta) \cosh \gamma d\,(L-n)}{Z_0 \cosh \gamma d\,(L-n) + (Z_L - \zeta) \sinh \gamma d\,(L-n)} \tag{8.40}$$

The physically interesting cases arise when $Z_L - \zeta = 0$ or $\infty$. If $\zeta = 0$, the first case corresponds to a short circuit and the second to an open circuit. In the two cases we have

$$Z_n - \zeta = Z_0 \tanh \gamma d\,(L-n), \text{ or } Z_0 \coth \gamma d\,(L-n). \tag{8.41}$$

To evaluate impedance constants $Z_{11}$, $Z_{12}$ etc., we note from Eq. (8.31) that

$$Z_n = Z_{11} - \frac{Z_{12}^2}{Z_{n+1} + Z_{22}}. \tag{8.42}$$

Comparison of Eqs. (8.40) and (8.42) with $n = L-1$ leads to the following values of $Z$

$$\begin{aligned} Z_{11} &= Z_0 \coth \gamma d + \zeta \\ Z_{22} &= Z_0 \coth \gamma d - \zeta \\ Z_{12} &= Z_0 \operatorname{csch} \gamma d \end{aligned} \tag{8.43}$$

The results above are for a single unit cell network. However, if identical networks are connected together than it is easy to see that for $n - L = s$,

$$\begin{aligned} (Z_{11})_{n-L} &= Z_0 \coth \gamma d\,(n-L) + \zeta \\ (Z_{22})_{n-L} &= Z_0 \coth \gamma d\,(n-L) - \zeta \\ (Z_{12})_{n-L} &= Z_0 \operatorname{csch} \gamma d\,(n-L) \end{aligned} \tag{8.44}$$

Now we switch on to a particular type of waveguide loading say a waveguide loaded with periodically spaced irises. Since a single iris acts like a shunt susceptance $jb_0 Y_0$, $Y_0$ becomes the characteristic admittance across the guide. The problem is simplified if the admittance matrix is considered instead of the impedance matrix considered above. Following the discussion as above, it can be shown that for the propagation of waves,

$$\begin{aligned} I_n' &= -V_n \, j \, Y_o \cot \beta d + V_{n+1} \, j \, Y_o \csc \beta d, \\ I_{n+1} &= -V_n \, j \, Y_o \csc \beta d + V_{n+1} \, j \, Y_o \cot \beta d \end{aligned}$$

Noting that $I_n$ equals the current in the iris susceptance plus $I_n'$, i.e. $I_n = I_n' + V_n\, j\, b_o Y_o$ then

$$I_n = -V_n\, j\, Y_o\, (\cot \beta d - b_o) + V_{n+1}\, j\, Y_o \csc \beta d,$$

$$I_{n+1} = -V_n\, j\, Y_o \csc \beta d + V_{n+1}\, j\, Y_o \cot \beta d \tag{8.45}$$

Hence
$$Y_{11} = -j\, Y_o\, (\cot \beta d - b_o),$$
$$Y_{22} = -j\, Y_o \cot \beta d,$$
$$Y_{12} = -j\, Y_o \csc \beta d \tag{8.46}$$

Using Eq. (8.34) in the $Y$'s we immediately find from Eq. (8.46) that

$$\cos \beta_o d = \cos \beta d - \frac{b_o}{2} \sin \beta d \tag{8.47}$$

$\beta_o$ is the phase constant for the loaded guide. Equation (8.47) may easily be transformed to

$$\cos \beta_o d = A \cos (\beta d + \theta), \tag{8.48}$$

where
$$A = \sqrt{1 + (b_o/2)^2},\ \theta = \tan^{-1} (b_o/2)$$

Let us now observe the behaviour of the function $A \cos (\beta d + \theta)$ as a function of frequency. When $\beta d + \theta$ approaches a value $n\pi$; $n$ being an integer, $\cos (\beta d + \theta)$ will approach $\pm 1$. In that case it follows from Eq. (8.48) that $A > 1$ unless $b_o = 0$, hence there will always be a range of frequencies where $A \cos (\beta d + \theta) > 1$. The First of Eqs. (8.48) cannot hold under these conditions unless $\beta_o$ is imaginary which in turn implies that there is only attenuation and no propagation. We are thus led to pass bands and they are determined by Eq. (8.48). In the limit of small $b_o$, the attenuation regions will be very near $\beta d = n\pi$ as shown in Fig. 8.6. For $b_o$ to be large, it follows from Eq. 8.47 that we must have $\sin \beta d$ to be very small so that $\beta d \simeq n\pi$ for the propagation or pass bands.

The pass band characteristics of the iris loaded waveguide can physically be understood as follows. For small $b_o$, we may consider each iris as a scatterer, so that a wave approaching an iris results in a reflected wave travelling back as well as a forward transmitted wave. When the iris spacing happens to be a half guide wavelength ($\beta d = n\pi$), the reflected wave from each successive iris will be a period out of phase from the reflected wave from the preceding iris. Therefore, all the reflected waves add in phase to produce a large effect. If the structure is lossless, the amplitude of the reflected wave (all reflected waves taken together), equals the amplitude of the incident wave. Consequently, a pure standing wave is set up. Thus, the phenomena resembles Bragg's reflection in the study of X-ray diffraction. Group velocity and hence the net power flow is zero at the frequencies $\beta d = n\pi$. However, if $b_o$ is large, i.e. iris opening is small, the unit cell may be treated as a resonant cavity with very small coupling, then condition $\beta d = n\pi$ implies the resonant oscillations of the cavity. As there is very small coupling, no power flow can be expected down such loaded line it has a stop band character.

In the design of waveguide filters, expression for the input admittance is desired which can be obtained in an analogy to Eq. (8.40).

$$Y_n-\zeta=Y'_o\frac{j\,Y'_o\sin\beta_o d\,(L-n)+(Y_L-\zeta)\cos\beta_o d\,(L-n)}{Y'_o\cos\beta_o d\,(L-n)+j\,(Y_L-\zeta)\sin\beta d\,(L-n)}$$

where,
$$\zeta=\frac{Y_{11}-Y_{12}}{2}=\frac{j\,Y_o b_o}{2},\quad Y'_o=jY_{12}\sin\beta_o d=Y_o\frac{\sin\beta_o d}{\sin\beta d} \tag{8.49}$$

The cases discussed in this section are very general and are applicable to any periodic structure; however, the pass and stop bands may be interchanged, e.g. in case when shunt susceptance is $-j\,b_oY_o$, i.e. in case of inductive loading. Moreover, pass and stop bands exist for each Hartree component of the Bloch wave and it is not difficult to see that only those components for which $\beta_n d=n\pi$ have appreciable amplitude.

## 8.5 HELIX

Consider a helical guide shown in Fig. 8.8(a). A TEM wave fed at the input continues to propagate along the conductor with a velocity very near to the velocity of light in free space. The time interval $\tau$ required for the wave to travel one turn of the helix is the ratio of turn circumference to the velocity of light $c$.

$$\tau=\frac{1}{c}\sqrt{(2\pi a)^2+p^2} \tag{8.50}$$

where $a$ is the radius of the helix and $p$ is the pitch defined as the distance travelled by the waves along the $z$-axis in time $\tau$ and $c$ is velocity of light. Thus, phase velocity is given by

$$v_p=c/\sqrt{1+(2\pi a/p)^2} \tag{8.51}$$

It follows from Eq. (8.51) that $v_p$ is always less than $c$ and hence slow waves are expected. Moreover, if $\psi$ be the angle of helix (Fig. 8.8b) then phase velocity is the component of $c$ along the $z$-axis, i.e.

$$v_p=c\sin\psi \tag{8.52}$$

Equations (8.51) and (8.52) yields

$$\sin\psi=1/\sqrt{1+(2\pi a/p)^2} \tag{8.53}$$

Equation (8.53) relates various parameters of a helix.

The approximate analysis* of a helix, disregarding spatial harmonics for the present, may be carried by assuming the helix to be a cylindrical waveguide whose walls are conducting only along the helical path, i.e. the conductivity of the wall is anisotropic with definite periodicity. If the wall

*Rigorous treatment may be found in D.A. Watkins, *Topics in Electromagnetic Theory*, John Wiley and Sons, Inc. (1958).

thickness in the structure of Fig. 8.8(b) is assumed to be infinitesimally small ($p \ll \lambda$) then such a helix is called a *sheath helix*.

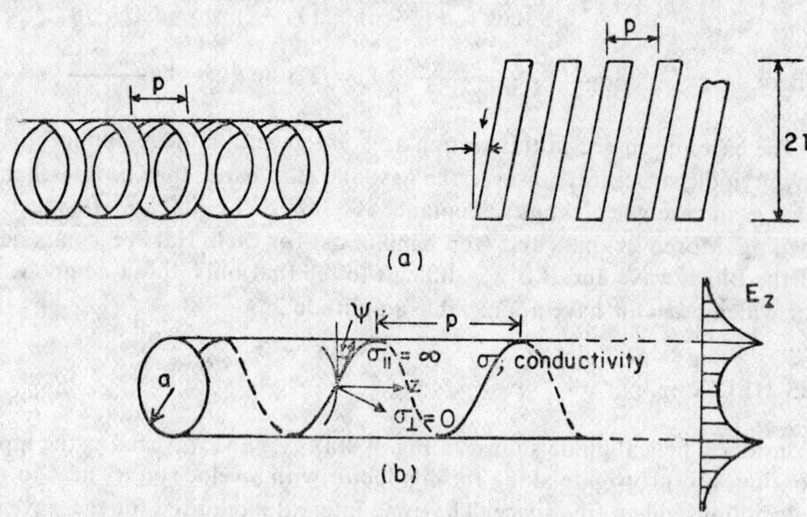


**Fig. 8.8** (*a*) Helix structure (*b*) Helix angle and the illustration of boundary conditions.

The fundamental boundary conditions to be observed (assuming the helix to be made of a perfect conductor) are that at $r=a$ current must be infinite along the direction of the helix tape and that it must vanish perpendicular to it. The infinite current implies infinite conductivity (assumed) and so the tangential electric field must vanish. The vanishing condition of current perpendicular to the tape implies continuity of the perpendicular electric field and the tangential magnetic field. Thus referring to Fig. 8.8(b) we have,

$$E_{\phi 1} \cos \psi + E_{z1} \sin \psi = E_{\phi 2} \cos \psi + E_{z2} \sin \psi = 0 \quad (8.54a)$$

$$E_{z1} \cos \psi - E_{\phi 1} \sin \psi = E_{z2} \cos \psi - E_{\phi 2} \sin \psi \quad (8.54b)$$

$$H_{z1} \sin \psi + H_{\phi 1} \cos \psi = H_{z2} \sin \psi + H_{\phi 2} \cos \psi \quad (8.54c)$$

where subscripts 1 and 2 refer to the field components in the two regions $\gamma \leqslant a$ and $\gamma \geqslant a$. The field solution for the helix consists of both TM and TE modes since these are coupled together by the boundary conditions (Eqs. 8.54) at $\gamma = a$. The expansions for these modes in the two regions outlined ($\gamma \lesseqgtr a$), may be obtained in terms of the axial components $E_z$ and $H_z$. The solutions of equations

$$\nabla_T^2 E_z(\gamma, \phi) + (k_o^2 - \beta^2) E_z(\gamma, \phi) = 0$$

and $$\nabla_T^2 E_z(\gamma, \phi) + (k_o^2 - \beta^2) H_z(\gamma, \phi) = 0$$

we are looking for, are slow waves, i.e. $\beta^2 > k_o^2$. Therefore, these can be written in terms of modified Bessel functions

$$I_n\left(r\sqrt{\beta^2 - k_o^2}\right) \text{ and } K_n\left(r\sqrt{\beta^2 - k_o^2}\right).$$

For small values of $r\sqrt{\beta^2-k_o^2}$ the functions go as

$$I_n\left(r\sqrt{\beta^2-k_o^2}\right) \sim \frac{1}{\sqrt{2\pi r\,(\beta^2-k_o^2)^{1/2}}}\quad e^{r\sqrt{\beta^2-k_o^2}}$$

$$K_n\left(r\sqrt{\beta^2-k_o^2}\right) \sim \sqrt{\frac{\pi}{r\,(\beta^2-k_o^2)^{1/2}}}\; e^{-r\sqrt{\beta^2-k_o^2}}.$$

Since we require a field that decays for large $r$, physically acceptable solutions in regions $r>a$ and $r<a$ respectively correspond to $K_n$ and $I_n$. Therefore, the field expressions are given by

$$E_z=\begin{cases}\sum\limits_{n=-\infty}^{+\infty} a_n\, e^{-jn\phi}\, I_n(hr) & r\leqslant a\\[2ex] \sum\limits_{n=-\infty}^{+\infty} b_n\, e^{-jn\phi}\, K_n(hr) & r\geqslant a\end{cases}$$

$$H_z=\begin{cases}\sum\limits_{n=-\infty}^{+\infty} c_n\, e^{-jn\phi}\, I_n(hr) & r\leqslant a\\[2ex] \sum\limits_{n=-\infty}^{+\infty} d_n\, e^{-jn\phi}\, K_n(hr) & r\geqslant a\end{cases}$$

where $h=(\beta^2+k_o^2)^{1/2}$ and $a_n$, $b_n$, $c_n$ and $d_n$ are unknown, amplitude constants. We are basically interested in the solutions of circular symmetry, i.e. $n=0$ in the expressions for $E_z$ and $H_z$. Using the waveguide theory of Chapter 4, it is easy to obtain other field components. These field components in the two regions of interest are given by:

(i) For $r < a$

$$E_z=a_o\, I_o(hr)\, e^{-j\beta z},\quad E_r=\frac{j\beta}{h}\, a_o\, I_1(hr)\, e^{-j\beta z}$$

$$H_z=c_o\, I_o(hr)\, e^{-j\beta z},\quad H_r=\frac{j\beta}{h}\, c_o\, I_1(hr)\, e^{-j\beta z},$$

$$E_\phi=-\frac{j\omega\mu_o}{h}\, c_o\, I_1(hr)\, e^{-j\beta z}$$

$$H_\phi=\frac{j\omega\epsilon_o}{h}\, a_o\, I_1(hr)\, e^{-j\beta z} \tag{8.55}$$

(ii) For $r > a$

$$E_z=b_o K_o(hr)\, e^{-j\beta z},\quad E_r=-\frac{j\beta}{h}\, b_o K_1(hr)\, e^{-j\beta z},$$

$$H_z=d_o K_o(hr)\, e^{-j\beta z},\quad H_r=-\frac{j\beta}{h}\, d_o K_1(hr)\, e^{-j\beta z},$$

$$E_\phi=\frac{j\omega\mu_o}{h}\, d_o K_1(hr)\, e^{-j\beta z}$$

$$H_\phi=-\frac{j\omega\epsilon_o}{h}\, b_o K_1(hr)\, e^{-j\beta z} \tag{8.56}$$

Unknown constants $a_o b_o c_o$ and $d_o$ are evaluated as usual by substituting the field components in the boundary conditions [Eqns. (8.54)] and obtaining a non-trivial solution for the equations so obtained, by putting the determinant equal to zero. This yields the following eigen value equation for $\beta$.

$$\frac{K_1\,(ha)\;I_1\,(ha)}{K_0\,(ha)\;I_0\,(ha)}=\frac{(ha)^2\tan^2\psi}{(K_0 a)^2}. \tag{8.57}$$

Ratio $K_1 I_1/K_0 I_0$ rapidly approaches unity for $ha>10$. In this region Eq. (8.57) yields $h=k_o\cot\psi$ which in turn gives a phase velocity

or $$\beta=(k_o^2+h^2)^{1/2}=k_o\csc\psi, \tag{8.58}$$

$$v_p=\frac{\omega}{\beta}=\frac{k_o}{\beta}\,c=c\sin\psi \tag{8.59}$$

which is the same result as was inferred before on purely qualitative grounds.

The field pattern for $E_z$ in the transverse plane is shown in Fig. 8.8(b). As expected it is confined near the structure and decreases rapidly, as we move away from the structure.

Figure 8.9 shows the dispersion characteristics of a sheath helix with angle of helix winding as a parameter, range of obtainable $c/v_p$ ratios under the

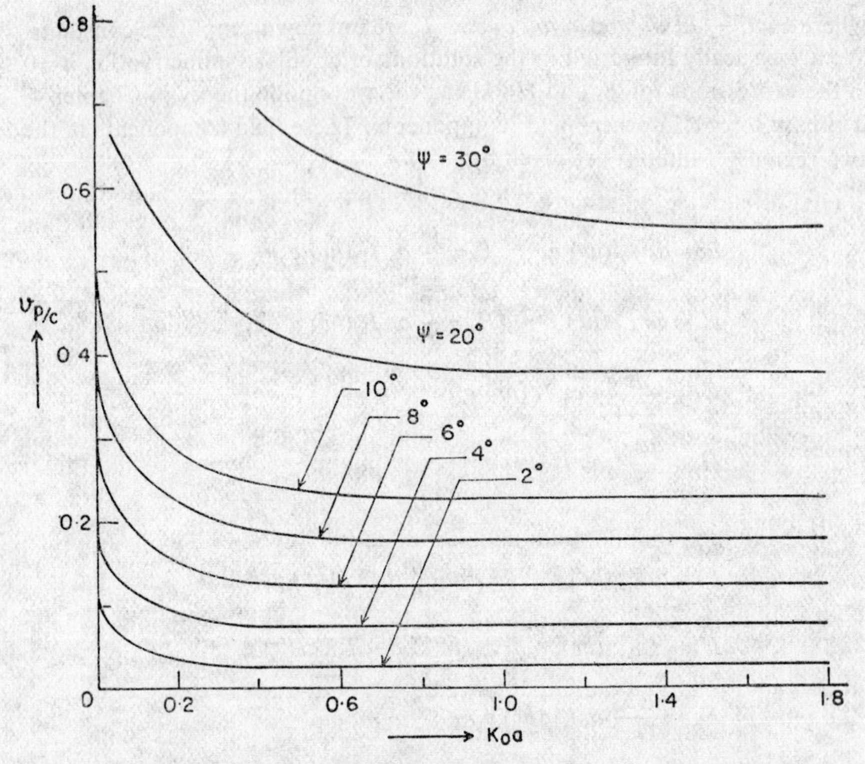


Fig. 8.9 Dispersion characteristics of a helical slow wave structure at various angles of helix winding.

stated conditions is from 2-3 to about 30. As seen from the graphs, the dispersion characteristics of a helix have an almost horizontal. Section in a very wide range of values of $K_o a$ ($=2\pi a/\lambda$) where $v_p$ is practically independent of operating frequency, i.e. a broad band operation is ensured. At higher frequencies, however, $v_p$ increases and approaches $c$ as $\omega \to \infty$. Thus, the fundamental slow wave of helix ($n=0$) features normal and positive dispersion. The phase velocity in the horizontal region is given by (for any $\psi$)

$$v_p \simeq c \sin \psi \, v \sqrt{1-\frac{\sin^2 \psi}{2\,(K_o a)^2}}. \tag{8.60}$$

The coupled impedance for the system in question can be obtained using the general relation (8.29). Figure 8.10 depicts the variation in coupled impedance with frequency for $\psi < 10°$. It is of the order of several ohms.

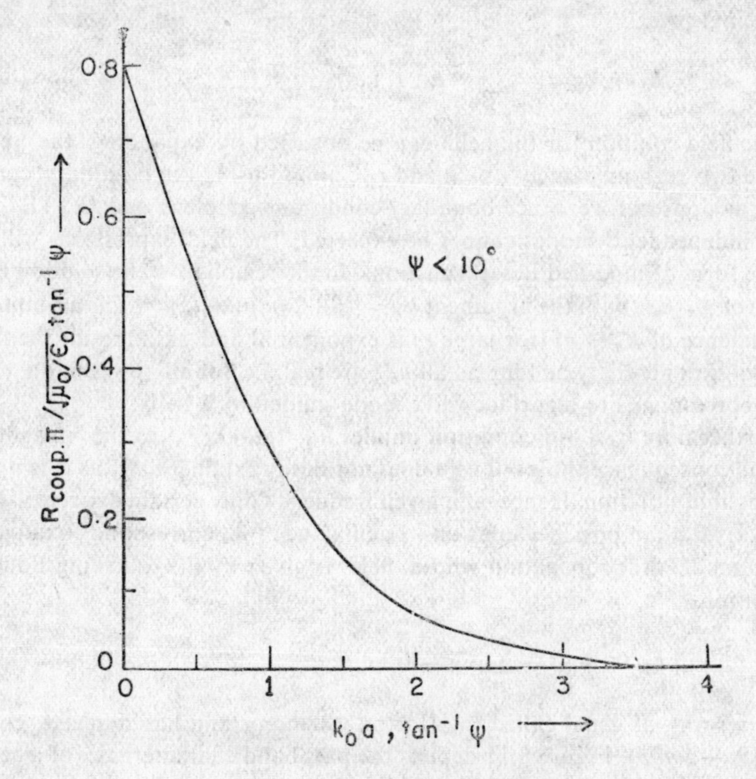


**Fig. 8.10** Variation in coupled impedance $\left(R_{\text{coup.}} \frac{\pi}{\sqrt{\frac{\mu_0}{\epsilon_0}}} \tan \psi\right)$ with frequency ($k_0 a$ tan $-1$ $\psi$).

We now return to the periodic character of the helix. A helix is not only periodic with respect to a translation by a distance $p$ along the axis of the helix but is also periodic with respect to rotation through an arbitrary angle $\theta$, followed by a translation $p\theta/2\pi$ along the axial direction. This implies that if $E_1(r, \phi, z)\, e^{-j\beta z}$ is a solution for the electric field then $E_1(r, \phi+\theta, z+p\theta/2\pi)$ $e^{-j\beta(p\theta/2\pi)}$ is also a solution because points* $(r, \phi, z)$ and $(r, \phi+\theta, z+p\theta/2\pi)$ are indistinguishable from one another. Consequently, $E_1(r, \phi, z)$ must be periodic in $\phi$ and besides a propagation factor must also be periodic in $z$ with period $p$. Therefore, $E_1$ can be expanded in double Fourier series,

$$E_1(r, \phi, z)=\sum_{m=\infty}^{\infty}\sum_{n=-\infty}^{\infty} E_{1,mn}(r)\, e^{-jm\phi-\frac{j2\pi nz}{p}}\, e^{-j\beta z} \tag{8.61}$$

However, with the noting that $E_1$ does not change when $\phi\rightarrow\phi+\theta$, $r\rightarrow r$, $z\rightarrow z+p\theta/2\pi$, i.e.

$$e^{-jm(\phi+\theta)-j2n\pi(z+p\theta/2\pi)/p}=e^{-jm(\phi+\theta)-jn\theta-jn\,2\pi z/p}=e^{-jm\phi-j2\pi nz/p},$$

which requires $m=-n$. Hence the double summation in Eq. (8.61) reduces to a single one

$$E_1(r, \phi, z)=\sum_{n=-\infty}^{\infty} E_{1n}(r)\, e^{-jn(2\pi z/p-\phi)-j\beta z} \tag{8.62}$$

The field solution for the helix can be obtained by expanding the series in the two regions namely $r > a$ and $r < a$ and using the boundary conditions as done before. Since boundary conditions couple $E$ and $H$, TE and TM independent modes cannot be expected. The field expressions will be in the form of modified Bessel functions. In the region $r>a$ the solution is in terms of $K_n(h_n r)$ with the argument $h_n r=[(\beta+2\pi n/p)^2-k_o^2]^{1/2}_{\Upsilon}$. The asymptotic dependence of $K_n(h_n r)$ (for large $r$) is exponential and as a result the field decays exponentially as long as all $h_n$'s are real, i.e. for all $(\beta+2\pi n/p)^2> k_o^2$. This corresponds to a surface wave mode guided by a helix.

Furthermore $k_o>\pm\beta$ condition implies $h_o=(\beta^2-k_o^2)^{1/2}$ to be imaginary and as consequence the $n=0$ field does not decay exponentially as it is not a permissible situation. Hence, at a given frequency only certain discrete values of $\beta$, say $\beta_m$, are possible solutions. Each value of $\beta_m$ corresponds to a particular mode of propagation whose field is given by the following Fourier expansion,

$$E_m(r, \phi, z)=\sum_{n=-\infty}^{+\infty} E_{mn}(r)\, e^{-j(2\pi z/p-\phi)-j\beta mz} \tag{8.63}$$

Each term is, as usual called the Hartree harmonic and has a phase constant $(\beta_m+2\pi n/p)$. Figure 8.11 depicts the pass band characteristics of a helix. The region above $k_o=\pm\beta$ in $K_o p-\beta p$ curve is a forbidden region. Moreover, it is required that $(\beta+2\pi n/p)>k_o$ hence all possible allowed values are restricted to the unshaded triangular region whose boundaries are determined by the lines,

$$k_o=\pm\left(\beta\pm\frac{2\pi n}{p}\right),\ n=0, 1, 2, \ldots \tag{8.64}$$

*Cylindrical coordinates are used here.

Note the uncommon feature that the forbidden region is characterized by complex propagation constant.

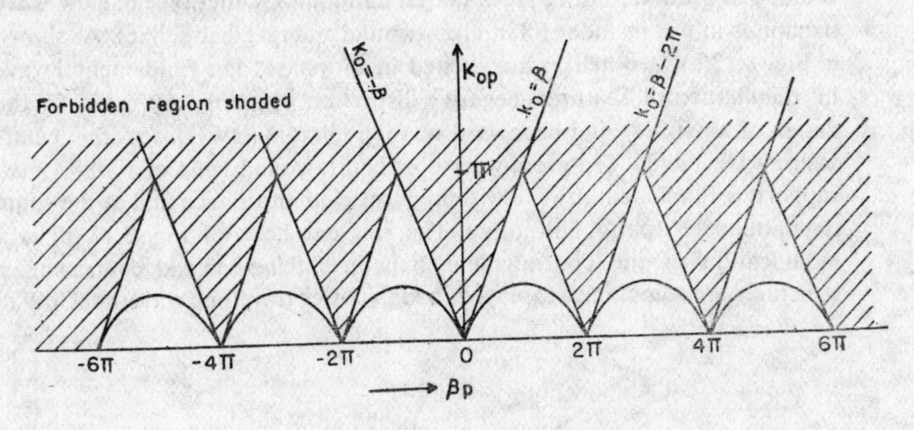


**Fig. 8.11** $K_0p - \beta p$ diagram for a helix: allowed and forbidden regions.

## 8.6 SLOW WAVE STRUCTURES IN USE

The helix was first put to use in microwave travelling wave tubes (TWT). Cur rently helical slow wave structures are commonly used in microwave TWT and other devices with non-resonant oscillatory systems. The main advantage of a helical slow wave systems is its wideband characteristics. The phase velocity of a slow wave is almost precisely equal to the group velocity and remains practically constant within a comparatively wide range of frequencies—of the order of an octave and even wider. However, the actual designs of slow wave helical structures differ from the model discussed in Section 8.6 because undesired spatial harmonics need to be reduced in amplitude. A typical structure is shown in Fig. 8.12. This mechanically rigid structure is obtained by mounting the helix on dielectric supports, rods or

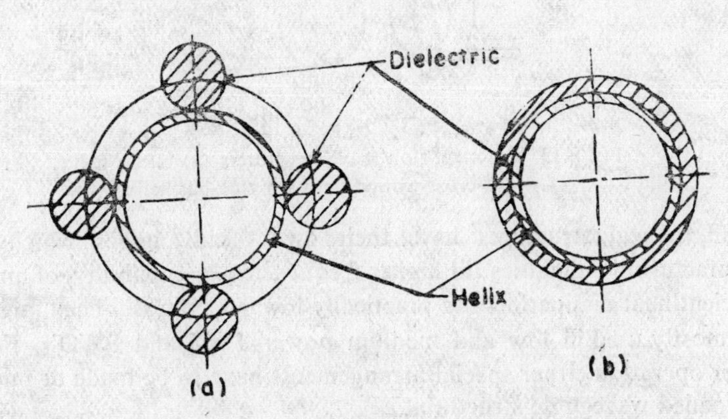


(a) with four dielectric rod support (b) Tabular support

**Fig. 8.12** Cross-section of a helix.

on a solid tube. The partial dielectric filling at the surface of the helix conductors increases slightly the $c/v_p$ ratio—and decreases $E_z$ and hence the coupled impedance, but reduces spatial harmonics. Other type of slow wave structures in use include bifilar, cross-wound and ring-bar helixes. As shown in Fig. 8.13(a), two helixes are excited in antiphase. The fundamental wave in this structure features negative dispersion which is very useful in the design of amplifiers and backward wave oscillators (BWO), for the centimetre wave band. The cross-wound or contrawound helix of Fig. 8.13 (b) supports a fundamental wave with high coupled impedance and at the same time attenuates spatial harmonics. The ring-bar helix of Fig. 8.13 (c) is a modification of the contrawound structure. Ring-bar and contrawound structures are commonly employed in high power travelling wavetubes TWT.

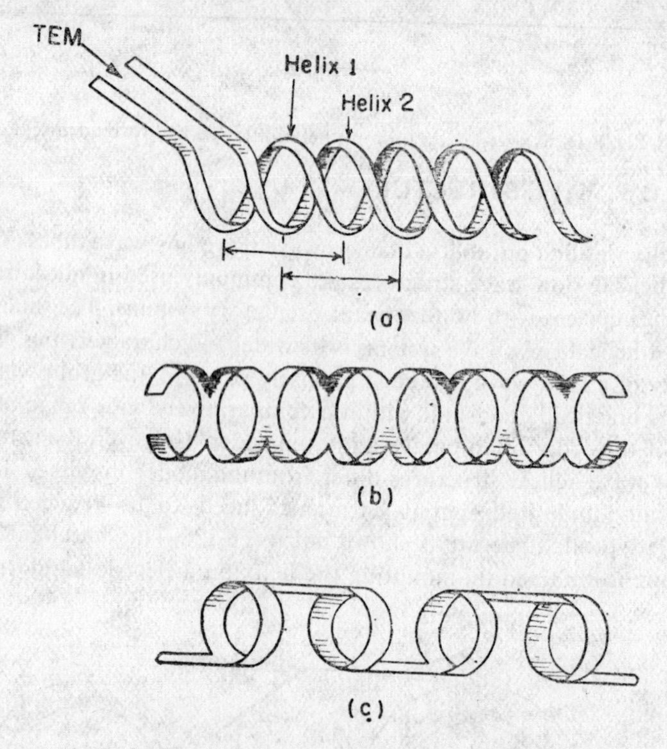


**Fig. 8.13** Practical slow wave structurers (*a*) bifilar helix. (*b*) cross-wound helix (*c*) ring-bar helix.

Still, helical structures have their own disadvantages, which include manufacturing difficulties (at higher frequencies), incapability of providing sufficient heat dissipation and practically low $c/v_p$ ratios. They are therefore mostly used in low and medium power TWT and BWO's. For high power operation either special arrangements have to be made or one has to use loaded waveguide structures.

Periodically loaded slow wave structures form a separate class of devices of special interest. A commonly used interdigital slow wave structure is shown in Fig. 8.14(a). The fundamental wave in this structure is a backward wave, i.e. the structure features negative dispersion as in the case of contrawound and ring bar helix structures. These structures are commonly employed in backward wave oscillators (BWO's). Figure 8.14(b) shows a circular waveguide periodically loaded with metal diaphragms with round central coupling holes. In general, loaded guide structures are mechanically rigid, have high heat dissipation and high coupled impedance at low $c/v_p$ ratios. These advantages make them specially suitable in the design of modern linear electron accelerators where $v_p$ is needed to be slowed down to $c$. However, the band width of these structures is low and can be widened by increasing coupling hole dimensions which of course lowers coupled impedance. Better results in a wider frequency band can be obtained by providing coupling apertures at the periphery of the diaphragms, thus utilizing magnetic coupling along with capacitive coupling due to the central coupling holes, as shown in Fig. 8.14(c). Each of the cylindrical resonant cavities so formed

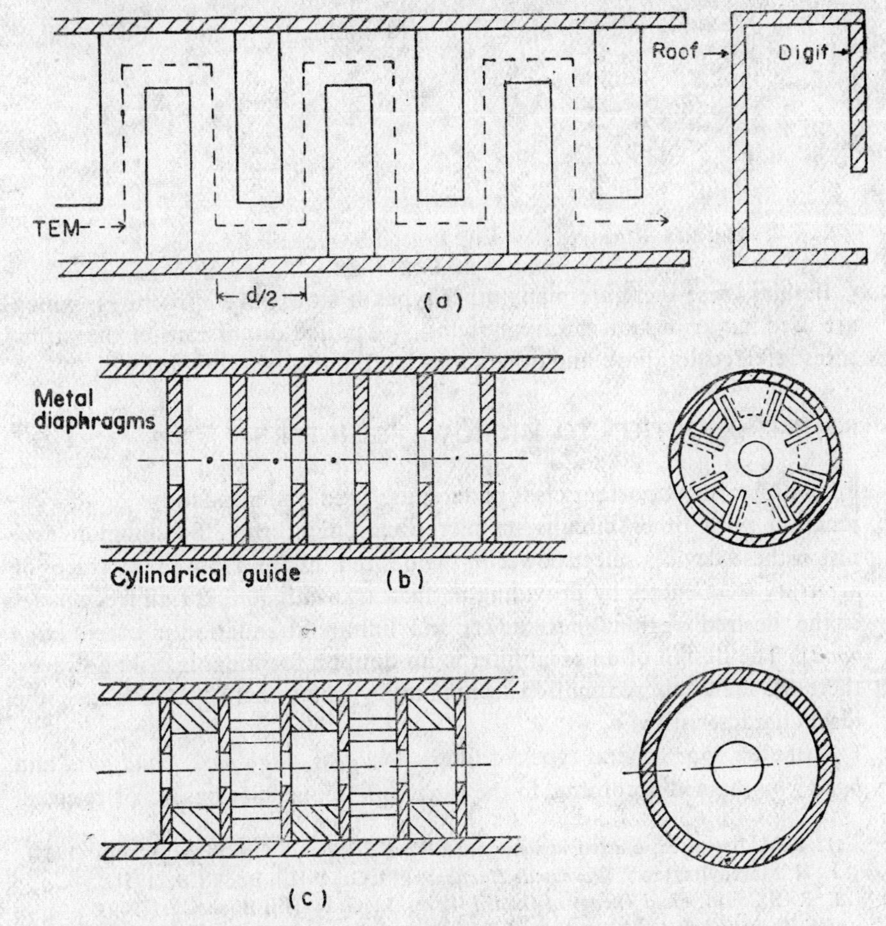


Fig. 8.14 Periodic slow wave structures (*a*) Interdigital (*b*) Periodically loaded circular waveguide (*c*) Overleaf slow wave circuit.

have four radial ridges displaced in adjacent cavities by an angle 45°. The coupling is provided by narrow radial slots and the central coupling hole. The fundamental wave features positive dispersion at a band width of about 20 percent of the central frequency with sufficiently high coupling impedance. In high power microwave tubes suitable modifications of the structure discussed above are used. At millimetre wavelength, the inductively loaded waveguides are more suitable.

Slow wave structures constituting a complete loop as shown in Fig. 8.15 are used in magnetron type electron tubes. These structures can also be analyzed by considering them as ring resonant cavities.

**Fig. 8.15** Ring type slow wave structures used in magnetrons.

Besides these there are many other types of slow wave structures which are used in various microwave systems. A detailed discussions of the structures referred to above and others may be found in the literature.*

## 8.7 INTRODUCTION TO MICROWAVE FILTERS

Filters form an important class of circuits, used independently or in conjunction with other circuits in microwave engineering. By common definition these devices select power in the desired range from a mixture of arbitrary frequencies by providing perfect transmission for all frequencies in the desired region (*pass band*) and infinite attenuation in others (*stop band*). The design of an ideal filter is no doubt a formidable task; however, there are many approximation techniques** which help to approach the ideal characteristics.

There are four general types of filters: *low pass*, *high pass*, *band pass* and *band stop*, named according to the range of frequencies passed or rejected.

*1. R.M. Bevensee, *Electromagnetic Slow Wave Systems*. John Wiley & Sons (1964).
2. N. Marcuvitz, (ed.), *Waveguide Handbook*, McGraw-Hill Book Co. (1951).
3. R.E. Collin, *Field Theory of Guided Waves*, McGraw-Hill Book Co. (1960).
**W.R. Daniels, '*Approximation Methods for Electronic Filter Design*' McGraw Hill Book Co. (1974).

Response curves of these filters are shown in Fig. 8.16. For a given response, the design of a filter is basically a problem of network synthesis. At low frequencies ideal inductors and capacitors are the base bricks of a filter network. Since the frequency response of these elements is precisely known, there exists a very general and complete filter synthesis procedure for low frequency filters. However, at microwave frequencies lumped elements lose their significance and if at all designed, have a very complicated

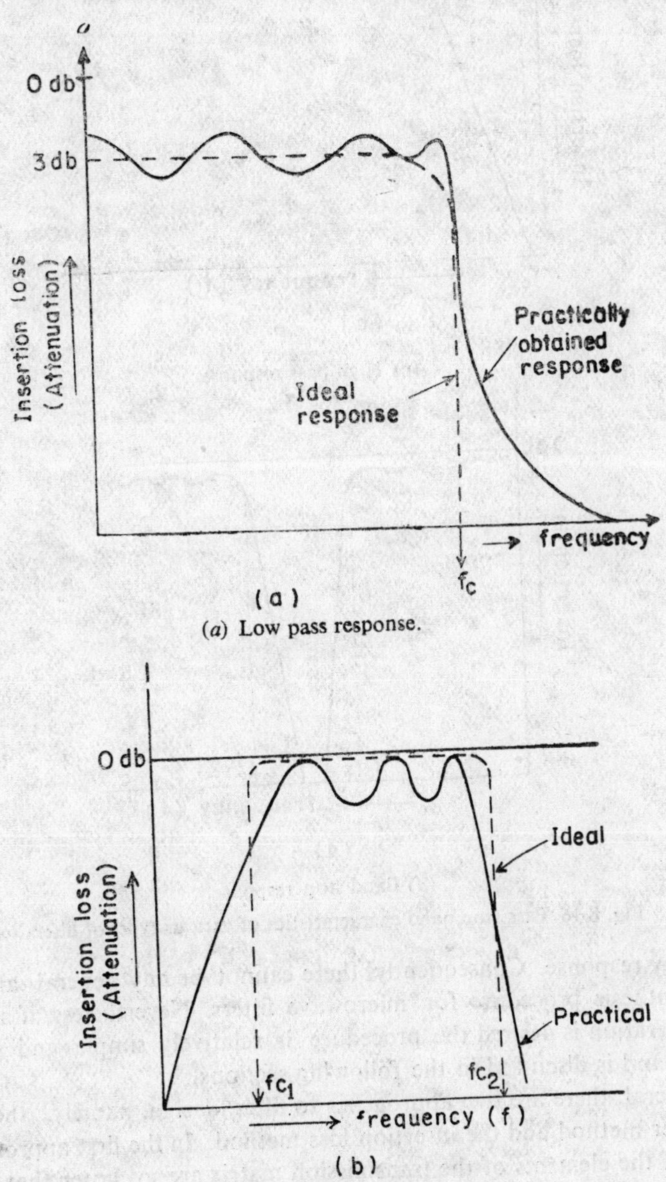


(*a*) Low pass response.



(*b*) Band pass response.

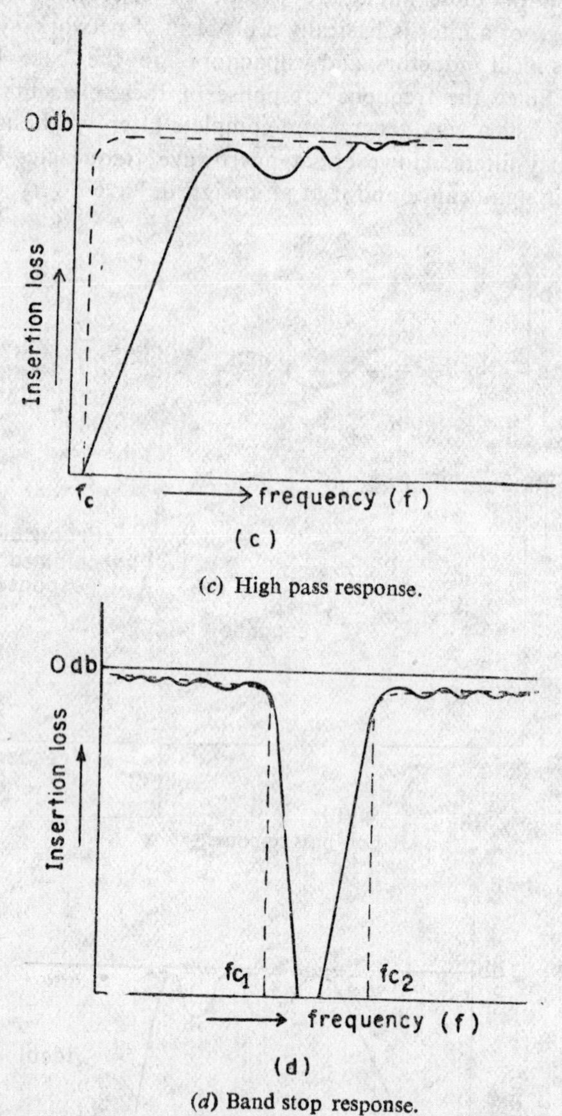


(*c*) High pass response.

(*d*) Band stop response.

**Fig. 8.16** Pass stop band characteristics of various types of filters.

frequency response. Consequently, there cannot be any general and complete synthesis procedure for microwave filters. Nevertheless, if a narrow band operation is desired the procedure is relatively simple and straightforward and is discussed in the following sections.

In general, there are two approaches to filter design, namely, the image parameter method and the insertion loss method. In the first approach, the nature of the elements of the transmission matrix are so chosen that the network provides the required pass band and stop band characteristics with the condition that image parameters are equal to the terminating impedances at the central frequency of the pass band. Clearly, the image parameter method does not specify the exact frequency characteristics over each region. The insertion loss approach, on the other hand, relies upon the complete specification of the frequency characteristics in determining the filter structure. Obviously, the second method is superior to the first and will be discussed in the following section. The details of the image parameter method may be found in the literature*. Conventionally, the insertion loss method comprises of two steps, viz. (*i*) determining a low pass "prototype" structure, of given insertion loss characteristics, normalized to 1 ohm termination and a cut-off frequency 1 rad/sec; and (*ii*) transforming the above structure into a high pass band stop or band pass structure using suitable frequency and impedance transformations.

## 8.8 BASIC MICROWAVE FILTER THEORY

The modern filter theory, developed by Darlington** and improved by Mathaei*** comprises the following steps:

(i) Assuming a low pass "prototype" lumped circuit (Fig. 8.17) and recognizing various elements of the circuit. (The structure is called "prototype" because other type of filters can be obtained from this structure.)

(ii) Expressing these elements in terms of known parameters such as characteristic impedance, maximum VSWR or insertion loss.

(iii) Finding out how many elements exist in a prototype circuit for a given insertion loss.

(iv) Determining frequency and impedance transformations that transform a low pass prototype network into a high pass, band pass or band stop network.

(v) Finding out rules that transform a lumped circuit into a distributed parameter circuit.

*1. S B. Cohn, *Very High Frequency Techniques*, Vol. 2, McGraw-Hill Book Co. (1947).
2. G.L. Ragan, *Microwave Transmission Circuits*, MIT Rad. Lab. Series Vol. 9. McGraw-Hill Book Co. (1948).
3. J.D. Rodes, *Theory of Electrical Filters*, John Wiley & Sons (1976).
4. H. Ozaki and J. Ishill, "Synthesis of transmission line networks and Design of UHF filters", *IRE, Trans.*, Vol. CT-2, (Dec. 1955).
5. J.O. Scanton and R. Levy, *Circuit Theory*, Vol. 2, Oliver and Boyd, Edinburgh (1973).
6. R.N. Ghose, *Microwave Circuit Theory and Analysis*, McGraw-Hill Book Co. (1963).

**S. Darlington 'Synthesis of reactance of four poles which produce prescribed insertion loss characteristics including, special applications to filter design', *J. Math. Phys*. 18. (1939).

***G.L. Mathaei, L. Young. and E.M.T. Jones, *Microwave Filters, Impedance Matching Networks and Coupling Structures*, McGraw-Hill Book Co. (1964).

(vi) Changing the network so developed into its microwave equivalent using either a lumped microwave equivalent or the distributed ones.

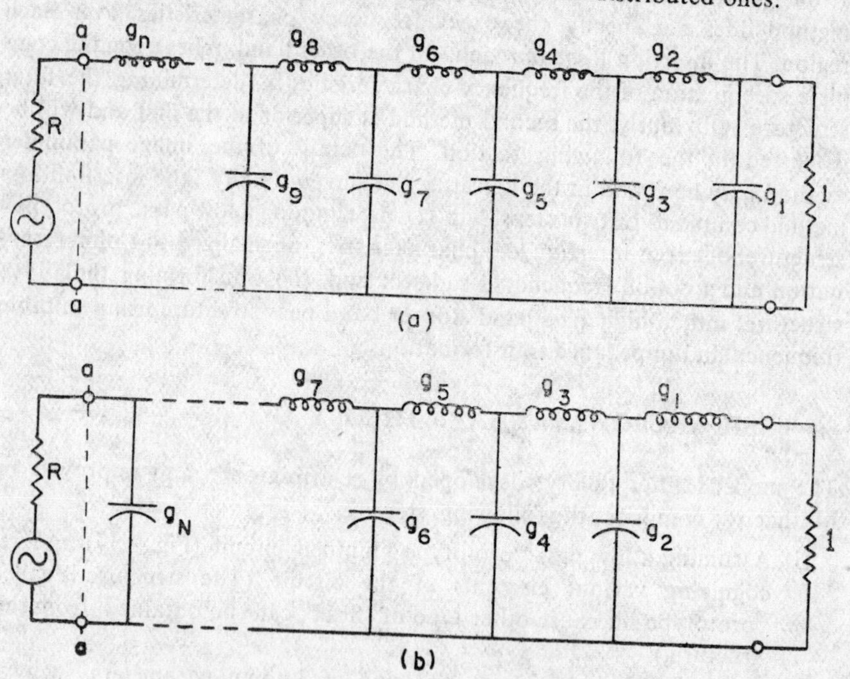


**Fig. 8.17** (*a*) Low pass ladder-prototype networks and (*b*) its equivalent circuit.

Basically, the theory relies on the development of a low pass prototype structure of given insertion loss (or scattering parameter) characteristics. However, before we switch on to such a development, it is instructive to state the following theorems which are frequently used in network synthesis.

**Theorem I*:** *The necessary and sufficient condition for a given network to be physically realizable is that the power loss ratio, defined as ratio of power input $P_i$ to the power delivered to the load $P_T$ when terminating impedances on both the sides of the network are equal, must be expressable in the following polynomial form*

$$P_{LR}=\frac{P_i}{P_T}=1+\frac{P(\omega^2)}{Q(\omega^2)} \tag{8.65}$$

where $P$ and $Q$ are polynomials in $\omega^2$. Using the definition of insertion loss, $I$, Eq. (8.65) may be written as

$$L=10\log_{10} P_{LR}=\log_{10}\left[1+\frac{P(\omega^2)}{Q(\omega^2)}\right]. \tag{8.66}$$

In terms of a scattering matrix if the network is matched at the input and output then the square of the scattering transfer coefficient, $|S_{21}|$, which

*G.L. Ragan, *op cit.*

in general is a function of complex frequency, equals the inverse of power loss ratio, i.e.

$$|S_{21}(j\omega)|^2 = \frac{P_l}{P_T} = \frac{1}{P_{LR}}. \tag{8.67}$$

From Eqs. (8.66) and (8.67), it follows that

$$\begin{aligned} L &= 10 \log_{10} P_{LR} = 10 \log_{10} (|S_{21}|^{-2}) \\ &= 10 \log_{10} \left[1 + \frac{P(\omega^2)}{Q(\omega^2)}\right]. \end{aligned} \tag{8.68}$$

For a loss less junction $[S]$ is unitary, hence for a two-port network we have

$$|S_{21}|^2 + |S_{11}|^2 = 1 \tag{8.69}$$

For a given insertion loss, $|S_{21}|^2$ is known from Eq. (8.68) and if it is substituted in Eq. (8.69), $|S_{11}|^2$ can be calculated. $|S_{11}|^2$ is connected to the input impedance of the network by the following standard relation.

$$Z_{in} = \frac{1}{Y_{in}} = \frac{1 + S_{11}(j\omega)}{1 - S_{11}(j\omega)}. \tag{8.70}$$

Thus, by properly normalizing $Z_{in}$, the filter network can be known. However, the basic requirement is the realizability of the network which is stated in Eq. (8.68). Hence the starting point of the filter theory will be the selection of the polynomials $P$ and $Q$ under various approximations.

**Theorem II:** This theorem, given by Foster* states: *The input or 'driving point' impedance of a network composed of a finite number of self-inductances, mutual inductances and capacitances, is equal to the impedance of a properly constructed series or parallel network with the number of parallel circuits equal to the number of natural frequencies at which the input terminal of the given network appears open (or the driving point impedance is infinite), and the number of series circuits is equal to the number of natural frequencies at which driving point impedance is zero (or the network appears to be shorted).*

The frequencies (in general complex) at which driving point impedance becomes zero and infinite are respectively called "zeros" and "poles".** Naturally, a lumped circuit has infinite zeros and poles corresponding to resonances and anti-resonances but because of unlimited resonances, a microwave circuit has infinite poles and zeros. This makes another basic difference between a low frequency and a microwave filter circuit.

However, using the above theorem for a given value of $Z_{in}$ (which can be calculated from Eq. (8.70)), for a narrow frequency band, a lumped circuit consisting of a finite number of inductances and capacitances can be drawn and knowing the pole-zero character of either $Z_{in}$ or $|S_{11}|$, the number of such elements can be calculated.

*R.M. Foster, "A Reactance Theorem" *Bell System Tech. J.* Vol. 3 (April 1924).

**Simplified discussions of poles and zeros may be found in J.K. Hardy, *High Frequency Circuit Design*, Reston Pub. Co. Reston (1979).

## 8.9 PROTOTYPE LOW PASS FILTER

The two most often used approximations for selecting $P$ and $Q$ polynomials appearing in Eq. (8.68) are the Butterworth and Chebychev approximations. In case of Butterworth's approximation polynomials $P$ and $Q$ respectively are chosen as $k^2\,(\omega/\omega_c)^{2n}$ and unity. Thus from Eq. (8.65), we have

$$P_{LR}=\frac{1}{|S_{12}|^2}=1+k^2\left(\frac{\omega}{\omega_c}\right)^{2n}. \tag{8.71}$$

A typical response curve [$P_{LR}$ vs $\omega/\omega_c$] for $n=5$ is shown in Fig. 8.18. The pass band (low $P_{LR}$ region) extends from $\omega=0$ to the cut-off value $\omega_c$. The maximum value of $P_{L_R}$ in the pass band is $1+k^2$ which is determined by $k$, therefore, $k$ is called *the pass band tolerance*. The response curve is maximally flat in the pass band region so this type of filter is also known as a maximally flat filter. The power loss ratio $P_{L_R}$ varies as $e^{2n}$ as we go outside the pass band ($\omega>\omega_c$). Thus an increase in $P_{LR}$ is a function of $n$, the number of sections employed.

In case of Chebychev or an equal ripple filter the polynomials $P$ and $Q$ are selected to give the power loss ratio

$$P_{LR}=\frac{1}{|S_{21}|^2}=1+k^2T_n^2\,(\omega/\omega_c) \tag{8.72}$$

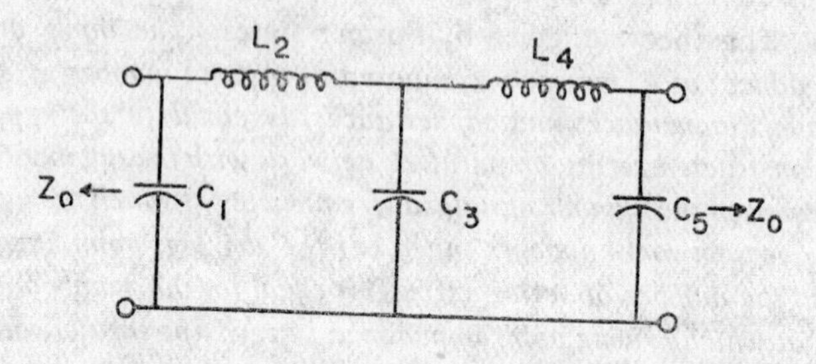


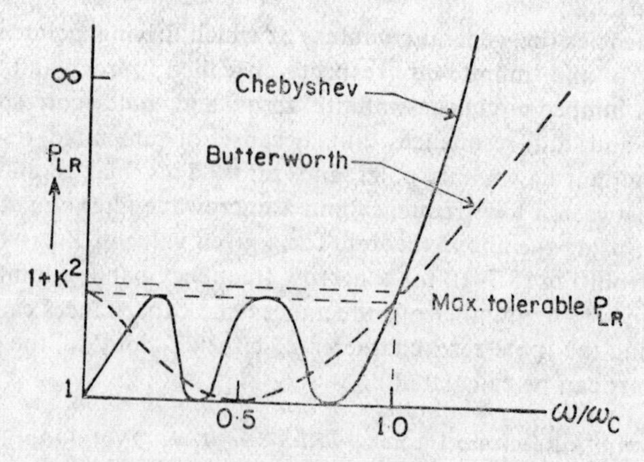


**Fig. 8.18** Low pass filter responses for Chebychev and Butterworth filters for $n=5$.

where $T_n(\omega/\omega_c)$ is the usual Chebyshev polynomial of degree $n$, discussed in Chapter 6. Recalling that

$$T_n(\omega/\omega_c)=\cos(n\cos^{-1}\omega/\omega_c) \tag{8.73}$$

we see that as long as $\omega<\omega_c$, $T_n(\omega/\omega_c)$ oscillates between $\pm 1$ and increases monotonically for $\omega>\omega_c$. This is the reason why this filter is called an *equal ripple filter*. From the response curve of Fig. 8.18 it can be seen that for $\omega>\omega_c$, $P_{LR}$ increases faster for Chebychev's approximation than for Butterworth's approximation. In fact, the Chebyshev polynomial gives a more rapid increase in $P_{LR}$ with $\omega$ for $\omega>\omega_c$ than any other polynomial. Therefore, for a given $P_{LR}$, the Chebyshev filter will have minimum tolerance or, in other words, the Chebyshev filter represents an optimum design.

Let us assume that Chebyshev's and Butterworth's low pass filter power loss ratios are realized in a ladder prototype filter network shown in Fig. 8.17(a). Assuming further that load impedance is unity and generator impedance is $R$, then, the reflection coefficient at the plane *aa* of Fig. 8.17(a) is given by

$$\Gamma=\frac{Z_{\text{in}}-R}{Z_{\text{in}}+R}$$

and the corresponding power loss ratio is given by

$$P_{LR}=1+\frac{|Z_{\text{in}}-R|^2}{2R\,(Z_{\text{in}}+Z_n^*)}. \tag{8.74}$$

The capacitive reactances are all infinite while inductive reactances are all zero at $\omega+\omega_o$. Hence we have $Z_{\text{in}}=1$. Moreover, for Butterworth's and Chebyshev's filters, for odd $n$, $P_L=1$ at $\omega=0$, therefore, for the above type of filters, generator impedance must be chosen to be unity. However, in case of Cheyshev's filter for $n$ even, $P_{LR}=1+k^2$ at $\omega=0$ and hence

$$1+k^2=1+\frac{(1-R)^2}{4R}$$

or

$$R=2k^2+1-\sqrt{4k^2\,(1+k^2)}=1/R' \tag{8.75}$$

where $R'$ is the generator impedance in the equivalent network shown in Fig. 8.17(b). Thus, both the structures of Fig. 8.17 are realizable subject to the condition that the generator impedance is selected in the manner discussed above.

The required values of the elements $g_k$ can be obtained by obtaining $Z_{\text{in}}$ and substituting it in Eq. (8.74). But the procedure becomes too complicated when $n>4$. Therefore, general solutions have been worked out* and these can be used. Moreover, the values of $g_k$ and other parameters may readily be found from the standard literature.**

*V. Belevitch, "Chebyshev filters and amplifier networks", *Wireless Eng.*, **29** (April 1952).

**1. T.S. Saad, (Ed.), *Microwave Engineers, Handbook*, Vol. I, Artech House Inc. Mass. (1971).

2. L. Weinberg, "Network design by use of modern techniques and tables", *Proc. Natt. Electron Conf.*, **12** (1956).

For a Butterworth filter with insertion loss ratio

$$P_{LR}=\frac{1}{|S_{21}|^2}=1+\omega^{2n}, \tag{8.76}$$

the values of elements $g_k$, which are either normalized shunt susceptances or normalized series reactances, are given by

$$\frac{\omega' C_k}{Y_o}=g_k=\frac{\omega' L_k}{Z_o} \tag{8.77}$$

$$R=1,$$

$$g_k=2 \sin \frac{2k-1}{2n}\pi \tag{8.78}$$

$\omega'$ is the frequency at which insertion loss equals the tolerable limit.

For a Chebyshev filter with $\omega_c=1$ these values are

$$\frac{\omega_1' C_k}{Y_o}=\frac{'L_k}{Z_o}=g_k$$

$$R=\begin{cases}1 & \text{for } n \text{ odd}\\ 2k^2+1-2k\sqrt{1+k^2} & \text{for } n \text{ even}\end{cases} \tag{8.79}$$

$$g_k=\frac{4a_{k-1}a_k}{b_{k-1}g_{k-1}}, \qquad g_1=\frac{2a_1}{\gamma} \tag{8.80}$$

where,
$$a_k=\sin\frac{2k-1}{2n}\pi \tag{8.81}$$

$$b_k=\gamma^2+\sin^2\left(\frac{k}{n}\pi\right) \tag{8.82}$$

$$\gamma=\sinh\left[\frac{1}{2n}\log\left(\coth\frac{L_m}{17.37}\right)\right] \tag{8.83}$$

$$L_m=20\log\left(\frac{S_m+1}{2\sqrt{S_m}}\right) \tag{8.84}$$

and $L_m$ is the maximum tolerable insertion loss.

The insertion loss below $\omega_c'$ is given by

$$L=10\log\{1+\epsilon\cos^2[n\cos^{-1}(\omega/\omega_c)]\} \tag{8.85}$$

while the isolation above $\omega_c'$ is given by,

$$I=10\log\{1+\epsilon\cosh^2[n\cosh^{-1}(\omega/\omega_c)]\} \tag{8.86}$$

$$\simeq 6(n-2)-10\log(1/L_m)+20\,n\log(\omega/\omega_c), \tag{8.87}$$

where
$$\epsilon=10L_m/10-1=\frac{(S_m-1)^2}{4S_m} \tag{8.88}$$

We note that for $n$ odd, $R=1$ in both Butterworth's and Chebyshev's filter. Thus, as long as the generator and load impedances are equal, the insertion loss will be zero at zero frequency and the structure will require an odd number of elements. The elements for a low pass prototype filter must alter between shunt capacitor and series inductor. Of course the structure may begin either with an inductor or with a condensor.

Though the value of $g_k$ for a given $n$ can be obtained from Eqs. (8.76) to (8.89), there is an alternative method by which $|S_{21}|^2$ is known for a given insertion loss. The poles of the transfer function $|S_{21}|^2$ are then obtained in the complex plane. Thus for the transfer function of Eq. (8.76) these poles are given by

$$1+(-s^2)^n=0$$

where $s=j\omega$ is a complex frequency function. Therefore the pole locations are

$$s_k=(-1)^{1/2}\,(-1)^{n/2}=e^{j\frac{2k+n-1}{n}\,\frac{\pi}{2}} \tag{8.89}$$

where $k=1, 2, 3, \ldots, 2n$.

These poles lie on a unit circle in the $s$-plane and are symmetrical with respect to both real and imaginary axes. For the impedance or admittance to be positive, the poles of the adopted solution must lie on the left half of the $s$-plane. Thus $S_{11}(s)$ is given by [using Eq. (8.69)],

$$S_{11}(s)\,S_{11}(-s)=\frac{(-s^2)^2}{1+(-s^2)^n} \tag{8.90}$$

$S_{11}$ is thus known and $Y_{\text{in}}$ may be evaluated from Eq. (8.70). To make the things more clear consider the case when $n=3$; then

$$S_{11}(s)\,S_{11}(-s)=\frac{s^3\,(-s^3)}{(1+2s+2s^2+s^3)\,(1-2s+2s^2-s^3)}$$

or

$$S_{11}(s)=\frac{\pm s^3}{1+2s+2s^2+s^3}.$$

Therefore,

$$Y_{\text{in}}=\frac{1-S_{11}}{1+S_{11}}=\frac{2s^3+2s^2+2s+1}{2s^2+2s+1}. \tag{8.91}$$

To obtain a canonical realization, the Cauer type ladder expansion, i.e. using division and inversion (inversion changes a series component into a shunt component and vice versa) of admittance (Eq. 8.91), has to be made so that the structure is normalized to a load of 1Ω at a cut-off frequency of 1 rad/sec.

$s\to$parallel capacitor (*1F*)

$2s^2+2s+1\,|\,\overline{2s^3+2s^2+2s+1}$

$\underline{2s^3+2s^2+s}$

$2s\to$series inductor (*2H*)

$s+1\,|\,\overline{2s^2+2s+1}$

$\underline{2s^2+2s}$

$s\to$parallel capacitor (*1F*)

$1\,|\,\overline{s+1}$

$\underline{s}$ ↓

Terminating resistor→1Ω

The resultant structure is depicted in Fig. 8.19. It may be mentioned that the values of $g_1$, $g_2$, $g_3$, if calculated using Eq. (8.78) come out to be the same as obtained above.

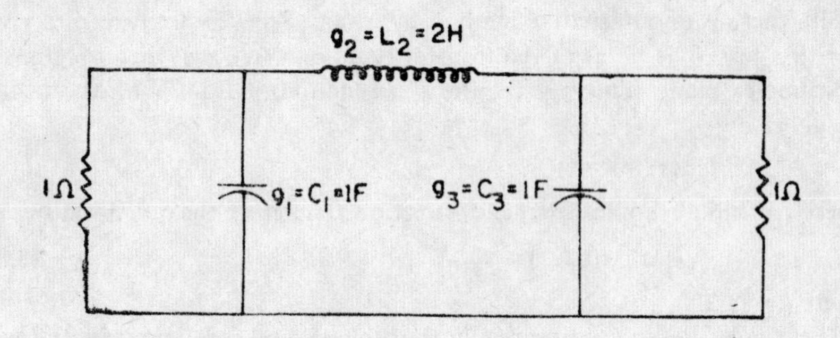


**Fig. 8.19** Low pass prototype filter network for $n=3$.

## 8.10 FREQUENCY TRANSFORMATIONS

Our objective, now, is to transform the low pass prototype structure into high pass, band pass and band stop filter structures. The prototype structure is characterized by $\omega_c=1$ and load$=1\ \Omega$. To have a cut-off frequency more than unity let us denote the frequency variable of prototype filter as $\omega'$ and replace it by a new frequency $\omega$ by a general transformation

$$\omega'=f(\omega).$$

The power loss ratio then becomes

$$P_{LR}=1+P(\omega'^2)=1+P[f^2(\omega)]$$

To have load other than unity, we assume that the structure is terminated in a load $R_L$. Then the prototype filter reactances can be viewed as normalized with respect to $R_L$. The new values of elements are then obtained by multiplying all reactances and the generator resistance by $R_L$, i.e. the new values $L_k'$, $C_k'$ and $R'$ are given by

$$L_k'=R_L L_k$$
$$C_k'=C_k/R_L,$$
$$R'=R_L R. \tag{8.92}$$

Which is nothing but *impedance transformation*.

It is not difficult to see from Eq. (8.92) that the response characteristics of a new filter will be different from those of the original prototype filter depending on the selection of the transformation function $f(\omega)$. The following transformations are commonly used.

### Frequency Expansion

Suppose $\quad f(\omega)=\omega/\omega_c,$

then $$P_{LR}=1+P(\omega^2/\omega_c^2). \tag{8.93}$$

Clearly the cut-off of the pass band occurs when $\omega/\omega_c=1$, i.e. at $\omega=\omega_c$. *The cut-off frequency is thus shifted from unity to* $\omega_c$ as shown in Fig. 8.20. The corresponding new series reactances and shunt susceptances are obtained by replacing $\omega$ by $\omega/\omega_c$. That is

$$jX'_k=j\left(\frac{\omega}{\omega_c}\right)L_k,\ jB'_k=j\left(\frac{\omega}{\omega_c}\right)C_k$$

where $$L'_k=L_k/\omega_c,\ C'_k=C_k/\omega_c. \tag{8.94}$$

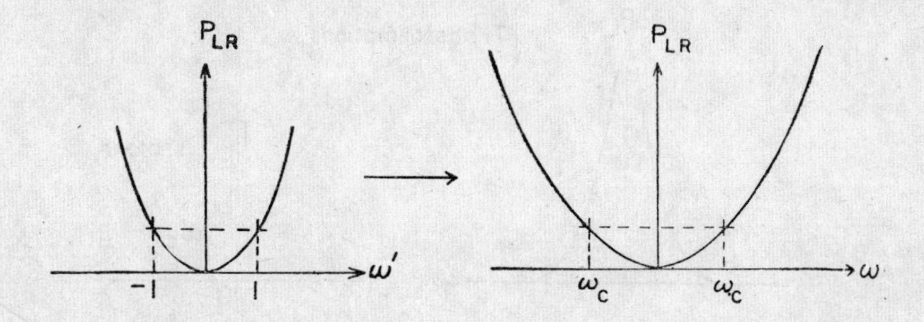


**Fig. 8.20** Frequency expansion transformation.

**Low pass to High Pass Transformation**

The low pass to high pass transformation is obtained by selecting $f(\omega)=-\omega/\omega_c$ or by defining the frequency variable $s$ as, $s=j\omega$ and replacing $s'$ by $\omega_c/s$ where $\omega_c$ is the cut-off frequency of a high pass filter. This transformation maps the segment $0<\omega'<1$ on the $s'$ plane to $\omega_c<|\omega|<\infty$ on the $s$-plane as shown in Fig. 8.21. Clearly this transformation interchanges the pass band and stop band. The elements of the new network are obtained by noting that immittance* is invariable under frequency transformation. For the inductance the impedance is

$$Z=j\omega' L$$

which under the transformation in question, goes to

$$Z=\left(-\frac{\omega_c}{\omega}\right)jL=\frac{1}{j\omega C_h} \tag{8.95}$$

where $$C_h=1/\omega_c L. \tag{8.96}$$

Similarly, the shunt admittance of the condenser $Y=j\omega C$, is transformed to

$$Y=j\,(\omega_c/\omega)\,C=1/j\omega\,L_h \tag{8.97}$$

where $$L_h=1/\omega_c\,C \tag{8.98}$$

*Common name for impedance and admittance.

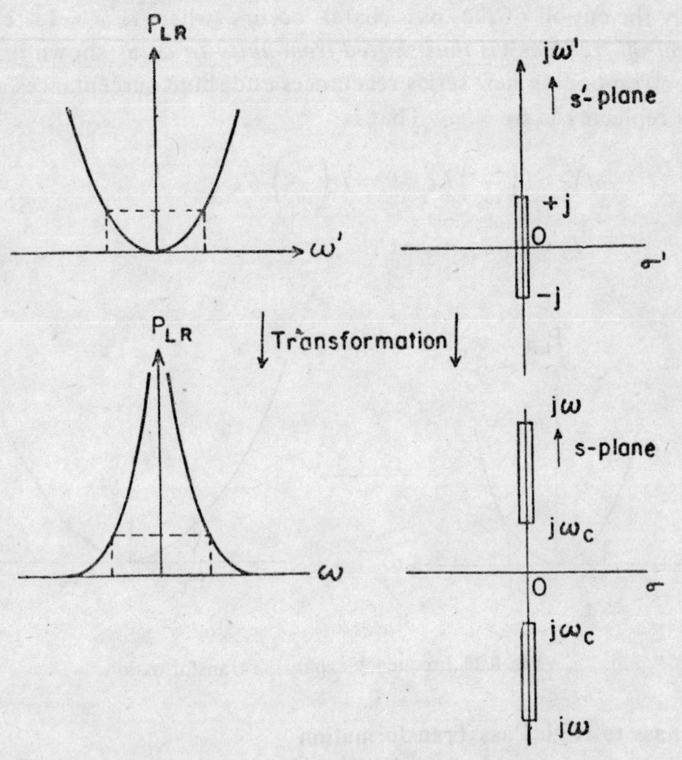


**Fig. 8.21** Low pass to high pass transformation.

*It thus follows that series inductance is transformed into series capacitance, and shunt capacitance is transformed into shunt inductance when frequency cut-off is $\omega_c$ and load is other than unity.* The transformed structure for $n=3$ is shown in Fig. 8.22.

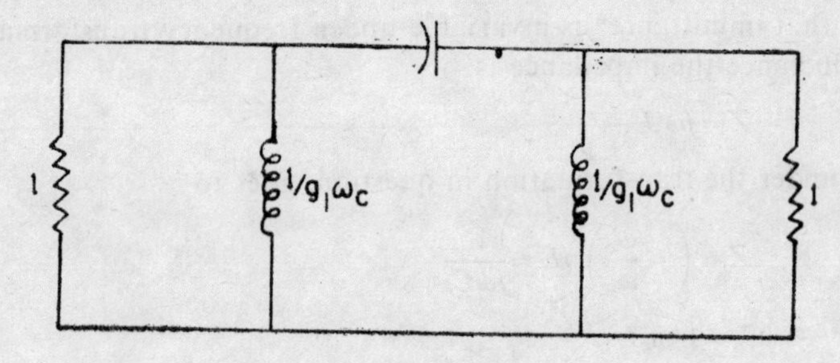


**Fig. 8.22** High pass prototype, $n=3$.

**Low Pass to Band Pass Transformation**

Low pass to band pass transformation is obtained by choosing the transformation function to be

$$\omega'=f(\omega)=\frac{\omega_0}{\omega_{c2}-\omega_{c1}}\left(\frac{\omega}{\omega_0}-\frac{\omega_0}{\omega}\right) \tag{8.99}$$

Taking $\omega_o = \sqrt{\omega_{c2}\,\omega_{c1}}$ and solving Eq. (8.99) we have,

$$\omega = \omega' \frac{\omega_{c2} - \omega_{c1}}{2} \pm \frac{1}{2}\sqrt{\omega'^2(\omega_{c2} - \omega_{c1})^2 + 4\,\omega_{c1}\,\omega_{c2}} \quad . \tag{8.100}$$

It can be seen fromEq. (8.101) that $\omega'=0$ and $\omega'=\pm 1$ respectively map into two and four points given by $\omega=\pm\omega_o$, and $\pm\omega_{c2}$ and $\pm\omega_{c1}$.
Thus segment $0<|\omega'|<1$ is mapped into $\omega_{c1}>|\omega|>\omega_{c2}$ as shown in Fig. 8.23. That is the prototype pass band between $\omega=\pm 1$ maps into passband extending from $\omega_{c1}$ to $\omega_{c2}$ and $-\omega_{c1}$ to $-\omega_{c2}$ which represent bandpass filter with band centered at $\pm\omega_o$, which is geometric mean of two cut-off frequencies $\omega_{c2}$ and $\omega_{c1}$.

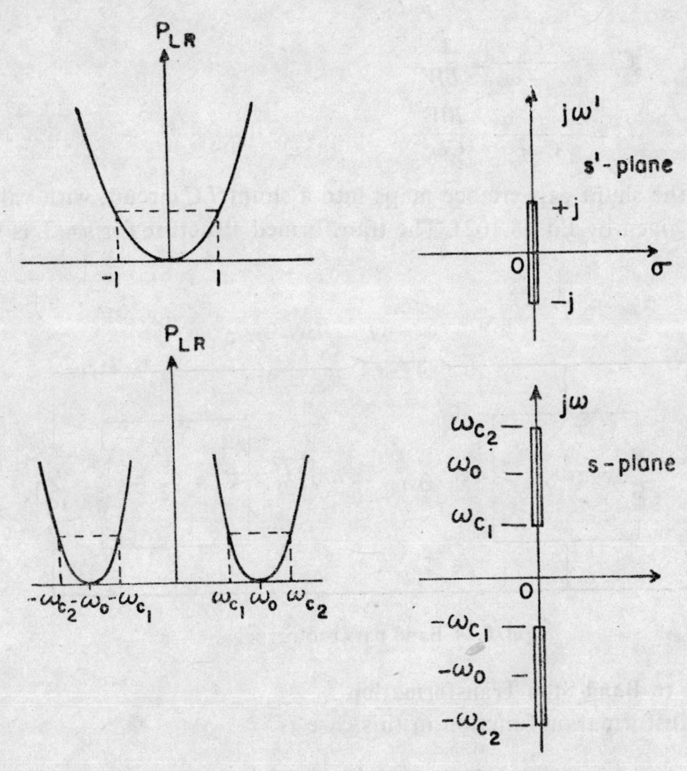


**Fig. 8.23** Low pass to band pass transformation.

To find the elements of a transformed filter we again apply the invariance property of the immittance under frequency transformation. Invariance of $Z$ requires

$$Z = j\omega' L = \left[\frac{\omega_o}{\omega_{c1} - \omega_{c2}}\left(\frac{j\omega}{\omega_o} + \frac{\omega_o}{j\omega}\right) L\right]$$

$$= j\omega L_s + \frac{1}{j\omega C_s}$$

where $$L_s=\frac{L}{\omega_{c2}-\omega_{c1}}=\frac{L}{\text{Band width}}=\frac{L}{BW},$$

and $$C_s=\frac{\omega_{c2}-\omega_{c1}}{\omega_o^2 L}=\frac{BW}{\omega_o^2 L} \tag{8.101}$$

*Thus, series inductance of the prototype low pass filter maps into a series LC resonator, with the values of L and C given by Eq. (8.101).*

Similarly, invariance of $Y$ requires

$$Y=j\omega' C=\left[\frac{\omega_o}{\omega_{c2}-\omega_{c1}}\left(\frac{j\omega}{\omega_o}+\frac{\omega_o}{j\omega}\right)\right]C$$
$$=j\omega C_p+\frac{1}{j\omega L_p}$$

where $$C_p=\frac{C}{\omega_{c2}-\omega_{c1}}=\frac{C}{BW}$$

and $$L_p=\frac{\omega_{c2}-\omega_{c1}}{C\omega_o^2}=\frac{BW}{C\omega_o^2} \tag{8.102}$$

That is, the shunt capacitance maps into a shunt $LC$ circuit, with values of $L$ and $C$ given by Eq. (8.102). The transformed structure for $n=3$ is shown in Fig. 8.24.

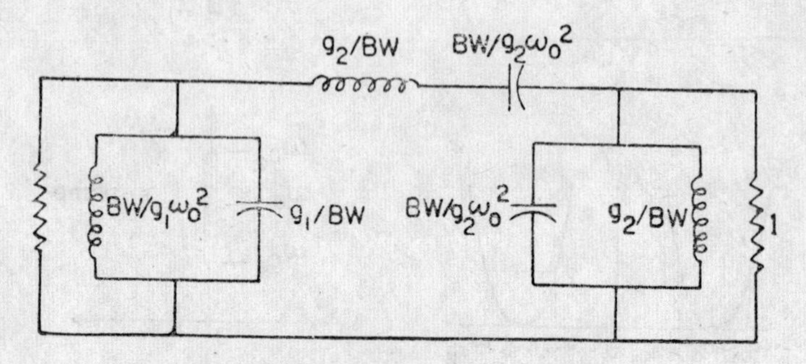


**Fig. 8.24** Band pass prototype, $n=3$.

### Low Pass to Band Stop Transformation

The transformation function in this case is

$$\omega'=f(\omega)=\frac{\omega_{c2}-\omega_{c1}}{\omega_o}\left(\frac{1}{\dfrac{\omega}{\omega_o}+\dfrac{\omega_o}{\omega}}\right). \tag{8.103}$$

It can be seen that this transformation maps the segment $|\omega'|<1$ into $\omega_{c2}>|\omega|>\omega_{c1}$ as shown in Fig. 8.25.

The invariance of the immittance under frequency transformation, in this case, leads to the result that *the series inductance maps into a shunt tank circuit* with elements volues following values of elements

$$C_p=\frac{1}{L\,(\omega_{c2}-\omega_{c1})}=\frac{1}{L\,(B.W)}$$

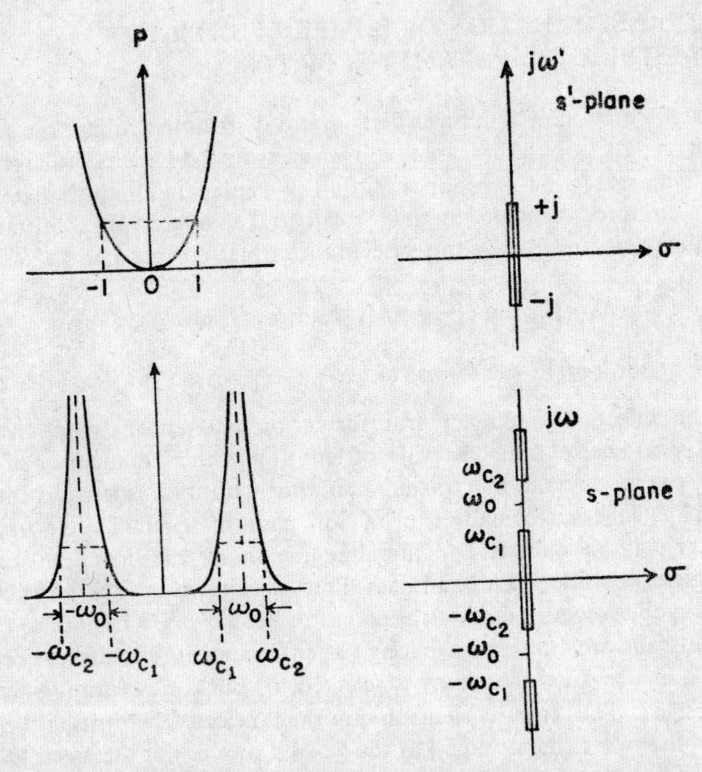


Fig. 8.25 Low pass to band stop transformation.

and $$L_p = L \cdot \frac{\omega_{c2} - \omega_{c1}}{\omega_o^2} = \frac{L\,(BW)}{\omega_o^2} \tag{8.104}$$

*whereas the shunt capacitance maps into a series tuned circuit with element values*

$$C_s = \frac{C\,(\omega_{c2} - \omega_{c1})}{\omega_o^2} = \frac{C\,(B.W)}{\omega_o^2}$$

and $$L_s = \frac{1}{C\,(\omega_{c2} - \omega_{c1})} = \frac{1}{C(BW)} \tag{8.105}$$

The transformed structure for $n=3$ is shown in Fig. 8.26.

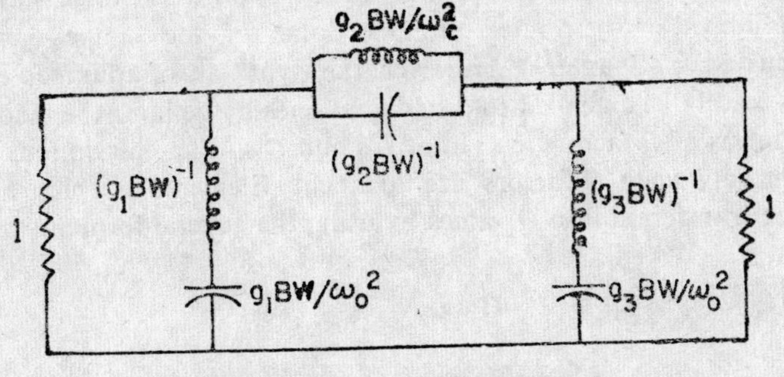


Fig. 8.26 Band stop prototype, $n=3$.

## 8.11 TRANSFORMATION OF LUMPED FILTER TO DISTRIBUTED PARAMETER FILTER

The transformation of a lumped filter into a distributed filter can be done by replacing all $L_k$ by short-circuited transmission line stubs of length $l$ ($l<\lambda/4$) and characteristic impedance $Z_k$ and by replacing all the capacitors $C_k$ by open-circuited stubs of length $l$ ($l<\lambda/4$) and characteristic admittance $Y_k$. The transformed reactance and susceptance functions become

$$j X'_k = jZ_k \tan\left(\frac{\omega l}{c}\right) = jZ_k \tan\theta = jg_k \tan\theta$$

$$j B'_k = jY_k \tan\theta = jg_k \tan\theta. \tag{8.106}$$

Thus, the effective frequency transformation $\omega' \to \tan \omega l/c$ maps the whole $\omega'$ axis periodically into intervals of length $\pi$ in the $\theta$ domain or of length $\pi c/l$ in the $\omega$ domain. A typical such filter with its response is shown in Fig. 8.27. The transformation of a low pass distributed filter to a high pass filter may be effected by interchanging short and open line stubs. In fact, a high pass filter is a band pass filter (as shown in Fig. 8.28) because of the cyclic repetition of the attenuation profile $P_{LR}=1+P\,[\tan^2(\omega l/c)]$. Also the stubs are quarter wave long (when resonant circuits are replaced by stubs) at $\omega_o$. This frequency is selected to put the isolation where it is most desired. The circuit elements are then selected by putting the reactances of the two elements equal at the desired ripple. For the given tolerable ripple the characteristic impedance of the stubs may be easily selected.

Often the realization of a microwave filter that involves alternate occurrence of series and parallel stubs becomes difficult particularly in strip line filters. It is therefore desirable to transform the filter of Fig. 8.28 into a filter that either has parallel stubs (Fig. 8.29a) or has only series stubs (Fig. 8.29b). The transformation from a series circuit to a parallel circuit can be made either by using Kurda's Identity (shown in Fig. 8.28) or using a quarter wave transformer as an impedance inverter which inverts the load impedance with respect to the characteristic impedance squared at the input (as discussed in Chapter 6).

The last step of the microwave filter theory requires the microwave equivalent to lumped or distributed circuits. Some of the lumped analogies were discussed in Chapter 4. However, the synthesis of a lumped element filter can readily be extended to distributed ones by defining the ladder branches in terms of $Q$ factors instead of $L$'s and $C$'s. The descriptions will be equivalent only when $Q$ factors are identical. Elements $g_r$ (may be $L_r$ or $C_r$) can be transferred into $Q$ values by using the general formula

$$Q_r = \frac{g_r\,\omega_o}{2BW} \tag{8.107}$$

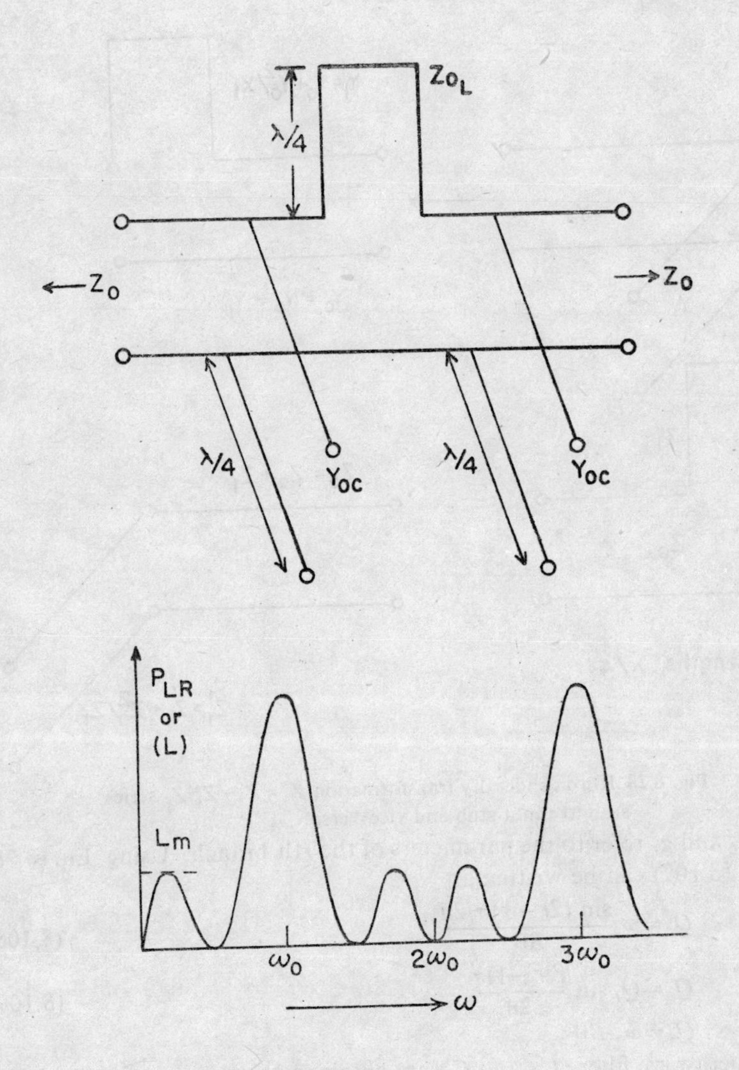


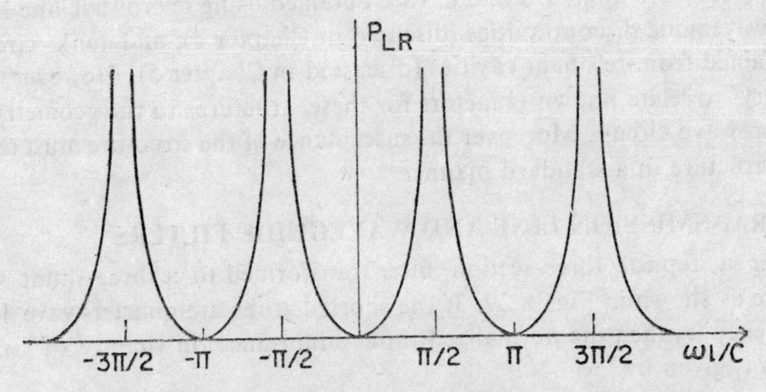


**Fig. 8.27** Transmission line filter (from frequency transformation obtained).

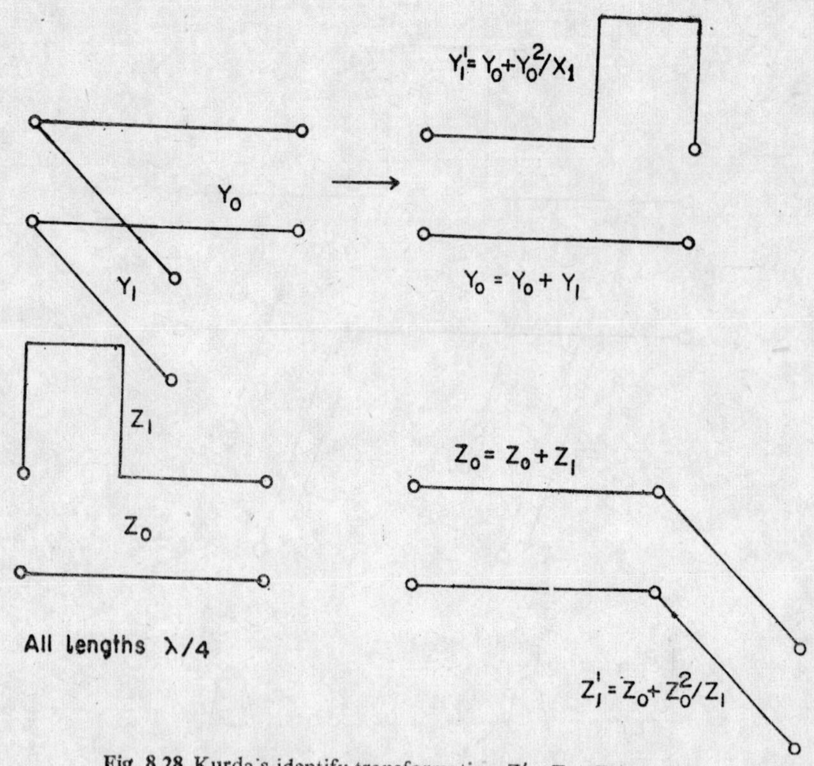


**Fig. 8.28** Kurda's identify transformation $Z_1' = Z_0 + Z_0^2/Z_1$ series stub to shunt stub and vice-versa.

where $Q_r$ and $g_r$ refer to the parameters of the $r$th branch. Using Eq. (8.78) and Eq. (8.107) can be written as

$$Q_r = \omega_o \frac{\sin (2r-1)\pi/2n}{BW} \tag{8.108}$$

or

$$Q_r = Q_t \sin \frac{(2r-1)\pi}{2n} \tag{8.109}$$

where $\quad Q_t = \omega_o/BW.$

In a microwave filter, $L$'s and $C$'s are obtained using microwave line stubs or the waveguide discontinuities (discused in Chapter 4); and tank circuits are obtained from resonant cavities (discussed in Chapter 5). However, it is necessary to relate known $Q$ factors for these structures to the geometry of the microwave circuit. Moreover the susceptance of the structure must relate to the structure in a standard manner.

## 8.12 TRANSMISSION LINE AND WAVEGUIDE FILTERS

Consider a typical three-section filter transformed to a three-shunt stub structure as shown in Fig. 8.29. If the shorted stubs are quarter-wave long at frequency $\omega_o$ then the normalized input admittance, in vicinity of $\omega_0$, to the stub is given by

$$y_{\text{in}} = j\, y\, (\omega - \omega_w) \frac{l}{c} = j\, y \frac{\omega - \omega_o}{\omega_o} \frac{\pi}{2} \tag{8.110}$$

where $l$ is the stub length, $y$ is its normalized characteristic admittance, and $\omega_o^l/c = \pi/2$. The input admittance of a parallel tuned circuit is given by

$$Y_{in} = j\omega C + \frac{1}{j\omega L} = \sqrt{\frac{C}{L}}\left(\frac{\omega}{\omega_o} - \frac{\omega_o}{\omega}\right)$$

$$\simeq 2j\sqrt{\frac{C}{L}}\,\frac{\omega - \omega_o}{\omega} = 2j\,C\,(\omega - \omega_o). \tag{8.111}$$

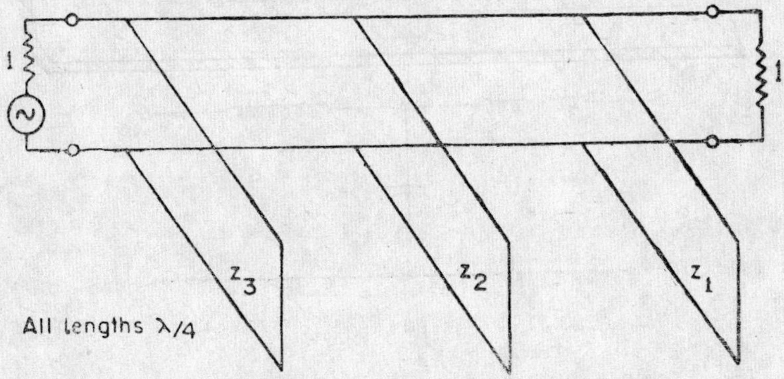

Fig. 8.29 Three-section transmission line filter.

It follows from Eqs. (8.110) and (8.111) that a shorted stub behaves as a parallel tank circuit with $\sqrt{C/L} = y\pi/4$ in the vicinity of the frequency for which it is quarter wave long. Thus shunt stubs may be used to replace parallel resonant circuits appearing in a filter circuit. The stub length may be calculated using Eq. (8.110). Typical available coaxial line filters along with their lumped equivalent circuits are shown in Fig. 8.30. A low pass filter (Fig. 8.30a) consists of two reentrant line sections separated by a concentric condensor. The structure behaves as a low pass filter as long as reentrant line sections are shorter than a quarter wavelength. At higher frequencies reactances of these sections have poles and zeros and so the existence of spurious pass bands cannot be ruled out. The predicted cut-off frequency of this structure is 2.86 GHz, but its actual observed value is 2.44 GHz.

High pass coaxial filter (Fig. 8.30b) is differentiated from a low pass filter so as to realize the suitable frequency transformations discussed in Section 8.10. Band stop and band pass filters may be similarly designed.

A waveguide is intrinsically a high pass filter because of its cut-off characteristics. The possibility of a low pass filter is thus excluded. A filter shown in Fig. 8.31 which is basically a waveguide version of the transmission line filter of Fig. 8.29 can be designed. But due care must be taken regarding the junction effects which normally create designing problems. A typical tapered waveguide high pass filter is shown in Fig. 8.32. The filter may be made variable* by placing it inside the guide, a thin longitudinal dielectric

*W.G. Wadey, *A Variable Cut-off High pass Filter* (for use with AN/APR/5A), RRL Report No. 411-161, (March 1945).

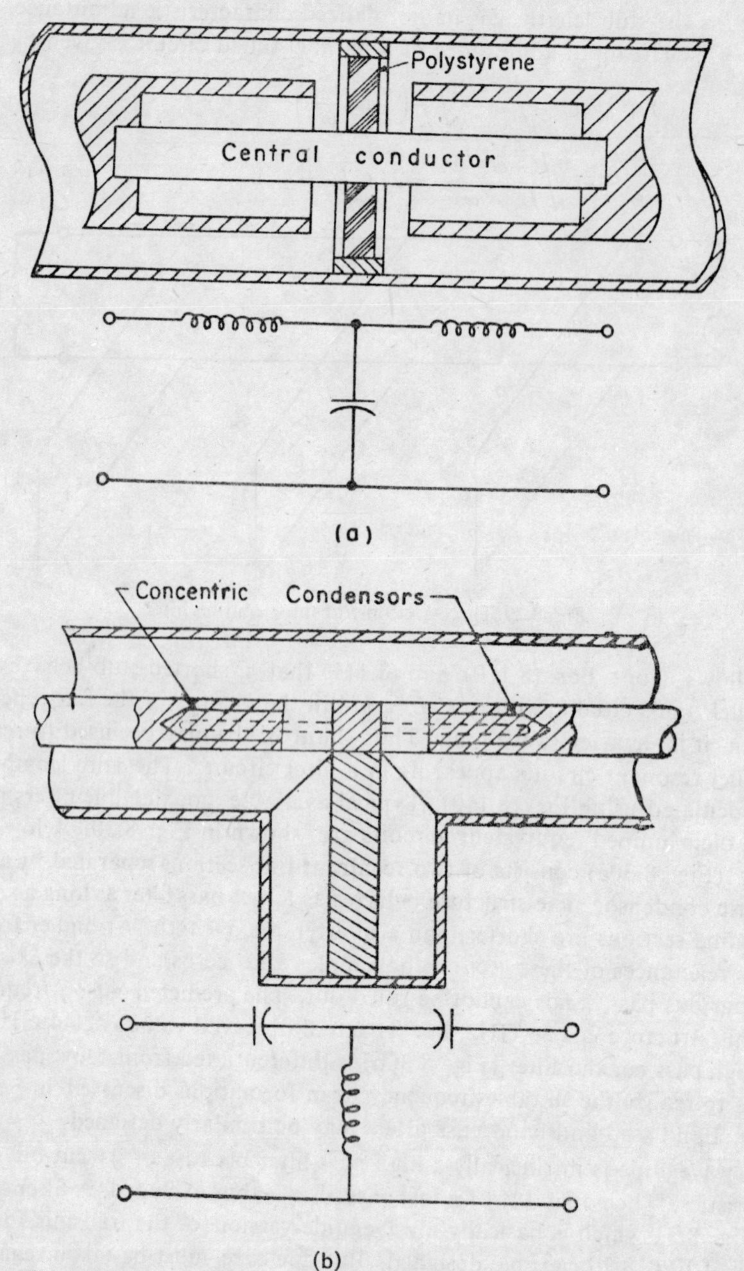


Fig. 8.30 Microwave coaxial line filters (*a*) low pass (*b*) high pass.

or metallic slab which can be moved across the waveguide.

Recently, waveguide low pass filters have been designed which consist of a single mode ridge waveguide constituting distributed shunt capacitors and an evanescent mode waveguide constituting series inductors. The use of a

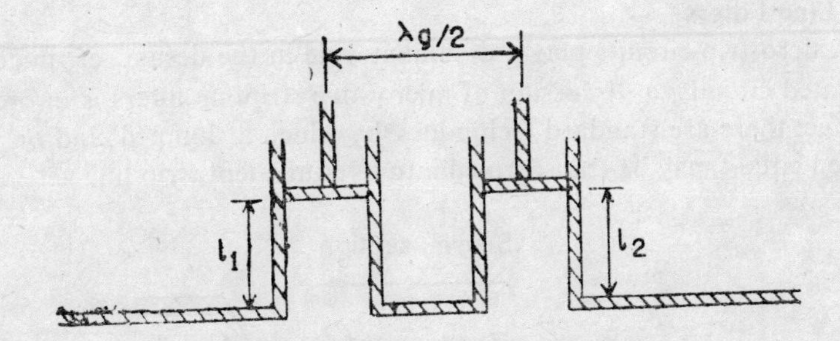

Fig. 8.31 Waveguide type microwave filter.

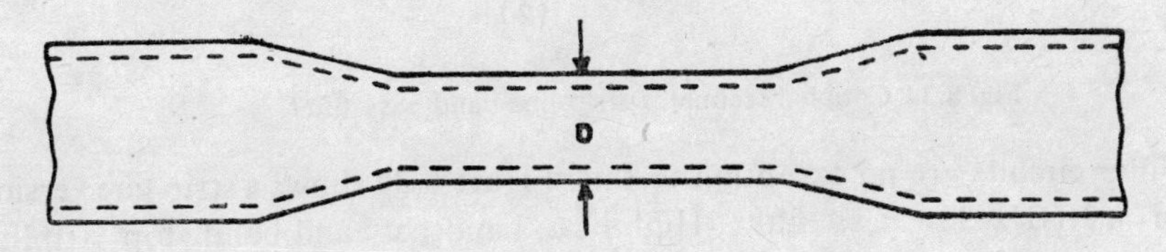


Fig. 8.32 Tapered waveguide high pass filter.

single mode ridge waveguide eliminates the possibility of higher order spourious responses and the evanescent inductances overcome high impedance limitations (which are present due to the use of a corrugated waveguide). Further details may be found in the literature.*

As mentioned in Section 8.4, a periodically loaded waveguide behaves as a band pass/band stop filter. Such a waveguide may be treated as being made of directly coupled cavity resonators. The discussion of such filters will be made in the following sections.

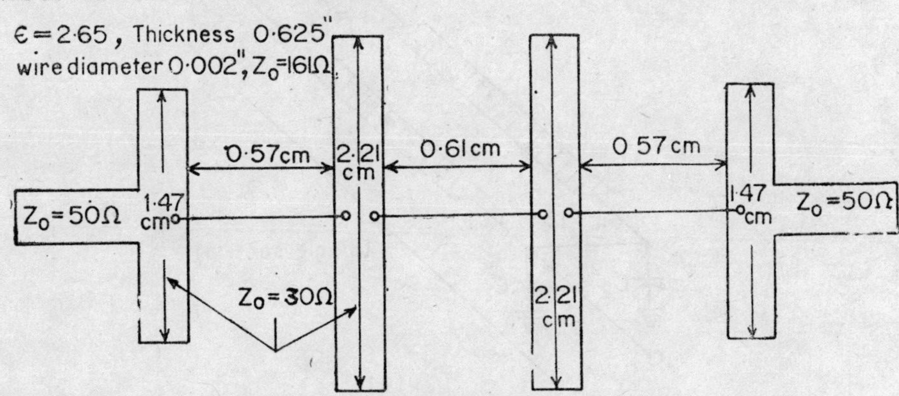


Fig. 8.33 2 GHz low pass stripline filter.

*H.F. Chappell, 'Waveguide low pass filter using evanescent mode inductors' *Microwave J.*, Vol. 21, No. 12 (Dec. 1978).

**Strip Line Filters**

As microstrip circuits play a dominant role in the design of microwave integrated circuits, a discussion of microwave stripline filters is in order.

In fact there are standard techniques*by which a lumped and/or a distributed circuit may be transformed into its equivalent strip line version. Fil-

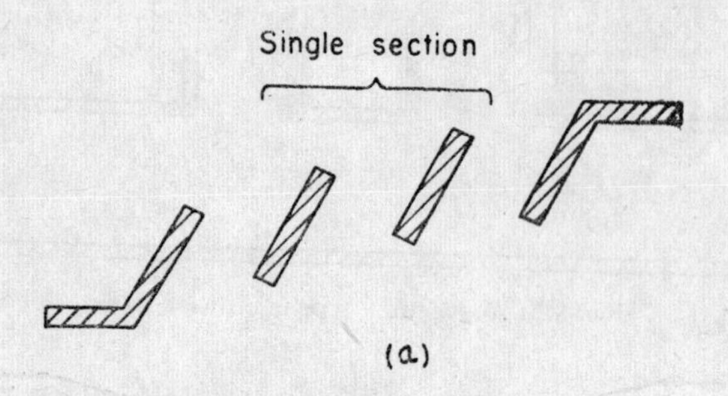


**Fig. 8.34** Coupled resonator strip line band pass filter.

Filter circuits are no exception to this. Figure 8.33 shows a strip line version of a 2 GHz low pass filter. High pass, band pass and band stop stripline versions may be similarly designed. However, the parallel coupled resonator configuration, shown in Fig. 8.34, is attractive for the realization of a band pass filter in microstrip configuration. The equivalent circuits of each strip section and that of a multisection filter are shown in Figs. 8.35 (a) and (b) respectively. The $\pi$-network of capacitors of either end arise from the consideration of the fringing fields associated with each of the sections. This structure is a narrow band one and can be made broad band by using multi-

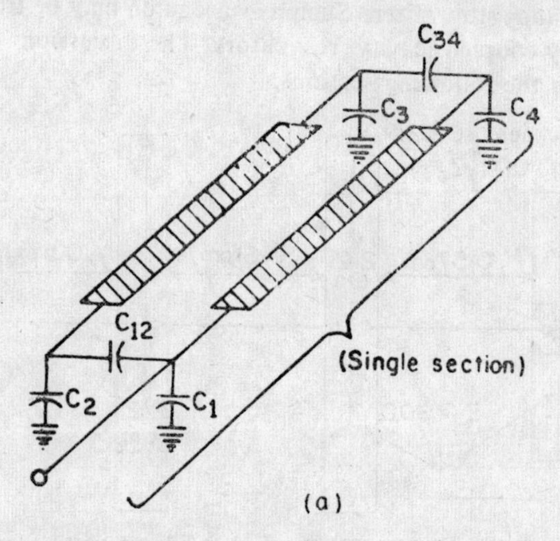


**Fig. 8.35** (*a*) Single section equivalent circuit

*H. Howe, *Strip line Circuit Design,* Artech House, Dedhan, Mass. (1974).

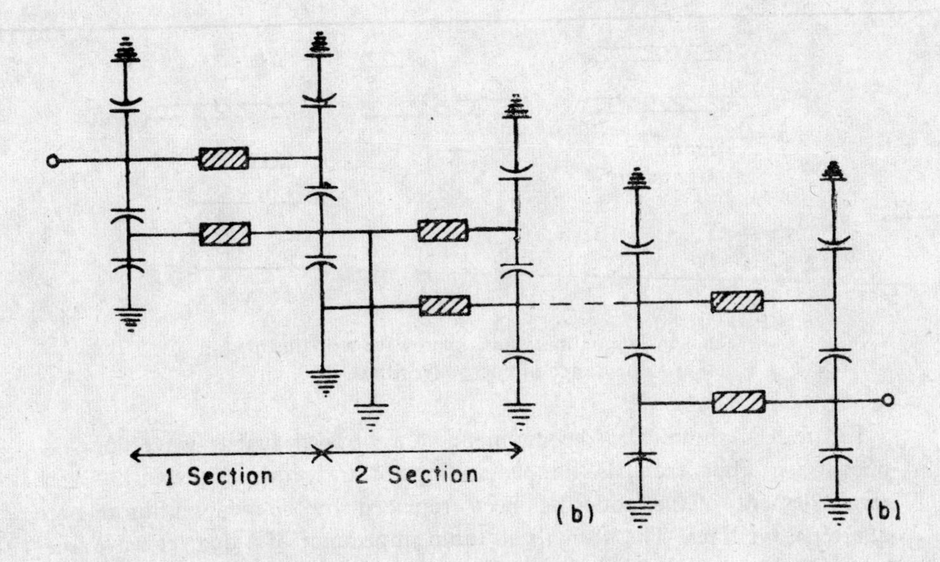


Fig. 8.35 (*b*) Multisection equivalent circuit.

line sections.* However, there are many problems in the design of such filters if the desired response is to be obtained. These problems are overcomed by introducing discontinuities** in the configuration as shown in Fig. 8.36. Moreover, the spacing of adjacent bands in these filters is small and may be increased† by properly selecting the stripline coupling parameters. Interdigital filters using elliptic functional responses have also been designed†† for wide band operation.

## 8.13 QUARTER WAVE COUPLED CAVITY FILTER

This type of filter is obtained by joining resonant cavities with quarter wave long microwave lines such as a coaxial line of waveguide. The chain-like structure obtained is approximately a microwave version of the low frequency lumped element ladder structure. However, a quarter-wave coupled structure is obtained if the resonant stubs of Fig. 8.30 are replaced by resonant cavities.

*J.F. Mara and J.B. Schappacher, "Broadband microstrip parallel-coupled filter using multiline sections", *Microwave J.*, Vol. 22, No. 4 (April 1979).

**C. Gupta, "Design of parallel coupled line filter with discontinuity compensation in microstrip", *Microwave J.* Vol. 22, No. 12 (Dec. 1979).

†G. Saulich "A simple method for spacing the adjacent passbands of a coupled line filter", *IEEE Trans. Microwave Theory and Techniques*, Vol. MTT-28, No. 4, (April 1980).

†† J.A.G. Matherbe, 'An etched digital elliptic filter", *Microwave J.*, Vol. 22, No. 12 (Dec. 1979).

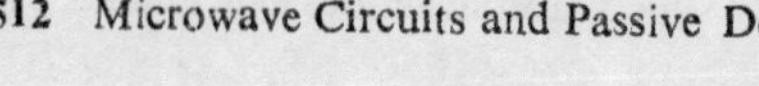

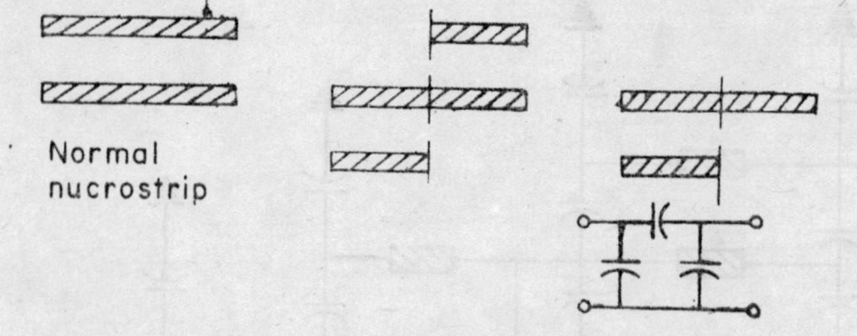


**Fig. 8.36** Discontinuity in a microwave strip line used in the design of microwave filters.

Figure 8.37 shows the development of a coaxial quarter-wave coupled microwave filter from its lumped analogue low frequency network. Each series element of the ladder has been replaced by the combination of two quarter-wave lines. The normalized input impedance of a quarter-wave line terminated in a normalized impedance $z$ is, $1/z$, therefore, series impedances ($Z_n$) in Fig. 8.37(a) have been replaced by shunt impedances of inverse value (i.e. $1/Z_n$) (Fig. 8.37b). The transformation from Fig. 8.37(b) to Fig. 8.37(c) is straightforward.

The analysis and design of such a filter may be carried out by considering cavities to be coupled by quarter-wave lines. Let us consider the $k$th section of the filter. It basically consists of two identical diaphragms separated by a distance $l_k$. This structure has an equivalent circuit shown in Fig. 8.38 (a). Mumford* has shown that this circuit has the same frequency response as Fig. 8.38 (b) so long as frequencies are very close to $\omega_o$. Thus, the circuit behaves as a filter. The structure so obtained by joining all such sections is nothing but a periodically loaded waveguide with spatial periodicity $\lambda_g/4$. Mumford showed that the diaphragm spacing $l_k$ for perfect transmission at $\omega=\omega_o$, is given by

$$\tan \beta_o l_k = -2/b_k \tag{8.112}$$

where $b_k$ is the normalized shunt susceptance of the $k$th cavity aperture. $b_k$ is related to $Q_k$ of the cavity as

$$Q_k = \frac{\tan^{-1}(2/b_k)}{2\sin^{-1} 2/(b_k^4+4b_k^2)^{1/2}} \simeq \frac{\sqrt{b_k^4+4b_k^2}}{4}\tan^{-1}\left(\frac{2}{b_k}\right) \tag{8.113}$$

The two line sections in Fig. 8.39 (a) should be chosen so that at $\omega_o$ frequency,

$$\beta_o l_k + 2\theta_{1k} = \theta_k + 2\theta_{1k} = \pi. \tag{8.114}$$

*W.W. Mumford, "Maximally flat filters in waveguides", *Bell System J.*, Vol. 27 (Oct. 1948).

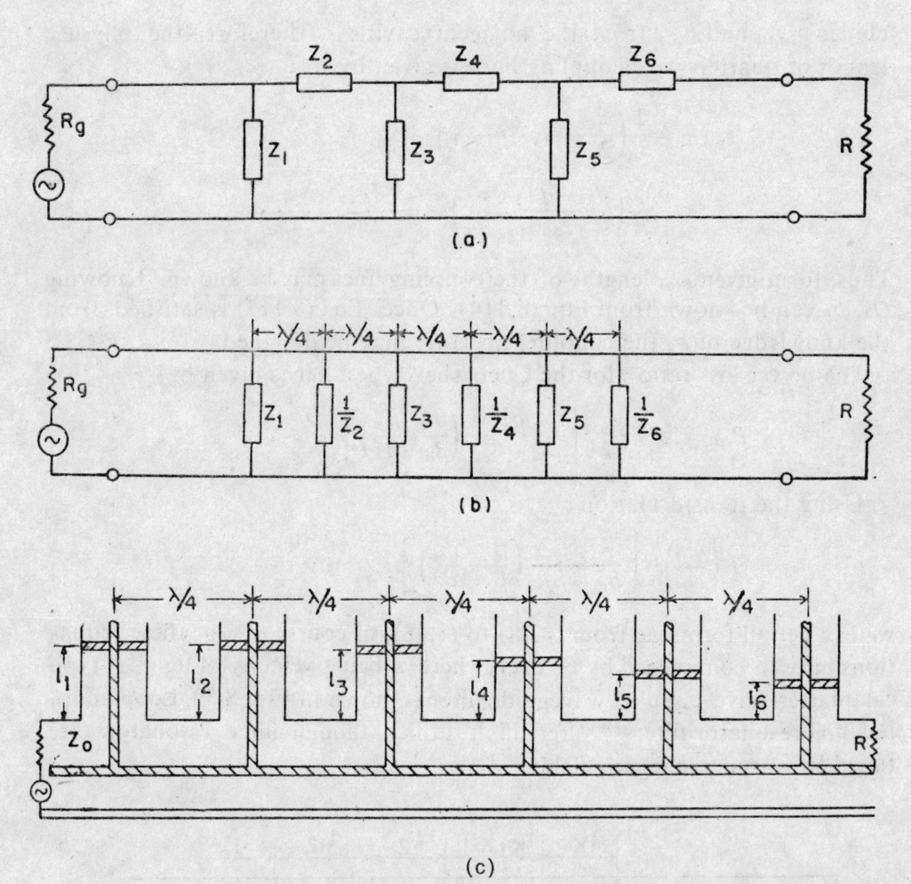


**Fig. 8.37** Transformation of a lumped element ladder structure into a microwave network (*a*) Lumped element ladder (*c*) structure (*b*) Microwave equivalent circuit (*c*) Transformed microwave network structure.

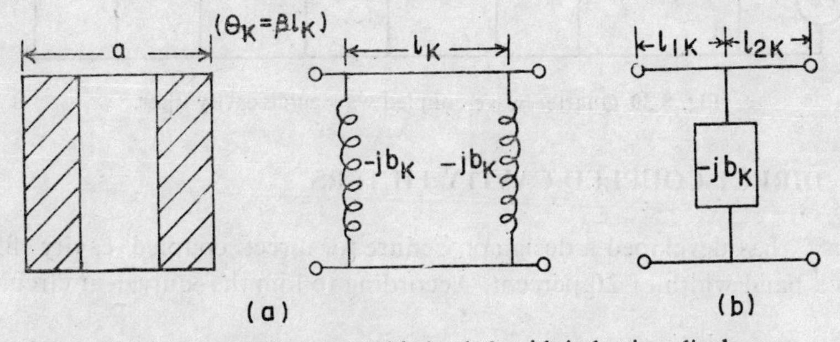


**Fig. 8.38** (*a*) Rectangular waveguide loaded with inductive diaphragms to form a cavity (*b*) transformed equivalent circuit.

The additional line lengths in the equivalent circuit of the cavity are absorbed into and make part of the quarter-wave coupling lines. The waveguide section between cavity $k$ and $k+1$ has an electrical length $\pi/2$ and also includes $\theta_{1k+1}$ and $\theta_{l,k}$ from the adjacent cavities. Therefore, the physical length of quarter-wave coupling lines is given by

$$l_{k,\,k+1}=\frac{1}{\beta_o}\left(\frac{\pi}{2}-\theta_{1k}-\theta_{1k+1}\right)$$

$$=\frac{\lambda_{go}}{2\pi}\left(\frac{\theta_k+\theta_{k+1}}{2}-\frac{\pi}{2}\right)=\frac{l_k+l_{k+1}}{2}-\frac{\lambda_{go}}{4}. \tag{8.115}$$

Thus, for a given $\omega_o$, lengths of the coupling lines may be known. Knowing $Q_k$, $b_k$ can be known from Eq. (8.114). Once Eq. (8 113) is satisfied, from the knowledge of $b_k$ the coupling aperture may be designed.

The power loss ratio* for the Chebyshev type filter is given by*

$$P_{LR}=1+k^2\,T_n^2\left[\frac{\beta_o}{\beta_{c_2}-\beta_{c1}}\left(\frac{\beta}{\beta_o}-\frac{\beta_o}{\beta}\right)\right].$$

Using the transformation

$$f\left(\frac{\omega}{\omega_c}\right)=\left[\frac{\beta_o}{\beta_{c_2}-\beta_{c1}}\left(\frac{\beta}{\beta_o}-\frac{\beta_o}{\beta}\right)\right],$$

we can get all formulae from (8.78) to (8.88), of course $\omega$'s in these equations have to be replaced by $\beta c$ everywhere, $c$ being velocity of light. A typical quarter-wave coupled waveguide filter is shown in Fig. 8.39. Loop coupled *YIG* resonator filters** offer high utility though filter resonators are tuned by varying magnetic field.

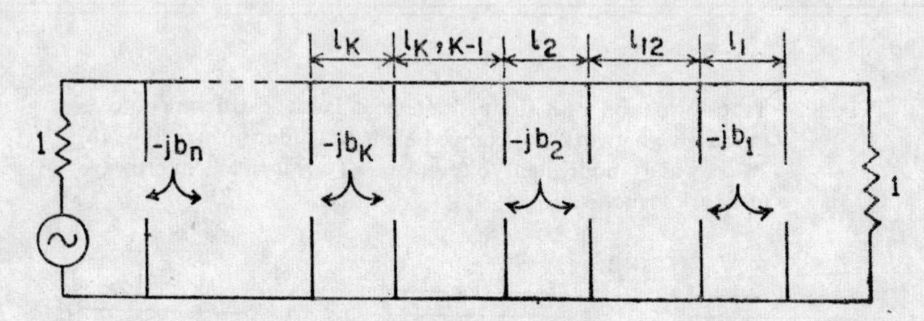


**Fig. 8.39** Quarter wave-coupled waveguide cavity filter.

## 8.14 DIRECT COUPLED CAVITY FILTERS

Cohn*** has developed a design procedure for direct coupled cavity filters upto a band width of 20 percent. According to him the equivalent circuit of

*Note that power loss ratio is expressed in $\beta$'s instead of $\omega$'s the reason being that a waveguide is a dispersive device and so good parameter is $\beta c$ and not $\omega$.

**J. Helszajn, *Passive and active Microwave Circuits*, John Wiley & Sons, New York (1978).

***S.B. Cohn, Direct coupled resonator filters', *Proc. IRE*, Vol, 45, (Feb.1957).

Fig. 8.38(b) can be represented by a $\pi$ network shunted with normalized inductive susceptances $b=-\cot \theta_k/2$, at each end as shown in Fig. 8.40(a). Since $\theta_k \simeq \pi$ hence $b$ may be neglected. So normalized series reactance replaces series resonant circuit in a prototype structure for a band pass filter. The shunt inductive susceptance and two waveguide sections, one on each side of the susceptance serve the purpose of impedance inversion. The line lengths of the impedance inverter are absorbed in a cavity and the filter becomes nothing but cavities placed in cascade as shown in Fig. 8.40(b.) The physical length of the $k$th cavity is given by

$$l_k=\frac{\lambda_{go}}{2}+\frac{\lambda_{go}}{2\pi}(\theta_{1k}+\theta_{1k+1}) \tag{8.116}$$

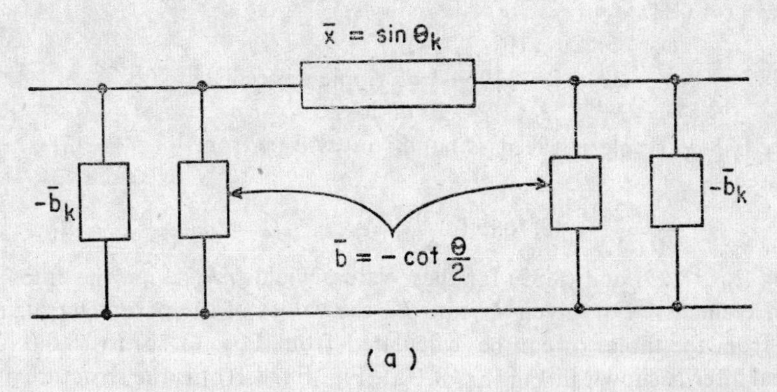


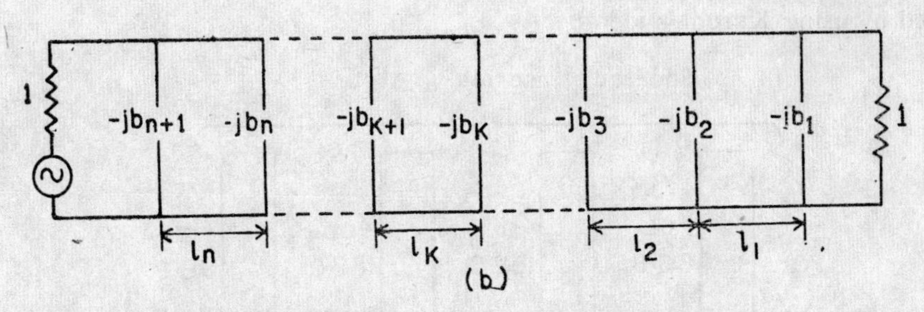


Fig. 8.40 A directly coupled waveguide cavity filter.

The normalized shunt susceptances required to decide cavity coupling are given by

$$\bar{b}_1=\left(1-\frac{\pi}{2g}\frac{\beta_{c2}-\beta_{c1}}{\beta_o}\right)\Big/\sqrt{\frac{\pi}{2g_1}\frac{\beta_{c2}-\beta_{c1}}{\beta_o}}$$

$$\bar{b}_2=\frac{2\beta_o\sqrt{g_1g_2}}{\pi(\beta_{c_2}-\beta_{c_1})}\left\{1-\frac{\pi^2}{4g_1\,g_2\,\beta_o^2}(\beta_{c_2}-\beta_{c_1})^2\right\}$$

$$\bar{b}_k=\frac{2}{\pi}\frac{\beta_o}{(\beta_{c_2}-\beta_{c_1})}\left\{1-\frac{\pi^2}{4g_k\,g_{k-1}}\frac{(\beta_{c_2}-\beta_{c_1})^2}{\beta_o^2}\right\}\sqrt{g_k\,g_{k-1}}$$

$$\bar{b}_n=\left\{1-\frac{\pi}{2}\frac{R}{g_n}\left(\frac{\beta_{c_2}-\beta_{c_1}}{\beta_o}\right)\right\}\Big/\sqrt{\frac{\pi}{2}\frac{R}{g_n}\frac{\beta_{c_2}-\beta_{c_1}}{\beta_o}} \tag{8.117}$$

where $g_k$ are the elements of the prototype filter.

Thus, a direct coupled resonator filter may be designed.

As many important topics such as phase characteristics, distortion, losses, etc. have not been covered, the reader may consult the references given at the end of this chapter.

## SOLVED EXAMPLES

**Example 8.1.** Design a microwave filter, given that:

(*i*) pass band is 1-2 GHz; (*ii*) tolerable ripple$=\frac{1}{2}$ *db*; and (*iii*) harmonic suppression 20 *db*.

*Solution*: The harmonics required to be filtered appear from 2 to 4 and 3 to 6 GHz. Therefore, the required band stop is 2-6 GHz, which has its centre at 4 GHz.

$$\omega_o=4\times 2\pi\times 10^9$$
$$\omega_{C_1}=1.9\times 2\pi\times 10^9 \rightarrow \tfrac{1}{2}\ db \text{ ripple desired}$$
$$\omega_n=2\pi\times 2.2\times 10^9 \rightarrow 20\ db \text{ loss.}$$

There is $\frac{1}{2}$ *db* ripple in the pass band so it will be from 1.1 – 1.9 GHz. Thus,

$$\frac{\omega_n}{\omega_c}=\frac{2.2}{1.9}=1.17085 \text{ for } L_m=\frac{1}{2}\ db. \quad L_n=20db.$$

Using Eqs (8.85) and (8.89) for these values yields $n=7$, i.e. the filter has seven elements. For given $L_m$ and $L_n$, the values of the parameters can be seen from the tables or can be calculated from Eqs. (8.76) to (8.89). The typical filter is shown in Fig. Ex. 8.1(a), Fig. Ex. 8.1(b) is the structure obtained by using Kuroda's identity.

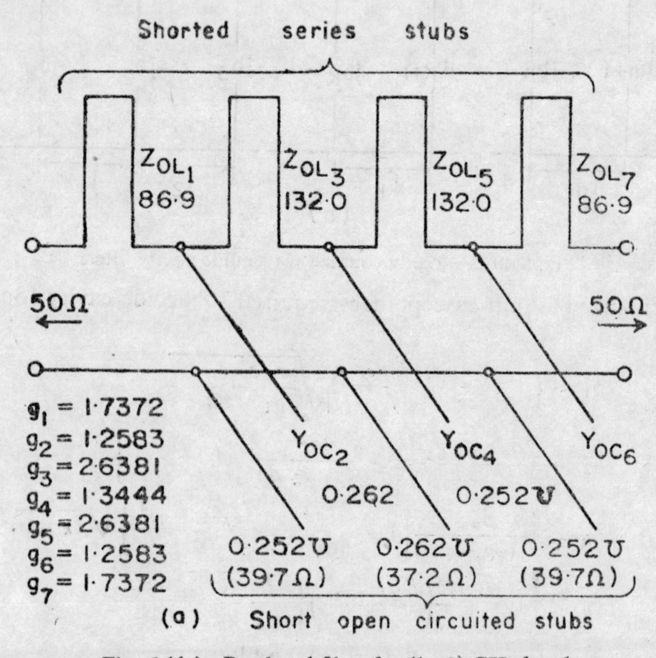


Fig. 6.1(*a*) Passband filter for (1—2) GHz band.

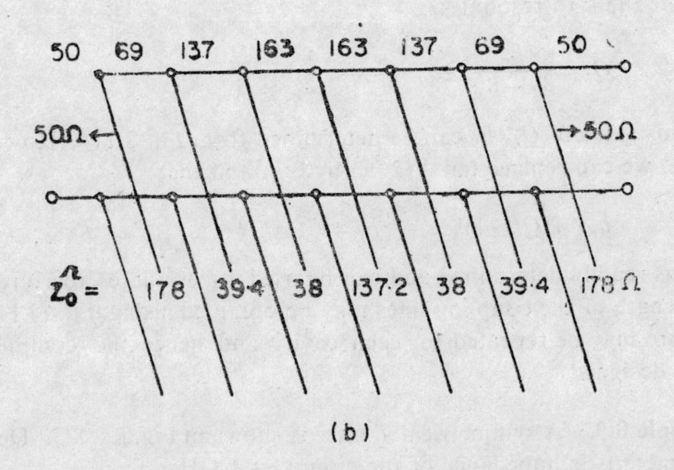


Fig. Ex. 6.1(*b*) Structure obtained after changing series stubs to shunt stube using Kuroda's identity.

**Example 8.2.** Design a five element quarter-wave cavity filter of Chebyschev response given that:

(*i*) Pass band tolerance $k^2=0.0233$.

(*ii*) Waveguide width $a=0.9$ inch.

(*iii*) Pass band $f_{c_1}=10$ GHz, $f_{c_2}=10.4$ GHz.

*Solution*: Values of wave vector $k_o$ corresponding to two cut-off frequencies is respectively

$$k_{oc_1}=\frac{\omega_{c_1}}{c}=2.1, \qquad k_{oc_2}=\frac{\omega_{c_2}}{c}=2.18 \text{ rad/cm.}$$

Thus, phase constants are

$$\beta_{c_1}=[(2.21)^2-(\pi/a)^2]=1.59$$

$$\beta_{c_2}=[(2.28)-(\pi/a)^2]=1.70.$$

The centre of the band will eventually be at

$$\beta_o=\sqrt{\beta_{c_1}\beta_{c_2}}=1.59\times1.70=1.64$$

or $\qquad f_o=10.2$ GHz.

Using Eq. (8.103)

$$C_p=\frac{C}{BW}=\frac{1}{\beta_{c2}-\beta_{c1}}=\frac{\beta_o}{\beta_{c_2}-\beta_{c_1}}\frac{C}{\beta_oC}=\frac{\beta_o}{\beta_{c_2}-\beta_{c1}}\frac{g_1}{\beta_oC}.$$

Here $C_{o1}=C_1$, the structure is prototype.

Since $\qquad Q_k=\frac{1}{2}\sqrt{\frac{C_{ok}}{L_{ok}}}=\frac{1}{2}\ (\beta_oC)\ C_{o1}=\frac{\beta_o}{\beta_{c_2}-\beta_{c_1}}\frac{g_1}{2}.$

Thus, for the $k$ th resonator

$$Q_k = \frac{\beta_o}{\beta_c - \beta_{c_1}} \frac{g_k}{2}.$$

From this value of $Q_k$, $b_k$ can be determined from Eq. 8.114. However, for large $b_k$, we can replace $\tan^{-1}(2/b_k)$ by $2/b_k$ and then,

$$b_k = 2(Q_k^2 - 1)^{1/2}, \qquad b_1 = 17.$$

$b_k$ is thus readily determined and can be used to evaluate aperture dimension. Length of the coupling lines may be obtained using Eq. (8.113). The procedure may be repeated for each cavity and hence the complete filter may be designed.

**Example 8.3.** A symmetrical $T$ filter is shown in Fig. Ex. 8.3. Determine the characteristic impedance of the circuit at 1 GHz.

*Solution*:

$$Z_1 = j\omega\,(0.5 \times 10^{-6}) = j2\pi \times 10^9 \times 0.5 \times 10^{-6} = .3.14 \times 10^3$$

$$Z_2 = \frac{1}{j\omega_c} = \frac{1}{j2\pi \times 10^9 \times 10^{-12}} = 159/j$$

$$Z_o = \sqrt{Z_1 Z_2 + \frac{Z_1^2}{4}}$$

$$= 998.52 \times 10^3 + 9.8 \times 10^6 = 3.295 \text{ k}\Omega.$$

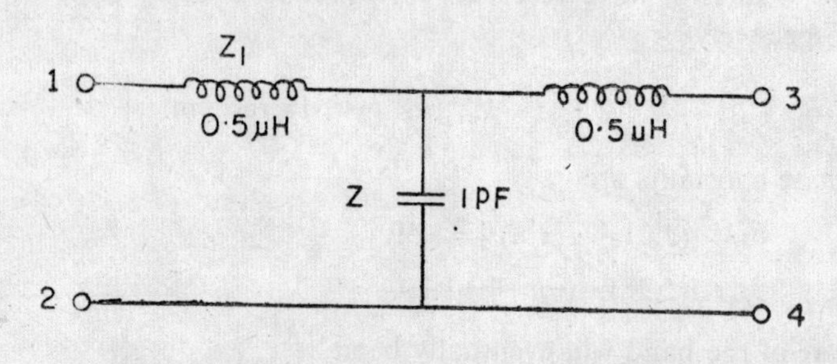


Fig. Ex. 8.3

**Example 8.4.** A five-section symmetrical $T$ filter has the following values. Determine the attenuation and phase shift of the filter.

$$\left.\begin{aligned} Z_1 &= 30 + j40\ \Omega \\ Z_2 &= 10 - j50\ \Omega \end{aligned}\right\} \text{ per section.}$$

*Solution*: $$\frac{Z_1}{2Z_2} = \frac{30 + j40}{2\,(10 - j50)} = -0.327 + j0.365$$

$$\therefore \quad Z_o = \sqrt{Z_1 Z_2 + \frac{Z_1^2}{4}}$$

$$=\sqrt{(30+j40)\ (10-j50)+(15+j20)^2}$$
$$=46\ 8\ \angle -6.6°.$$

Again $\dfrac{Z_0}{Z_2}=\dfrac{46.8\ \angle -6.6°}{51.2\ \angle -78.7°}=0.905\ \angle 72.1°$

$$=0.294+j0.857$$

$\alpha$, the attenuation is given by

$$=\ln\left[1+\frac{Z_1}{2Z_2}+\frac{Z_0}{Z_2}\right]$$
$$=\ln\ (1.55\angle 51.4)$$
$$=\ln\ (1.55)=0.437\ \text{neper/sec.}$$
$$\beta=\frac{51.4°}{57.3°}=0.895\ \text{rad/sec.}$$

**Example 8.5.** Design a high pass filter that will attenuate all frequencies upto 10 GHz and will work into a 50 kΩ load. Draw the circuit diagram.

*Solution.* $L=\dfrac{R_L}{4\pi f_{co}}=\dfrac{50000}{4\pi\times 10^{10}}=3.975\times 10^{-7}$ H

$$C=\frac{1}{4\pi f_{co}\ R_L}=\frac{1}{4\pi\times 10^{10}\times 5000}=1.59\times 10^{-10}\ \text{F}$$

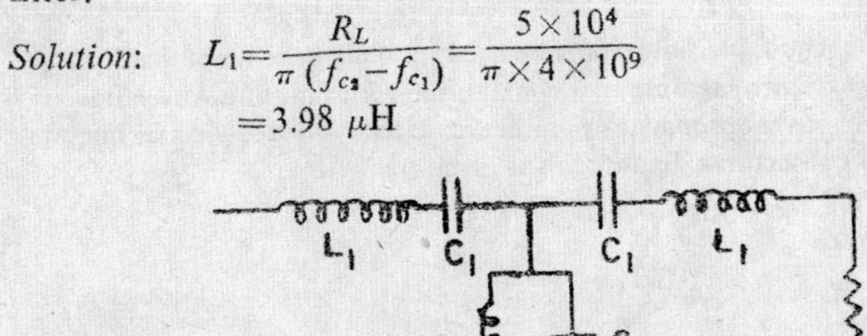


**Example 8.6.** Design a band pass filter that will pass all frequencies from 1 to 5 GHz. The load impedance is 50 kΩ. Specify all circuit components and draw the circuit. Draw equivalent series and or shunt stub microwave filter.

*Solution*: $L_1=\dfrac{R_L}{\pi\ (f_{c_2}-f_{c_1})}=\dfrac{5\times 10^4}{\pi\times 4\times 10^9}$

$$=3.98\ \mu\text{H}$$

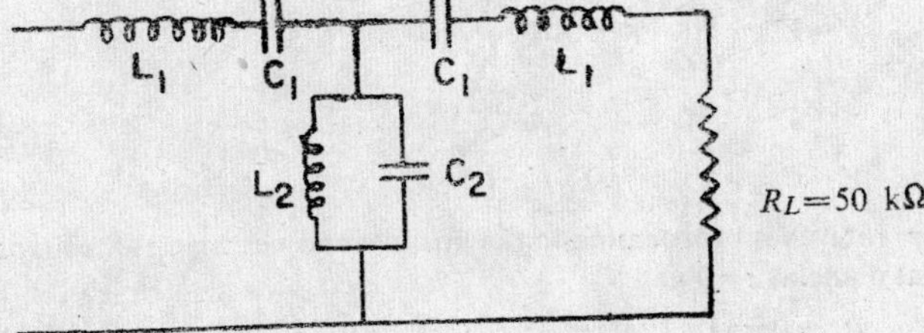

$$C_1=\frac{f_{c2}-f_{c1}}{4\pi R_o f_{c1} f_{c2}}$$

$$=\frac{4\times 10^9}{4\pi\times 5\times 10^4\times 5\times 10^{18}}$$

$$=12.7 \text{ pF}$$

$$L_2=\frac{R_L (f_{c2}-f_{c1})}{4\pi f_{c1} f_{c2}}$$

$$=\frac{50000\times 4\times 10^9}{4\pi\times 5\times 10^{18}}=31.8\ \mu\text{H}$$

$$C_2=\frac{1}{\pi R_L (f_{c2}-f_{c1})}=\frac{1}{\pi\times 50\times 10^3\times 4\times 10^9}$$

$$=0.318 \text{ pF}.$$

The corresponding microwave filter is

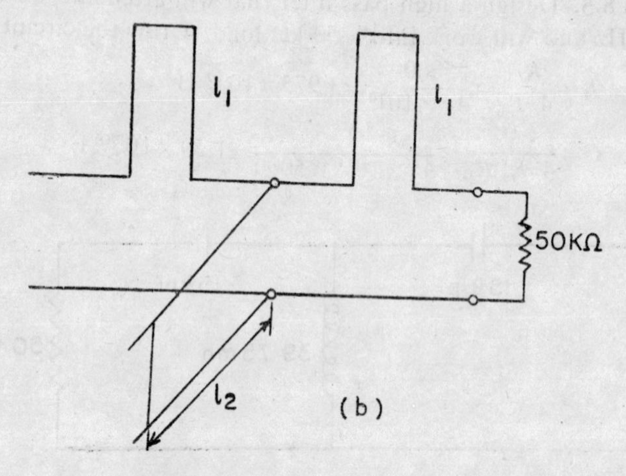


The stub positions and lengths may be calculated as discussed in Chap. 2.

## PROBLEMS

1. What is negative dispersion and how can it be used in microwave amplification?
2. Compare the slow waves propagating in a coaxial line with that in a waveguide.
3. Show that for TEM waves propagating in a capacitively loaded coaxial line the voltage between any two consecutive diaphragms is given by

$$V=V_n^+ \ e^{-jk_o (z-nd)}+V_n^- \ e^{+jk_o (z-nd)}$$

$$=V_n^+ e^{-jk_o (z-nd)}+\Gamma_B^+ \ v_n^+ \ c^{jk_o (z-nd)}$$

and
$$V=V_n^+ \ e^{-jk_o (z-nd)}+\Gamma_B^- \ V_n^+ e^{jk_o (z-nd)}$$

respectively, for waves propagating in the positive and negative $z$-directions. Assume the 0th terminal plane at $z=0$ and

$$V_B^+= V_n^+ (1+\Gamma_B^+), V_B^-=V_n^+ (1+\Gamma_B^-).$$

4. Obtain the values of the elements of high pass, band pass and band stop filters for $n=5$ using elements of the low pass Butterworth prototype filter.

5. Synthesize a low pass filter for $|S_{21}|^2=\frac{1}{1+\omega^4}$ o.

6. Design band pass filters with tolerable ripple of ½ *db* and harmonic suppression 20 *db* for bands 10-12 GHz and 10 to 12 MHz. Comment on the practical aspects of these filters.

7. Design a three cavity quarter-wave filter in a waveguide structure ($a=0.9$ in). The pass band extends from 10 GHz to 12 GHz with pass band tolerance $k^2=0.0233$ and Chebyshev response is desired. Determine the hole radii and diaphragm spacings if inductive diaphragms with circular holes are used.

8. Design a four cavity direct coupled cavity filter for the ratings of Prob. 8.7. Specify the diaphragm dimensions in this case and compare these with those in Prob. 8.7.

9. Find the elements of a Butt erworth low pass filter specified as follows: (*a*) attenuation at=GHz: max 1 *db*; (*b*) attenuation at=2GHz: min 47 *db* and (*c*) the filter is inserted between a 1 Ω generator and a ½ Ω load.

10. Design a Chebyshev low pass filter with the following requirements: maximum ripple in the pass band 0.5 *db*; minimum attenuation at double cut-off frequency 40 *db*; and normalized resistance of both generator and load is one.

11. Design a Chebyshev band pass filter to meet the following specifications: (*a*) ½ *bd* ripple in the pass band extending from 6 GHz to 10.8 GHz; (*b*) minimum attenuation of 60 *db* in the stop bands, e.g., below 5.4 GHz and 12 GHz. Find the microwave transmission line equivalent circuit.

## *REFERENCES*

MOSCHYTZ, G.S. AND HORN, P. Active Filter Design Handbook John Wiley & Sons 1981.

BEVENSEE, R.M., *Electromagnetic Slow Wave Systems* John Wiley & Sons. Inc. New York (1964).

BHUYAN, L. AND CHATTERJEE, B.N., 'Very-Low Resistivity Recessive Digital Filter Structures' Eelectronics Letters 17 (10) (1981).

COLLIN, R.E., *Field Theory of Guided Waves* McGraw-Hill Book Company, New York (1960).

DAVIDSON, C.W., *Transmission Lines for Communication* McMillan Press, London (1978).

GARVER, R.V., 'Basic Microwave Filter Theory' Microwave *J.* 23 (9) 1979.

GHOSE, R.N., *Microwave Circuit Theory and Analysis*, McGraw-Hill Book Co., New York (1963).

GOLDSMITH, P.F. AND SCHLOSSBERG, H., 'A Quasi-Optical Single Side band Filter Employing a Semiconductor Resonators' *IEEE Trans.* MTT-28 (10) 1980.

GREGORIAN, R., 'Filtering Techniques with Switched Capacitor Circuits' Microelectronics *J.* 11 (2) 1980.

HELSZAJN, J., *Passive and Active Microwave Circuits* John Wiley & Sons, New York (1978).

HUMPHERYS, D.S., *The Analysis, Design and Synthesis of Electrical Filters* Prentice-Hall, Inc. Englewoodcliffs, N.J. (1970).

IEEE TRANS. Special Issue on Microwave Filters, MTT-13 (9) 1965.

JOHNSON, D.E., *Introduction to Filter Theory* Prentice-Hall Inc. Englewoodcliffs, N.J. (1976).

JOKETA, K.T., 'Narrow-band Strip line or Microstrip Filters with Transmission Zeros at Real and Imaginary Frequencies' *IEEE Trans.* MTT-28 (6) 1980.

KUNIEDA, H., 'Simplified Equivalent Representation for Multimode Lines and Their Applications to Filter Design' *IEEE Trans.*, MTT-28 (8) 1980.

KUWANO, S. AND KAKUBUN, K., 'Waveguide Filters Using Dielectric Slabs Instead of Irises' Trans. IECEJ J.62-B (12) 1979.

MAKIMOTO, M. AND YAMASHITA, S., 'Band pass Filters Using Parallel Coupled Strip-line Stepped Impedance Resonators' *IEEE Trans*. MTT-28 (12) 1980.

NAVARRO, M.S., *et. al.*, 'Propagation in Rectangular Waveguide Periodically Loaded with Resonant Irises' *IEEE Trans*. MTT-28 (8) 1980.

PERINI, J., 'Periodically Loaded Transmission Lines' *IEEE Trans*. MTT-28 (9) 1980.

RHODES, J.D., *Theory of Electrical Filter*, John Wiley & Sons, New York (1976).

RICHARD, S.V., 'Microwave Transistor Filter and Design' *IEEE Trans*. MTT-18 (1) 1970.

RIZZOLI, V. AND LIPPORINI, A., 'Bloach-Wave Analysis of Strip line and Microstrip Array Slow-wave Structures', *IEEE Trans*, MTT-29 (2) 1981.

ROONEY, J.P. AND UNDERKOFLER, L.M., 'Printed Circuit Integration of M.W. Filters', *Microwave J.* 21 (9) 1978.

SANDER, K.F., *Transmission and Propagation of Electromagnetic Waves* Cambridge University Press Cambridge 1978.

SAULICH, G., 'A Simple Method for Spacing the Adjacent Pass bands for a Coupled Line Filter', *IEEE Trans*. MTT-28 (4) 1980.

SLATER, J.C., *Microwave Electronics* D. Van Nostrand Co. Inc., Princeton, N.J. 1950.

STORER, J.E., *Passive Network Synthesis* McGraw-Hill Book Co. New York (1957).

SURSMAN-FORT, S.E., 'Microwave All-Pass Network Using FETS', *IEEE Trans*. MTT-27 (12) 1979.

TEMES, G.C. AND LAPATRA, W.J., *Introduction to Circuit Synthesis & Design* McGraw-Hill Book Co. New York (1977).

TOYOTA, S. AND MAKIMOTO, T., 'Waveguide-Type Variable Band Pass Filters Using Varacter Diodes' Trans. IECEJ J62-B (6) 543, 1979.

VERBITSKII, I.L., 'Dispersion Relation for Comb-Type Slowwave Structures', *IEEE Trans*. MTT-28 (1) 1980.

WATANABE, R. AND NAKAJIMA, N., 'Reflection-Type Periodic Filter with Two Resonators Using Gaussian' Beam at Millimeter-Wave Region', Trans. IECEJ J62-B (11) 1979.

WILHELM, J., 'Transversal Filter Using Mesfets as Coupling Elements' *IEEE Trans*. MTT-19 (9) 1971.

YOUNG, L. (ed.), *Advances in Microwaves* Vol. IV '*Selected Topics in Microwave Filters*', Academic Press, New York (1970).

# INDEX